Economic, Social, and Environmental Insights on Energy Transition

Panos Dimitrakopoulos
https://orcid.org/0009-0000-8188-4255
University of the Peloponnese, Greece

Dimitrios Salampasis
Swinburne University of Technology, Australia

Ilias Makris
https://orcid.org/0000-0002-4276-2375
University of the Peloponnese, Greece

Sotiris Apostolopoulos
Neapolis University, Pafos, Cyprus

IGI Global
Scientific Publishing
Publishing Tomorrow's Research Today

Vice President of Editorial — Melissa Wagner
Director of Acquisitions — Mikaela Felty
Director of Book Development — Jocelynn Hessler
Production Manager — Mike Brehm
Cover Design — Jose Rosado

Published in the United States of America by
IGI Global Scientific Publishing
701 East Chocolate Avenue
Hershey, PA, 17033, USA
Tel: 717-533-8845 | Fax: 717-533-7115
Website: https://www.igi-global.com E-mail: cust@igi-global.com

Library of Congress Cataloging-in-Publication Data

Names: Dimitrakopoulos, Panos, 1982- editor | Salampasis, Dimitrios editor | Makris, Ilias, 1974- editor | Apostolopoulos, Sotiris, 1984- editor
Title: Economic, social, and environmental insights on energy transition / edited by: Panos Dimitrakopoulos, Dimitrios Salampasis, Ilias Makris, Sotiris Apostolopoulos.
Description: Hershey PA : IGI Global Scientific publishing, [2026] | Includes bibliographical references. | Summary: "This book plans to explore of local, as well as national and international energy initiatives within renewable energy options, the deployment schemes and available technologies, without overlooking the impact of decarbonization and energy transition in society and the environment, will be the backbone of this collective volume"-- Provided by publisher.
Identifiers: LCCN 2025034603 (print) | LCCN 2025034604 (ebook) | ISBN 9798337353029 hardcover | ISBN 9798337353036 paperback | ISBN 9798337353043 ebook
Subjects: LCSH: Renewable energy sources--Social aspects | Renewable energy sources--Economic aspects | Renewable energy sources--Environmental aspects
Classification: LCC TJ808 .E26 2026 (print) | LCC TJ808 (ebook)
LC record available at https://lccn.loc.gov/2025034603
LC ebook record available at https://lccn.loc.gov/2025034604

British Cataloguing in Publication Data
A Cataloguing in Publication record for this book is available from the British Library.

Table of Contents

Detailed Table of Contents

Comparative analysis shows significant policy divergence due to institutional and economic factors. The chapter concludes that "policy instruments must be adapted to national contexts," highlighting the need for coordinated strategies, flexible indicators, and stronger ESCO markets to support a more effective energy transition across diverse EU realities.

This chapter focuses on the energy transition and explores its connection with sustainability accounting and the performance management of economic organizations. This transition requires strategic actions from economic entities, aiming to promote the adoption of renewable energy sources. A bibliometric analysis was conducted using the VOSviewer software, based on 129 papers retrieved from the Scopus database. The analysis aimed to explore emerging research trends in the areas of sustainability accounting, corporate social responsibility (CSR), and performance measurement indicators. The results reveal significant thematic clusters and recurring scholarly interests. The findings underscore the strategic challenges faced by key stakeholders - particularly small and medium-sized enterprises (SMEs)- in aligning their operations with sustainability objectives. Finally, the literature highlights that integrating emerging technologies such as artificial intelligence and blockchain constitutes a critical factor for improving the quality and transparency of sustainability accounting systems.

This study operates at the nexus between decentralised energy systems and just energy transition, studying how digital technologies such as blockchain and IoT transform energy governance with socio-economic inequalities in mind. While there are technological advances, integration still has not taken place between decentralised models and inclusive policy frameworks. Using a qualitative thematic analysis of recent literature (2019–2025), the study identifies how digital tools increase efficiency and transparency yet require regulation that can respond to these challenges. The findings present opportunities, from empowering communities to P2P trading, but also risks, including the emergence of regional inequalities and labour displacement. The study puts forward integrative governance models that stretch the enabling potential of digital innovation toward justice-aligned transition frameworks, therefore providing critical knowledge for policy and practice.

Cyprus is considered an EU Member State that is isolated from the rest of Europe in terms of energy, and its energy mix continues to be dependent on imports of petroleum and fuel oil. Simultaneously, it faces challenges in achieving its sustainable development goals, a fact that places the country low on the ranking of the European Sustainable Development Report 2025. However, the prevailing climatic conditions, combined with Cyprus's geostrategic position in the Eastern Mediterranean, provide Cyprus with the energy advantage to become a green energy producer. The current study examines Cyprus' sustainable development through energy transition and energy hubs. Specifically, it explores Cypriot energy planning and strategy and examines whether there is potential to transform Cyprus into a climate-neutral green country by 2050 and, at the same time, into an energy hub for Europe. Employing a multi-level approach, this study examines and evaluates the policies implemented for the energy transition and energy hubs whilst drawing policy recommendations.

The current study conducted across Greece's lignite regions – which are experiencing the consequences of the energy transition – and essentially aims to examine how the potential establishment of incentives could foster green entrepreneurship. It has ultimately indicated the necessity of pursuing measures in order to prevent desertification. On top of utilising a Just Energy Transition plan of social legitimacy, building a concrete sustainable development model is required, and therefore, it must be implemented without any delays or deviations. Additionally, the energy crisis that erupted as a result of the war in Ukraine and the political decision to abandon the use of lignite in Greece by 2028, which is considered an unrealistic goal so far, have already put pressure on project implementation timeline. By conducting qualitative interviews with power plant workers and entrepreneurs located in lignite areas, this empirical research has specifically revealed that environmental goals should be supported by the revival of the local green economy.

Wind energy is central to the renewable transformation in Greece since this is among the main resources for fulfilling the National Energy and Climate Plan by achieving 70% renewable penetration by 2030. The majority of wind farms are situated within grassland habitats containing high cultural, ecological, and agricultural values often coincide with Natura 2000 protected areas. However, creation and operation of a wind farm could cause fragmentation accompanied by a long-term reduction in plant richness, soil degradation supporting invasive species establishment for a long-term effect on biodiversity, ecosystem function as well as landscape connectivity. This study discusses the impacts of wind energy development on Greek grassland ecosystems, observed through the forest road network that connect with the construction of wind farms. Therefore, brings out a call for the involved infrastructure-related impacts in renewable energy planning and management to achieve climate goals without a loss in ecological integrity.

India's energy transition reflects a delicate balance between rapid economic expansion, increasing energy consumption, and the need for a cleaner environment. Even though coal still occupies a dominant position in the overall national energy portfolio, renewables, led by solar and wind, have made significant inroads as costs have dropped and policy support, along with technological advances, has strengthened. But that vast renewable potential has so far gone largely unused, slowing the path to decarbonization. The country's dual identity as both a leading coal consumer and a rapidly emerging renewable power highlights the complexity of its energy transition. It also calls for accelerating the deployment of renewables, expanding the grid infrastructure, and investing in green finance. As India proceeds along this trajectory, its decisions will not just shape domestic energy security but also have a large bearing on the success of global climate and sustainable development goals.

This chapter explores the multisectoral implications of the energy transition by examining India–Gulf Cooperation Council (GCC) trade and the sustainability challenges of hospitals. While hydrocarbons have long defined India–GCC ties, renewable cooperation in solar and hydrogen is emerging, supported by policies such as the India–UAE Comprehensive Economic Partnership Agreement. Using a gravity model, the analysis finds that GDP, population, and distance remain central drivers of trade, but renewable commitments and institutional frameworks increasingly shape flows. Hospitals highlight the micro-level dimension: they depend on continuous energy for care yet face pressure to cut emissions and adopt efficiency standards. India's Energy Conservation Building Code and the UAE's Net Zero 2050 strategy mark progress, though uneven implementation and financing gaps persist. The chapter argues that linking renewable trade with hospital sustainability can reinforce bilateral ties, advance resilience, and show how energy policy can deliver broader social benefits.

A moral reorientation based on equality and justice is necessary as the world moves away from extraction-driven energy systems and toward sustainable alternatives. This shift calls for more than simply technological advancement. Indigenous knowledge challenges prevailing patterns of exploitation by providing important insights into resource management, community-centered sustainability, and land stewardship. This chapter explores how Indigenous epistemologies are changing the moral terrain of energy transformation, emphasizing the moral necessity of incorporating traditional ecological knowledge into contemporary practice and policy. The study highlights how Indigenous perspectives offer avenues to resilience, climate justice, and intergenerational responsibility by contrasting extractive paradigms with equity-centered methods. In order to achieve equitable and inclusive energy futures, it is not only legally required but also morally necessary to acknowledge and institutionalize Indigenous rights within energy governance.

In today's data-driven world, modeling and forecasting environmental, energy, and demographic systems has become crucial for addressing some of the most pressing global challenges such as pollution, energy management, climate change, and disaster risk reduction. These systems are inherently dynamic, evolving across both space and time, and are influenced by a complex interplay of geographic, economic, and social factors. Traditional forecasting techniques often struggle to account for such spatio-temporal complexity, making room for the emergence of artificial intelligence (AI) and machine learning as powerful tools in this domain. By integrating spatial and temporal dimensions, AI-powered systems can detect patterns, predict trends, and offer timely, location-specific insights that combine the strengths of different time series and LSTM algorithms. This advancement is particularly valuable for governments, researchers, and industries aiming to make proactive decisions for sustainable development, efficient resource allocation, and disaster preparedness.

This chapter explores the concept of the Triple Dividend—economic prosperity, social equity, and climate resilience in the context of the global energy transition. It examines how inclusive and sustainable energy policies can generate synergistic benefits across economic growth, social justice, and environmental sustainability. Through global case studies, policy analysis, and theoretical frameworks, the chapter provides actionable recommendations for integrating the triple dividend into energy and climate governance.

Preface

INTRODUCTION

The global energy transition occupies a critical juncture between necessity and opportunity, transforming the ways in which energy is produced, distributed, and consumedon a worldwide scale. As Editors of the, *Economic, Social, and Environmental Insights on Energy Transition* volume, we have convened a diverse group of international contributors to explore the multifaceted dimensions of this transformation. Each chapter provides a distinct analytical perspective through which the drivers, challenges and impacts of this transformation are explored.

The impetus for this book stems from a recognition of the increasing urgency to address climate change, enhance energy security, and pursue sustainable development goals, by critically engaging with the complex forces shaping this global energy transitions. These pressures converge with rapid advancements in frontier and emerging technologies—decentralized energy systems, digitalization, sectoral electrification, and circular economy integration. All these factors collectively contribute to a dynamic and continuously evolving global energy landscape.. This complexity is accompanied by a corresponding proliferation of data, systems and models that demand rigorous analysis, contextual understanding and interpretation..

Our intention with this collective volume is to highlight how these changes unfold not in isolation, but within deeply interconnected economic, social, and environmental contexts. The energy transition is not a singular or uniform process; it varies significantly across geographies, governance systems, and socio-economic realities. Accordingly, this book presents empirical studies, conceptual approaches, and practical insights drawn from local, national, and international contexts. These include analyses of renewable energy initiatives, deployment frameworks, policy mechanisms, and the integration of technologies such as hydrogen and digital systems.

Beyond the technical and policy dimensions, we put emphasis on the socio-economic and environmental implications of decarbonization. Understanding these broader impacts is essential if we are to design inclusive, equitable, and resilient energy futures. The chapters in this volume engage with such questions, reflecting an interdisciplinary effort to map both the current state and the trajectory of energy transitions worldwide.

We hope that this book will not serve only as a reference for researchers, policy-makers, and practitioners, but also as a catalyst for further dialogue and innovation. As the global energy transition continues to accelerate, we must collectively navigate its uncertainties and possibilities, guided by robust evidence, diverse perspectives, and a commitment to long-term sustainability.

STRUCTURE OF THE VOLUME

This volume brings together interdisciplinary insights and empirical research from diverse global contexts to illuminate the multifaceted nature of energy transition. The chapters collectively examine the economic, societal, and environmental dimensions of the transition towards sustainable energy systems. Below, we provide an overview of the thematic contributions across the 15 chapters.

Chapter 1 investigates the complexities of energy transition in geographically and institutionally constrained island regions, focusing on the Aegean Islands. It critically analyzes European Union funding programs, identifying implementation challenges and policy gaps that impede progress. The chapter provides practical recommendations to overcome structural barriers and enable fossil fuel independence, thereby positioning islands as frontrunners in achieving green autonomy.

Chapter 2 offers a macroeconomic perspective on energy efficiency in the European Union, highlighting differences in national policy approaches across selected member states. The chapter distinguishes efficiency from savings, examining the role of key directives and instruments such as EEOS and green finance. Through comparative analysis, it underscores the importance of tailoring policy instruments to national contexts, supported by flexible indicators and robust ESCO markets.

Chapter 3 bridges corporate sustainability and energy transition by exploring performance management and accounting frameworks. A bibliometric analysis reveals critical research trends in sustainability accounting, CSR, and performance measurement. The findings emphasize the strategic challenges faced by SMEs and the transformative role of technologies like AI and blockchain in enhancing transparency and decision-making within sustainability reporting.

Chapter 4 focuses on the intersection of decentralized energy systems and digital governance, exploring the implications for a just energy transition. Through a thematic literature analysis, the chapter reveals both the potential of technologies like blockchain and IoT to empower communities and the risks they pose—such as labor displacement and regional inequality. It advocates for integrated governance models that align digital innovation with social justice imperatives.

Chapter 5 examines Cyprus's dual challenges and opportunities in the energy transition. Although isolated in terms of infrastructure and still reliant on fossil fuels, Cyprus's climatic and geopolitical positioning offers potential for the country to become a leading regional energy hub. The chapter assesses national strategies and energy policies, proposing pathways for the island nation to achieve climate neutrality by 2050, while playing a critical role in Europe's broader energy ambitions.

Chapter 6 focuses on the socio-economic impacts of energy transition within Greece's lignite-dependent regions. Drawing from qualitative interviews, it emphasizes the urgent need for socially legitimate transition plans and targeted incentives to foster green entrepreneurship. The findings highlight the fragility of the transition timeline in light of external shocks such as the Ukraine war and call for adopting inclusive models of sustainable regional development.

Chapter 7 brings ecological considerations to the forefront, analyzing the unintended impacts of wind farm infrastructure—particularly forest road networks—on Greek grassland ecosystems. As Greece aims for high renewable energy penetration, this chapter raises critical concerns about biodiversity loss, landscape fragmentation, and invasive species. It calls for integrative planning that balances climate goals with ecological stewardship.

Chapter 8 presents India's energy transition as a case of contrasts—between coal dependency and renewable potential, economic growth and environmental constraints. While solar and wind investments have gained ground, structural and financial bottlenecks persist. This chapter advocates for faster deployment of clean technologies, strengthened grid infrastructure, and expanded green finance to ensure India's energy choices support both domestic security and global climate goals.

Chapter 9 expands the lens of analysis by exploring the multisectoral implications of energy transition through the case of India–GCC trade and healthcare infrastructure. It reveals how renewable energy trade, institutional policy, and hospital sustainability converge in shaping resilient systems. The chapter offers a novel perspective on linking macro trade dynamics with micro-level social infrastructure to deliver inclusive energy transition benefits.

Chapter 10 introduces a moral and epistemological rethinking of energy systems through Indigenous knowledge systems. It challenges extractive paradigms by foregrounding values of equity, stewardship, and intergenerational justice. The chapter advocates for embedding Indigenous perspectives within institutional energy governance, not only as a legal imperative but as a pathway to more just and resilient energy futures.

Chapter 11 explores the application of AI-driven frameworks in modeling and forecasting spatio-temporal systems related to energy, environment, and demographics. By leveraging machine learning and deep learning algorithms, the chapter offers solutions for improving decision-making in climate adaptation, disaster risk reduction, and resource allocation—paving the way for data-driven governance in energy and environmental planning.

Chapter 12 articulates the concept of the *Triple Dividend*, emphasizing how energy policies can simultaneously advance economic growth, social equity, and climate resilience. Through global case studies and policy frameworks, the chapter provides a compelling argument for holistic governance approaches that harness synergies among sustainability objectives.

Chapter 13 closes the volume by examining the role of marketing analytics in green marketing strategy. As consumer awareness and environmental standards evolve, organizations must integrate sustainability into their marketing mix. The chapter unpacks how analytical tools can enhance decision-making around green products, pricing, distribution, and promotion—highlighting the opportunities and challenges in aligning market strategies with environmental responsibility.

Collectively, these chapters provide a comprehensive andinterdisciplinary exploration of the global energy transition. They underscore the necessity of context-specific strategies, inclusive governance, technological innovation, and moral/ethical accountability. As Editors, our objective is to equip readers with the analytical tools and conceptual frameworks essential for understanding and shaping the evolving global energy future.

CONCLUSION

As the world stands at a critical juncture in its pursuit of a sustainable energy future, this volume has been assembled to reflect the urgency, complexity, and multidimensional nature of the energy transition. The chapters included offer a diverse array of perspectives—from local to global, theoretical to empirical, technical to moral—demonstrating that energy systems cannot be meaningfully understood in isolation from their broader economic, social, and environmental contexts.

One of the defining messages that emerges from this collective work is the recognition that no singular path exists for achieving a just and effective energy transition. Each context—whether an island region, a post-industrial community, or a rapidly developing economy—brings with it unique challenges, capacities, and opportunities. The transition must therefore be approached not as a uniform technical shift, but as a deeply systemic transformation requiring coordinated policy, inclusive governance, informed financial instruments, ecological awareness, and social engagement.

We further underscore the importance of interdisciplinarity in effectively navigating this transformation. The insights drawn from economics, engineering, environmental science, sociology, public policy, and Indigenous knowledge systems must inform one another. Only by integrating these diverse strands of inquiry can we generate solutions that are not only technologically sound, but also socially just and environmentally viable.

It is our hope that this volume serves as a valuable resource for scholars, practitioners, policymakers, and students alike—those who are actively engaged in shaping the contours of energy transition in their respective fields and communities. We invite readers to engage critically with the contributions herein and to view them as starting points for continued dialogue, research, and action.

The energy transition is not merely a technological upgrade; it is a societal shift. As editors, we believe this shift must be guided by a commitment to equity, resilience, and long-term sustainability. This book is a small contribution to that broader effort.

Chapter 1
European Funding Programs for the Energy Transition of the Aegean Islands: A Critical Approach

Maria Karamanoli
https://orcid.org/0000-0001-8319-9358
University of the Aegean, Greece

Georgios Papadopoulos
https://orcid.org/0009-0006-8407-258X
University of the Peloponnese, Greece

Sotiris Apostolopoulos
https://orcid.org/0000-0002-1065-891X
Neapolis University, Pafos, Cyprus

Nikolaos Apostolopoulos
https://orcid.org/0000-0003-2810-5886
University of the Peloponnese, Greece

Amvrosios Prodromou
https://orcid.org/0009-0008-4787-6292
Neapolis University, Pafos, Cyprus

Nicholas Theodorakopoulos
https://orcid.org/0000-0002-2262-0054
Neapolis University, Pafos, Cyprus

ABSTRACT

The physical isolation of island regions, combined together with their geographical facets and institutional structures, creates additional constraints compared to mainlands, including their continued energy reliance on fossil fuels. Energy dependency, instability, and energy isolation hinder the sustainable development of island regions. EU policies aim to remove these obstacles through the just energy transition. This study examines the EU funding programs supporting the energy transition of island regions and their overall impact, respectively. It identifies challenges and obstacles towards the implementation of the islands' transition plan and concludes

DOI: 10.4018/979-8-3373-5302-9.ch001

with policy recommendations. The latter includes concrete measures to overcome the aforementioned issues in order for the islands to become independent from polluting fossil fuels and achieve green energy autonomy.

INTRODUCTION

Energy transition of island regions faces challenges that in many cases are differentiated from continental regions (Peñalvo-López et al., 2024; Kaldellis et al., 2012; Kavadias et al., 2007). Within EU Member States there are more than 2.200 inhabited islands that despite funding for Renewable Energy Sources (RES) and interconnections with the countries' main energy grids, many islands continue to use polluting fossil fuels for their energy needs (Kotzebue & Weissenbacher, 2020). Greece is home to a variety of islands and island clusters covering 16.28% of the country's total territory, of which only 165 are inhabited (Hellenic Statistical Authority, 2025; Katsoulakos, 2019; Gillespie, 2009). There are inhabited islands in Greece that, despite the funding provided so far, continue to use polluting fossil fuels for their energy needs (Southeastern European Energy Institute, 2020). Electricity generation in the islands not connected to the country's central energy system is based on twenty-nine autonomous systems (Southeastern European Energy Institute, 2020). There are also Greek islands that have moved to decentralized energy systems such as the island of Astypalaia (Tsagkari & Roca Jusmet, 2020; Hatziargyriou et al., 2017). In addition, the planned interconnection of islands with the central energy grid is scheduled to be completed in 2030 (Katsaprakakis et al, 2022). Despite the delays and time overruns in planning, significant steps have been made in the development of renewable energy to the extent that some islands are meeting their electricity needs (Southeastern European Energy Institute, 2020). In order to overcome obstacles and address existing problems to realize the decarbonization of islands from fossil fuels, the EU has taken a number of initiatives (Kotzebue & Weissenbacher, 2020; Spilanis et al., 2013). In 2017, the European Commission together with its fourteen member states with island regions proceeded to the "Valletta Declaration" (Europa.eu, 2017). The signing of the declaration provided for the gradual decarbonization of islands from fossil fuels, their energy autonomy, and the support of renewable energy for inhabited islands that the EU has (Kotzebue & Weissenbacher, 2020). The technical secretariat that was created supported the pilot transition strategy of six islands including the Greek island of Sifnos and twenty pioneer islands including the Greek islands of Zakynthos, Samothrace, Samos, Samos, Ikaria, Patmos, Amorgos, Symi, Kasos, and Crete (IEA, I., 2023; Katsaprakakis et al., 2022; Tsagkari, 2020). This was preceded by the Agreement of Mayors in 2008 for the promotion and implementation of sustainable energy policies in which thirty-two Greek island municipalities participated and the Agreement of Islands in 2011 in which one hundred and seventeen islands participated, of which thirty are Greek (Melica et al., 2022; Santopietro & Scorza, 2021; Kona et al., 2021; Pablo-Romero et al., 2015). It was also preceded by the Smart Islands Initiative in 2016 in which more than two hundred islands from fifteen EU Member States participated (Peñalvo-López et al., 2024; Tranberg et al., 2019; Priano et al., 2016).

To support its policies in promoting the energy transition of islands, EU has established financial instruments such as the Just Transition Fund, where for islands it has established a special treatment entitled "Special funds for the outermost regions and islands" (European Commission, 2020a). It also directed funds for the decarbonization of islands from fossil fuels from the Sectoral and Regional Programs of the Partnership for Regional Development 2021-2027 (Ministry of Development and Investment, 2021). The main objective was, and remains to be, is that where oil and diesel power generation facilities exist, these should be replaced by LNG, and at the same time accelerate the development of renewable energy (Makris et al., 2024; Southeastern European Energy Institute, 2020). Furthermore, attention was paid towards the planning of the interconnection of the islands with the country's central energy system (Ministry of Environment and Energy, 2021). The provision for the creation of energy storage systems, for the storage of electricity regardless of the interconnection with the central energy system of the country constitutes a major priority which can provide additional operational benefits (Southeastern European Energy Institute, 2020). Successful energy transition requires policies to support both the sustainable development and the creation of local revenue streams which is achieved through the availability and accessibility of several financial tools (Tandon, 2021; Cheung et al., 2019; Carley et al., 2018). European funding for islands, enable reducing, or even, eliminating the impact of energy transition on human resources and green investments (Apostolopoulos et al., 2024; Ministry of Environment and Energy, 2021). Securing the relevant financial tools is essential in order to attract new investment in clean energy technologies, and therefore, to support workers to acquire the necessary knowledge and skills that clean energy technologies require (Dimitrakopoulos et al., 2024; Apostolopoulos et al., 2023; Official Journal of the European Union, 2021; European Commission, 2020b).

This research examines the impact of European funding programs oriented towards the energy transition of the islands located to the North Aegean Region and the South Aegean Region. It examines the implementation of the transition plan of these islands based on European funding by identifying obstacles and problems. The article concludes with policy recommendations in order to help energy policy makers and local communities to overcome such obstacles and problems which will enable the islands to wean themselves from polluting fossil fuels by gaining their green energy autonomy.

LITERATURE REVIEW

In recent years, the research interest related to European policies concerning island regions has been continuously growing (Mauger et al., 2024; Kotzebue & Weissenbacher, 2020; Armstrong et al., 2012). Importantly, the research interest focuses on the energy transition of islands, since a large part of the islands use polluting fossil fuels for their energy needs, have increased energy cost, grid instability and energy dependence (Sakari & Roca Jusmet, 2020). The particular economic, social and environmental conditions prevailing in island regions in relation to their energy mix are a field of increased research interest (Gabderakhmanova & Marinelli, 2022; Jaramillo-Nieves & Del Río, 2010). A large number of islands are dependent from foreign energy sources, which makes these islands socially and economically vulnerable (Jaramillo-Nieves & Del Río, 2010). Comparative analyses of decentralized electricity systems in islands regions are largely absent from the international literature (Katsoulakos, 2019; Uche-Soria & Rodríguez – Monroy, 2018). According to research by Katsoulakos (2019), the average energy cost on Greek islands that are not connected to the continental grid is 2.5 times higher than in islands that are interconnected with the continental grid. The international literature focuses on the energy sustainability of islands (Curto et al., 2020; Groppi et al., 2018; Ioannidis & Chalvatzis, 2017; Franzitta & Curto, 2017). In order for island sustainable energy systems to support cooling, heating and electricity, there is a high need for the development of hybrid renewable energy resources systems (Lassalle et al., 2022; Kougias et al., 2019; Meschede et al., 2018; Chua et al., 2014). Such actions are important in order to support the efficient utilization of energy self-sufficiency of islands through the utilization of storage systems (Fotopoulou et al., 2024; Groppi et al., 2018; Notton, 2015; Blechinger et al., 2014). The renewable energy resources of islands have been found to have an impact on the economy, society and the environment (Meschede et al., 2022; Milone et al., 2022; Barone et al., 2021; Weisser, 2004), as well as the energy dependence of islands on fossil fuels and the effort to replace them with RES (Sánchez et al, 2023; Caldera et al, 2023; Groppi et al, 2018). Furthermore, aspects such as the interconnection of islands with national and international energy systems have also been extensively examined (Ramos-Real et al., 2018; Zafeiratou & Spataru, 2017; Georgiou et al., 2011), the energy vulnerability of islands due to their isolation from energy markets (Tzanes et al., 2023; Ioannidis et al., 2019; Genave, 2019), the energy security of islands (Rajaeian & Davarani, 2023; Lal & Kumar, 2022; Wolf et al, 2016), and the European policies for the energy transition of islands (Mauger et al, 2024; Mauger, 2021; Kotzebue & Weissenbacher, 2020) have also been fields that have attracted extensive research. However, there is a research gap on the European funding mechanisms for the energy transition in islands regions and how effective they are. The current research focuses on filling this gap.

THE SPECIAL CHARACTERISTICS OF ISLANDS AND THEIR ENERGY SAMPLE

According to Spilanis et al. (2012) in the Aegean Sea there are aver 3.050 Greek islands and rocky islets, as well as 253 islands and rocky islets that belong to Turkey. The islands of Aegean Sea that belong to Greece are administratively divided into the North and South Aegean Regions. In the South Aegean region two large island complexes are administratively integrated, those belonging to the Cyclades and those belonging to the Dodecanese. Figure 1 illustrates the islands of the Aegean Sea included in the North Aegean Region and the South Aegean Region.

Figure 1. Islands of the North and South Aegean regions (Ministry of Environment and Energy, 2025)

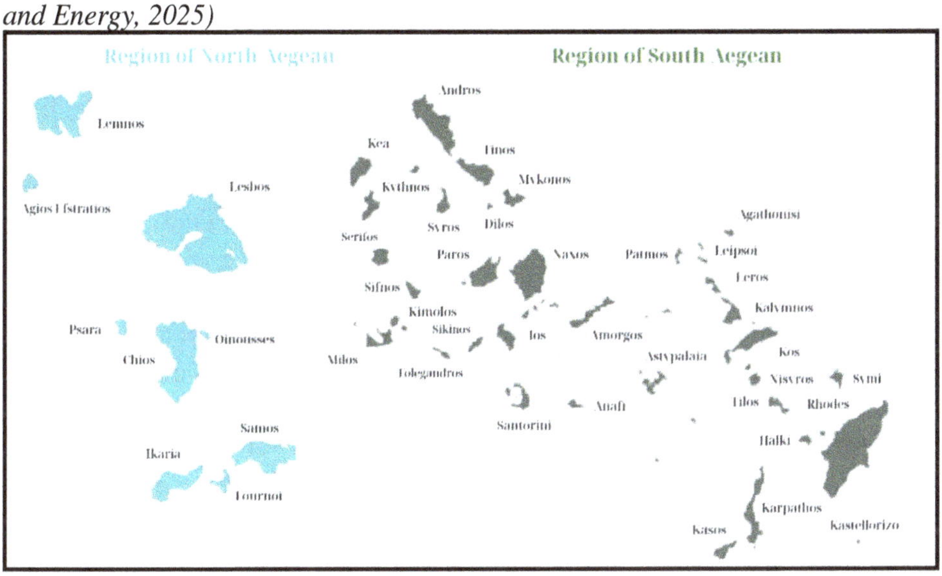

It should be noted that there are also islands of the Aegean Sea that administratively belong to Regional Units of mainland Greece such as the Sporades. The islands Rhodes and Lesvos gather the largest number inhabitants (Hellenic Statistical Authority, 2025; Katsoulakos, 2019). Some of the islands of the Aegean Sea are popular tourist destinations, such as Santorini, Mykonos, and Rhodes, and which have increased seasonal demand for electricity (Katsoulakos, 2019). Characteristics of the Aegean Sea islands are their small size, difficulty in accessing them, high transport cost, the high cost and long time required for services not offered on the islands difficulties in accessing the health system (Ballis et al., 2018; Karampela et al., 2014; Spilanis et al., 2003). The majority of these islands are connected by air from Athens and Thessaloniki and by ferry from Piraeus, while the air and ferry grid between the islands is limited (Spilanis et al., 2012). In level of electricity many islands continue to be dependent on fossil fuels, despite the continuous development of RES, and there is a plan to interconnect them with the central grid of country (John, 2024; Iliopoulou et al., 2018; Georgiou et al., 2011). Many islands in the Aegean Sea are located close to continental Greece, such as Cyclades and some are located very far from continental Greece and very close to the Turkish coasts, such as Dodecanese, Lesvos, Chios and Samos. For those close to continental Greece there is an easy interconnection with the central electricity grid of the country, and in many the interconnection has already been established. On the contrary, in remote islands there are existing difficulties in implementing their interconnection project, and the national plan foresees exceeding the existing problems and connecting with the central grid of country by 2030. However, frequent revisions have been created affecting the timetable for completing the islands interconnection project. In January 2025, there were forty-four islands unconnected to the national grid that recommended twenty-eight autonomous electrical systems. Thirty thermal stations and 696 RES stations are currently operating in these islands (Stergiopoulou, 2025). However, the solar and wind energy production, including existing hybrid plants in Ikaria and Kalymnos do not exceed 15% of the overall produced power (Stergiopoulou, 2025). In regards of islands that are not interconnected with the central grid of the country, the energy cost is excessively high and therefore with the law passed by the Greek parliament has imposed a special fee on all electricity consumers in the country (Katsoulakos, 2019). The implementation of plan for the interconnections of the islands with the country's central grid, which is delayed, is expected to solve the energy problem on the islands of Aegean Sea. However, there are islands of Aegean Sea such us Agios Efstratios, Astypalaia and Megisti that the plan of 2019 has not included them in the interconnection with the country's central grid, and thus, will remain autonomous in terms of electrical systems on these islands due to cost (Stergiopoulou, 2025). Figure 2 illustrates the electrical interconnections of the North and South Aegean islands that are planned to be implemented. A section on the Cyclades islands has already been established.

Figure 2. Independent Electricity Transmission Operator's (ITTO) plan for the interconnection of the Aegean Islands' electricity system by 2030 (Ministry of Environment and Energy, 2025)

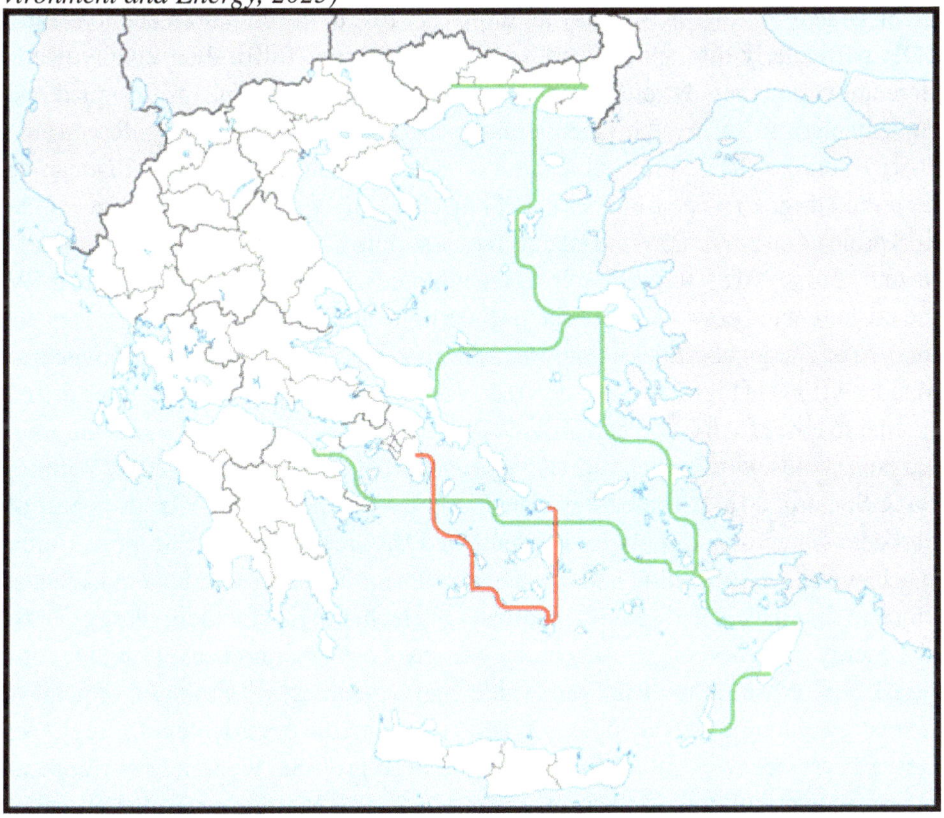

ENERGY TRANSITION PLAN OF AEGEAN ISLANDS

The territorial plan for just and sustainable transition of the islands of the North and South Aegean Sea was formulated in 2021 (Ministry of Environment and Energy, 2021). The concrete plan also incorporates the EU policy on the environment and energy transition (Ministry of Environment and Energy, 2021). Based on the energy system of the islands of the North and South Aegean Regions, the islands are divided into two categories. More specifically, the first group includes islands that are already interconnected to the mainland grid via high-voltage (150 kV) and medium - voltage interconnections, and the second group includes islands that have autonomous and remote energy systems, yet still not interconnected to the mainland grid. The Non–Interconnected Islands (NIIs) are the 10% country's total electricity demand, making them important to be taken into consideration for the national energy policy (John, 2024; Karystianos et al., 2021; Zafeiratou & Spataru, 2017). Greek islands of Aegean Sea play an important role to the country's economy and GDP, particularly through tourism demand. In order to fulfill their energy needs there are various regulations, legislations, policies and mechanisms designed and implemented, to secure the islands energy needs. More specifically, Renewable Energy Sources (RES) projects located on non-interconnected islands that can be supported through a Feed-in Tariff (FiT) mechanism regardless of their size, while RES projects on interconnected islands have a ceiling of 400kW (Greek Government Gazette, 2016, 2018). Regarding RES technologies, Greece supports both solar PV and onshore wind power. Greek islands have high energy demand which creates the concern of energy cost alleviation. Because energy costs of the non-interconnected islands (NII's) are high, the State created a levy charged to all electricity consumers, in order to cover such costs, and provide help to families with four or more children, and households with low income (Kougias et al, 2019; Katsoulakos, 2019; Zafeiratou & Spataru, 2017). The energy transition plan for the Aegean islands is part of a broader framework of policies implemented by Greece and the European Union to achieve energy transition and decarbonization, with a focus on isolated islands, which to date have relied heavily on oil and old technologies for their energy needs. The energy transition of the Aegean islands involves a comprehensive strategy for the islands' transition to clean, renewable energy sources, with the aim of achieving energy autonomy, reducing costs, and protecting the environment through two axes. Experience from pilot projects ("lighthouse projects") and the development of energy communities (Katsaprakakis et al., 2022). More specifically, the pilot projects concern wind and photovoltaic parks, hybrid energy storage systems local smart grids, as well as the use of pumped storage systems and hydrogen production (Katsaprakakis et al, 2022). At the same time, energy communities developed on the islands of the Aegean Sea (Papapostolou et al, 2019). Energy communities are cooperative organisations created and managed by citizens, municipalities or local businesses.

Planning and Schedules

The non-interconnected islands of the Aegean belong to autonomous electricity systems consisting of one or more islands and supplied by one or more power stations. Unlike most continental electricity systems, these autonomous insular networks are not connected to the country's primary transmission or distribution networks, which presents distinctive operational, technical, and economic challenges (Ministry of Environment and Energy, 2021). The management of the electric systems on these islands is a complex task involving two main responsibilities. First, there is the operational management of electricity production facilities and associated infrastructure. Second, there is the management of the electricity market and distribution systems within the islands. Both functions fall under the jurisdiction of the Hellenic Electricity Distribution Network Operator-HEDNO S.A (Ministry of Environment and Energy, 2021).

Figure 3. The Non-connected islands' systems and their respective peak annual demand (for 2022) and average energy production from thermal units and renewable energy resources (as of July 2023) (HEDNO (2023) & Hellenic Association for Energy Economics (2023))

NII Electricity System		Peak Annual Demand (MW)	Average Energy Production (MWh)
Small	Antikythera	0.12	24.47
	Arki	0.16	35.04
	Gavdos	0.19	47.60
	Agathonisi	0.23	61.32
	Othonoi	0.25	54.02
	Agios Efstratios	0.3	84.55
	Ereikousa	0.49	69.55
	Donousa	0.52	84.32
	Anafi	0.63	116.12
	Megisti	1.31	381.50
	Astypalaia	2.36	546.81
	Amorgos	3.23	947.77
	Serifos	3.58	755.37
	Symi	3.93	1129.83
	Kythnos	4.05	942.36
	Skyros	4.18	1245.81
	Patmos	5.18	1509.81
	Sifnos	6.64	1678.03
	Ikaria	6.74	2168.62
Medium	Karpathos	10.94	3300.08
	Milos	13.59	4536.06
	Lemnos	14.1	4771.28
	Samos	28.6	10337.09
	Chios	44.9	15384.80
	Thira	58.5	20844.01
	Lesbos	63.25	22561.66
	Kos - Kalymnos	97.5	31243.51
Large	Rhodes	218	65449.94

It should be noted that the energy market with regard to the non-interconnected islands of the Aegean Sea has changed since 2023 and now consists of twenty-eight autonomous systems. (Hellenic Association for Energy Economics, 2023). These systems do not always correspond to single islands; in some cases, they include multiple islands grouped as one operational unit. This reflects the geographic and logistical realities of island clusters and the need for efficient management of interconnected energy resources within these clusters (Hellenic Association for Energy Economics, 2023). According to the official Ten-Year Network Development Plan 2025–2034 by Independent Power Transmission Operator this interconnection process involves linking each island's autonomous power system directly to Greece's national grid via submarine electricity cables (IPTO, 2024). IPTO's The aforementioned plan refers to investments approximately EUR 4.4 billion towards Greece's electricity transmission infrastructure. The main dimesions feature the interconnection of islands, modernization of the domestic grid, and the expansion of cross-border connections. In Crete, the interconnection of the Crete – Peloponnese was completed in December 2020 while in May 2025, the Crete – Attica electric interconnection to the mainland grid was finalized (ADMIE, 2025a). Regarding the transition timetable, the latest Ten – Year Development Plan of the Hellenic Electricity Transmission System Operator (June 2023), was submitted to the Regulatory Authority for Energy (RAE) for approval, providing a detailed timeline for the completion of the interconnection of Aegean islands within the period 2024-2033, in three different stages. Between 2024-2025, the completion of phase D of the interconnection of the southern and western Cyclades. The Thira-Naxos interconnection projects are actively underway, while contracts for the cable interconnections of Folegandros, Milos, and Serifos were signed in 2023. By 2027–2028, the interconnection of the Dodecanese islands is scheduled. Finally, the North Aegean islands are targeted for interconnection by 2028–2029.

At the same time, actions were launched for the implementation of holistic energy interventions on smaller islands that will not be interconnected, such as Agios Efstratios and Astypalaia. It is planned to fulfill projects that will give the opportunity to locals to produce their own energy through renewable resources, to build energy efficiency and improve it at the same time, while electric mobility is also promoted so that fossil fuel consumption and emissions will be reduced (ADMIE, 2025b). To conclude, the autonomous electricity systems in non – interconnected islands in Greece are of special characteristics that create difficulties through the fulfillment process. Greece has set a concise strategy for electric interconnection of islands, that includes innovative projects, in order not only to fulfill demand but also to transform the energy landscape of islands. The completion of the interconnection plan in Greek islands, will promote clean energy practices in Greece, creating a sustainable energy future for islands communities.

Challenges

The Aegean islands face severe challenges in terms of their energy supply and energy sufficiency because of their high dependence on imported fossil fuels (Kotzebue & Weissenbacher, 2020; Spilanis et al, 2013). Fossil fuels, which are imported to meet the needs of Greek consumers, are expensive and polluting, increase the price of electricity, contribute negatively to air pollution, and limit the competitiveness of small and medium-sized businesses operating in island regions

(John, 2024; Iliopoulou et al., 2018; Georgiou et al., 2011). Their impact to air pollution deteriorates public health of locals (Greek Ministry of Energy, 2021). Additionally, pollution affects tourism, which is critical for all island economies (Greek Ministry of Energy, 2021). On the contrary of these negative characteristics, the recent announcement of the establishment of an Island Decarbonization Fund by the Ministry of Environment and Energy, the European Commission, and the European Investment Bank constitutes a positive step towards making the islands energy independent from fossil fuels (European Commission, 2024).

Most Aegean islands operate as non-interconnected islands, with autonomous power systems disconnected from the mainland grid. This independence introduces several technical and operational constraints (Karystianos et al., 2021). No alternative power supply exists should the local Autonomous Power Station (APS) fails, risking blackouts. Electricity generation costs are significantly higher than on the mainland. Low system inertia causes frequency fluctuations, threatening grid stability. Voltage fluctuations occur due to limited short-circuit power capacity. Overall system stability is fragile due to a narrow operational margin. Integration of renewable energy sources (RES) is limited as a result of the small size and isolated nature of the grids. These operational challenges add to the existing complicated path towards the transition for cleaner energy and impose economic penalties on island communities.

Beyond technical issues, the Aegean islands face entrenched socioeconomic difficulties linked to their geography, morphology, and small size. These factors create structural inequalities that hinder sustainable development (Greek Ministry of Energy, 2021). There are two major problems. The first one refers to the limited access and the second refers to the economic dependence from the tourism industry. (Katsoulakos, 2019; Ballis et al., 2018; Karampela et al., 2014; Spilanis et al., 2003). In light of the above, it is obvious that the interconnectivity of islands, and in particular those facing double insularity, is necessary for their development and the preservation of territorial cohesion. The most significant challenges confronted by the islands are related to population decline, the low competitiveness of their economies, geographical isolation, and the potential for geopolitical instability. Contrary to popular belief, unemployment rates on many islands, especially the Cyclades, are relatively low. However, low unemployment masks structural weaknesses. Island economies generally have low gross value added per capita. Economic output closely correlates with tourism intensity, highlighting vulnerability to seasonal and external factors. Islands lack economies of scale and diversified industries, making them susceptible to economic shocks (Zisopoulos et al., 2023). The economic vitality of these islands depends heavily on the success and sustainability of tourism. Geographic isolation is a defining characteristic of the islands, resulting in higher transport and logistics costs that raise the price of goods and services. Delays or failures in island interconnection perpetuate reliance on polluting thermal power plants, which have severe environmental and public health impacts. Emissions of sulfur dioxide (SO_2), nitrogen oxides (NOx), particulate matter (PM), and greenhouse gases (CO_2, CH_4, N_2O) contribute to local air pollution. Thermal plants near urban areas, such as Mytilene on Lesbos, cause significant pollution and visual degradation, affecting tourism and quality of life. Islands without interconnection lack redundancy in power supply, risking blackouts during station failures. Power shortages are projected in the future on islands like Kos, Lesbos, and Lemnos due to growing demand. These factors diminish the islands' ability to transition to a sustainable economy. The ongoing green energy momentum creates opportunities to foster island-based businesses in renewable energy, circular economy, and sustainable tourism sectors. This "green economy" can attract environmentally conscious visitors and generate local jobs (Ministry of Environment and Energy, 2021).

Based on the territorial plan for just and sustainable transition of the islands of the North and South Aegean Sea, the challenge of energy transition is twofold. Initially, it must ensure and maximise the socio-economic benefits for local communities and, and contrarily, it must limit the uneven sustainable development of the islands (Ministry of Environment and Energy, 2021, Ministry of Development and Investment, 2020). The aim is for the policies implemented to bring about the greatest possible benefits from the green transition of the islands, offering prospects in the sectors of energy, tourism and the blue economy. Another objective is to eliminate inequalities between islands resulting from the green transition, with an emphasis on islands that are lagging behind in terms of interconnection or for which no interconnection is planned and which face inherent socio-economic challenges (Ministry of Environment and Energy, 2020, 2021). After finishing consulting with social partners and assessing the socio-economic impact of the energy transition, the designers of the aforementioned major plan have quantified this impact into two categories. Firstly, towards the loss of jobs due to a decline in economic activity, and secondly, towards the opportunity for growth in local economies through the green transition (Ministry of Environment and Energy, 2021). Policies and financial instruments were focused on these two areas (North Aegean Region, 2021; South Aegean Region, 2021; Hellenic Republic, 2020). National and regional smart specialisation strategies have been developed, as well as regional national development plans (North Aegean Region, 2021; South Aegean Region, 2021; Hellenic Republic, 2020; Ministry of Finance, 2020). The aim of all these national and regional strategies is to strengthen and promote entrepreneurship, support the energy transition with a fair labour transition, developing transport infrastructure, integrated territorial investments and supporting the circular economy (Ministry of Environment and Energy, 2021; Ministry of Environment and Energy, 2018).

EUROPEAN FUNDING PROGRAMMES

The decoupling of islands from crude oil and fuel oil in their electricity generation and the production of green energy is one of the main objectives of European funding (Mauger et al., 2024; Marczinkowski et al., 2022; Skjølsvold et al., 2020; Kotzebue & Weissenbacher, 2020). In this light, the EU has directed resources through the European Just Transition Fund, the Island Decarbonisation Fund and the 2021-2027 Regional Development Partnership for the North and South Aegean Regions (Ministry of Development and Investment, 2021). The GReco Islands Initiative constitutes an additional initiative of particular importance towards the sustainable development and climate-neutral economy of the Greek islands is (Apostolopoulos et al., 2025; Ministry of National Economy and Finance, 2024).

European Just Transition Fund

The emergence of climate change and the ongoing degradation of the environment pose a huge threat to Europe and the whole planet. In recent years, to address these challenges, the transition to a climate neutral and circular economy is one of the most important policy objectives for the European Union. The European Union has committed to becoming the first climate-neutral region in the world by 2050 (Tkachuk, 2024). In this light, in December 2019 the European Commission issued a communication on the European Green Deal which aims to reduce greenhouse gas emissions by 55% by 2030 and achieve climate neutrality by 2050 (Theodoropoulou et al., 2025). However, achieving this target leads to wider social and economic impacts in regions that employment and income are heavily dependent on fossil fuels or emissions intensive industries. Significant challenges in the transition to climate neutrality are presented the islands that cover their energy needs mainly from autonomous electrical systems (Katsoulakos, 2019). For this reason, the inclusion of the North and South Aegean islands and Crete in the overall plan for decarbonization is a prerequisite for the approval of the new Just Development Transition Operational Program 2021-2027. According to Burrow (2017), the just transition will affect the whole of society, which is currently dependent on the consumption of coal and oil this is why investments are needed in new forms of energy and new industrial models, as well as in the creation of new jobs. In this context, in January 2020, the European Commission published its proposal for a Just Transition Mechanism which aims to provide support to regions facing serious socio-economic challenges related to the transition towards climate neutrality (Cameron et al., 2020). The Just Transition Mechanism provides targeted support to help mobilise around €55 billion over the period 2021-2027 in the most affected regions, to alleviate the socio-economic impact of the transition (Tserkezos et al., 2025). The Just Transition Mechanism consists of the following three distinct funding pillars, a) Just Transition Fund, b) Special scheme under the InvestEU Programme, and c) European Investment Bank Public Sector Loan Facility 2027. The main pillar of the Just Transition Mechanism plan is the Just Transition Fund, which aims to support the people, economies and environment of territories that face serious socioeconomic challenges stemming from the shift towards a climate-neutral (European Union, 2023). The Just Transition Fund (JTF) is a financial instrument of the European Union, with a total budget of €17.5 billion, aimed at supporting the region's most affected by the transition towards climate neutrality. It aims to reduce regional disparities and address structural changes in the EU, helping these regions adapt to changing labour markets and diversify their economies. Support from the Just Transition Fund will be conditional on the approval of territorial just transition plans, which must include, in particular, a description of the Member State's commitment to the transition process in line with its national energy and climate plans and the EU's 2050 climate neutrality objective. According by Territorial Just Development Transition Plans, Greece will mobilise a total investment of €1.63 billion to alleviate the impact of the energy and climate transition on the local economy and society (European Commission, 2022). In the territories most affected by the energy and climate transition, the Programme will support decarbonisation in Western Macedonia, Megalopolis and adjacent municipalities, and the phasing out of fossil fuel power stations in the islands of North-South Aegean and Crete. The JTF support will foster the diversification and modernization of the economy, including the creation of new jobs and the upskilling and reskilling of people affected by the transition (European Commssion, 2022). According to the Just Development Transition Program in Greece, the allocation of resources per territorial Just Transition Plan is: West Macedonia 1.026 billion Euros, Megalopolis 391 million Euros and North-South Aegean and Crete 211 million Euros.

The 2021-2027 Regional Development Partnership

The 2021-2027 Regional Development Partnership (RDP) is supported by the European Regional Development Fund (ERDF), the Cohesion Fund (CF), the European Social Fund (ESF+), the European Union's Fair Transition Fund (FTT) and the European Maritime, Fisheries and Aquaculture Fund (EMFF). Its public expenditure with regard to Greece for the period 2021-2027 amounted to EUR 26.2 billion of which 20.9 billion was the EU contribution and EUR 5.3 billion was the Greek participation (Ministry of Development and Investment, 2021). The NSRF 2021-2027 was divided into eight Sectoral Programmes amounting to €17.6 billion and thirteen Regional Programmes amounting to €8.1 billion. The Fisheries, Aquaculture and Maritime Programme worth €0.5 billion (Ministry of Development and Investment, 2021). Resources for climate change projects and actions were sourced from the ERDF (30%) and the TA (37%). For the first time, the Discretionary Programme for Environment, Climate Change and Energy was included in the NSRF 2021-2027 (Ministry of Development and Investment, 2021). The EU contribution together with the Greek input to the Environment and Climate Change Programme amounted to €3,607 million and the Fair Energy Transition Programme amounted to €1,629 million. The Regional Programme of the NSRF 2021-2027 of the North Aegean Region reached 394 million euro and the South Aegean Region reached 285 million euro (North Aegean Region, 2022; South Aegean Region, 2021). Compared to the resources of the previous NSRF 2014-2020, the resources have been increased towards the North and South Aegean Region by 196 million euros. However, the allocation of funds to the Regions of Greece should have been more enhanced for the Aegean islands due to their specificities. The policy objectives for the Aegean Sea islands through the NSRF 2021-2027 for the energy sector was the transition of the islands to clean energy. In this light, the promotion of energy autonomy of the islands through the use of renewable energy sources and energy storage systems was supported, respectively. Additionally, among the objectives was the strengthening of the GReco Islands Initiative (Ministry of National Economy and Finance, 2024).

The Initiative of "GReco Islands"

The 'GReco Islands' initiative aiming at socio-economic sustainable development and the transition to a climate neutral economy was funded through the NSRF-2021-2027 (Ministry of National Economy and Finance, 2024). Specifically, it targets 39 islands in the Aegean and Ionian Sea with a population of 100 to 3,500 inhabitants. The GReco Islands Initiative and its timeframe is shown in Table 1 and the objectives of the Initiative in Table 2.

Table 1. "Greco islands" initiative

Region	No expected interconnection or potentially under consideration	Interconnected islands by 2028	Interconnected islands
North Aegean	Agios Efstratios	Oinousses, Fish, Fournoi, Thymena	
South Aegean	Astypalaea, Symi, Megisti, Agathonisi	Kassos, Halki, Pserimos, Lipsi, Telendos, Nisiros, Tilos	Amorgos, Anafi, Donousa, Therasia, Kythnos, Serifos, Sifnos, Heraklia, Antiparos, Schinoussa, Ios, Sikinos, Koufonissi, Folegandros

Source: Ministry of National Economy and Finance (2024)

Table 2. Objectives and scheduling of the Greco Islands Initiative

Goals by 2027	Goal by 2030
• Exclusive use of solar energy for water heating in tourist accommodation • Replacement of the rental fleet with electric vehicles • Replacement of the fleet of vehicles of small and medium-sized enterprises by electric vehicles. • Separate collection in seven different waste streams and recycling • Sustainable waste water management • Smart management of water resources • Recruiting digital nomads	• Penetration of Renewable Energy Sources by 60% • Transforming electricity grids into smart ones

Source: Ministry of National Economy and Finance (2024)

The funding for the projects of the GReco Islands Initiative derives from the Sectoral Programmes "Environment and Climate Change" and "'Just Development Transition"' as well as from the Regional Programmes "North Aegean" and "South Aegean" of the NSRF 2021-2027. Other programmes of the NSRF 2021-2027 may also contribute to the achievement of the objectives of the GReco Islands Initiative should it be required (Government Gazette, 2025). The inclusion and funding of certain projects under the GReco Islands Initiative requires the establishment of a label to eligible islands as a reward for the achievement of specific goals. The GRecoISLAND label features three levels (Government Gazette, 2025). The GReco Candidate label is awarded at the introductory stage and indicates the island's commitment to implement the Local Action Plan. The intermediate label "GReco in Transition" which certifies the positive progress in achieving the objectives of the Local Action Plan. The particular label recognises the island's progress in implementing the work achieved with the Initiative's funding. Lastly, the 'GRecoISLAND' label certifies the successful fulfilling of the final targets.

Island Decarbonisation Fund

The Island Decarbonisation Fund emerged from the tripartite treaty between the EU, the Ministry of Environment and Energy of Greece and the European Investment Bank in November 2024 on the Aegean island of Naxos (European Commission, 2024), which provided Greece with twenty-five million CO2 emission allowances from the EU's greenhouse gas emissions trading scheme for auctioning (Ministry of Environment and Energy, 2024). The resource resulting from the auctioning is allocated towards the carbonization of the Greek islands (HEA, 2024). The available resources of the Fund for the Decarbonisation of the Greek islands amount to €1.6 billion, which in total is estimated to mobilise investments of €3.8 billion for the period 2025-2032 (Ministry of Environment and Energy, 2024). Within these amounts it is projected that the new auctioning of CO2 emissions from the EU's greenhouse gas emission allowance trading scheme will add over €1.5 billion in grants and €3.5 billion in additional investment (Ministry of Environment and Energy, 2024). The structure of the Island Decarbonisation Fund is organised on three axes, namely; new renewable energy installations, interconnection of islands, and installations to increase the penetration of renewable energy (European Commission, 2024). With regard to the first dimension, more than 50% of the fund's budget will be invested on projects in order to develop renewable energy systems and energy storage on islands (Ministry of Environment and Energy, 2024). Eligible projects include investment in energy self-consumption for households and enterprises as well as hybrid green energy production systems including offshore wind parks (Ministry of Environment and Energy, 2024). Concerning the second dimension, projects of electrical interconnection of the islands are funded, respectively (Athens Chamber of Commerce and Industry, 2024). The third dimension includes projects related to the development of cold-ironing systems in the ports of the islands in order to ensure uninterrupted energy supply to ships (KEDE, 2024; Athens Chamber of Commerce and Industry, 2024). Moreover, this dimension includes projects for the charging of electric vehicles and the construction of reservoirs for the production of electricity, storage and management of water resources (European Commission, 2024; Ministry of Environment, 2024).

The function of the Fund and the evaluation of its effectiveness shall be assessed in the future as the process is still at the planning stage and the starting point for the inclusion of concrete projects. According to the Greek government, the management of the Islands' Resilience Fund is undertaken by the Strategic Contracts Unit of the Public Sector Private Property Development Fund (ICTPLUS, 2024). The Strategic Contracts Unit will manage the funded projects and will also be responsible for tendering actions (ICTPLUS, 2024). In 2024, the Greek Electricity Transmission System Operator has announced a new grid line connecting mainland Greece to the island of Kos in the Dodecanese and for this project there will be mobilisation of funding from the Island Decarbonisation Fund (ADMIE, 2024). This is the first project that the government seeks to financially support through the Island Uninterconnection Fund. Via the interconnection of the island of Kos with mainland Greece, the six autonomous electricity systems of the islands of Karpathos, Rhodes, Symi, Kalymnos, Patmos and Arkia will then be connected (ADMIE, 2024).

CONCLUSIONS AND POLICY PROPOSALS

The large number of islands scattered across the Aegean Sea do not present the same characteristics. The further away from the mainland and the smaller in area and population they are, the more the economic, social and energy inequalities worsen. The specific characteristics of the islands may be recognised by Article 174 of the Treaty on the Functioning of the European Union (Official Journal of the European Union, 2016) however; at a practical level, the European and national development plan for the islands, have not yet been able to eliminate existing and unfair inequalities to date. Many of the remote island areas are less developed than mainland Greece. The physical disconnection of the remote islands from mainland Greece and the great distance from the country's capital, Athens, create additional constraints on these islands, including their energy dependence on fossil fuels. However, despite the challenges faced by the islands of the Aegean Sea, as well as their different geographical characteristics and institutional structures, they are vital for the overall sustainable development of Greece. In addition, the formation of a common and well developed European and national vision is required in order to efficiently exploit the several existing comparative advantages. However, this exploitation of advantages will be difficult to be achieved as long as the islands face energy dependence, instability and energy isolation. The impact of the energy transition requires additional compensatory measures such as attracting investment, loan facilities, additional investment and tax incentives. Support for human resources is also required, as well as pursuing sustainable development through circular economy, new technologies, research and innovation, the strengthening of energy communities, and the acceleration of island interconnection. The implementation of energy transition-related projects of the Aegean islands is based on the availability of European funding. The energy transition plans that have been prepared for these islands require the availability of European funding with the amount provided for by European regulations, as well as the percentage that will be contributed by the Greek authorities. European funding aims to eliminate the polluting fossil fuels that many islands use for their electricity. Projects related to the expansion of RES, energy storage systems and the electrical interconnection of the islands with the country's central grid are funded. Interventions are funded in order to support workers affected by the energy transition as well as to enhance the green entrepreneurship. Interventions for the circular economy, the reuse of materials, electromobility as well as the energy upgrade of buildings are also funded. Island regions, due to their specificities, have separate energy transition plans. The EU pays exceptional attention to the sustainable development of islands as well as their energy autonomy and has in recent years directed significant financial resources towards this direction through European funds. In addition, the Island Decarbonization Fund has been directed for the island regions of Greece. However, despite the funding from the Just Transition Fund, the Corporate Regional Development Pact (ΕΣΠΑ) and the Island Decarbonization Fund, there are large discrepancies in the project implementation schedules due to the complexity of the projects and bureaucracy. In many cases, there is an absence of local the societies with a simultaneous large involvement of private initiative.

There are often reactions from local societies, especially for the installations of wind farms. However, there are also islands where local authorities have taken advantage of European resources and carried out pioneering projects, such as the islands of Tilos and Skyros. Additionally, there are also islands where private initiatives have carried out important projects so that these islands can become examples of best practices. Astypalaia and Halki are among these islands. The European support mechanisms for hybrid power plants on the islands are yielding positive results. In addition to all the above, initiatives such as the "EU Clean Energy 2030" and "GReco" ISLANDS" play a crucial role for the Greek islands. Further, once the "EU program Clean Energy 2030" is fully completed, the islands of Astypalaia, Ikaria, Lesvos, Megisti, Tilos, and Psara their electricity will be 100% produced from RES (Papachristou, 2025). Finally, when the "GReco program ISLANDS" is completed in 2030, greenhouse gas emissions will also be reduced by over 55% on thirty-nine islands on which there are 218 settlements (Papachristou, 2025).

In order to reach social consensus on energy projects, it is primarily necessary to institutionalize energy spaces by state bodies and local governments through effective social dialogue. The institutionalisation of energy areas will facilitate investments, alleviate contradictions and activate social participation. On the other hand, energy democracy requires eliminating the differences and discrepancies observed between the energy support of mainland and island Greece. On a practical level that the energy interconnection of the islands with the country's central energy network must be accelerated and, if necessary, additional funding must be directed towards this goal. The projects to connect the islands to the country's main electricity grid must be implemented within the planned timelines. The delays observed favor the oil industry, increase energy production costs at the expense of consumers, the environment, and the climate.

The University of the Aegean operates on the islands of the North and South Aegean Regions, spread across six islands (Lesbos, Lemnos, Rhodes, Samos, Syros, and Chios). It has six schools and eighteen departments. The University of the Aegean was actively involved in the planning and consultation for the energy transition of the islands. One of its departments is the Department of Environment. It has developed postgraduate programs related to energy and climate, such as the postgraduate program "Planetary Environmental Management and Technology." It participates in building energy transition programs and provides professional knowledge and skills in the fields of the environment and related technologies. Its contribution to the islands is significant, which is why its work should receive further support from both the state and local authorities.

The absence of statistical data behalf the EU on its islands, hinders the formulation and implementation of specific targeted policies. The creation of a European institute for island regions would facilitate island-directed financial policies for citizens, businesses and the energy transition of islands, especially the small and remote islands. At the same time, it is necessary to establish specific rules for financial support for islands that are oriented towards achieving climate goals through and through the energy transition. This targeting must take into account the additional costs associated with energy systems and transport. Legislative measures taken for island regions must be consistent with the economic, social and territorial cohesion of islands. Island regions face structural problems due to their dependence on sea and air transport, which increases the cost of living for island residents. The transport of goods, especially energy raw materials, increases the cost of services such as solid waste management. This cost is particularly high on small and remote islands. To address this problem, the European Parliament has called for a reassessment of the subsidies given to maritime and air transport companies and for compensatory measures to be put in place (European Parliament, 2022).

Green maritime infrastructure in relation to the decarbonisation of transport should be a European and national priority. The aim should be twofold; on the one hand, to address the challenges posed by energy supply with fossil fuel imports for electricity generation, transport and heating, and on the other hand, to prioritise the development of RES and the interconnection of islands with the central systems of the countries. However, the development of RES should also take into account the islands' local habitats, as well as their traditional architecture in order to preserve and develop their unique character. It is important that the development of offshore RES, where required, is based on scientific research, taking into account the specific characteristics of the sea basins. The effective strengthening of the administrative capacity of local bodies constitutes a key element in order to facilitate the islands' energy transition. The cooperation of local actors and initiatives for research and innovation in the production of green energy should be at the forefront of European funding programs.

REFERENCES

ADMIE. (2024). Tender for the Corinth-Cos electricity interconnection begins, Independent Electricity Transmission System Operator, https://www.admie.gr/kentro-typoy/deltia-typoy/xekina-o-diagonismos-gia-tin-ilektriki-diasyndesi-korinthoy-ko

ADMIE. (2025a), Ten – year Network Development Plan, Independent Electricity Transmission System Operator, https://www.admie.gr/en/grid/development/ten-year-development-plan

ADMIE. (2025b), Electrical Interconnection between Crete and Attica in operation mode, https://www.admie.gr/kentro-typoy/deltia-typoy/se-leitoyrgia-i-ilektriki-diasyndesi-kritis-attikis

Apostolopoulos, N., Liargovas, P., & Apostolopoulos, S. (2025). European Transnational Cooperation Programmes Involving the Academic Community and Their Contributions to Sustainable Entrepreneurship. In *Real-World Tools and Scenarios for Entrepreneurship Exploration* (pp. 1–24). IGI Global.

Apostolopoulos, N., Makris, I., Apostolopoulos, S., & Dimitrakopoulos, P. (2023). Resilience of rural micro-businesses in an adverse entrepreneurial environment: Adapting to the energy crisis. *Journal of Enterprising Communities: People and Places in the Global Economy*, *18*(5), 1023–1044. DOI: 10.1108/JEC-08-2023-0144

Apostolopoulos, N., Makris, I., Apostolopoulos, S., & Dimitrakopoulos, P. (2024). Resilience of rural micro-businesses in an adverse entrepreneurial environment: adapting to the energy crisis. *Journal of Enterprising Communities: People and Places in the Global Economy*, (ahead-of-print).

Armstrong, H. W., Giordano, B., Kizos, T., Macleod, C., Olsen, L. S., & Spilanis, I. (2012). The European Regional Development Fund and island regions: An evaluation of the 2000-06 and 2007-13 programs. *Island Studies Journal*, *7*(2), 177–198. DOI: 10.24043/isj.267

Athens Chamber of Commerce and Industry, (2024). Island Fragmentation Fund: €5 billion investment in interconnections, renewable energy, water supply and reduction of electricity bills, Athens Chamber of Commerce and Industry (in Greek)

Ballis, A., Paravantis, J. A., & Moschovou, T. (2018). Assessing the tourism potential of the Greek islands of South Aegean. In *2018 9th International Conference on Information, Intelligence, Systems and Applications (IISA)* (pp. 1-7). IEEE. DOI: 10.1109/IISA.2018.8633653

Barone, G., Buonomano, A., Forzano, C., Giuzio, G. F., & Palombo, A. (2021). Increasing renewable energy penetration and energy independence of island communities: A novel dynamic simulation approach for energy, economic, and environmental analysis, and optimization. *Journal of Cleaner Production*, *311*, 127558. DOI: 10.1016/j.jclepro.2021.127558

Blechinger, P., Seguin, R., Cader, C., Bertheau, P., & Breyer, C. (2014). Assessment of the global potential for renewable energy storage systems on small islands. *Energy Procedia*, *46*, 325–331. DOI: 10.1016/j.egypro.2015.01.071

Burrow, S. (2017). *Just Transition. A Report for the OECD*. Just Transition Center.

Caldera, U., Gulagi, A., Jayasinghe, N., & Breyer, C. (2023). Looking island wide to overcome Sri Lanka's energy crisis while gaining independence from fossil fuel imports. *Renewable Energy*, *218*, 119261. DOI: 10.1016/j.renene.2023.119261

Cameron, A., Claeys, G., Midões, C., & Tagliapietra, S. (2020). *How good is the European Commission's Just Transition Fund proposal?* (No. 2020/04). Bruegel Policy Contribution.

Carley, S., Evans, T. P., & Konisky, D. M. (2018). Adaptation, culture, and the energy transition in American coal country. *Energy Research & Social Science*, *37*, 133–139. DOI: 10.1016/j.erss.2017.10.007

Cheung, G., Davies, P. J., & Bassen, A. (2019). In the transition of energy systems: What lessons can be learnt from the German achievement? *Energy Policy*, *132*, 633–646. DOI: 10.1016/j.enpol.2019.05.056

Chua, K. J., Yang, W. M., Er, S. S., & Ho, C. A. (2014). Sustainable energy systems for a remote island community. *Applied Energy*, *113*, 1752–1763. DOI: 10.1016/j.apenergy.2013.09.030

Curto, D., Franzitta, V., Trapanese, M., & Cirrincione, M. (2020). A preliminary energy assessment to improve the energy sustainability in the small islands of the Mediterranean Sea. *Journal of Sustainable Development of Energy. Water and Environment Systems*, *8*(4), 735–753.

Dimitrakopoulos, P., Makris, I., & Apostolopoulos, S. (2024, July). The Impact of Major Energy Indexes on the Economic Growth of EU Countries, According to the Contribution of Industry to GDP. In *International Conference on Applied Economics* (pp. 59-74). Cham: Springer Nature Switzerland.

Europa.eu. (2017). Political declaration on clean energy for EU islands, https://energy.ec.europa.eu/system/files/2017-05/170505_political_declaration_on_clean_energy_for_eu_islands-_final_version_16_05_20171_0.pdf

European Commission. (2020a) Regulation of the European Parliament and of the Council establishing a European Union Just Transition Fund. Available at https://eur-lex.europa.eu/legal-content/EN/TXT/?uri=CELEX%3A52020PC0022 (accessed 14 January 2020).

European Commission. (2020b) The European Green Deal Investment Plan and Just Transition Mechanism explained. Available at https://ec.europa.eu/commission/presscorner/detail/el/qanda_20_24 (accessed 14 January 2020)

European Commission. (2022). *Press release: European Green Deal – Just Transition Mechanism and funding sources*. European Commission. https://ec.europa.eu/commission/presscorner/detail/en/ip_22_3711

European Commission. (2024). Executive Vice-President Šefčovič signs agreement between Greece, European Investment Bank and the Commission on a Decarbonisation Fund for Greek islands, https://climate.ec.europa.eu/news-your-voice/news/executive-vice-president-sefcovic-signs-agreement-between-greece-european-investment-bank-and-2024-11-21_en

European Parliament. (2022). Report on EU islands and cohesion policy: current situation and future challenges, Report - A9-0144/2022, https://www.europarl.europa.eu/doceo/document/A-9-2022-0144_EN.html

Fotopoulou, M., Pediaditis, P., Skopetou, N., Rakopoulos, D., Christopoulos, S., & Kartalidis, A. (2024). A review of the energy storage systems of non-interconnected European islands. *Sustainability (Basel), 16*(4), 1572. DOI: 10.3390/su16041572

Franzitta, V., & Curto, D. (2017). Sustainability of the renewable energy extraction close to the Mediterranean Islands. *Energies, 10*(3), 283. DOI: 10.3390/en10030283

Gabderakhmanova, T., & Marinelli, M. (2022). Multi-energy system demonstration pilots on geographical Islands: An overview across Europe. *Energies, 15*(11), 3908. DOI: 10.3390/en15113908

Genave, A. (2019). Energy vulnerability in the Southwest Indian Ocean islands. *Journal of the Indian Ocean Region, 15*(1), 40–57. DOI: 10.1080/19480881.2019.1560760

Georgiou, P. N., Mavrotas, G., & Diakoulaki, D. (2011). The effect of islands' interconnection to the mainland system on the development of renewable energy sources in the Greek power sector. *Renewable & Sustainable Energy Reviews, 15*(6), 2607–2620. DOI: 10.1016/j.rser.2011.03.007

Gillespie, R. G. (2009). Encyclopedia of islands. *University of California, 1074.*

Government Gazette, (2016). Law 4414 "New support scheme for power plants using renewable energy sources and high-efficiency cogeneration of electricity and heat - Provisions for the legal and operational separation of the supply and distribution sectors in the natural gas market and other provisions," Government Gazette A' 149/9-8-2016, (in Greek).

Government Gazette. (2025). GReco Islands Charter: Determining the conditions for co-financing GReco Islands projects under Article 21 of Law No. 4936/2022 from the NSRF Programmes 2021-2027 and establishment of the "GRecoIS-LAND" label, Sheet No. 2119/5.5.2025, (in Greek), https://www.espa.gr/elibrary/YA6422715042025-FEK2119B05052025_Charta-GReco-islands.pdf

Government Gazette, (2018). Law 4546 "Transposition into Greek law of Directive 2014/89/EU "establishing a framework for maritime spatial planning" and other provisions", Government Gazette A 101/12.06.2018, (in Greek).

Greek Ministry of Energy. (2021), Territorial plan for a just and developmental transition of the North and South Aegean Islands & Crete (SDAM), sdam.gr, https://sdam.gr/sites/default/files/2021-05/ΕΔΑΦΙΚΟ%20ΣΧΕΔΙΟ%20ΝΗΣΩΝ%20_ΔΙΑΒΟΥΛΕΥΣΗ%2027042021.pdf

Groppi, D., Garcia, D. A., Basso, G. L., Cumo, F., & De Santoli, L. (2018). Analysing economic and environmental sustainability related to the use of battery and hydrogen energy storages for increasing the energy independence of small islands. *Energy Conversion and Management*, *177*, 64–76. DOI: 10.1016/j.enconman.2018.09.063

Hatziargyriou, N., Margaris, I., Stavropoulou, I., Papathanassiou, S., & Dimeas, A. (2017). Noninterconnected Island systems: The Greek case. *IEEE Electrification Magazine*, *5*(2), 17–27. DOI: 10.1109/MELE.2017.2685739

HEDNO. (2023). The Electricity Market of Non-Interconnected Islands in Greece, Islands Network Operation Department, Argiro Magkanioti Head of Electricity Generation Development and Management Section, https://clean-energy-islands.ec.europa.eu/system/files/2023-01/HEDNO%20-%20Operator%20of%20NIIs.pdf

Hellenic Association for Energy Economics. (2023). Assessing the energy transition of the Greek island, *IN FOCUS,* Issue 4, at https://www.haee.gr/media/5374/haee_2023_infocus_issue_4_standard_final.pdf]

Hellenic Republic. (2020). *National Development Program 2021-2025*. Hellenic Republic - Ministry of Development and Investment. (in Greek)

Hellenic Statistical Authority. (2025). Greece in numbers (in Greek) https://www.statistics.gr/documents/20181/18573742/GreeceInFigures_2025Q1_GR.pdf/871e88cc-0188-a27b-0a6a-3bc8e03f56b3

ICTPLUS. (2024). Hellenic Privatisation Fund: how the Island Resettlement Fund will work (in Greek) https://www.ictplus.gr/taiped-pos-tha-leitourgei-to-tameio-apanthrakopoiisis-nision/

IEA. I. (2023). CO2 Emissions in 2022. *IEA, Paris.*

Iliopoulou, C., Kepaptsoglou, K., & Schinas, O. (2018). Energy supply security for the Aegean islands: A routing model with risk and environmental considerations. *Energy Policy, 113*, 608–620. DOI: 10.1016/j.enpol.2017.11.032

Ioannidis, A., & Chalvatzis, K. J. (2017). Energy supply sustainability for Island Nations: A study on 8 Global Islands. *Energy Procedia, 142*, 3028–3034. DOI: 10.1016/j.egypro.2017.12.440

Ioannidis, A., Chalvatzis, K. J., Li, X., Notton, G., & Stephanides, P. (2019). The case for islands' energy vulnerability: Electricity supply diversity in 44 global islands. *Renewable Energy, 143*, 440–452. DOI: 10.1016/j.renene.2019.04.155

IPTO. (2024). Public consultation: Preliminary draft of the Ten Year Development Plan of HETS, for the period 2025-2034, https://www.admie.gr/en/anakoinoseis/announcement/public-consultation-preliminary-draft-ten-year-development-plan-hets

Jaramillo-Nieves, L., & Del Río, P. (2010). Contribution of renewable energy sources to the sustainable development of islands: An overview of the literature and a research agenda. *Sustainability (Basel), 2*(3), 783–811. DOI: 10.3390/su2030783

John, V. (2024). Energy Islands in Greece: Are they Feasible in Aegean Sea? *International Journal of Advanced Multidisciplinary Research and Studies, 4*(2), 148–155. DOI: 10.62225/2583049X.2024.4.2.2454

Kaldellis, J. K., Gkikaki, A., Kaldelli, E., & Kapsali, M. (2012). Investigating the energy autonomy of very small non-interconnected islands: A case study: Agathonisi, Greece. *Energy for Sustainable Development, 16*(4), 476–485. DOI: 10.1016/j.esd.2012.08.002

Karampela, S., Kizos, T., & Spilanis, I. (2014). Accessibility of islands: Towards a new geography based on transportation modes and choices. *Island Studies Journal, 9*(2), 293–306. DOI: 10.24043/isj.307

Karystianos, M. K., Pitas, C. N., Efstathiou, S. P., Tsili, M. A., Mantzaris, J. C., Leonidaki, E. A., Voumvoulakis, E. M., & Sakellaridis, N. G. (2021). Planning of Aegean Archipelago Interconnections to the Continental Power System of Greece, Energies, 14, 3818, Katsaprakakis, D. A., Proka, A., Zafirakis, D., Damasiotis, M., Kotsampopoulos, P., Hatziargyriou, N., ... & Xevgenos, D. (2022). Greek Islands' energy transition: From lighthouse projects to the emergence of energy communities. *Energies*, *15*(16), 5996.

Katsoulakos, N. M. (2019). An overview of the Greek Islands' autonomous electrical systems: Proposals for a sustainable energy future. *Smart Grid and Renewable Energy*, *10*(04), 55–82. DOI: 10.4236/sgre.2019.104005

Kavadias, K. A., Zafirakis, D., Kondili, E., & Kaldellis, J. K. (2007). The contribution of renewables on reducing the electricity generation cost in autonomous island networks. In *2007 International Conference on Clean Electrical Power* (pp. 777-783). DOI: 10.1109/ICCEP.2007.384302

KEDE. (2024) New 2 billion fund for the carbonization of the Greek islands, Central Union of Greek Municipalities (in Greek) https://kede.gr/neo-tameio-2-dis-gia-tin-apanthrakopoiisi-ton-ellinikon-nision/

Kona, A., Monforti-Ferrario, F., Bertoldi, P., Baldi, M. G., Kakoulaki, G., Vetters, N., Thiel, C., Melica, G., Lo Vullo, E., Sgobbi, A., Ahlgren, C., & Posnic, B. (2021). Global Covenant of Mayors, a dataset of greenhouse gas emissions for 6200 cities in Europe and the Southern Mediterranean countries. *Earth System Science Data*, *13*(7), 3551–3564. DOI: 10.5194/essd-13-3551-2021

Kotzebue, J. R., & Weissenbacher, M. (2020). The EU's Clean Energy strategy for islands: A policy perspective on Malta's spatial governance in energy transition. *Energy Policy*, *139*, 111361. DOI: 10.1016/j.enpol.2020.111361

Kougias, I., Szabó, S., Nikitas, A., & Theodossiou, N. (2019). Sustainable energy modelling of non-interconnected Mediterranean islands. *Renewable Energy*, *133*, 930–940. DOI: 10.1016/j.renene.2018.10.090

Lal, R., & Kumar, S. (2022). Energy security assessment of small Pacific Island Countries–Sustaining the call for renewable energy proliferation. *Energy Strategy Reviews*, *41*, 100866. DOI: 10.1016/j.esr.2022.100866

Lassalle, J. M., Martínez, D. F., & Fernández, L. V. (2022). Optimisation of hybrid renewable energy systems on islands: A review. *Island Studies Journal*, *17*(1), 221–242. DOI: 10.24043/isj.167

Makris, I., Apostolopoulos, S., & Anastasopoulou, E. E. (2024). An Entrepreneurial Perspective on the Transition of Lignite Rural Areas to a New Regime within a Suffocating Timeframe. *Administrative Sciences*, *14*(4), 64. DOI: 10.3390/admsci14040064

Marczinkowski, H. M., Østergaard, P. A., & Mauger, R. (2022). Energy transitions on European islands: Exploring technical scenarios, markets and policy proposals in Denmark, Portugal and the United Kingdom. *Energy Research & Social Science*, *93*, 102824. DOI: 10.1016/j.erss.2022.102824

Mauger, R. (2021). Cast Away? How EU Energy Law Provides for a Just Transition for EU Islands. *OGEL*, *2021*(1), 1–16.

Mauger, R., Diestelmeier, L., & Nieuwenhout, C. (2024). Harnessing EU legal concepts for the energy transition on islands. *Journal of World Energy Law and Business*, *17*(3), 167–183. DOI: 10.1093/jwelb/jwae002

Melica, G., Treville, A., Franco De Los Rios, C., Baldi, M., Monforti-Ferrario, F., Palermo, V., & Bertoldi, P. (2022). Covenant of Mayors: 2021 assessment. *Climate change mitigation and adaptation at local level (European Commission. Joint Research Centre, JRC130957*, 2022.

Meschede, H., Bertheau, P., Khalili, S., & Breyer, C. (2022). A review of 100% renewable energy scenarios on islands. *Wiley Interdisciplinary Reviews. Energy and Environment*, *11*(6), e450. DOI: 10.1002/wene.450

Meschede, H., Child, M., & Breyer, C. (2018). Assessment of sustainable energy system configuration for a small Canary island in 2030. *Energy Conversion and Management*, *165*, 363–372. DOI: 10.1016/j.enconman.2018.03.061

Milone, D., Curto, D., Franzitta, V., Guercio, A., Cirrincione, M., & Mohammadi, A. (2022). An economic approach to size of a renewable energy mix in small islands. *Energies*, *15*(6), 2005. DOI: 10.3390/en15062005

Ministry of Development and Investment (2020). National Development Program 2021–2025, (in Greek).

Ministry of Development and Investment. (2021) Just Transition Development Plan 2021-2027. NSDP 2014-2020. https://2014-2020.espa.gr/el/Documents/Kalliris_17-6.pdf

Ministry of Environment and Energy. (2018). *National Strategy for the Circular Economy*. Hellenic Republic. (in Greek)

Ministry of Environment and Energy, (2021). Territorial Just Transition Development Plan for the North and South Aegean Islands and Crete (in Greek)

Ministry of Environment and Energy. (2024). Establishment of an Island Resettlement Fund, (Greek), https://ypen.gov.gr/systinetai-tameio-apanthrakopoiisis-ton-nision/

Ministry of Finance. (2020). *Explanatory Report on the 2021 Budget, Hellenic Republic*. Ministry of Finance. (in Greek)

Ministry of National Economy and Finance. (2024). Presentation of the GReco Islands initiative and Integrated Spatial Investments at the 15th Panhellenic Conference of Small Islands, (in Greek), https://www.espa.gr/el/Pages/NewsFS.aspx?item=1810

North Aegean Region. (2021). *Strategic Operational Program for the North Aegean Region*. North Aegean Region. (in Greek)

North Aegean Region, (2022). North Aegean Regional Programme 2021-2027, (in Greek)

Notton, G. (2015). Importance of islands in renewable energy production and storage: The situation of the French islands. *Renewable & Sustainable Energy Reviews*, *47*, 260–269. DOI: 10.1016/j.rser.2015.03.053

Official Journal of the European Union. (2021) Regulation (EU) 2021/1056 of the European Parliament and of the Council of 24 June 2021 establishing the Just Transition Fund. 30 June. Brussels: Publications Office of the European Union Official Journal of the European Union, (2016). Consolidated Versions of the Treaty on European Union and the treaty on the Functioning of the European Union (2016/C 202/01), https://eur-lex.europa.eu/resource.html?uri=cellar:9e8d52e1-2c70-11e6-b497-01aa75ed71a1.0006.01/DOC_1&format=PDF

Pablo-Romero, M. D. P., & Sánchez-Braza, A. (2015). Productive energy use and economic growth: Energy, physical and human capital relationships. *Energy Economics*, *49*, 420–429. DOI: 10.1016/j.eneco.2015.03.010

Papachristou, D., (2025). Presentation at the event organized by the Waste, Energy and Water Regulatory Authority in Thessaloniki in the framework of the 'Helexpo Dialogues' conference, the timetable for the electrical interconnection of the islands in March 2025 (in Greek).

Papapostolou, C. M., Tzanes, G. T., & Kondili, E. M. (2019). Optimisation of the integrated water–energy systems in small group of islands communities. []. Elsevier.]. *Computer-Aided Chemical Engineering*, *46*, 1735–1740. DOI: 10.1016/B978-0-12-818634-3.50290-3

Peñalvo-López, E., Andrada-Monrós, C., León-Martínez, V., & Valencia-Salazar, I. (2024). Assessing energy transition in Mediterranean islands. A review. *E-Prime-Advances in Electrical Engineering. Electronics and Energy, 9*, 100719.

Priano, F. H., Armas, R. L., & Guerra, C. F. (2016). A model for the smart development of island territories. In *Proceedings of the 17th international digital government research conference on digital government research* (pp. 465-474). DOI: 10.1145/2912160.2912187

Rajaeian, E., & Davarani, R. Z. (2023). Supplying the energy of islands with geopolitical positions using distributed generations to increase energy security. *International Journal of Hydrogen Energy, 48*(42), 16057–16071. DOI: 10.1016/j.ijhydene.2023.01.114

Ramos-Real, F. J., Barrera-Santana, J., Ramírez-Díaz, A., & Perez, Y. (2018). Interconnecting isolated electrical systems. The case of Canary Islands. *Energy strategy reviews, 22*, 37-46.

Sánchez, A. S., Junior, E. P., Gontijo, B. M., de Jong, P., & dos Reis Nogueira, I. B. (2023). Replacing fossil fuels with renewable energy in islands of high ecological value: The cases of Galápagos, Fernando de Noronha, and Príncipe. *Renewable & Sustainable Energy Reviews, 183*, 113527. DOI: 10.1016/j.rser.2023.113527

Santopietro, L., & Scorza, F. (2021). The Italian experience of the covenant of mayors: A territorial evaluation. *Sustainability (Basel), 13*(3), 1289. DOI: 10.3390/su13031289

Skjølsvold, T. M., Ryghaug, M., & Throndsen, W. (2020). European island imaginaries: Examining the actors, innovations, and renewable energy transitions of 8 islands. *Energy Research & Social Science, 65*, 101491. DOI: 10.1016/j.erss.2020.101491

South Aegean Region. (2021). *Strategic Operational Program for the South Aegean Region*. South Aegean Region. (in Greek)

South Aegean Region, (2021). South Aegean Regional Programme 2021-2027 (in Greek)

Southeastern European Energy Institute. (2020). *Just Development Transition Plan: Current situation and perspectives for regions in energy transition in Greece*. Southeastern European Energy Institute. (in Greek)

Spilanis, I., Kizos, T., & Petsioti, P. (2012). Accessibility of Peripheral Regions: Evidence from Aegean Islands (Greece). *Island Studies Journal, 7*(2), 199–214. DOI: 10.24043/isj.268

Spilanis, I., Kizos, T., Vaitis, M., & Koukourouvli, N. (2013). Measuring the economic, social and environmental performance of European island regions: Emerging issues for European and regional policy. *European Planning Studies*, *21*(12), 1998–2019. DOI: 10.1080/09654313.2012.722970

Spilanis, I., Kondili, J., & Gryllaki, S. (2003). Measuring the attractiveness of small islands. A tool for sustainability. In *International conference on sustainability indicators. Malta.*–http://www. aegean. gr/lid/internet/elliniki_ekdosi/Dimosieuseis/Malta_paper. pdf *[2015 02 21]*.

Stergiopoulou, E. (2025). Presentation at the event organized by the Waste, Energy and Water Regulatory Authority in Thessaloniki in the framework of the conference 'Helexpo Dialogues', the timetable for the electrical interconnection of the islands, March 2025, (in Greek)

Tandon, A. (2021) Transition finance: Investigating the state of play: A stocktake of emerging approaches and financial instruments. Report no. 179, OECD, 3 August. Paris: OECD Publishing

Theodoropoulou, S., Sabato, S., & Akgüç, M. (2025). National eco-social policies in the framework of EU just transition: The cases of Greece, Italy, Portugal and Spain. *Global Social Policy*, *25*(1), 86–111. DOI: 10.1177/14680181241291122

Tkachuk, O. (2024). The European Union in its pursuit of being a global leader in climate neutrality. *Global Policy*, *15*(S8), 69–74. DOI: 10.1111/1758-5899.13480

Tranberg, B., Corradi, O., Lajoie, B., Gibon, T., Staffell, I., & Andresen, G. B. (2019). Real-time carbon accounting method for the European electricity markets. *Energy Strategy Reviews*, *26*, 100367. DOI: 10.1016/j.esr.2019.100367

Tsagkari, M. (2020). Local energy projects on islands: Assessing the creation and upscaling of social niches. *Sustainability (Basel)*, *12*(24), 10431. DOI: 10.3390/su122410431

Tsagkari, M., & Roca Jusmet, J. (2020). Renewable energy projects on isolated islands in Europe: A policy review. *International Journal of Energy Economics and Policy*, *10*(5), 21–30. DOI: 10.32479/ijeep.9683

Tserkezos, D., Apostolopoulos, S., & Prodromou, A. (2025). Sustainable Entrepreneurship in Under-Resourced Rural Areas in Relation to The EU Rural Vision: The Greek Case. In Entrepreneurial Opportunities in Disadvantaged Rural Communities (pp. 1-22). IGI Global Scientific Publishing.

Tzanes, G., Zafirakis, D., Makropoulos, C., Kaldellis, J. K., & Stamou, A. I. (2023). Energy vulnerability and the exercise of a data-driven analysis protocol: A comparative assessment on power generation aspects for the non-interconnected islands of Greece. *Energy Policy*, *177*, 113515. DOI: 10.1016/j.enpol.2023.113515

Uche-Soria, M., & Rodríguez-Monroy, C. (2018). Special regulation of isolated power systems: The Canary Islands, Spain. *Sustainability (Basel)*, *10*(7), 2572. DOI: 10.3390/su10072572

Weisser, D. (2004). Power sector reform in small island developing states: What role for renewable energy technologies? *Renewable & Sustainable Energy Reviews*, *8*(2), 101–127. DOI: 10.1016/j.rser.2003.10.002

Wolf, F., Surroop, D., Singh, A., & Leal, W. (2016). Energy access and security strategies in Small Island Developing States. *Energy Policy*, *98*, 663–673. DOI: 10.1016/j.enpol.2016.04.020

Zafeiratou, E., & Spataru, C. (2017). Potential economic and environmental benefits from the interconnection of the Greek islands. *International Journal of Global Warming*, *13*(3-4), 426–458. DOI: 10.1504/IJGW.2017.087211

Zisopoulos, F. K., Noll, D., Singh, S. J., Schraven, D., De Jong, M., Fath, B. D., Goerner, S., Webster, K., Fiscus, D., & Ulanowicz, R. E. (2023). Regenerative economics at the service of islands: Assessing the socio-economic metabolism of Samothraki in Greece. *Journal of Cleaner Production*, *408*, 137136. DOI: 10.1016/j.jclepro.2023.137136

Chapter 2
Macroeconomic Approaches to Energy Efficiency:
The Case of Indicative EU Member States

Panos Dimitrakopoulos
https://orcid.org/0009-0000-8188-4255
University of Peloponnese, Greece

Andreas Masouras
https://orcid.org/0000-0001-8989-3930
Neapolis University, Pafos, Cyprus

ABSTRACT

This chapter explores the macroeconomic approach to energy efficiency, focusing on EU policy and its implementation in selected member states. Efficiency is defined as "the ratio of energy services delivered to energy consumed" and differs from savings. EU policy is structured around the Ecodesign, Buildings, and Energy Efficiency Directives. Instruments like the EEOS contribute 34% of required savings under Article 7. It is emphasized that "the transition to a clean energy system requires mobilizing private capital," while green finance and tools like green bonds gain importance. Despite progress, the Energy Intensity Index is limited by structural differences. Comparative analysis shows significant policy divergence due to institutional and economic factors. The chapter concludes that "policy instruments must be adapted to national contexts," highlighting the need for coordinated strategies, flexible indicators, and stronger ESCO markets to support a more effective energy transition across diverse EU realities.

DOI: 10.4018/979-8-3373-5302-9.ch002

INTRODUCTION

The Concept of Energy Efficiency

Energy efficiency represents a method of managing and limiting the growth of energy consumption. A process or system is considered more energy efficient if it provides more services for the same energy input, or the same or more services for less energy input. In recent years, energy policy makers have remained focused on improving energy performance, and several studies have underscored the significance of enhancing energy efficiency (Lovins, 2017). These studies, from an economic perspective, concentrate on ways to improve energy performance without adversely affecting economic growth.

Specifically, in 2014, the "International Energy Efficiency Scorecard" published by the American Council for an Energy-Efficient Economy (ACEEE) emphasized that "Countries can conserve their resources, address global warming, stabilize their economies, and reduce the cost of their economic outputs by using energy more efficiently – a highly feasible goal" (Molina, 2014). Improvements in energy efficiency are widely accepted, as they provide the same energy service and are often accompanied by an increase in supply and an enhancement in quality (Beccali et al., 2015). Numerous modeling techniques have been developed to address energy efficiency, as it is a concept that is difficult to define.

This concept is often confused with energy savings, although savings imply using less energy, whereas efficiency entails achieving a given output through more moderate and rational use of resources (Jebali et al., 2017; Seidl et al., 2017). Energy efficiency, defined as the ratio of energy services delivered to the energy consumed, can be understood as the quotient of energy services over energy inputs. Such a relationship implies that every unit of purchased energy is being utilized to its fullest potential. It primarily constitutes a technical (and historically rooted) process driven by capital turnover, in which outdated equipment is replaced by newer and more efficient technologies (Herring, 2006).

Despite the significant progress in the design and adoption of energy efficiency policies across the 27 EU Member States, a notable variation persists among countries in terms of their macroeconomic approach, policy instruments, and achieved outcomes. This study seeks to address this gap by focusing on the comparative documentation and assessment of energy efficiency policies implemented in selected EU Member States.

POLICY INSTRUMENTS AND STRATEGIC FRAMEWORKS FOR ENERGY EFFICIENCY

EU Energy Efficiency Policy Design

The European Union's policy on energy efficiency has been implemented through three main directives: Ecodesign Directive (2009/125/EC), Energy Performance of Buildings Directive (EPBD - 2010/31/EU), and Energy Efficiency Directive (EED - 2012/27/EU). As their titles suggest, the Ecodesign Directive addresses the energy performance of products, the EPBD covers aspects of energy use in buildings, and the EED is a broader directive. Additionally, energy efficiency policy or energy-saving measures focused on larger organizations include the requirement for energy audits by high energy users (Article 8 of the Energy Efficiency Directive) and participation in the EU Emissions Trading Scheme (Fawcett & Hampton, 2020).

EU countries are required to use energy more efficiently at all stages of the energy chain from production to final consumption. Under the EED, Article 3 obliges Member States to set an indicative national energy efficiency target. This target includes resource savings from all actions at both the national and EU levels, including new actions under other provisions of the EED. The directive includes several provisions and new pathways for achieving resource savings—including energy audits and energy management systems, the Energy Efficiency Obligation Scheme (EEOS), and measures targeting public sector buildings (Community, 2014).

Furthermore, Article 7 of the EED defines national energy savings targets for the 2014–2020 period, to be achieved through the EEOS and/or "alternative measures." The EEOS is a key policy instrument: analysis of Member States' reports shows it is expected to deliver 34% of the savings under Article 7—the largest contribution of any single policy instrument. Other savings are expected from funding programs or grants (19%), taxes (14%), regulations/voluntary agreements (11%), standards (9%), with smaller contributions from education, national energy efficiency funds, labeling, and other policy measures (Fawcett et al., 2019).Experts and researchers in energy policy must indeed identify exogenous factors and help design policy instruments capable of exerting environmental influence. Conversely, governments' responsibility primarily lies in implementing policy instruments and providing the financial resources needed by the stakeholders involved in policy research, design, enforcement, and evaluation. According to modern policy approaches, individuals are increasingly seen as the agents responsible for energy sustainability and climate change mitigation (Labanca & Bertoldi, 2018).

Many countries worldwide have implemented policies to improve energy efficiency. Regardless of the progress made over time, there remains substantial potential for additional energy savings across various industrial sectors through public policy interventions that remove previously identified energy efficiency barriers. According to various scholars and policymakers, a significant portion of this potential can be successfully realized through the development of a competitive and sustainable ESCO (Energy Service Companies) market (Barta, 2008).Previous studies support the view that ESCOs reduce energy use by establishing diffusion pathways for promoting energy-efficient technologies. The relationships between ESCO indicators and country-specific factors suggest that either ESCO inefficiency or underdeveloped ESCO markets may hinder improvements in energy efficiency in many countries worldwide (Barta, 2008).

At the EU level, efforts have been made to frame or reframe energy efficiency as a supranational issue, linked to internal market competences, energy security concerns, and environmental objectives. It is estimated that at the current rate of savings, a transition to a 100% renewable energy system appears unfeasible before the end of the century. Keeping energy demand at today's levels would drastically shorten the transition period, but is not sufficient. This is due to the higher capital requirements of renewable energy technologies, which may crowd out non-energy sectors in capital allocation. Such a shift could increasingly burden economic growth, slowing and eventually blocking the transition as the economy moves toward stagnation (Dupont et al., 2021). Given that these barriers reduce the attractiveness of new measures, an effective financial framework must be developed to ensure optimal interaction between private and public actors (Andersson, 2020). The EU has increased the volume of public funding to meet its targets and support the transition to a clean energy system. However, there is still a need to further unlock private financing, particularly with respect to investments. To this end, risk-sharing instruments, technical assistance programs, and the optimal use of public funds—including the European Structural and Investment Funds, in conjunction with financing through the European Fund for Strategic Investments—are critical parameters. It is estimated that an additional €177 billion annually will be required during the 2021–2030 period to achieve the EU's 2030 energy and climate targets (Gkonis et al., 2020).

Moreover, efforts to (re)frame the EU's energy policy strategy have entailed consolidating authority and shaping the framework within which this policy will be designed, setting the tone for subsequent negotiations. Within the broader development of the energy policy domain, the energy efficiency framework plays a long-term role in reinforcing or legitimizing authority and in preparing and justifying future policy developments by limiting the scope for (legitimate) contestation. Framing energy efficiency measures in specific ways (e.g., linking them to climate change efforts and the EU's global role) may reduce opposition to the rationale behind such policy proposals. The framing process can also form part of a strategy to (de)politicize energy efficiency, depending on whether the frame demands high-politics or low-politics decision-making solutions (Dupont et al., 2020). EU policy sets efficiency requirements for organizations of all sizes. Energy efficiency is one of the five closely connected and mutually reinforcing dimensions of the EU's Energy Union strategy, which was developed to achieve a secure, affordable, and climate-friendly European energy system.

Innovation and Green Energy

Annual public energy R&D funding must be doubled, public institutions for energy innovation must evolve, and energy innovation and its diffusion must be accelerated to enable future reductions in energy emissions (Meckling et al., 2022). The financial crisis appears to be linked to an increase in total energy expenditures but not in new expenditures for clean energy R&D, as governmental responses—especially in major countries—have not aligned with the notion of a green recovery (Nahm et al., 2022). For the expansion and governance of energy efficiency, governments must promote global competition, which, at the local level, requires new government agencies to design and manage industrial policy. This would build broader political support for energy R&D, which is directly linked to economic policy-making and promotes the commercialization of technology while simultaneously mobilizing larger private R&D expenditures (Atkinson, 2022). On the other hand, at the international level, global competition can be promoted through collaborative standards and rules, as well as open trade regulations. For these to function reciprocally, they must be linked to public commitments and be reinforced to attract investment (Trinh et al., 2023).

Green Investment Resources

Mobilizing green financial resources is key to meeting the growing energy demand in Southeast Asia, fueled by economic growth, population increase, and enhanced access to energy resources. A significant shift in investment patterns is required to ignite green financing, which is an increasing focus of governmental policies. More and more, new measures are being implemented in Asia to support green finance, such as green bond standards, green bond grant programs, and sovereign green bonds. Green bonds are attracting increasing interest across Asia and globally as an alternative financing source for low-carbon investments. The green bond market has grown rapidly, from USD 3.4 billion in 2012 to USD 156 billion in 2017 (In et al., 2020). To attract private financing for low-carbon projects, the European Investment Bank and thc World Bank were the first to issue green bonds in 2007 and 2008, respectively. After entering the green bond market in 2015, China is now the largest green bond issuer in the world. In 2016 and 2017, China issued green bonds worth USD 34 billion and USD 31 billion, respectively (Azhgaliyeva et al., 2020).

TECHNOLOGICAL INNOVATION AND ITS SIGNIFICANCE FOR ENERGY EFFICIENCY

Currently improving energy efficiency through industrial restructuring is challenging, and more advanced technologies are required to address problems of high pollution and high energy consumption in industrial development (Wang et al., 2020). As a result, innovation-driven development strategies have become a reliable method for reducing energy intensity. When promoting such a strategy, technological innovation capability and energy efficiency are important complementary indicators for measuring the success of an energy strategy (Jin et al., 2019). As one of the largest emerging economies, China is accelerating its transformation through innovation, and its innovation capacity and performance have attracted attention from scholars and policymakers worldwide. In 2017, China's R&D intensity reached 2.12%, ranking second globally after the United States. However, there has always been debate regarding the outcome and energy efficiency in the process of transforming R&D input into output. Efficiency-based innovation strategies emphasize the degree of innovation in products or technologies brought to market. These strategies focus on speed and cost, reflecting first-mover and cost-advantage benefits over competitors (Yuan, 2019).

China's increased capacity for technological innovation, as explored in panel data collected from 284 cities, shows that technological innovation exerts a consistent impact on energy efficiency across different types of cities in China (Zhang et al., 2022). Several industry-related review studies have attempted to propose managerial energy policies, e.g., state intervention policies, taking into account industry characteristics and the scale of recovery. Distributed Energy Systems (DES) offer better performance, flexibility, and cost-effectiveness compared to centralized generation systems. Given these advantages, decentralization through DES is regarded as one of the key pillars of the 21st-century energy transition, as renewable energy sources take a leading role in these systems (Nadeem et al., 2023). Supportive policies have increased solar PV capacity in the UK, Germany has been a pioneering country in electricity supply and sustainable energy transition, China introduced such systems primarily in transmission and emissions management, while developing countries like Nigeria and Pakistan focused on rural electrification. Overall, 165 countries have active renewable energy policies across electricity generation, transportation, heating, and cooling. DES based on renewable sources account for nearly 26% of the global energy supply and operate in around 32 countries (Nadeem et al., 2023). Furthermore, a research gap remains regarding the evaluation of Energy Efficiency Programs (EEPs) and the comparison between barriers and benefits internationally. For industrial EEPs, overall outcome analysis is more complex than for commercial and residential programs. Studies indicate that mandatory penalty-based policies are more effective than subsidies in motivating traditional suppliers (Safarzadeh et al., 2022), CO2 emissions are positively correlated with economic growth and GDP, but negatively correlated with technological innovation and renewable energy use (Wang & Wang, 2020). Technological advancements in machinery, equipment, and energy structure ratio have improved energy efficiency in China's manufacturing sector by an average of 7.1% annually from 1997 to 2014 (Zhu et al., 2019). From an environmental perspective, reduced energy use remains a cornerstone in global greenhouse gas mitigation. However, without fully internalizing external costs, mitigation may not be a high strategic priority for industries. Production costs and financial revenues alone do not reflect the success or failure of such policies and practices. Nevertheless, reduced energy costs can significantly impact the profitability of even small enterprises (Johansson & Thollander, 2018).

THE ROLE OF THE ENERGY INTENSITY INDEX AND MEASURING ENERGY EFFICIENCY

Analyses describing various dimensions for quantifying cross-country energy efficiency parameters led to the creation of the energy intensity index, defined as the ratio of energy input to output (e.g., GDP). This can reflect a country's energy efficiency while accounting for sectoral and structural characteristics (Schipper, 2004). The EEI has been criticized for not measuring underlying technical efficiency, as its value reflects the specific energy mix of the expanding sector in the economy. The EEI is also affected by changes in energy input composition and energy-for-labor substitution. None of these factors can be adequately captured through a simplistic measurement of energy efficiency (Velasco-Fernández et al., 2020). The lack of contextualized information reveals that while groups of countries may share similar EEI values, they can differ substantially in economic development and wealth distribution. Thus, at the national level, the EEI may obscure the role of the service sector, which can increase energy efficiency but at the cost of energy intensity, altering each sector's energy mix at the microeconomic level (Giampietro et al., 2011). For example, in technical performance terms, electricity generation is relatively inefficient due to the significant energy and capital required to extract and convert primary energy sources into a smaller quantity of secondary energy. Despite a significant increase in total energy consumption accompanying the growth of the emerging electricity industry in the US between 1910 and 1950, official statistics showed a sharp decrease in the economy's economic energy intensity, measured as real GDP per unit of energy input (Velasco-Fernández et al., 2020).

Measuring energy efficiency pertains to the energy required to deliver the same service, which does not necessarily align with energy produced from renewables.

Although the use of renewable energy can increase energy efficiency in the long term, it also requires innovations, technologies, and financial resources that may not make it immediately energy/economically efficient (Dong et al., 2020). The concept of a socio-economic society's metabolic pattern identifies the set of expected functions needed to reproduce societal structures. The metabolic pattern can be represented as the combination of final uses of energy carriers per unit of human activity (i.e., metabolic rates). These expected functions must ensure: (i) a desired quality of life compatible with expectations and values to stabilize the social fabric; (ii) a sustainable set of integrated processes in the technosphere compatible with technical, institutional, and economic constraints to provide stability and adaptability; and (iii) a feasible profile of flows exchanged with the environment (trade with other societies and collection/disposal from/to the local biosphere), thus preventing the loss of favorable boundary conditions that provide stability and adaptability (Velasco-Fernández et al., 2020). Naturally, the energy intensity index cannot serve as an unquestionable basis for assessing an economy's energy efficiency, as it is determined by many factors including not only energy use effectiveness but also economic structure, type of industrial base, country size, climate, living standards, energy service availability, and consumer behavior. These factors explain, for example, the low EEI scores for Nigeria, Egypt, or Colombia. However, comparing Russia's GDP energy intensity with similar indicators for developed countries and countries with similar climatic and geographic conditions also reveals significant underperformance in energy efficiency in the domestic economy.

Based on Russia's example as one of the most energy-intensive economies in the world, technological factors contributed to a 0.6% reduction in GDP energy intensity between 2015 and 2021 (Bashmakov et al., 2023). Conversely, globally, although structural factors increased energy demand, technological factors reduced it by an average of 1.4%. EU data show that from 2013 to 2019, technological factors contributed to an annual 0.9% reduction. It is thus evident that Russia faces a gap in meeting energy efficiency targets, and this gap presents challenges in international comparisons. A recent study suggests that excluding non-energy use from GDP energy intensity can help reduce this gap and make cross-country comparisons more consistent with the information conveyed by energy intensity (Bashmakov et al., 2023). Using appropriate indicators and setting necessary targets are crucial for monitoring and assessing success in energy efficiency. Since EU member states may choose national targets based on different indicators—such as primary energy consumption (PEC), final energy consumption (FEC), primary or final energy savings, or energy intensity—policies often fail to meet global targets. Therefore, improved design is required, and decomposition analysis could support this process by providing insights unaffected by structural or activity changes that might result from energy efficiency monitoring (Deloitte, 2016). In an effort to approach energy efficiency from a macroeconomic perspective across selected EU countries, this study undertakes an analytical investigation of the energy policies implemented and assimilated by these countries in recent years.

CASE STUDIES OF INDICATIVE EU MEMBER STATES

Germany's Energy Policy

Germany is the largest energy consumer in Europe and was the seventh-largest country in 2015. The country's energy policy began to take shape in the 1990s through the decentralization of East Germany's energy industries and the establishment of stronger local utilities in the West—a step that reduced lignite's share and elevated the role of natural gas (Becker et al., 2018). After 1992 and the global environmental agreements, Germany developed a policy aiming to balance environmental goals with the interests of the established energy industry by adopting a CO2 emissions trading system with generous allocation policies (Corbach, 2012). The Kyoto Protocol in 1997 led to a search for low-emission energy technologies and their replacement in 1,660 German facilities from 2007 onwards. Legislation for feed-in tariffs from renewable energy sources further stimulated the development and use of renewables, aiming to meet 5% of the country's energy needs.

Later, the energy gap caused by the closure of nuclear power plants was partly filled by coal industries. However, Germany's energy commitments in 2008/2009 and the pressure to reduce emissions led to the renovation of lignite units and funding for investment programs in power plants to sustain East Germany's industry. Regarding nuclear energy, the 2009 elections marked a policy shift to extend the life of nuclear reactors and revive the nuclear sector with a 12-year plan (von Hirschhausen, 2018a). In 2000, Germany continued to support renewable energy through mechanisms initially targeting wind, solar, and geothermal energy. New measures were adopted to include bioenergy, raising concerns among traditional energy industries. In the following years, renewables increased their share to 16% in 2009. Between 2010 and 2011, the announcement of the 2050 Energy Concept accelerated the energy transition towards low-carbon policies without nuclear power and focused on renewables (Meeus et al., 2011). This transition included coal, provided that carbon capture, transport, and storage technologies—still unavailable—would be used. However, the 2011 Fukushima nuclear accident changed Germany's policy direction towards a gradual nuclear phase-out by 2022 (von Hirschhausen, 2018a, 2018b). Given Germany's goals to reduce primary energy consumption by 20% by 2020 and by 50% by 2050 compared to 2008 levels, the government committed to a 20% reduction by 2020 and a 27% reduction by 2030 relative to EU energy forecasts.

Consequently, a new set of energy efficiency policy measures was proposed under the National Action Plan on Energy Efficiency (NAPE) and the Climate Action Program 2020 (APC) (Kuhn, 2017; Ringel et al., 2016).

Today, fossil fuels still dominate Germany's energy mix, accounting for 80% of primary energy consumption. Oil represents 35.3%, and natural gas 25%. Renewables cover 14.7%, primarily from biomass (7.6%). Nuclear energy—expected to be phased out by 2022—accounts for 6.4% of primary energy (Telli et al., 2021). In 2016, Germany's primary energy demand exceeded 3,700 TWh, of which only 68% was used as final energy, half of which was for heating, one-third for fuels, and only one-fifth for electricity (Energiewende, 2018). These figures highlight the need for more integrated policies, such as sector coupling, referring to the technical and economic interdependence between sectors like electricity, heating, and transportation, with a greater share of renewables (Weibezahn, 2018). Policy support primarily comes through low-interest loan systems, such as the CO2 Building Rehabilitation Program (CBRP), funded by public resources managed by the KfW development bank. Germany's energy upgrade policy lacks targeted social focus and does not provide funding for specific social groups. The loan mechanism allows significant leveraging of private investment, a result not usually achieved with grants (Kerr et al., 2017). Over the past 10–15 years, Germany's energy efficiency policy mix has included regulations (e.g., building codes, end-use products, electricity tariff structures), price-affecting measures, financial incentive schemes (e.g., energy taxation, KfW subsidized loan and grant programs for building renovation), and information campaigns and energy audits promoted by national and regional energy agencies. Financing tools rely exclusively on federal government funds and reflect its reluctance to mandate utilities to support energy efficiency development (e.g., through energy-saving quotas or "white certificates") (Markandya et al., 2014). Although most investment support and awareness measures are funded by an energy efficiency fund, its scope remains limited, with a funding volume of €1.68 billion (2015). The potential role of an expanded energy efficiency fund is only considered within NAPE, which will launch a resource-saving competition and be replenished by the general budget (Ringel et al., 2016).

Among recent instruments, two measures deviate from conventional policy paths: First, the introduction of a competitive bidding system for energy efficiency, bringing quantity-based mechanisms into the German policy mix. Second, an initiative to establish 500 energy efficiency networks aimed at accelerating efficiency improvements in industry and services. Evaluation findings indicate that while these measures do not fully meet all needs, they can yield multiple benefits, as there are synergies and interactions—sometimes resulting in policy overlaps of up to 10% (Ringel et al., 2016). Energy efficiency, as shown, is a widely recognized concept since energy savings are directly linked to cost savings. Standard, horizontal energy efficiency measures are thoroughly evaluated, and extensive research has explored why potential measures are not adopted. Well-developed approaches identify barriers to adopting energy-efficient practices, categorized into economic, behavioral, and organizational obstacles, along with limited information on benefits and risks, hidden costs, access to capital, split incentives, and bounded rationality. Over 45% of service sector firms were engaged in energy efficiency activities. According to a modeling approach for efficiency gains in Germany's tertiary sector, a 16% reduction in final energy demand is projected by 2030 compared to 2008, assuming the realization of "nearly cost-effective" potentials—26% at the EU level (Wohlfarth et al., 2020).

Italy's Energy Policy

Italy is a relatively energy-intensive country in Europe, heavily dependent on fossil fuels, with a particularly high energy dependence on imports (87%) and some of the highest electricity and natural gas prices (Trotta, 2020). Despite significant obstacles to energy efficiency, Italy, along with Germany, was identified in 2016 by the American Council for an Energy-Efficient Economy as one of the countries with the most effective energy efficiency policies (Berg et al., 2016). According to the EU agreements and Italy's National Renewable Energy Action Plan (NAPRE) presented in 2011, the country committed to meeting 19–20% of its energy needs from renewable sources, surpassing the EU-set target of 17%. Furthermore, Italy's recent National Energy Strategy (NES) highlighted that 28% of total energy consumption and 55% of electricity use should come from renewable sources by 2030 (MISE, 2017). In 2012, Italy introduced the concept of regional energy burden sharing, setting energy targets for each region (Magoni, 2018). This success was attributed to the white certificate policy, tax relief schemes, and thermal incentives implemented by the country (Trotta, 2018). Under the Italian white certificate system, electricity and gas distribution companies were required to achieve annual energy efficiency targets of around 1 toe, either through implementing efficient solutions or purchasing white certificates from other organizations. In the short term, this provided a significant incentive to achieve energy goals, such as the adoption of fluorescent lamps, and it facilitated information dissemination at the local level (OECD, 2016), with an expected energy savings of 16 Mtoe between 2014 and 2020.

The tax relief scheme was primarily targeted at the residential sector, offering income tax deductions for individuals and companies undertaking building energy efficiency improvements. The deduction rate ranged between 50% and 85%, depending on the type of intervention, and was estimated to result in energy savings of 3.92 Mtoe. Similarly, the thermal incentive scheme was designed to support public entities and non-industrial private entities (e.g., housing cooperatives, farmers, and individuals) in implementing energy efficiency projects and small-scale renewable thermal energy production. It covered 65% of total expenditures incurred by beneficiaries, disbursed in annual installments over two to five years, with expected savings of 5.88 Mtoe (Trotta, 2020). Data on Italy's energy consumption indicate a steady decline, with 2015 figures showing lower consumption levels than the EU target (Clemente, 2014). This decline coincides with the implementation of most energy efficiency policies. However, increased energy consumption in the freight transport sector emerged as a key constraint on reducing energy intensity during the period 2006 to 2015 (Trotta, 2020). Regarding energy efficiency policies for SMEs in Italy, these focused on white certificates, support schemes for heating and cooling systems based on renewable energy, the provision of national energy efficiency funds, the obligation for electricity-intensive companies to undergo energy audits, and a three-year educational and information campaign entitled "Italy in Class A," aimed at promoting energy efficiency (Trotta, 2020).

France's Energy Policy

France's energy system is characterized by particular stability due to strong state control and a longstanding preference for large-scale projects driven by the central government. As of 2019, France's energy mix consisted of 40% nuclear energy, 29% oil, 15% natural gas, just 3% coal, and 12% renewables (mainly hydro, wind, biofuels, and waste). Additionally, the high level of electricity generation from nuclear energy (67.1%) has kept the share of renewables in electricity production relatively low and stable over the last 25 years, at around 25.3% (13% hydro and 7.9% wind). Thus, France shows the slowest transition to renewables among European countries, underutilizing its geographic potential for solar and wind energy due to its reliance on nuclear power (Bocquillon & Evrard, 2022). This "all-electric, all-nuclear" energy mix renders the French energy system relatively "clean," positioning France as the second-lowest country in terms of greenhouse gas emissions intensity in electricity generation in the EU (56 g CO2e/kWh), and the eighth-lowest in per capita greenhouse gas emissions (6.8 tons CO2e per capita). However, these goals provide limited incentives for a broader energy transition. Combined with France's traditional energy policy focus on energy security, social equity, and both domestic and international competitiveness, this defines its strategic energy direction (EEA, 2017). Regarding access to energy, France traditionally offered "social tariffs," which were replaced in 2018 by the "energy voucher," a financial aid program for low-income households. Furthermore, France's nuclear industry holds a dominant position in global energy markets, competing for contracts in Europe, the Middle East, and China, thus significantly contributing to its national economy. Recent efforts have focused on "greening" nuclear energy, which has become a focal point of national policy since the “Grenelle de l’environnement” consultations ((Bocquillon & Evrard, 2022).

In 2010, organizations like NegaWatt brought forward the debate on phasing out nuclear power and limiting fossil fuel use by 2050. The Fukushima accident in 2011 and subsequent international concerns prompted the French government to commit to reducing the nuclear share in electricity production to 50%, modernizing the nuclear industry, and decentralizing renewable energy deployment. This shift was formalized in the 2015 Energy Transition for Green Growth Act (LTECV), which set targets to reduce greenhouse gas emissions by 40%, cut energy consumption by 50% by 2030, and increase the renewable share to 32% of total consumption and 40% of electricity, while reducing the nuclear share to 50% and closing one nuclear power plant (Chailleux & Hourcade, 2021). In 2019, the Climate and Energy Law followed, with commitments to reduce fossil fuel use by 40% by 2030, phase out the last four coal-fired power plants by 2022, and implement various measures to accelerate renewable energy development and energy-efficient housing renovation. Currently, key institutions include the High Council for Climate and the Citizens’ Convention on Climate, which propose measures to address global warming. These proposals have been legislated through the 2021 Climate and Resilience Law (Gougou, 2020). However, the concentration and vertical integration of energy industries, along with traditionally centralized and technocratic governance structures, have made the country less flexible in its transition to greener and more renewable energy sources (Bocquillon & Evrard, 2022).

Belgium's Energy Policy

Belgium's energy policy is divided between the central government and the local authorities of its three regions, resulting in four energy policy authorities. The federal government is responsible for policies related to oil and gas supply security, nuclear matters, tariff regulation, electricity transmission, and offshore wind energy. Each region governs electricity and gas distribution, energy efficiency, renewable energy (excluding offshore wind), and non-nuclear energy research and development (Van de Graaf et al., 2022). Belgium is a small but economically open EU member state, highly dependent on fossil fuel and crude oil imports due to a lack of domestic energy resources. In 2015, fossil fuels accounted for 77.9% of total energy (oil 44.6%, natural gas 27.1%, coal 6.2%), nuclear energy represented 13.2%, and renewables 8.9%. Renewable energy primarily comprised biofuels and waste (6.8%), with a small share (2%) from geothermal, wind, and solar sources (IEA, 2015). However, Belgium is a net exporter of refined products from the Port of Antwerp and a key gas transit hub due to its LNG terminal in the Port of Zeebrugge. Nuclear energy from its seven reactors produces half of its electricity (EEA, 2017). Under a 2003 law, Belgium aimed to phase out nuclear energy between 2022 and 2025. A more recent 2012 law, the "Wathelet Plan," proposed maintaining the Tihange 1 reactor for another decade and decommissioning Doel 1 and 2 by 2015. However, some reactors had to be shut down earlier due to safety issues, while others were refurbished and restarted to maintain energy security. Therefore, the nuclear phase-out plan remains unachieved, as renewable energy sources have not yet reached market competitiveness to support the country's energy demands, and reliance on external sources would cause economic instability. Continued operation of nuclear facilities beyond planned timelines has thus been recommended (Kunsch & Friesewinkel, 2014). Despite a slow achievement of the EU's 15% GHG emissions reduction target by 2020, Belgium showed good progress in renewable energy, increasing its share from 2.4% in 2004 to 7.6% in 2014. By 2017, it ranked third globally in per capita use of photovoltaic energy. However, the financial burden of this achievement shifted to distribution service operators, as in the case of Flanders (Van de Graaf et al., 2022). Renewable energy promotion was supported by the creation of tradable green certificates (TGC) in 2000 by the Flemish government. However, cost and effectiveness led other regions to develop alternative certificate systems with minimum price levels, defined support durations, and technology-specific support, mainly targeting households and small-scale users. Although industrial and technological innovation frameworks were limited, the average support level for renewable energy production was relatively high (IEA, 2016).

Belgium's energy policy began under EU influence. Before liberalization, electricity markets were controlled by private firms, but EU directives opened electricity markets in all regions by 2007 and pushed gas market liberalization. The main utility, Electrabel, held a dominant 66% production share in 2014 (IEA, 2016). Efforts to improve building energy efficiency included subsidies, mandatory inspections, and awareness programs. Energy-saving measures in industry focused on mobility initiatives and the creation of green systems and CHP certificate systems. The 2018 Energy Pact signed by Belgium's political entities outlined strategic goals for the coming decades but lacked detailed measures, leading to continued reliance on nuclear capacity and existing energy agreements (Belgian et al., 2018).

Spain's Energy Policy

Spain is a country that follows a pattern of high energy consumption, heavily reliant on fossil fuels and highly dependent on foreign sources, which makes it vulnerable to negative impacts on its economic development (Fernandez, 2018). Energy consumption has continued to rise, having doubled between 1990 and 2007, even without a corresponding increase in population. On the other hand, Spain's energy efficiency and intensity levels are moderate compared to the European average, which is attributed to the contribution of the service sector to its GDP. Given the high dependency on imported oil, natural gas, uranium, and coal, the underdevelopment of the oil distribution sector, and low emission taxes in transportation, Spain faces serious constraints in reducing its reliance on external energy sources (Campos-Martín et al., 2022). A key factor in mitigating this dependency has been the development of the electricity sector through renewable energy production. This was initially promoted in 1997 with the help of a feed-in tariff system, which significantly increased renewable energy installations. Later, the introduction of incentives and rewards through Royal Decree 661/2007 further boosted renewable electricity production (Fernandez, 2018), placing Spain among the most advanced nations in this regard. However, the 2008 economic crisis halted these subsidies, and investment in renewables fell sharply. The liberalization of the Spanish energy sector came through decentralization, where energy production and sales were treated as unregulated activities, while private companies were subjected to legislation concerning the transmission and distribution of natural gas and electricity. These regulations were repeatedly revised over the years, with recent changes introducing obstacles to the development of renewables in favor of more traditional energy mixes. Generous subsidies for certain technologies, such as photovoltaics, destabilized the electricity system and led to financial losses for foreign investors, who were partially compensated in 2017 (Energy, 2017).

Following the adoption of the "Winter Package," Spain designed an energy and climate plan aligned with EU policies. Nonetheless, investment debt in combined cycle backup technologies and electricity tariffs, along with limited development of new transport and trading technologies, hindered the package's implementation. The establishment of the National Energy Efficiency Fund to financially support technical assistance, training, information, and other measures aimed at increasing energy efficiency across consumption sectors marked a continuation of Spain's efforts to align with the EU (Gürtler et al., 2019). Spain had set the target of achieving 20% of primary energy from renewables by 2020, mainly through wind power installations, and further aims to achieve 100% penetration of renewables by 2050 (Campos-Martín et al., 2022).

These goals are to be pursued through improvements in energy efficiency across all sectors, aided by population decline and corresponding reductions in energy consumption, along with the use of new technologies and more efficient energy standards. With the transport sector having peaked in demand in 2009, adjustments became necessary, and it is projected to maintain two-thirds of consumption through 2030. Households are expected to maintain moderate growth, while freight transport is projected to improve its efficiency by 15%. The greatest technological integration and implementation of horizontal technologies and new processes occurred in the industrial sector between 2004 and 2005, which also experienced the inclusion of more productive and energy-intensive sub-sectors in the EU Emissions Trading System (ETS) (González, 2018). Nevertheless, policies related to bioclimatic buildings remain inadequate, and the service sector (offices, tourism, shopping centers) is expected to play a more prominent role in the coming years (Campos-Martín et al., 2022). Spain's political landscape regarding energy demonstrates a degree of instability that hampers investments in low-carbon energy sources. However, small steps in the right direction are emerging, which must be linked to the role of national and local policies and their effective incorporation of EU commitments.

The Netherlands' Energy Policy

The governance of the Netherlands in terms of energy transition is characterized by distinct approaches to policy coordination, expert participation, strategic planning, consensus building, and mutual self-commitments. According to the Energieakkoord of 2013, the country aimed to ensure that 16% of its energy consumption would come from sustainable sources, that 74.4% of its electricity would be generated from renewables, and that CO2 emissions would be reduced by 49% by 2030 (Musch, 2022). The Dutch energy mix is largely composed of fossil fuels, with natural gas representing the primary source of energy consumption at around 40%, followed by oil and coal. The Netherlands is the second-largest producer of natural gas in Europe and an important supplier to Northwestern Europe. However, due to seismic activity linked to gas extraction, the government decided in 2018 to phase out gas exports by 2030, signaling significant changes in the country's energy landscape (Musch, 2022). The share of renewable energy sources is gradually increasing, with biomass, wind, and solar energy being the main contributors. The creation of the "wander fund" aims to finance the development and implementation of renewable energy sources, promote energy savings in households and the tourism sector, enable the storage and production of green fuels, and support the application of blue energy (tidal and wave) and biomass (IEA, 2021).

Policies regulating the production, distribution, and management of energy—such as electricity in 1998, natural gas in 2000, heating in 2014, and the environment in 2017—set limits, laws, and goals that have been redefined over time. These policies included environmental plans, quality programs, compensation mechanisms, handling procedures, taxation, and insurance schemes, all aimed at supporting energy and economic prosperity standards. Concerning the energy tax, which imposes a carbon levy on consumers, the government offered exemptions in 2016 to entities involved in electricity generation. Simultaneously, the "Stimulation of Sustainable Energy Production" (Stimulering Duurzame Energieproductie, SDE+), introduced in 2011 and funded through household consumption taxes, links subsidies to market energy prices—when market prices rise, subsidies decrease. Disbursements are prioritized for low-cost technologies. Tax relief is available to consumers of renewable energy or users of environmentally friendly facilities, and households can apply for green loans to implement energy-efficient measures. The Dutch effort to increase competitiveness through the development of new technologies created a gap that was addressed in 2011 by promoting other economic sectors such as agriculture, food and chemical industries, life sciences, and water management (OECD/IEA, 2014). Given that the Netherlands managed to reduce greenhouse gas emissions by 6.4% between 2008 and 2012 and has set ambitious emission reduction goals of 49% by 2030 and 60% by 2050, further development of biomass and wind energy production is essential. The country also plans to increase natural gas imports from Norway and the North Sea while reducing exports to compensate for the domestic production shortfall. Through participatory governance and new climate change agreements, the Netherlands will continue to move toward a more sustainable and carbon-independent future (Musch, 2022).

Denmark's Energy Policy

Denmark's energy dependence is primarily on oil, followed by biomass, waste, solar and wind energy, and natural gas. Domestic oil production has been a significant revenue source, financing infrastructure projects like electrified transport and high-speed trains until 2014, when falling oil prices caused a budget deficit. Biomass, having replaced coal in the energy mix under a phase-out plan by 2030, is the main heating source, but 43% of it is imported, shifting the emission burden to exporting countries. Likewise, while waste is a decentralized energy source, improved waste management and recycling policies have significantly reduced its volume (Dyrhauge, 2020). As of 2017, 50% of Denmark's electricity was generated from renewable sources. With demand expected to increase by 2030, Denmark risks becoming a net electricity importer unless it makes significant investments in offshore wind farms and adopts self-sufficiency policies (Christensen et al., 2020). Research and development in green technology play a crucial role and require government support. Although Denmark lacks heavy industry compared to other nations, it hosts the world's largest shipping company, Maersk. In 2018, Maersk announced its goal of achieving zero emissions by 2050, a major policy shift for Denmark. This move supports the Energy Technology Development and Demonstration Program funded by the Innovation Fund, which promotes floating renewables and introduces new models of energy organization, storage, and distribution. Legislation governing heating and electricity traditionally followed non-profit principles, with prices regulated by the government and profits reinvested in production and consumers. The liberalization of the electricity market in 2016 introduced more actors into production and development, expanding opportunities for investment in renewables and wind energy. Since power plants have been banned since 2000, local energy cooperatives evolved into national and international players distributing energy across Scandinavia (Larsen, 2019).

Separating the energy and climate policy plans in 2018 contributed positively to Denmark's multilevel governance structure, characterized by a wide range of public and private actors, though coordination depends heavily on political and legal commitments. The Ministry of Energy is responsible for energy efficiency, with annual reports indicating that successful transition requires inclusion of the transport and agriculture sectors and investment in specialized carbon storage and capture technologies. The shift to electric vehicles has led to new charging infrastructure, and battery sector maturity suggests complete replacement of combustion vehicles within the next decade. However, this transition poses risks related to energy security and cyberattacks. In response, the Ministry of Energy, Utilities, and Climate announced strategic protection measures in 2019 (Ministry for Energy et al., 2019). These developments signal a new era of energy policy that must address emerging challenges and actors involved in managing and integrating green technologies across all societal structures and in coordinating these efforts internationally.

Poland's Energy Policy

Compared to other EU member states, Poland is often characterized as the least ambitious regarding energy and climate policy. Most literature suggests that energy security and economic development hold greater priority for the Polish government. Political parties tend to show limited awareness of climate issues and lack support for the EU's decarbonization vision. Changes in legislation concerning renewable energy have been particularly slow to implement. This attitude appears grounded in Poland's perception of the coal industry as a symbol of economic independence and state sovereignty, making it difficult to adopt and promote climate-oriented reforms (Vasev, 2017).

Poland is the second-largest coal consumer in Europe after Germany, with coal accounting for around 20% of electricity generation and home to Europe's largest lignite plant, which returns only 8% of the energy as electricity. Recent investments in modernizing coal-fired power stations have not yielded expected benefits, and the share of electricity from renewables dropped to 12.7%, the lowest since 2014 (Formuszewicz & Gawlikowska-Fyk, 2014). Poland also has the worst air quality in the EU, largely due to widespread use of low-quality coal in households and outdated technologies in industry due to financial constraints. The strong reliance on coal for economic stability has entrenched employment and income structures dependent on the industry. However, the sector is increasingly unprofitable, faces environmental challenges, and struggles to remain competitive (Gawlik, 2018). Natural gas has begun to fill the energy gap but raises concerns over Poland's energy autonomy since it is largely imported. Until recently, Russia held a monopoly on this supply, with which Poland has historically had strained relations. This situation prompted efforts to diversify the energy mix, including nuclear energy. However, the continued reliance on imports remained problematic. The attempt to exploit unconventional shale gas (fracking) was short-lived due to unsatisfactory results and was quickly abandoned (Jankowska, 2016).

According to Poland's Energy Policy 2040, coal is expected to remain the primary energy source, contributing 60% of the energy mix by 2030, with a reduction to 32% by 2040. Meanwhile, nuclear energy is projected to increase to 18%. These goals are likely unachievable given the rising electricity demand and the inadequacy of existing nuclear reactors to meet current technological and environmental standards. Moreover, there is a widespread belief among Poles that clean energy is unaffordable and that climate change mitigation via neutral technologies is not a national priority (Leiren et al., 2019). Some Polish cities and municipalities, however, are developing progressive environmental agendas and promoting renewable energy investments. These efforts often conflict with central government policies, which have imposed higher taxes and minimum distance regulations that significantly reduced wind energy investments (Skjærseth, 2018). In conclusion, Poland demonstrates significant resistance to developing an environmentally friendly energy policy. The national approach oscillates between EU compliance pressures and the enduring influence of the coal industry. As a result, the energy transition is expected to proceed slowly, with the vision of sustainable development still largely marginalized (Zoll, 2022).

Greece's Energy Policy

Greece is an economy predominantly based on tourism, shipping, and the public sector, which together account for about 80% of its GDP. Between 2002 and 2008, the country experienced rapid economic growth, driven by its entry into the European Monetary Union. However, the lack of structural reforms to enhance productivity left Greece vulnerable to the financial crisis, which by 2009 significantly increased its fiscal deficit and led to a cycle of austerity measures and economic turmoil. The imposed bailout programs also included a set of energy sector reforms, which opened up the electricity market, reduced subsidies for renewable energy, increased fuel prices, and promoted privatization efforts—such as the partial privatization of the Public Power Corporation (PPC)—to attract investment (Tsagkari, 2022). Primary energy in Greece remains largely dependent on lignite, followed by imports of oil and natural gas, making the country energy-dependent. In 2008, Greece was the second-largest lignite producer in Europe, with lignite accounting for over half of its electricity generation. However, a 2014 report highlighted the sector's high costs, and under EU guidelines and post-crisis pressures, lignite use dropped from 63% in 2004 to 32% in 2016, despite the establishment of two new lignite units. Today, lignite accounts for 23% of energy needs, with renewable energy sources expanding and natural gas imports rising. Additionally, the lack of clean energy infrastructure and the use of traditional power methods on islands increases national energy costs and CO2 emissions due to the unified energy pricing system (Ioannidis et al., 2019). Oil consumption has declined due to economic constraints and higher taxes, but overall energy demand has increased, resulting in energy poverty for many households. During winter, this has led to increased wood burning and smog-related issues. The liberalization of the electricity market attempted to reduce the dominance of PPC by auctioning shares to private entities (Fafaliou & Polemis, 2010). Nevertheless, PPC remains dominant, while a lack of structural reforms and high capital costs hinder the emergence of competitive firms. The liberalization of the natural gas market appears more promising, with two new supply and distribution companies entering the Greek market. Additionally, agreements for natural gas pipelines—initially with Russia, and later with the Trans Adriatic and EastMed pipelines—are expected to enhance energy security, a previously vulnerable aspect of Greek policy (Vlados et al., 2021).

Renewable energy sources currently meet only 11% of the country's energy needs, significantly lower than other Mediterranean nations such as Spain, Portugal, and Italy. Greece has set a target to increase this share to 20% by 2020 and 30% by 2030, in line with EU emission reduction goals. However, despite the strategic potential for a sustainable energy transition, changes remain insufficient, with barriers including the high cost of renewables and low public awareness about integrating them into all sectors (Neofytou et al., 2020). The "Energy Saving at Home" program, launched in 2011 and revised in 2018, continues to offer incentives for building renovations and energy efficiency upgrades. Similarly, the "Green Roofs on Public Buildings" initiative aims to enhance energy performance and reduce CO2 emissions. The transition to electromobility and transport decarbonization is progressing slowly, although the sector accounts for 40% of national energy use. The most successful measure—vehicle retirement incentives—ended in 2016. Despite high taxation on emissions and luxury, combined with subsidies for electric and hybrid vehicles, these efforts have not yet delivered the desired outcomes (Neofytou et al., 2020). The recent "Move Electric" program, which subsidizes electric vehicles, bikes, and home chargers, has been absorbed mainly by wealthier, more educated urban areas, while many rural regions lag behind EU targets (Shaban et al., 2023). Greece's efforts to align with EU energy transition goals are often overshadowed by economic crisis, political instability, and a lack of long-term strategic planning. Many energy-related policies have been prematurely abandoned or replaced, undermining their impact. In some cases, attempts to comply with EU mandates have led to energy poverty.

Bureaucratic inefficiencies, limited incentives, and public information gaps remain the most pressing challenges. As Greece continues to face economic hardship, aligning energy policies with current conditions may delay the expected benefits until the country fully recovers (Tsagkari, 2022).

DISCUSSION

Energy efficiency is a key strategy for managing energy consumption while maintaining or improving service delivery, which has over the past decade gained significant attention from policymakers and organizations, for its potential in resource conservation, climate action, and economic stability (Jebali et al., 2017). While various modeling techniques have been developed to address energy efficiency challenges, implementation varies considerably across EU Member States (Dupont et al., 2024). Many of these techniques point towards R&D funding, industrial policy management, collaborative networks, green investment and monitoring key energy efficiency indicators that will leverage private investment, local and international competitiveness, clean energy innovation and essential information in policy making and implementation (Velasco-Fernández et al., 2020).

Despite widespread policy adoption, member states have followed their own path revealing distinct convergences and divergences in their national approaches. Germany focused on structural energy transition, through nuclear phase out, efficiency loans and large-scale renewable projects, whereas Italy emphasized on policy incentives such as white certificates, tax relief and regional burden sharing, highlighting its successful household and SME efficiency. In accordance, France, Belgium and Poland are all retaining their nuclear dominance, but have adopted individual levels in their renewable ambitions. Spain, the Netherlands and Denmark are all highly supportive of renewable progress, though their support schemes especially for Spain and the Netherlands shows instability. Whereas Greece hindered by economic crisis and structural inefficiencies, exhibits a slow uptake of renewables and is struck by energy poverty more than the other members. Taken together, these cases illustrate a fragmented but converging European energy landscape: northern and western countries generally push faster toward renewables and innovation, while southern and eastern states move more cautiously, balancing transition goals with economic, social, and geopolitical constraints.

Nevertheless, when examining the differences in energy efficiency among OECD countries, it is evident that the implementation of uniform policies may not always be appropriate (Iram et al., 2020). Specifically, total efficiency scores—based on primary energy consumption (PEC), CO_2 emissions, and GDP—indicate that Norway, Switzerland, Poland, Israel, Australia, and Japan performed as energy-efficient countries during the period 2013–2017. On the other hand, Hungary, Turkey, and Mexico were found to be the least efficient, while New Zealand, Hungary, Mexico, Austria, South Korea, Germany, the United Kingdom, and France showed relatively stable rankings across the same period. In contrast, the Netherlands, Belgium, and Canada demonstrated notable improvements, whereas Finland, the USA, and Chile experienced consistent declines between 2013 and 2017 (Iram et al., 2020).

Energy efficiency is a cornerstone of international agreements on sustainable development and the preservation of lifestyles and the services that support them. Expanding the current policy focus from supply-side parameters—centered on cheaper, cleaner, and more sustainable energy—toward demand-side factors, such as food, transportation, comfort, convenience, and productivity presents a considerable challenge (Brown, 2015). Such frameworks consider the capacity to deliver services that meet the needs of respective populations, including institutional support and the skills or qualifications of the workforce. Consequently, the benefits of energy efficiency policies may vary by context, and some countries may require alternative criteria for promoting green energy (Tietenberg, 2010). Sectors with high energy intensity often necessitate the use of renewable sources and advanced technologies to improve efficiency, but these principles must be supported by the local implementation environment. Therefore, in addition to the parallel use of diverse policy instruments—informational, financial, cooperative, technical, voluntary, regulatory—there is also a need for a proper sequencing of implementation to yield the intended benefits. Moreover, conditions such as financial, informational, regulatory, socio-economic, cultural, and geopolitical contexts—both nationally and sectorally—can significantly shape decision-making. Studies that explore the practical application of policy tools in light of barriers and the coexistence of multiple instruments can generate actionable resources for policymakers (Aboltins & Blumberga, 2019)

Macroeconomic approaches to energy efficiency highlight both the economic rationale and the systemic challenges of the energy transition, emphasizing efficiency not only as a tool for reducing emissions but also as a driver of competitiveness, innovation, and social welfare. For EU member states, the future of energy efficiency will hinge on their capacity to integrate decarbonization targets with economic growth strategies, while addressing asymmetries in resources, infrastructures, and institutional frameworks. Sustained progress will therefore require coordinated EU-wide mechanisms that balance common objectives with national specificities, ensuring that efficiency policies evolve as engines of resilience, inclusiveness, and long-term sustainability.

CONCLUSION

The establishment of transnational energy efficiency requires collaboration and coordinated progress among countries and organizations, as globalization renders energy a cross-border issue whose challenges must be addressed collectively. Practices involving cooperation in research and the development of new programs, technologies, and alternative energy sources; the exchange of best practices that optimize energy efficiency programs at institutional, technological, and operational levels; and the implementation of joint projects and infrastructures for developing more energy-efficient applications—such as cross-border energy transmission networks—can significantly contribute to both the temporal and functional transition of nations. However, these outcomes can only be achieved through the adoption of common policies and regulations, along with the implementation of financial incentives and joint investments. For example, the Intergovernmental Panel on Climate Change (IPCC) identifies limited access to financing and the lack of banking channels for projects, as the main barriers to widespread implementation of new measures in the EU context (Andersson, 2020).

Overall, the heterogeneity in energy efficiency trajectories across OECD countries underscores the limitations of adopting standardized policy prescriptions, reinforcing the need for nuanced, country-specific approaches that account for structural, economic, and technological contexts in order to achieve sustained improvements in energy efficiency. This insight affirms that energy efficiency is not merely a technical or economic concern, but a multidimensional policy domain, whose success depends on alignment with national structures, sectoral characteristics, and institutional capacities. The comparative evidence across European and OECD countries illustrates that while common goals of sustainability and decarbonization are widely shared, the pathways to achieving them remain deeply context-dependent, thereby reinforcing the imperative for tailored, flexible, and well-sequenced policy frameworks, that can adapt to diverse socio-economic and geopolitical realities.

Collectively, the analysis of indicative EU Member States confirms the heterogeneity of energy efficiency approaches across Europe. Differences in macroeconomic conditions, institutional structures, and implementation capacities contribute to fragmented policy outcomes. This highlights the need to move beyond uniform targets and toward policy instruments that reflect context-specific dynamics. The chapter contributes to the literature by framing energy efficiency, through a macroeconomic lens, drawing attention to policy design constraints and barriers to implementation. In this context, the development of harmonized yet flexible indicators, the strengthening of national ESCO markets, and the promotion of coordination between public and private funding mechanisms are critical to ensuring a more coherent and effective energy efficiency trajectory across the EU.

REFERENCES

Aboltins, R., & Blumberga, D. (2019). Key factors for successful implementation of energy efficiency policy instruments: A theoretical study and the case of Latvia. *Environmental and Climate Technologies*, *23*(2), 187–206. DOI: 10.2478/rtuect-2019-0063

Andersson, M. B. Claudio; Morgan, Julian (2020). Climate change and the macro economy. In (Vol. 243). ISBN 978-92-899-4249-2: European Central Bank (ECB) Occasional Paper.

Atkinson, R. D. (2022). *Why America Should Compete to Win in Advanced Industries*.

Azhgaliyeva, D., Kapoor, A., & Liu, Y. (2020). Green bonds for financing renewable energy and energy efficiency in South-East Asia: A review of policies. *Journal of Sustainable Finance & Investment*, *10*(2), 113–140. DOI: 10.1080/20430795.2019.1704160

Barta, C. (2008). *ENSCO: The First 20 Years*. Elsevier.

Bashmakov, I., Myshak, A., Bashmakov, V. A., Bashmakov, V. I., Borisov, K., Dzedzichek, M., Lunin, A., Lebedev, O., & Shishkina, T. (2023). Russian energy balance, energy efficiency, and energy-related GHG emission accounting system. *Energy Efficiency*, *16*(6), 67. DOI: 10.1007/s12053-023-10132-6

Beccali, M., Bonomolo, M., Ciulla, G., Galatioto, A., & Brano, V. L. (2015). Improvement of energy efficiency and quality of street lighting in South Italy as an action of Sustainable Energy Action Plans. The case study of Comiso (RG). *Energy*, *92*, 394–408. DOI: 10.1016/j.energy.2015.05.003

Becker, S., Naumann, M., & Moss, T. (2018). Between coproduction and commons: understanding initiatives to reclaim urban energy provision in Berlin and Hamburg. In *Coproducing Water, Energy and Waste Services* (pp. 63–85). Routledge. DOI: 10.4324/9781351238526-4

Belgian, Federal, & Government. (2018). Visienota: Belgisch Interfederaal Energiepact. In.

Berg, W., Nowak, S., Kelly, M., Vaidyanathan, S., Shoemaker, M., Chittum, A., DiMascio, M., & Kallakuri, C. (2016). The 2016 state energy efficiency scorecard.

Bocquillon, P., & Evrard, A. (2022). Energy Governance in France. In Knodt, M., & Kemmerzell, J. (Eds.), *Handbook of Energy Governance in Europe* (pp. 647–666). Springer International Publishing., DOI: 10.1007/978-3-030-43250-8_10

Campos-Martín, J. M., Crespo, L., & Fernandez, R. M. (2022). Energy Governance in Spain. In Knodt, M., & Kemmerzell, J. (Eds.), *Handbook of Energy Governance in Europe* (pp. 1121–1156). Springer International Publishing., DOI: 10.1007/978-3-030-43250-8_29

Chailleux, S., & Hourcade, R. (2021). Introduction. Local energy policies: Renewal under constraint. *Nature Sciences Sociétés*, *29*(1), 3–12. DOI: 10.1051/nss/2021018

Community, E. (2014). Energy Efficiency Directive in the Energy Community. In *Justification Note*.

Corbach, M. (2012). *Die deutsche Stromwirtschaft und der Emissionshandel* (Vol. 5). ibidem-Verlag/ibidem Press.

Dong, K., Dong, X., & Jiang, Q. (2020). How renewable energy consumption lower global CO2 emissions? Evidence from countries with different income levels. *World Economy*, *43*(6), 1665–1698. DOI: 10.1111/twec.12898

Dupont, C., Moore, B., Boasson, E. L., Gravey, V., Jordan, A., Kivimaa, P., Kulovesi, K., Kuzemko, C., Oberthür, S., Panchuk, D., Rosamond, J., Torney, D., Tosun, J., & von Homeyer, I. (2024). Three decades of EU climate policy: Racing toward climate neutrality? *Wiley Interdisciplinary Reviews: Climate Change*, *15*(1), e863. https://doi.org/https://doi.org/10.1002/wcc.863. DOI: 10.1002/wcc.863

Dupont, E., Germain, M., & Jeanmart, H. (2021). Feasibility and Economic Impacts of the Energy Transition. *Sustainability (Basel)*, *13*(10), 5345. DOI: 10.3390/su13105345

Dupont, E., Koppelaar, R., & Jeanmart, H. (2020). Global available solar energy under physical and energy return on investment constraints. *Applied Energy*, *257*, 113968. DOI: 10.1016/j.apenergy.2019.113968

Dyrhauge, H. (2020). Energy Governance in Denmark. In Knodt, M., & Kemmerzell, J. (Eds.), *Handbook of Energy Governance in Europe* (pp. 1–26). Springer International Publishing., DOI: 10.1007/978-3-319-73526-9_7-1

EEA. (2017). *Climate change mitigation: Trends and projections in Europe*. European Environment Agency.

Energiewende, A. (2018). *Energiewende 2030: The Big Picture; Megatrends, Targets, Strategies and a 10-Point Agenda for the Second Phase of Germany's Energy Transition. Impulse*. Agora Energiewende.

Energy, M. o. (2017). 2017–2020 National energy efficiency action plan. In. Spain.

Fafaliou, I., & Polemis, M. L. (2010). Liberalisation of the European natural gas market: Myth or reality? Evidence from Greece. *Research Topics in Agricultural and Applied Economics*, *1*, 168–183.

Fawcett, T., & Hampton, S. (2020). Why & how energy efficiency policy should address SMEs. *Energy Policy*, *140*, 111337. https://doi.org/https://doi.org/10.1016/j.enpol.2020.111337. DOI: 10.1016/j.enpol.2020.111337

Fawcett, T., Rosenow, J., & Bertoldi, P. (2019). Energy efficiency obligation schemes: Their future in the EU. *Energy Efficiency*, *12*(1), 57–71. DOI: 10.1007/s12053-018-9657-1

Fernandez, R. (2018). Crisis on Spain's Environmental Policy. *The impact of the economic crisis on European environmental policy*, 152.

Formuszewicz, R., & Gawlikowska-Fyk, A. (2014). New EU energy and climate framework: challenges for Poland and Germany.

Gawlik, L. (2018). The Polish power industry in energy transformation process. *Mineral Economics*, *31*(1-2), 229–237. DOI: 10.1007/s13563-017-0128-5

Giampietro, M., Mayumi, K., & Sorman, A. (2011). *The metabolic pattern of societies: Where economists fall short*. Routledge. DOI: 10.4324/9780203635926

Gkonis, N., Arsenopoulos, A., Stamatiou, A., & Doukas, H. (2020). Multi-perspective design of energy efficiency policies under the framework of national energy and climate action plans. *Energy Policy*, *140*, 111401. https://doi.org/https://doi.org/10.1016/j.enpol.2020.111401. DOI: 10.1016/j.enpol.2020.111401

González, F. J. P. (2018). El céntimo sanitario: Potencial recaudatorio frente a recaudación real. *Anales de ASEPUMA*, (26), 15.

Gougou, F., & Persico, S. (2020). Deciding together: The Citizens' convention on climate and the democratic challenge. In. Books & Ideas.net.

Gürtler, K., Postpischil, R., & Quitzow, R. (2019). The dismantling of renewable energy policies: The cases of Spain and the Czech Republic. *Energy Policy*, *133*, 110881. DOI: 10.1016/j.enpol.2019.110881

Herring, H. (2006). Energy efficiency—A critical view. *Energy*, *31*(1), 10–20. DOI: 10.1016/j.energy.2004.04.055

IEA. (2015). Policy pathway-Accelerating energy efficiency in small and medium-sized enterprises. In: International Energy Agency Paris, France.

IEA. (2016). *Energy policies of IEA countries: Belgium 2016 review*. OECD/IEA.

IEA. (2021). Energy system of The Netherlands. In.

In, S. Y., Monk, A. H. B., & Knox-Hayes, J. (2020). Financing Energy Innovation: The Need for New Intermediaries in Clean Energy. *Sustainability (Basel)*, *12*(24), 10440. DOI: 10.3390/su122410440

Ioannidis, A., Chalvatzis, K. J., Li, X., Notton, G., & Stephanides, P. (2019). The case for islands' energy vulnerability: Electricity supply diversity in 44 global islands. *Renewable Energy*, *143*, 440–452. DOI: 10.1016/j.renene.2019.04.155

Iram, R., Zhang, J., Erdogan, S., Abbas, Q., & Mohsin, M. (2020). Economics of energy and environmental efficiency: Evidence from OECD countries. *Environmental Science and Pollution Research International*, *27*(4), 3858–3870. DOI: 10.1007/s11356-019-07020-x PMID: 31823262

Jankowska, K. (2016). Poland's clash over energy and climate policy: green economy or grey status quo? In *The European Union in International Climate Change Politics* (pp. 169–182). Routledge.

Jebali, E., Essid, H., & Khraief, N. (2017). The analysis of energy efficiency of the Mediterranean countries: A two-stage double bootstrap DEA approach. *Energy*, *134*, 991–1000. https://doi.org/https://doi.org/10.1016/j.energy.2017.06.063. DOI: 10.1016/j.energy.2017.06.063

Jin, W., Zhang, H., Liu, S., & Zhang, H. (2019). Technological innovation, environmental regulation, and green total factor efficiency of industrial water resources. *Journal of Cleaner Production*, *211*, 61–69. https://doi.org/https://doi.org/10.1016/j.jclepro.2018.11.172. DOI: 10.1016/j.jclepro.2018.11.172

Johansson, M. T., & Thollander, P. (2018). A review of barriers to and driving forces for improved energy efficiency in Swedish industry–Recommendations for successful in-house energy management. *Renewable & Sustainable Energy Reviews*, *82*, 618–628. DOI: 10.1016/j.rser.2017.09.052

Kerr, N., Gouldson, A., & Barrett, J. (2017). The rationale for energy efficiency policy: Assessing the recognition of the multiple benefits of energy efficiency retrofit policy. *Energy Policy*, *106*, 212–221. DOI: 10.1016/j.enpol.2017.03.053

Kuhn, T. (2017). The revision of the German Fertiliser Ordinance in 2017.

Kunsch, P. L., & Friesewinkel, J. (2014). Nuclear energy policy in Belgium after Fukushima. *Energy Policy*, *66*, 462–474. https://doi.org/https://doi.org/10.1016/j.enpol.2013.11.035. DOI: 10.1016/j.enpol.2013.11.035

Labanca, N., & Bertoldi, P. (2018). Beyond energy efficiency and individual behaviours: Policy insights from social practice theories. *Energy Policy*, *115*, 494–502. https://doi.org/DOI. DOI: 10.1016/j.enpol.2018.01.027

Larsen, T. (2019). The Hegemony of Green Capitalism: A critique of the imaginary that the environmental crisis can be solved by capitalism.

Leiren, M. D., Szulecki, K., Rayner, T., & Banet, C. (2019). Energy security concerns versus market harmony: The Europeanisation of capacity mechanisms. *Politics and Governance*, *7*(1), 92–104. DOI: 10.17645/pag.v7i1.1791

Lovins, A. (2017). Energy efficiency. *Energy Economics*, *1*, 234–258.

Magoni, M. (2018). European energy policies and regional burden sharing: What are the effects on the territory of Lombardy? *City. Territory and Architecture*, *5*(1), 1–7. DOI: 10.1186/s40410-018-0086-4

Markandya, A., Labandeira, X., & Ramos, A. (2014). Policy instruments to foster energy efficiency. In *Green Energy and Efficiency: An Economic Perspective* (pp. 93–110). Springer.

Meckling, J., Galeazzi, C., Shears, E., Xu, T., & Anadon, L. D. (2022). Energy innovation funding and institutions in major economies. *Nature Energy*, *7*(9), 876–885. DOI: 10.1038/s41560-022-01117-3

Meeus, L., Hafner, M., Azevedo, I., Marcantonini, C., & Glachant, J.-M. (2011). *Transition Towards a Low Carbon Energy System by 2050: What Role for the EU?* Ministry for Energy, Utilities, & Climate. (2019). *Cyber- og informationssikkerhedsstrategi for energisektorerne.*

MISE. M. (2017). Strategia Energetica Nazionale. In: SEN.

Molina, M. (2014). The best value for America's energy dollar: a national review of the cost of utility energy efficiency programs. American Council for an Energy-Efficient Economy, Musch, E. (2022). Energy Governance in the Netherlands. In Knodt, M., & Kemmerzell, J. (Eds.), *Handbook of Energy Governance in Europe* (pp. 863–895). Springer International Publishing., DOI: 10.1007/978-3-030-43250-8_20

Nadeem, T. B., Siddiqui, M., Khalid, M., & Asif, M. (2023). Distributed energy systems: A review of classification, technologies, applications, and policies. *Energy Strategy Reviews, 48*, 101096. https://doi.org/https://doi.org/10.1016/j.esr.2023.101096

Nahm, J. M., Miller, S. M., & Urpelainen, J. (2022). G20's US $14-trillion economic stimulus reneges on emissions pledges. *Nature*, *603*(7899), 28–31. DOI: 10.1038/d41586-022-00540-6 PMID: 35236968

Neofytou, H., Sarafidis, Y., Gkonis, N., Mirasgedis, S., & Askounis, D. (2020). Energy Efficiency contribution to sustainable development: A multi-criteria approach in Greece. *Energy Sources. Part B, Economics, Planning, and Policy*, *15*(10-12), 572–604. DOI: 10.1080/15567249.2020.1849449

OECD, I. (2016). Energy and air pollution: world energy outlook special report 2016.

Ringel, M., Schlomann, B., Krail, M., & Rohde, C. (2016). Towards a green economy in Germany? The role of energy efficiency policies. *Applied Energy*, *179*, 1293–1303. DOI: 10.1016/j.apenergy.2016.03.063

Safarzadeh, S., Hafezalkotob, A., & Jafari, H. (2022). Energy supply chain empowerment through tradable green and white certificates: A pathway to sustainable energy generation. *Applied Energy*, *323*, 119601. https://doi.org/https://doi.org/10.1016/j.apenergy.2022.119601. DOI: 10.1016/j.apenergy.2022.119601

Schipper, L. (2004). International Comparisons of Energy End Use: Benefits and Risks. In Cleveland, C. J. (Ed.), *Encyclopedia of Energy* (pp. 529–555). Elsevier., https://doi.org/https://doi.org/10.1016/B0-12-176480-X/00169-8 DOI: 10.1016/B0-12-176480-X/00169-8

Seidl, R., Moser, C., & Blumer, Y. (2017). Navigating behavioral energy sufficiency. Results from a survey in Swiss cities on potential behavior change. *PLoS One*, *12*(10), e0185963. DOI: 10.1371/journal.pone.0185963 PMID: 29016642

Shaban, F., Siskos, P., & Tjortjis, C. (2023). Electromobility Prospects in Greece by 2030: A Regional Perspective on Strategic Policy Analysis. *Energies*, *16*(16), 6083. DOI: 10.3390/en16166083

Skjærseth, J. B. (2018). Implementing EU climate and energy policies in Poland: Policy feedback and reform. *Environmental Politics*, *27*(3), 498–518. DOI: 10.1080/09644016.2018.1429046

Telli, A., Erat, S., & Demir, B. (2021). Comparison of energy transition of Turkey and Germany: Energy policy, strengths/weaknesses and targets. *Clean Technologies and Environmental Policy*, *23*(2), 413–427. DOI: 10.1007/s10098-020-01950-8

Tietenberg, T. (2010). *Emissions trading: principles and practice*. Routledge. DOI: 10.4324/9781936331284

Trinh, H. H., Trinh, K., Trinh, N., & Haouas, I. (2023). Energy Technology RD&D Budgets, sustainable transition, carbon risk, and climate change: Fresh insights from IEA member countries under policy uncertainty. *Available at SSRN* 4475237. DOI: 10.2139/ssrn.4475237

Trotta, G. (2018). Factors affecting energy-saving behaviours and energy efficiency investments in British households. *Energy Policy*, *114*, 529–539. https://doi.org/https://doi.org/10.1016/j.enpol.2017.12.042. DOI: 10.1016/j.enpol.2017.12.042

Trotta, G. (2020). Assessing drivers of energy consumption and progress toward energy targets in Italy. *Energy Sources. Part B, Economics, Planning, and Policy*, *15*(3), 137–156. DOI: 10.1080/15567249.2020.1778817

Tsagkari, M. (2022). Energy Governance in Greece. In *Handbook of Energy Governance in Europe* (pp. 709–736). Springer. DOI: 10.1007/978-3-030-43250-8_12

Van de Graaf, T., Laes, E., & Verbruggen, A. (2022). Energy Governance in Belgium. In Knodt, M., & Kemmerzell, J. (Eds.), *Handbook of Energy Governance in Europe* (pp. 511–532). Springer International Publishing., DOI: 10.1007/978-3-030-43250-8_4

Vasev, N. (2017). Governing energy while neglecting health–The case of Poland. *Health Policy (Amsterdam)*, *121*(11), 1147–1153. DOI: 10.1016/j.healthpol.2017.09.008 PMID: 28958707

Velasco-Fernández, R., Dunlop, T., & Giampietro, M. (2020). Fallacies of energy efficiency indicators: Recognizing the complexity of the metabolic pattern of the economy. *Energy Policy*, *137*, 111089. https://doi.org/https://doi.org/10.1016/j.enpol.2019.111089. DOI: 10.1016/j.enpol.2019.111089

Vlados, C., Chatzinikolaou, D., & Kapaltzoglou, F. F. (2021). Energy market liberalisation in Greece: Structures, policy and prospects. *International Journal of Energy Economics and Policy*, *11*(2), 115–126. DOI: 10.32479/ijeep.10804

von Hirschhausen, C. (2018a). German Energy and Climate Policies: A Historical Overview. In von Hirschhausen, C., Gerbaulet, C., Kemfert, C., Lorenz, C., & Oei, P.-Y. (Eds.), *Energiewende "Made in Germany": Low Carbon Electricity Sector Reform in the European Context* (pp. 17–44). Springer International Publishing., DOI: 10.1007/978-3-319-95126-3_2

von Hirschhausen, C. (2018b). Nuclear Power in the Twenty-First Century [Part I]. *Assessment*.

Wang, H., & Wang, M. (2020). Effects of technological innovation on energy efficiency in China: Evidence from dynamic panel of 284 cities. *The Science of the Total Environment*, *709*, 136172. https://doi.org/https://doi.org/10.1016/j.scitotenv.2019.136172. DOI: 10.1016/j.scitotenv.2019.136172 PMID: 31905588

Wang, R., Mirza, N., Vasbieva, D. G., Abbas, Q., & Xiong, D. (2020). The nexus of carbon emissions, financial development, renewable energy consumption, and technological innovation: What should be the priorities in light of COP 21 Agreements? *Journal of Environmental Management*, *271*, 111027. https://doi.org/https://doi.org/10.1016/j.jenvman.2020.111027. DOI: 10.1016/j.jenvman.2020.111027 PMID: 32778307

Weibezahn, J. (2018). Sector Coupling for an Integrated Low-Carbon Energy Transformation: A Techno-Economic Introduction and Application to Germany. In von Hirschhausen, C., Gerbaulet, C., Kemfert, C., Lorenz, C., & Oei, P.-Y. (Eds.), *Energiewende "Made in Germany": Low Carbon Electricity Sector Reform in the European Context* (pp. 217–237). Springer International Publishing., DOI: 10.1007/978-3-319-95126-3_9

Wohlfarth, K., Worrell, E., & Eichhammer, W. (2020). Energy efficiency and demand response–two sides of the same coin? *Energy Policy*, *137*, 111070. DOI: 10.1016/j.enpol.2019.111070

Yuan, B. (2019). Effectiveness-based innovation or efficiency-based innovation? Trade-off and antecedents under the goal of ecological total-factor energy efficiency in China. *Environmental Science and Pollution Research International*, *26*(17), 17333–17350. DOI: 10.1007/s11356-019-05082-5 PMID: 31012077

Zhang, L., Mu, R., Zhan, Y., Yu, J., Liu, L., Yu, Y., & Zhang, J. (2022). Digital economy, energy efficiency, and carbon emissions: Evidence from provincial panel data in China. *The Science of the Total Environment*, *852*, 158403. https://doi.org/https://doi.org/10.1016/j.scitotenv.2022.158403. DOI: 10.1016/j.scitotenv.2022.158403 PMID: 36057314

Zhu, W., Zhang, Z., Li, X., Feng, W., & Li, J. (2019). Assessing the effects of technological progress on energy efficiency in the construction industry: A case of China. *Journal of Cleaner Production*, *238*, 117908. https://doi.org/https://doi.org/10.1016/j.jclepro.2019.117908. DOI: 10.1016/j.jclepro.2019.117908

Zoll, M. (2022). Energy governance in the Republic of Poland. In *Handbook of energy governance in Europe* (pp. 923–958). Springer. DOI: 10.1007/978-3-030-43250-8_22

Chapter 3
Performance Management and Accounting Frameworks for Assessing Corporate Sustainability in the Energy Transition Era

Ioannis Ch. Lampropoulos
https://orcid.org/0000-0002-8956-2178
University of Patras, Greece

Eleftherios Aggelopoulos
https://orcid.org/0000-0002-4090-9417
University of Patras, Greece

Vasileios Giannopoulos
University of Peloponnese, Greece

Angeliki Tsameti
https://orcid.org/0000-0002-2741-2297
University of Thessaly, Greece

Spyridon G. Kariofyllas
https://orcid.org/0009-0004-8325-7617
University of Peloponnese, Greece

DOI: 10.4018/979-8-3373-5302-9.ch003

ABSTRACT

This chapter focuses on the energy transition and explores its connection with sustainability accounting and the performance management of economic organizations. This transition requires strategic actions from economic entities, aiming to promote the adoption of renewable energy sources. A bibliometric analysis was conducted using the VOSviewer software, based on 129 papers retrieved from the Scopus database. The analysis aimed to explore emerging research trends in the areas of sustainability accounting, corporate social responsibility (CSR), and performance measurement indicators. The results reveal significant thematic clusters and recurring scholarly interests. The findings underscore the strategic challenges faced by key stakeholders - particularly small and medium-sized enterprises (SMEs)- in aligning their operations with sustainability objectives. Finally, the literature highlights that integrating emerging technologies such as artificial intelligence and blockchain constitutes a critical factor for improving the quality and transparency of sustainability accounting systems.

INTRODUCTION

In a world that is constantly changing and where the global economy and society stand at the center of these changes, the energy transition constitutes one of the most significant challenges, but at the same time, a great opportunity. In order to shift to clean, renewable energy sources and reduce greenhouse gas emissions, we need to radically change both the way businesses operate and how we measure and report their performance. At this point, sustainability accounting and corporate performance management come into a central role through the tools that businesses use to record, evaluate, and communicate their performance on economic, environmental, and social levels. This need is evidenced by the adoption of integrated performance measurement systems such as the Sustainability Balanced Scorecard (Hubbard, 2009), which combines traditional financial indicators with sustainability metrics, as well as advanced True Cost Accounting systems that capture the full cost of environmental and social impacts (Gusc et al., 2022). Gebhardt et al. (2023) emphasize that the development and use of ESG (Environmental, Social, Governance) indicators is particularly important because these tools allow us to evaluate the sustainable practices of businesses. According to Zharfpeykan and Akroyd (2022), as well as Thomas, Maheswari, and Kavitha (2025), the implementation of such tools faces significant challenges, especially for small and medium-sized enterprises, which often lack the necessary know-how and resources.

A critical factor for the reliability and comparability of sustainability data is the need for internationally accepted reporting standards and transparency (Jankalová & Jankal, 2024). Furthermore, it appears that new innovative solutions derived from the impact of technology, with an emphasis on artificial intelligence and blockchain, seem to constitute the means that will enhance quality and transparency in the field of sustainability accounting (Gusc et al., 2022).

This chapter attempts to present and analyze the main theoretical and practical approaches regarding sustainability accounting and performance management, particularly in the context of the energy transition. The aim of this chapter is to understand how measurement and reporting systems adapt to the requirements of sustainable development. Also, it explores the obstacles and prospects so that economic organizations can be led toward a green economy.

THE ROLE OF PERFORMANCE MANAGEMENT, ESG REPORTING, AND SUSTAINABILITY ACCOUNTING IN THE NEW ENERGY LANDSCAPE.

The energy systems in view of the energy transition include not only changes but also various shifts in issues related to energy security, geopolitical power, and governance. This energy transition leads to significant socio-economic and environmental impacts because it is a process embedded in our society and is influenced by technological innovations, market mechanisms, policy interventions, and even cultural habits. According to Yang et al. (2024), it is important to ensure a "just transition," taking into account that differences exist between social groups and geographic regions. Achieving such a transition necessitates a fundamental reconfiguration of business operations, where performance management and accounting frameworks are repositioned at the core of organizational strategy.

Krechovská and Procházková (2014) examined how companies integrate sustainability into corporate governance, specifically within the context of performance management and non-financial reporting. The authors found that although many firms recognize the importance of sustainability and its long-term benefits (good performance), they do not truly integrate it into their strategic management. Especially SMEs continue to emphasize traditional financial indicators, and sustainability reporting remains limited and unsystematic.

Particular attention is also given to the significant difficulties SMEs face when incorporating sustainability into their operations. Among the most critical barriers are the absence of robust financial reporting systems and the limited availability of financial, human, and technological resources. It is considered especially important for companies to build a unified and sustainable corporate culture, which requires the development of integrated measurement and reporting systems along with employee training and awareness to align with international ESG standards. Therefore, systematic and reliable non-financial reporting becomes essential to enhance transparency, accountability, and long-term performance of businesses.

Also, Zharfpeykan and Akroyd (2022) confirm in their study that although companies are issuing increasingly more sustainability reports, the indicators contained in them are not integrated into their internal Performance Management Systems (PMS). This finding is particularly significant because it constitutes an obstacle to implementing sustainability in practice, thus highlighting the gap between external sustainability reporting and the internal integration of these indicators. The authors also report that factors affecting the extent of this gap between external and internal integration of indicators are the company size and industry sector. This observation is especially meaningful for the energy sector, because effective performance management with embedded real sustainability targets is a critical factor for companies to achieve the energy transition and reduce environmental impacts. The importance of actual integration of sustainability targets into performance management systems is emphasized in the research of Maheswari et al. (2025), where the authors state that when socio-economic and environmental goals are combined, companies achieve enhanced efficiency, accountability, and simultaneously create long-term corporate value.

In turn, Kantabutra (2024) highlights a theoretical framework for sustainability performance management, which is based on four key factors related to organizational sustainability culture, sustainability strategy, corporate practices, and sustainability outcomes. These four mechanisms are interconnected through a feedback mechanism. Saratun (2016) also identifies another crucial factor for implementing sustainability strategy-namely, the role of employee engagement-arguing that traditional performance management is deemed insufficient and needs redesign to promote trust and participation. Manalu et al. (2023), as they describe stakeholder theory, provide the appropriate theoretical basis for sustainability accounting in order to ensure transparency and strategic alignment. This need is reinforced by Pratiwi & Edeh (2024), showing that ESG disclosures increase investor trust while simultaneously reducing the cost of capital, despite the challenges that exist regarding the reliability and harmonization of data.

On the other hand, Abou Taleb et al. (2015) focus on the fact that sustainability accounting focuses largely on costs without taking into account all the potential sources of revenue that can arise from sustainable practices. In order to address this imbalance, they propose a comprehensive tool, namely Environmental Management Accounting (EMA), which would enable organizations to capture both the risks and the benefits associated with sustainability initiatives. This new tool would allow for balanced information.

In parallel, Thomas, Maheswari & Kavitha (2025) argue that performance management systems (PMS) have emerged as critical mechanisms for the integration of sustainability. It is estimated that significant benefits such as cost savings and increased trust may arise if ESG indicators are linked with tools such as the Balanced Scorecard. It is particularly important that these benefits occur despite the difficulties that exist either due to the lack of unified indicators or the resistance that businesses may encounter.

A broader academic consensus further highlights the evolving role of sustainability accounting within corporate governance and reporting frameworks. Hopwood, Unerman, and Fries (2010) emphasize that sustainability accounting must move beyond marginal considerations and become a core component of internal strategies and decision-making processes. Jankalova & Jankal (2024) reinforce the view that sustainability accounting should have a commonly accepted terminology so that the accounting profession can have a future and focus on non-financial information. Pratiwi & Edeh (2024), for their part, enrich existing perspectives by proposing the use of technology in order to enable reliable ESG reporting. Equally important is the view of Genc & Kosempel (2023), who place particular emphasis on the financial dimensions of the energy transition, introducing a new perspective on the creation of business models and policies designed to promote sustainable investments.

Green accounting, at this point, appears to become critical in order to enhance transparency and sustainability strategy (Soraya, Nurrochmah & Hwihanus, 2024), while sustainability performance management can be achieved through an integrated framework that combines socio-economic and environmental data (Schaltegger & Wagner, 2006). To confirm this, Zhan (2023) finds that good ESG practices contribute to business performance by enhancing reputation, innovation, and risk management.

Finally, Lamberton (2005) presents a framework based on the principles of accountability and transparency, designed for sustainability accounting. The purpose of this conceptual framework is to integrate environmental and social goals into the economic goals of organizations. However, it is particularly useful for there to be institutional and organizational incentives in order for sustainability accounting to be applied on a large scale. Indeed, Martinez & Vazquez (2024) report that through access to supply chains and more favorable financing conditions, the mandatory disclosure of ESG data is associated with improved performance. Therefore, the creation of effective sustainability management and reporting frameworks becomes a necessary condition in order to strengthen corporate resilience and address the challenges of the energy transition.

THEORETICAL FRAMEWORKS AND MEASUREMENT INDICATORS FOR SUSTAINABLE CORPORATE PERFORMANCE AND THE ENERGY TRANSITION

From the early 21st century, Rotmans et al. (2001) had already proposed the concept of transition management as a way to handle large socio-technical changes, such as the shift toward energy-sustainable paradigms. This transition management approach focused on the process itself, promoting long-term thinking, multi-level collaboration among stakeholders, and emphasized collective learning and experimentation. Within this framework, organizations' performance management systems progressively incorporate long-term objectives that are guided not only by a technocratic framework but also by socio-ecological visions.

Zou & Wang (2024) emphasize that stricter environmental policies have a positive effect on both energy efficiency and the expansion of renewable energy sources, with carbon intensity and energy efficiency indicators showing significant improvement in areas where strict regulatory frameworks have been applied. Such regulations, as the authors state, can serve as a foundational pillar for the strategic integration of energy transition goals, thereby indirectly enhancing the environmental performance management of businesses.

The main theoretical frameworks and tools used in the business sector related to the measurement and management of sustainability are summarized in Table 1.

Table 1. Theoretical frameworks and sustainability tools in the energy sector

Year	Reference	Framework/Tool	Description	Key Features / Objective
2009	Hubbard, G.	Sustainability Balanced Scorecard (SBSC)	Extension of the Balanced Scorecard with integration of environmental and social dimensions	Multidimensional performance measurement with 5 dimensions
2009	Hubbard, G.	Organizational Sustainability Performance Index (OSPI)	Indicator that summarizes an organization's sustainability performance	Quantification and easily understandable depiction of sustainability
2022	Gusc et al.	True Cost Accounting (TCA)	Accounting that incorporates the total cost (economic, social, environmental)	Accurate depiction of the cost of the energy transition
2025	Damtoft et al.	Sustainability Performance Measurement Systems (SPMS)	Integrated sustainability measurement systems, tailored to business needs	Combination of economic, social, and environmental KPIs
	GRI, CFA Institute, EFFAS	ESG Reporting Standards	Standards for reporting environmental, social, and governance indicators	Comparability and transparency of ESG reports

To understand sustainable performance, it is essential to integrate both financial and non-financial ESG (Environmental, Social, Governance) indicators into the evaluation systems of businesses. The use of Key Performance Indicators (KPIs) gives significant weight to guiding decision-making and strategic planning within corporate actions, enabling firms to measure their progress not only toward sustainability targets but also in assessing their socio-economic and environmental impacts. Table 2 below illustrates the international organizations responsible for developing these indicators, with the primary objective of ensuring comparability and credibility in sustainability reporting. However, these indicators alone do not carry inherent weight - they must be applied in consideration of how a business operates, the nature of its activities, and the expectations of its investors and stakeholders.

Table 2. International organizations for sustainability indicators (indicative table)

International Organization	Description	Website
Global Reporting Initiative (GRI)	Develops widely used standards for sustainability reporting to ensure comparability and credibility of ESG data.	https://www.globalreporting.org/
CFA Institute	Provides guidance and frameworks on ESG integration and reporting, especially relevant for financial analysts and investors.	https://www.cfainstitute.org/
European Federation of Financial Analysts Societies (EFFAS)	Develops ESG standards and key performance indicators, focusing on harmonization and quality of sustainability reporting.	https://effas.com/

A recent study by Dimitrakopoulos, Makris & Apostolopoulos (2024) examines the factors influencing the Gross Domestic Product (GDP) in EU member states. This study, which classifies countries according to the level of their industrial activity, helps us understand how macroeconomic indicators are linked to energy indicators within the European Union. Their findings showed that the production of renewable and nuclear energy is particularly important for economies with a weak industrial base, which supports the view that customized sustainability indicators tailored to the energy profile of each country are necessary.

Already in the early 2010s, in the Czech Republic, a study of small and medium-sized enterprises (Kocmanova et al., 2012) found that energy consumption, waste generation, and emissions played a decisive role in operational efficiency and investor attraction. Similar findings were recorded for social indicators, including employee relations and workplace safety, as well as for governance indicators. Therefore, to achieve corporate targets and strengthen shareholder trust, adopting Codes of Good Practice and transparent reporting was of crucial importance.

The social dimension of the energy transition plays a particularly important role, especially in regions dependent on fossil fuels. In their study, Apostolopoulos et al. (2023) state that in Greece, Cyprus, and Bulgaria, policies for a just transition will only succeed if we manage to redefine skills while simultaneously strengthening local entrepreneurship. However, there are obstacles to this effort, namely the lack of mechanisms for skills upgrading and insufficient support for green business initiatives. According to the authors, and in order for the transition to be sustainable, it is necessary to incorporate social indicators - such as participation in training programs and the creation of new green jobs - into the Key Performance Indicators (KPIs) of the relevant policies.

Building on these insights, Kocmanová & Dočekalová (2011) proposed a comprehensive performance measurement framework grounded in the three pillars of environment, society, and corporate governance. These indicators are selected using statistical methods and are grouped into categories (resource use, emissions, social responsibility, human rights, working conditions, and compliance with corporate governance principles). To enhance the capacity for monitoring and evaluation, thereby strengthening the ability to track and assess sustainable performance both internally and towards external stakeholders, the proposed framework allows for the adaptation of indicators with the main goal of aligning them with the strategic directions of companies.

Thomas, Maheswari & Kavitha (2025) recognize that the integration of sustainability goals into performance management systems is considered a fundamental strategy to support the energy transition. Performance management systems that are linked to socio-economic and environmental goals enhance operational efficiency, promote transparency and accountability, and create long-term value. In this context, and in order to turn global challenges (such as climate change and biodiversity loss) into opportunities for strategic planning and innovation, it is very important to integrate sustainability into business frameworks so that the outcome is the creation of opportunities for strategic design and innovation. To ensure an effective mechanism for alignment between organizational performance and sustainable development goals, connecting ESG indicators with performance management tools such as the Balanced Scorecard is essential.

A key tool for measuring companies' progress toward sustainable development goals is performance indicators (Key Performance Indicators – KPIs). The integration of both financial and non-financial KPIs into internal management systems is considered particularly important in order to link them with the strategic goals of companies, assessing their environmental and socio-economic performance. As previously mentioned, it is important when selecting appropriate indicators to take into account the specific characteristics of the company while also facilitating both internal and external sustainability reporting. It is also emphasized that the existence of measurable indicators enhances transparency, accountability, and organizational performance improvement, and this arises from the causal relationships between activities and results (Kocmanová and Dočekalová, 2011).

Hubbard (2009) proposes an improved framework aimed at addressing the increasing complexity and shortcomings of existing measurement methods. Through this framework, it is possible to synthesize financial and non-financial performance elements from the earlier Triple Bottom Line (TBL) approach, which relates to measuring organizational performance. This specific framework introduces the Sustainable Balanced Scorecard (SBSC), combining it with the Organizational Sustainability Performance Index (OSPI). According to the author, the use of the SBSC helps improve transparency due to its coherent way of monitoring sustainability objectives across five dimensions (financial, customer, internal processes, learning and development, and environmental/social). On the other hand, the OSPI aims to quantitatively summarize the sustainability performance of businesses so that it is understandable to all stakeholders. This new approach by Hubbard highlights the need to create multidimensional and adaptable measurements rather than simple one-dimensional indicators. Essentially, effective sustainability measurement must be aligned with the organization's strategy and values, providing access to all stakeholders.

Therefore, the integration of ESG indicators into corporate governance positively affects performance, especially in the environmental and social domains. Gebhardt et al. (2023) emphasize that embedding ESG KPIs into internal management processes enhances ESG performance - an effect that is further amplified when these indicators are not directly tied to incentives such as executive reward systems. Although the use of these indicators does not immediately impact performance, when combined with compensation mechanisms they provide substantial benefits for achieving strategic sustainability goals. Hence, true strategic integration yields greater returns than mere compliance alone.

In order to move towards net-zero carbon emissions, it is necessary to adopt decarbonization practices that rely on green technologies and renewable energy sources. Accounting and accountability practices are of crucial importance for the integration of these technologies, as they help enhance transparency and legitimize corporate strategies in the eyes of stakeholders. Connecting decarbonization with clear accounting practices is deemed essential, as it not only aligns net-zero targets with sustainability reporting, but also facilitates the acceptance of these practices by both institutional and social actors (Di Vaio et al., 2025).

Baumgartner (2014) proposes a holistic model that shows how companies can effectively manage sustainability and Corporate Social Responsibility (CSR). Essentially, his model explains how the values, strategies, and tools of a company are interconnected in order to lead to sustainable development. In this model, Baumgartner divides management into three levels (normative, strategic, and operational) and through this provides the practical tools required to align the daily operations of companies with social and environmental goals. Essentially, through this specific model, companies can turn sustainability into a driving force for innovation and economic success, bridging the gap between their core corporate values and the effective implementation of their strategy.

A new framework is also introduced by the study of Jędrych et al. (2021), which examines how energy companies set their goals. In this framework, capital management is connected with sustainability, and the authors emphasize that if the traditional criterion of profit is replaced by investment-oriented goals, then energy companies can become more effective in the long term. Although it is estimated, according to the article, that this change is slow to come, it is considered necessary to support and achieve the energy transition. This new strategy, which is based on this new framework, requires that performance management be based on capital balance and not only on profitability, radically changing the way companies are managed in the energy sector.

Gusc et al. (2022), in a relatively recent article, argue that the integration of technologies such as artificial intelligence (AI), big data, and blockchain technology offers a new framework for measuring corporate sustainable performance, especially in the field of the energy transition, particularly in the "True Cost Accounting" (TCA) approach. Through the use of these technologies, it becomes possible to record, evaluate, and make decisions more accurately, which means that not only economic costs but also all environmental and social costs that were often overlooked in traditional performance systems will be taken into account, thus changing the path towards the transition. More specifically, according to the authors, the proposed TCA framework supports sustainable performance management by incorporating cost data related to emissions, energy consumption, and the use of natural resources. This integration appears to directly contribute to achieving the goals of the energy transition and additionally enhances transparency and accuracy in the measurement of ESG (Environmental, Social, Governance) indicators, providing the possibility for dynamic updating of indicators through blockchain networks, significantly improving the ability to control and monitor performance in real-time.

In order to move toward sustainable energy systems, it is deemed necessary to study energy management in industrial environments, as it represents a critical factor. Energy-management practices are not merely about reducing consumption or upgrading technology; they encompass a complex set of administrative, educational, and policy parameters that can strengthen organizations' capacity to adopt sustainable solutions and achieve long-term energy goals. Companies that implement integrated energy-management systems benefit substantially - not only from immediate reductions in energy consumption, but also from fostering organizational learning and enhancing their innovation capability. Performance management in the energy sector acts as a tool to optimize both technical and strategic objectives, aiming ultimately for a sustainable energy future. Moreover, it is vital to emphasize that effective policies supporting these practices and reinforcing knowledge flows within organizations are particularly important. The implementation of performance-evaluation and monitoring mechanisms - directly linked to sustainability strategies - enhances the ability of organizations to respond to the challenges of the energy transition (Andrei, 2023).

A decisive role in driving organizational change in companies can be played by sustainability accounting, as it serves as a bridge between strategic sustainability objectives and everyday business operations. In their study, Garcia-Torea, Larrinaga & Luque-Vílchez (2022) present the connection between sustainability accounting and organizational change as a complex process that requires interdisciplinary approaches bridging accounting and organizational management. Specifically, they refer to the interaction of three elements - creation of new accounting frameworks, management of resistance within organizations, and emergence of new accountability practices - which are essential for effective implementation of sustainability accounting. The interplay of these three elements confirms the dynamic relationship between sustainable accounting and the energy transition, especially due to resistance to change, which necessitates institutional and organizational readjustments supported by accounting. Also, a significant aspect of their research is the importance of action research as a methodology that facilitates exploration of sustainability accounting in practice, thereby enhancing its transformative role.

As previously mentioned, the integration of sustainability indicators (especially environmental and social ones) into internal management and decision-making systems enables businesses to monitor their progress toward sustainable development with greater accuracy and consistency. Adams & Frost (2008) argued that such indicators not only contribute to strategic management but also enhance organizational transparency toward stakeholders. Their study, based on empirical data from organizations located in Australia and the United Kingdom, shows that focusing on social and environmental issues leads to improved risk management and strategic performance, with significant implications for corporate accountability and sustainability.

In the context of the energy transition, efforts toward decarbonization in the waste management sector are increasingly being integrated into the strategies of European energy companies. According to Di Vaio et al. (2024), through the analysis of non-financial reports, companies are incorporating practices of green technology and renewable energy sources. A particularly important finding is the lack of quantitative data to assess the real impact of the initiatives on greenhouse gas emissions, which highlights the need to strengthen the accounting of environmental outcomes so that we can effectively monitor progress toward energy efficiency targets.

A recent study (Cunha et al., 2025) differentiates the connection between ESG indicators and corporate performance so that we can better understand how these indicators affect the financial and non-financial aspects of businesses. The authors argue in their paper that non-financial aspects, regardless of when they are evaluated, provide a more positive relationship with ESG performance, in contrast to financial indicators, which should be examined in the long term in order to eliminate possible misinterpretations of their results or various negative correlations. In order to strengthen the role of sustainability and play a decisive role in decision-making, by combining ESG indicators and corporate performance, companies can benefit by understanding and aligning their ESG strategies with their overall performance.

Stauropoulou and Sardianou (2019) put forward a composite sustainability index designed to assess how sustainable companies are within the banking sector. This index incorporates socio-economic and environmental dimensions according to the guidelines of the GRI (Global Reporting Initiative). Additionally, Aggelopoulos, Lampropoulos, & Paraschi (2023) used Data Envelopment Analysis (DEA) to evaluate the efficiency of retail stores in a crisis environment, thereby highlighting the usefulness of the method in measuring organizational resilience. Meanwhile, in a subsequent study, Aggelopoulos & Lampropoulos (2023) also employed DEA and the Malmquist Productivity Index and found that although businesses may initially lag in efficiency, their performance improves in the long term.

Meanwhile, Saeed & Kersten (2020) developed an analytical framework for assessing sustainable performance based on empirical data from industrial firms across different sectors. They identified 89 indicators covering critical dimensions of sustainability intended to assess performance across supply chains. The research highlighted the gap between the theoretical approach to the subject and practical implementation, especially regarding the stabilization of standardized measurement methods. This lack of universal systems limits the completeness and comparability of data. It was also found that companies that publish separate sustainability reports provide higher-quality information compared to those that incorporate them into their annual financial reports.

Integrating sustainability indicators (Sustainability KPIs) into corporate strategies thus provides us with a comprehensive tool to measure performance, linking socio-economic and environmental dimensions. Hristov & Chirico (2019) propose a theoretical model that allows companies to integrate sustainability into their core strategic processes, enhancing value creation in ESG terms. This framework not only emphasizes the need to select appropriate KPIs depending on the type and goals of the company but also highlights the importance of flexibility in their application in order to achieve truly sustainable performance.

In summary, understanding theoretical frameworks and sustainability indicators is the foundation for evaluating corporate performance within the context of the energy transition.

Towards an Integrated Model of Environmental Accounting and Emissions Transparency in the Energy Sector

Environmental accounting, based on climate change policy - which significantly affects corporate environmental strategies - appears to be entering a necessary stage of transformation into a strategic planning tool that will enable businesses to respond to the demands of policies designed for environmental issues and align with the functions and objectives of sustainability. Evangelinos, Nikolaou & Filho (2015) introduce into the ongoing discussion of recent years a new regulatory framework that explains how environmental policies influence corporate strategic decisions. The implementation of the new model proposed by the authors is estimated to contribute to businesses by improving the accuracy and reliability of environmental information and enhancing the transparency of environmental performance. In this way and based on data that reflects the actual performance of businesses, it will be possible to make strategic decisions supported by accounting practices that are based on the goals of the energy transition, thus providing a practical implementation tool for sectors with high energy intensity, while simultaneously promoting transparency and accountability.

Another significant role of environmental accounting is the evaluation and reporting of the carbon footprint in the energy sector in order to increase transparency and achieve the goals of sustainable development. Kurniawan & Utami (2025) emphasize the importance of environmental accounting as a strategic management tool that will allow businesses to improve their environmental performance by making informed decisions. Investors now seem to increasingly seek companies with high transparency and sustainability, and thus the integration of carbon footprint reporting as a way of corporate governance increases their value through the reliability and accuracy of the reports they produce. According to the authors, it appears that large companies compared to smaller companies in the energy sector apply better the principles of environmental accounting and are subject to audits by third-party accredited bodies in order to obtain certification of compliance with international standards such as the GHG Protocol and GRI standards. On the other hand, small and medium-sized enterprises in the energy sector seem to have lower systematic recording due to limited resources and because they have not fully understood the importance of transparency. In order to address all these issues, the authors highlight the integration of blockchain and big data technologies as an important step towards improving the accuracy and security of emissions, although they acknowledge that their high cost and technical difficulties limit the access of companies. Therefore, according to Kurniawan & Utami (2025), the effective implementation of environmental accounting can lead companies not only to align with the goals of sustainable development and the energy transition but also to enhance their reputation, compliance with regulatory frameworks, and provide them with long-term resilience.

Accounting now appears to be expanding, and alongside financial accounting, cost accounting, managerial accounting, etc., there is now another key field that seems likely to be of particular importance in the coming years. This new field has emerged from the recognition of climate change as a critical factor of economic and social risk that we are and will continue to be called upon to address in the coming years. Thus, based on this recognition of climate change as an economic risk, the content and practices of accounting are changing by incorporating environmental and social values into traditional reporting, thereby creating accounting for sustainable development (ASD). This new branch of accounting incorporates, on the one hand, practices related to carbon footprint accounting, social and environmental accounting, as well as ESG and sustainability reporting, and, on the other hand, broadens the role of accountants, who are now called to actively participate in risk analysis and policy design to ensure transparent information to all stakeholders. In this new branch of accounting, and in order for sustainable accounting practices to exist, the principles of materiality, accuracy, and stakeholder participation are important (Kosarkoska & Joldeska, 2024). Also, Gulluscio et al. (2020), through a systematic literature review, conclude that there are still gaps in understanding how companies manage accounting processes, despite the increased interest in integrating climate change into accounting and reporting, suggesting strengthening research on accounting processes that support emission reduction and adaptation to climate change.

As early as 2013, Ramírez and González González had highlighted that climate change has brought significant challenges to the field of accounting science, mainly due to new obligations related to carbon emissions and the creation of carbon markets. The management of these obligations, such as the need for accounting recording of assets and liabilities related to emissions, had already created significant issues in financial accounting. The lack of universal accounting standards for carbon-related assets led the community to adopt different methods that did not offer comparability among them. The need to complete projects that would include a common framework covering both mandatory and voluntary carbon markets was deemed imperative, with climate change accounting considered the field that must incorporate the complexity of these markets and support businesses with consistency and transparency in recording and assessing climate risks.

To gain a deeper understanding of the scientific trends, research gaps, and evolving discourse surrounding sustainable performance management in the context of the energy transition, the following chapter presents a bibliometric analysis. This analysis focuses on key thematic areas, including energy transition, organizational performance, and ESG reporting, with the aim of mapping the intellectual structure of the field and identifying emerging research directions.

Uncovering Research Pathways in Sustainability and Energy Transition: Insights from VOSviewer

In this section of the chapter, we used VOSviewer (Van Eck & Waltman, 2013), a popular tool for bibliometric analysis of scientific articles. Our main objective is to map scientific trends, visualize connections between different research areas, and identify various research gaps within the scope of corporate sustainability and energy transition. We focused particularly on themes related to the development and evaluation of performance management frameworks, as these are increasingly important and help guide organizations through their complex transition toward sustainable energy models.

An excellent example of VOSviewer's use comes from the study by Shafi et al. (2025), who conducted an extensive bibliometric analysis covering the period from 2010 to 2024 using VOSviewer - the same tool employed in our current chapter. Their research focused on how environmental and socio-economic factors are integrated in the evaluation of corporate sustainable performance, recognizing this area as a critical research priority.

Similarly, Harichandan et al. (2022) used bibliometric techniques to explore the academic discussion around the energy transition, especially the shift from fossil fuels to renewable energy sources. Their findings confirm how central this transition is for addressing climate change and promoting sustainable development. In their research, the authors identified key areas of interest such as energy governance, renewable energy technologies, and policy instruments, as well as significant gaps in developing strong, unified policy frameworks and the necessary infrastructure to support this transition. Their results clearly indicate the need for further research in these specific areas.

Additionally, Jadhav and Sarangi (2025), in their recent study, found a strong empirical link between corporate sustainability and operational performance, while also highlighting the urgent need for further research in the context of emerging economies. Their study suggests an increased focus on strategies based on innovation and ESG criteria, as these factors are critical for long-term competitiveness and resilience in developing markets.

Bibliometric Analysis Using VOSviewer

In this section of our study, we employed the VOSviewer tool to analyze scientific literature at the intersection of performance management, accounting, and the energy transition. The goal of the analysis was to identify the dominant research trends in order to extract emerging thematic associations as well as potential research gaps arising from the existing literature. The analysis using VOSviewer can contribute to the development of an organized framework for performance evaluation within the context of the energy transition. Figure 1 illustrates the methodological approach we followed to conduct the bibliometric analysis.

Figure 1. Design of the methodological flow of the bibliometric analysis

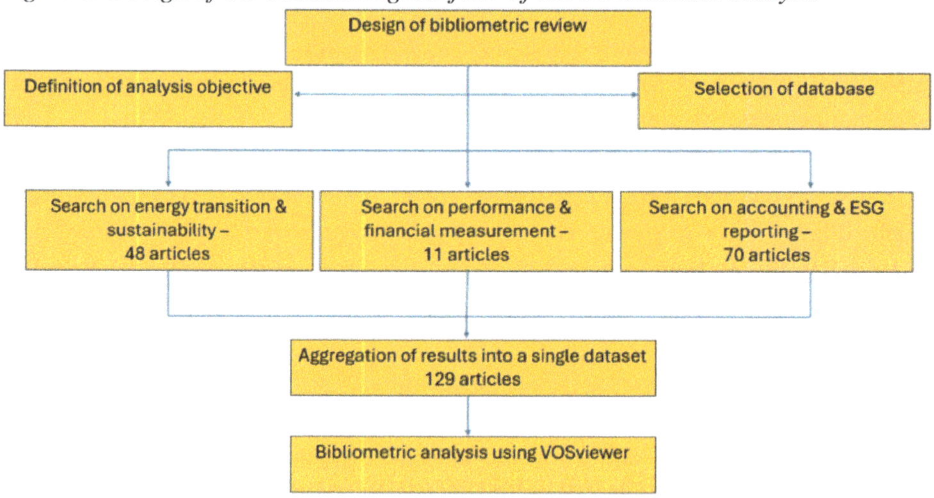

For the execution of the present bibliometric analysis, we sourced scientific articles from the Scopus database, widely acknowledged as a reliable platform housing a comprehensive collection of peer-reviewed literature. Initially, we conducted a structured search using keywords that combine the following three thematic axes:

1. Energy transition and sustainability
2. Performance and financial measurement
3. Accounting and ESG reporting

For these three axes, the following searches were performed without applying any restrictions, to ensure a comprehensive overview of the available literature. Accordingly, the keyword searches below did not include exclusions based on publication year, language, subject area, or document type, and include articles up to May 2025:

(A) ("energy transition" OR "sustainable development") AND ("performance management" OR "corporate performance") AND ("ESG" OR "sustainability accounting" OR "green accounting") – **48 articles**

(B) ("energy transition" OR "climate change") AND ("sustainability accounting" OR "green accounting") AND ("key performance indicators" OR "non-financial reporting" OR "ESG") - **11 articles**

(C) ("energy transition") AND ("financial performance" OR "KPIs" OR "balanced scorecard") – **70 articles**

The results of each of the three thematic searches were exported as CSV files and subsequently merged into a single consolidated dataset to facilitate a comprehensive bibliometric A total of 129 articles remained after removing duplicates from the three datasets (retrieved up to May 2025). This unified dataset was analyzed using the VOSviewer software, which enabled the visualization of bibliometric networks based on keyword co-occurrence. Through this approach, we enhanced the understanding of research activity within the studied field, identifying key thematic clusters and their interrelations. Through this approach, we enhanced the understanding of research activity within the studied field, identifying key thematic clusters and their interrelations. The aggregated results following the merging of the datasets are presented in Table 3.

Table 3. Distribution of articles by year

Year	Article
2011	1
2014	2
2018	4
2019	4
2020	12
2021	6
2022	13
2023	27
2024	37
2025	23

Within the framework of the bibliometric analysis, three analyses were carried out. Initially, Figure 2 shows the network visualization analysis where the keywords from the 129 articles of our sample are correlated. In the figure, we can see each keyword as a node, with the size of the node varying depending on how frequently the keyword co-occurs in the set of articles. We also observe lines connecting the nodes to each other, thus displaying the co-occurrence of keywords in relation to how often the keywords appear together in the same articles, thereby depicting the thematic relevance and the interconnection of scientific fields.

The network visualization analysis presents the structure and clustering of the main topics in the research we conducted on the energy transition and sustainable corporate performance. In the figure above, we can discern that the literature leads us to five different thematic clusters, which are identified as follows:

- The green cluster focuses on sustainability issues, ESG, and corporate performance.
- The red cluster focuses on economic analyses and investments in renewable energy sources.
- The blue cluster focuses on terms related to decision-making, energy efficiency, and performance measurement methods.
- The yellow cluster focuses on environmental issues such as greenhouse gas emissions, decarbonization, and climate change.
- Finally, the orange cluster focuses on the energy market and cost-benefit analysis.

Based on the above clusters, we conclude that the field under study, according to the thematic clusters, indicates the need for interdisciplinary collaborations among scientists, since economic, environmental, and managerial issues coexist and are governed by common themes, thematic convergences, and the emergence of research trends in energy transition and sustainability.

Figure 2. Network visualization analysis

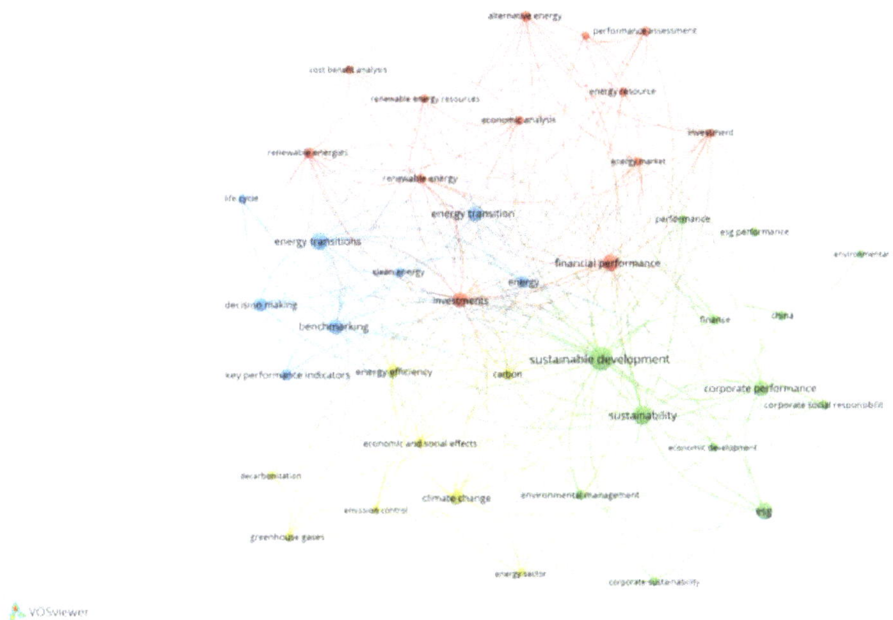

Figure 3 displays the Overlay Visualization Analysis, which incorporates a temporal dimension by mapping keywords according to their year of appearance and publication. The colors represent these years: cooler tones (blue – green) indicate keywords used earlier by researchers, while warmer colors (yellow) highlight terms that have drawn scientists' attention in recent years.

This visualization clearly displays a shift in research interest - from traditional topics such as performance management and financial sustainability to contemporary approaches centered on ESG reporting, the energy transition, and innovative technologies for sustainable development. Additionally, a notable finding is the increased emphasis on social aspects and risk management in recent years. This analysis serves as a dynamic tool that helps us understand the evolution of research fields, map emerging trends, and highlights the continuous change and expansion of scientific interest, thereby illustrating the trajectory of developments in the field.

Figure 3. Overlay visualization analysis

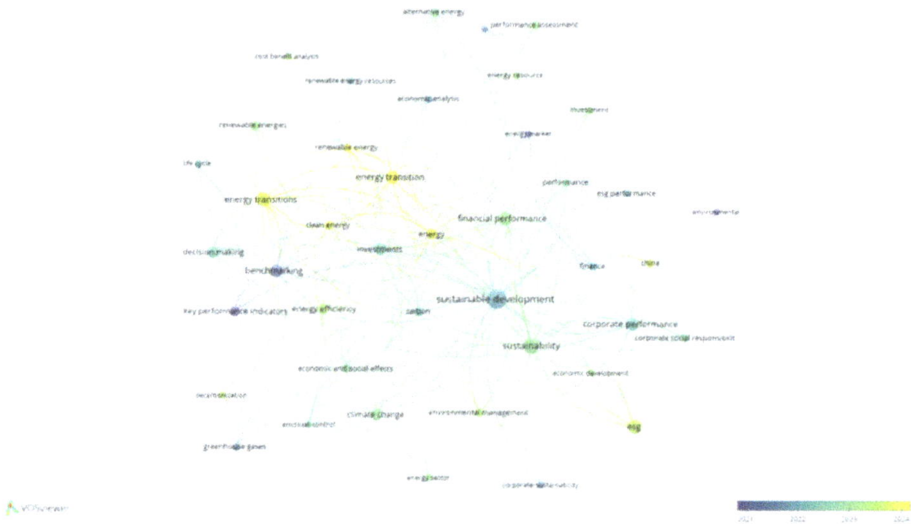

Figure 4 presents the Density Visualization Analysis, illustrating the frequency and distribution of keyword occurrences across the 129 articles included in the study. The regions in yellow highlight the most central and active research themes in the field of sustainability and energy performance, while the darker shades represent older research terms. According to the figure, the core topics identified are corporate social responsibility, ESG indicators, energy performance, sustainability, and climate change. These terms appear most frequently, indicating that they have been the most extensively studied. Additionally, the image reveals words with moderate density that highlight emerging new concepts, such as green economy, non-financial indicators, and sustainability governance. These terms signal their growing importance and the need for further focused research.

The Density Visualization analysis serves as a tool that identifies research cores and peripheral thematic areas, enabling a clear understanding of the structure and priorities of the literature over time.

Figure 4. Density visualization analysis

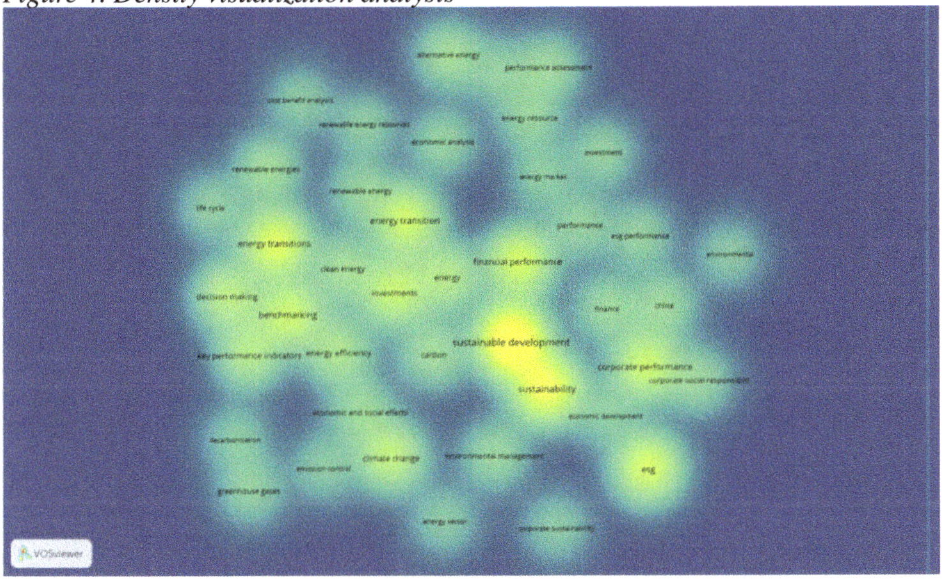

In this part of our study, we used the VOSviewer tool to examine scholarly articles that combine performance management, accounting, and the energy transition. The aim of the analysis was to identify dominant research trends, extract emerging thematic associations, and uncover potential research gaps arising from the documented literature. The VOSviewer analysis contributes to the development of a structured performance evaluation framework within the context of the energy transition. Figure 5 illustrates the methodological approach adopted to reach the bibliometric analysis stage.

Figure 5. Visualization of specific keywords

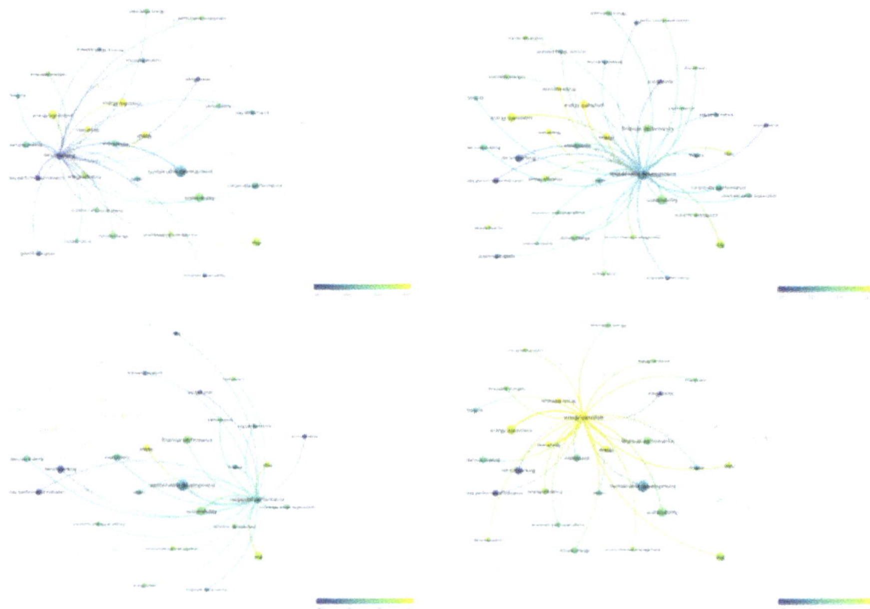

In the above figure, we initially isolated the term "Benchmarking" to investigate its connections with sustainability and the energy transition. From the keyword Benchmarking, it appears that there is a connection with numerous important terms, something that highlights its crucial significance in performance management, sustainability, and the energy transition. Benchmarking represents the process of comparison with standards or best practices and constitutes a key mechanism for evaluating performance and continuous improvement in companies. Based on the 129 articles, this specific keyword is connected with terms such as energy transition, sustainable development, energy efficiency, decision making, key performance indicators (KPIs), corporate sustainability, financial performance, climate change, ESG (environmental, social, and governance performance). These connections with the central term benchmarking show that the term functions as a critical tool for measuring and comparing results in relation to the challenges and requirements of the energy transition and corporate sustainability. We also observe a large number of connections which indicates that Benchmarking is a timeless concept that helps connect various thematic fields, such as economic performance, environmental management, and social responsibility. Therefore, based on this visualization of the connection of Benchmarking with other terms, it shows us that this timeless concept bridges theory with the practical management of performance, offering a framework through which companies can evaluate and improve their performance, particularly in relation to the requirements needed for sustainable development and the energy transition.

The next word chosen is “sustainable development,” which also constitutes a central axis in contemporary scientific discussions on issues related to energy transition and corporate sustainability. As shown by the connections created, sustainable development is not only linked to economic growth but focuses more on balanced progress that takes into account environmental and social dimensions. This includes the need to reduce pollutant emissions, manage natural resources, and improve social well-being. The concept of sustainable development serves as a node connecting performance, accounting, and KPIs in a constantly changing and demanding environment of corporate responsibility and energy transition. Sustainability accounting and related KPIs function not only as measurement tools but also as strategic guides that allow companies to transform environmental and social challenges into opportunities for growth and innovation.

The next concept depicted in Figure 5a is associated with the node created by the keyword **corporate performance**. This node likewise represents a central point in the scholarly dialogue on sustainable development and energy transition. Its pivotal position highlights its interdisciplinary nature and broad significance within the field. We observe four main connections associated with the following terms:

a) **Sustainable development**: The connection emphasizes that business evaluation is no longer limited to financial metrics but also incorporates social and environmental criteria. This confirms that sustainable development defines the framework within which corporate performance is assessed based on long-term value and impact.
b) **Financial performance** and **investments**: These links show that traditional financial performance remains fundamental but is now examined alongside environmental and social indicators. Corporate performance requires measures not only from financial outcomes but also from how investments support sustainable development.
c) **ESG** and **corporate social responsibility**: These associations indicate that corporate performance now includes the management of social and environmental matters essential for building value and competitiveness in business.
d) **Performance**, **decision making**, and **benchmarking**: These connections underscore the importance of measurement, evaluation, and comparison of performance within the corporate/management framework, with benchmarking and performance management forming the basis for enhancing corporate performance management in the evolving context of energy transition.

Based on the above analysis, it appears that the concept of corporate performance in accounting is evolving into a more holistic approach, one that integrates both traditional financial metrics and non-financial sustainability indicators. This evolution is essential for capturing a comprehensive picture of corporate value and its contribution to energy transition. Therefore, the keyword "corporate performance" highlighted in the figure serves as a connecting node between financial and non-financial measurements, management practices, and sustainability strategies. Its position in the visualization underlines the shift from traditional financial accounting toward more integrated, multidimensional frameworks that reflect the demands of energy transition and sustainable development.

Moreover, the figure reveals that the keyword "energy transition" functions as a central axis in our bibliometric analysis and within contemporary accounting discourse. This term bridges traditional and modern approaches to measuring business performance and accountability. The connections identified in this node show a link between energy transition, accounting concepts, and performance. Specifically, we observe a holistic depiction of corporate performance, with businesses now required to integrate environmental parameters into their performance measurements. It appears necessary to record and report non-financial data such as carbon emissions, energy consumption, and the use of renewable sources. We also notice that accounting frameworks and management systems are evolving and incorporating sustainability indicators (KPIs) related to energy transition. Additionally, there is a particularly important connection between energy transition and ESG requirements and accountability at the corporate governance level. Accounting appears to be moving toward supporting, recording, and disclosing environmental and social indicators that also affect financial performance.

Furthermore, the financial dimension - particularly investments in clean technologies - emerges as a critical component in driving long-term value creation through the energy transition. Energy transition seems to be central to the study of risk management and decision making, with accounting playing an especially significant role. Therefore, the energy transition node stands as a core element of modern corporate sustainability and is directly connected to the evolution of accounting frameworks and corporate performance measurement systems.

CONCLUSION

In conclusion, this study combined a theoretical approach with a bibliometric analysis performed using the VOSviewer software on 129 articles retrieved from the Scopus database. The results showed that the energy transition is a pivotal field of research and practice for sustainable development and corporate strategy. The analysis of the theoretical framework of the study highlights that the integration of sustainability into business strategy is achieved through performance indicator systems (KPIs) covering dimensions related to the economy, environment, society, organizational, and cultural aspects. The effective use of KPIs facilitates decision-making and strategic alignment, thereby strengthening sustainable practices across all areas of operations, with corporate culture and organizational structures playing a crucial role in the success of this integration, with significant obstacles identified in cultural and organizational challenges (Hristov & Chirico, 2019).

Benchmarking emerges as a key tool necessary for systematic measurement and continuous improvement of the sustainable performance of energy companies. KPIs through benchmarking are linked to sustainability accounting and strategic decision-making (Xu et al., 2024). While ESG indicators function as a critical measure of benchmarking despite short-term pressures on financial performance and contribute to risk management and achieving long-term value (Makridou et al., 2024; Fometescu et al., 2024). ESG strategy enhances companies' adaptation to market investor demands but the lack of unified ESG standards and the risk of greenwashing remain challenges. Meanwhile, KPIs within the ESG framework function as tools for monitoring and improving performance, enhancing transparency and decision-making (Wanday & El Zein, 2022). At the same time, the development of KPIs requires collaboration with stakeholders, with transparency and participation remaining key issues. Sustainable development is introduced through the integration of non-financial data into accounting systems and performance management, such as balanced scorecards (Adams & Frost, 2008).

Sustainability benchmarking proves to be a robust, long-term mechanism for comparing and enhancing organizational performance across multiple dimensions (Maltz et al., 2018). The Sustainability Balanced Scorecard (SBSC) specifically integrates critical sustainability dimensions, linking sustainability strategies with operational goals, thereby improving corporate reputation, reducing costs, and facilitating access to green financing (Hristov & Chirico, 2019). Empirical evidence suggests a positive relationship between ESG indicators and the corporate financial performance of energy companies, particularly regarding profitability and corporate value, with results typically manifesting after a 1–2-year lag and varying by market context (Cortez & Kelly, 2025).

The bibliometric analysis performed via VOSviewer revealed key themes including corporate social responsibility, ESG strategies, performance measurement tools, and emerging technologies such as blockchain, big data, and artificial intelligence (Gusc et al., 2022; Kurniawan & Utami, 2025). The analysis also highlighted the challenges faced by SMEs in integrating sustainability, the importance of international reporting standards and the chronological evolution of research in the energy transition field (Zharfpeykan & Akroyd, 2022; Krechovská & Procházková, 2014; Pratiwi & Edeh, 2024; Shafi et al., 2025; Harichandan et al., 2022).

In summary, performance management, sustainability accounting, and the integration of ESG metrics and KPIs constitute fundamental pillars for advancing the energy transition, managing risks, strengthening corporate resilience, and creating long-term value. These findings are strongly supported by both contemporary literature and the results of the bibliometric analysis presented in this chapter.

SUGGESTIONS FOR FURTHER RESEARCH

From the extensive theoretical analysis we conducted, we identified that although significant progress has been made in the development and application of performance management and sustainability accounting frameworks, several gaps and challenges remain that require further investigation. Firstly, it is necessary to conduct studies that examine non-financial indicators, specifically those related to how ESG KPIs are integrated into internal performance management systems. There appears to be a clear gap between what companies present (to their external environment) regarding sustainability and the ways they adopt these practices within their actual internal environment (Zharfpeykan & Akroyd, 2022; Thomas, Maheswari & Kavitha, 2025). A second point we propose for further research is conducting studies that focus on small and medium-sized enterprises, as the existing literature results remain limited, despite the frequent mention of this gap.

A fourth point we identified concerns the development of common, internationally accepted reporting standards and the harmonization of sustainability data, which are critical in order to improve their comparability and reliability (Jankalová & Jankal, 2024; Martinez & Vazquez, 2024). Furthermore, a fifth important area requiring future research is the use of modern technologies such as artificial intelligence (AI) and blockchain. At the same time, the development of innovative benchmarking approaches and the use of advanced analytical tools could strengthen sustainability management and its connection with financial performance, especially in the energy sector (Xu et al., 2024; Makridou, Doumpos & Lemonakis, 2024).

Finally, another point that would be particularly useful for further study is the emphasis that should be placed on strengthening interdisciplinary research, where the main objective will be bridging sustainability accounting with organizational management, taking into account organizational and cultural parameters in order to facilitate improvements in sustainable performance (Garcia-Torea, Larrinaga & Luque-Vílchez, 2023; Adams & Frost, 2008).

REFERENCES

Abou Taleb, M., Gibson, B., & Hovey, M. (2015). Fifty years of Sustainability Accounting: Does accounting for income in business sustainability really exist?. *International Journal of Accounting and Financial Reporting*.

Adams, C. A., & Frost, G. R. (2008)., Integrating sustainability reporting into management practices. [). No longer published by Elsevier.]. *Accounting Forum*, *32*(4), 288–302. DOI: 10.1016/j.accfor.2008.05.002

Aggelopoulos, E., & Lampropoulos, I. (2023). The effect of organic growth and acquisitions on performance of retailing networks in an alternating external operating environment. *Benchmarking*, *31*(1), 1–28. DOI: 10.1108/BIJ-02-2022-0087

Aggelopoulos, E., Lampropoulos, I., & Paraschi, E. P. (2023). An empirical analysis of evaluating the operational and economic efficiency of retail stores under an adverse operating environment. *MDE. Managerial and Decision Economics*, *44*(5), 2713–2741. DOI: 10.1002/mde.3844

Andrei, M. (2023). *The role of industrial energy management in the transition toward sustainable energy systems: Exploring practices, knowledge dynamics and policy evaluation*. Linkopings Universitet.

Apostolopoulos, N., Kakouris, A., Liargovas, P., Borisov, P., Radev, T., Apostolopoulos, S., Daskou, S., & Anastasopoulou, E. E. (2023). Just transition policies, power plant workers and green entrepreneurs in Greece, Cyprus and Bulgaria: Can education and retraining meet the challenge? *Sustainability (Basel)*, *15*(23), 16307. DOI: 10.3390/su152316307

Baumgartner, R. J. (2014). Managing corporate sustainability and CSR: A conceptual framework combining values, strategies and instruments contributing to sustainable development. *Corporate Social Responsibility and Environmental Management*, *21*(5), 258–271. DOI: 10.1002/csr.1336

Cortez, M. A. A., & Kelly, A. (2025). The Impact of ESG on Corporate Financial Performance: The Case of Global Energy Companies. *Asia-Pacific Social Science Review*, *25*(1), 2. DOI: 10.59588/2350-8329.1555

da Cunha, Í. G. F., Policarpo, R. V. S., de Oliveira, P. C. S., Abdala, E. C., & do Nascimento Rebelatto, D. A. (2025). A systematic review of ESG indicators and corporate performance: Proposal for a conceptual framework. *Future Business Journal*, *11*(1), 106. DOI: 10.1186/s43093-025-00539-1

Damtoft, N. F., van Liempd, D., & Lueg, R. (2025). Sustainability performance measurement–a framework for context-specific applications. *Journal of Global Responsibility*, *16*(1), 162–201. DOI: 10.1108/JGR-05-2023-0082

Di Vaio, A., Chhabra, M., Zaffar, A., & Balsalobre-Lorente, D. (2025). Accounting and Accountability in the Transition to Zero-Carbon Energy for Climate Change: A Systematic Literature Review. *Business Strategy and the Environment*, *34*(5), 5925–5946. DOI: 10.1002/bse.4282

Di Vaio, A., Van Engelenhoven, E., Chhabra, M., & Garofalo, A. (2024). Decarbonization of waste management practices and GHG accounting for energy transition: Evidence from European electricity corporations' reporting. *Environment, Development and Sustainability*, *27*(7), 1–25. DOI: 10.1007/s10668-024-04629-y

Dimitrakopoulos, P., Makris, I., & Apostolopoulos, S. (2024). The Impact of Major Energy Indexes on the Economic Growth of EU Countries, According to the Contribution of Industry to GDP. In *International Conference on Applied Economics* (pp. 59-74). Cham: Springer Nature Switzerland.

Evangelinos, K., Nikolaou, I., & Leal Filho, W. (2015). The effects of climate change policy on the business community: A corporate environmental accounting perspective. *Corporate Social Responsibility and Environmental Management*, *22*(5), 257–270. DOI: 10.1002/csr.1342

Fometescu, A., Hategan, C. D., & Pitorac, R. I. (2024). How Responsible Are Energy and Utilities Companies in Terms of Sustainability and Economic Development? *Energies*, *17*(23), 6209. DOI: 10.3390/en17236209

Garcia-Torea, N., Larrinaga, C., & Luque-Vílchez, M. (2023). Bridging the understanding of sustainability accounting and organizational change. *Organization & Environment*, *36*(1), 17–38. DOI: 10.1177/10860266221083339

Gebhardt, M., Thun, T. W., Seefloth, M., & Zülch, H. (2023). Managing sustainability - Does the integration of environmental, social and governance key performance indicators in the internal management systems contribute to companies' environmental, social and governance performance? *Business Strategy and the Environment*, *32*(4), 2175–2192. DOI: 10.1002/bse.3242

Genc, T. S., & Kosempel, S. (2023). Energy transition and the economy: A review article. *Energies*, *16*(7), 2965. DOI: 10.3390/en16072965

Gulluscio, C., Puntillo, P., Luciani, V., & Huisingh, D. (2020). Climate change accounting and reporting: A systematic literature review. *Sustainability (Basel)*, *12*(13), 5455. DOI: 10.3390/su12135455

Gusc, J., Bosma, P., Jarka, S., & Biernat-Jarka, A. (2022). The big data, artificial intelligence, and blockchain in true cost accounting for energy transition in Europe. *Energies, 15*(3), 1089. DOI: 10.3390/en15031089

Harichandan, S., Kar, S. K., Bansal, R., Mishra, S. K., Balathanigaimani, M. S., & Dash, M. (2022). Energy transition research: A bibliometric mapping of current findings and direction for future research. *Cleaner Production Letters, 3*, 100026. DOI: 10.1016/j.clpl.2022.100026

Hopwood, A. G., & Unerman, J. (Eds.). (2010). *Accounting for sustainability: Practical insights*. Earthscan. DOI: 10.4324/9781849776332

Hristov, I., & Chirico, A. (2019). The role of sustainability key performance indicators (KPIs) in implementing sustainable strategies. *Sustainability (Basel), 11*(20), 5742. DOI: 10.3390/su11205742

Hubbard, G. (2009). Measuring organizational performance: Beyond the triple bottom line. *Business Strategy and the Environment, 18*(3), 177–191. DOI: 10.1002/bse.564

Jadhav, D. S., & Sarangi, S. (2025). Corporate sustainability and performance: A bibliometric review for two decades. *Benchmarking, 32*(4), 1459–1483. DOI: 10.1108/BIJ-07-2023-0467

Jankalová, M., & Jankal, R. (2024). Review of Sustainability Accounting Terms. *Administrative Sciences, 14*(7), 137. DOI: 10.3390/admsci14070137

Jędrych, E., Klimek, D., & Rzepka, A. (2021). Principles of Sustainable Management of Energy Companies: The Case of Poland. *Energies, 14*(8), 2042. DOI: 10.3390/en14082042

Kantabutra, S. (2024). Toward a sustainability performance management framework. *Heliyon, 10*(13), e33729. DOI: 10.1016/j.heliyon.2024.e33729 PMID: 39071584

Kocmanová, A., & Dočekalová, M. (2011). Corporate sustainability: Environmental, social, economic and corporate performance. *Acta Universitatis Agriculturae et Silviculturae Mendelianae Brunensis, 59*(7), 203–208. DOI: 10.11118/actaun201159070203

Kocmanova, A., Nemecek, P., & Docekalova, M. (2012). Environmental, social and governance (ESG) key performance indicators for sustainable reporting.

Kosarkoska, D., & Joldeska, I. (2024). The impact of climate change and climate risk on the Accounting and Accounting professions.

Krechovská, M., & Procházková, P. T. (2014). Sustainability and its integration into corporate governance focusing on corporate performance management and reporting. *Procedia Engineering*, *69*, 1144–1151. DOI: 10.1016/j.proeng.2014.03.103

Kurniawan, A., & Utami, Z. R. (2025). Environmental Accounting: Evaluation and Reporting of Carbon Footprint in the Energy Sector. *FIDUSIA: JURNAL KEUANGAN DAN PERBANKAN, 8*(1).

Lamberton, G. (2005). Sustainability accounting - a brief history and conceptual framework. []. No longer published by Elsevier.]. *Accounting Forum*, *29*(1), 7–26. DOI: 10.1016/j.accfor.2004.11.001

Makridou, G., Doumpos, M., & Lemonakis, C. (2024). Relationship between ESG and corporate financial performance in the energy sector: Empirical evidence from European companies. *International Journal of Energy Sector Management*, *18*(4), 873–895. DOI: 10.1108/IJESM-01-2023-0012

Maltz, E., Bi, H. H., & Bateman, M. (2018). Benchmarking sustainability performance: The next step in building sustainable business models. *Journal of Public Affairs*, *18*(3), e1606. DOI: 10.1002/pa.1606

Manalu, C. I. B., Manurung, M., Muda, I., & Kesuma, S. A. (2023). Implementation of stakeholder theory in sustainability accounting: A literature review. *Brazilian Journal of Development*, *9*(12), 31853–31871. DOI: 10.34117/bjdv9n12-087

Martinez, S., & Vazquez, A. B. (2024). Mandatory ESG reporting and corporate performance. *Available at SSRN* 4358256.

Pratiwi, A. P., & Edeh, F. O. (2024). The Role of ESG Disclosure in Corporate Performance and Investment Decision-Making. *Sinergi International Journal of Accounting and Taxation*, *2*(1), 1–14. DOI: 10.61194/ijat.v2i1.476

Rotmans, J., Kemp, R., & Van Asselt, M. (2001). More evolution than revolution: transition management in public policy. *foresight, 3*(1), 15-31.

Saeed, M. A., & Kersten, W. (2020). Sustainability performance assessment framework: A cross-industry multiple case study. *International Journal of Sustainable Development and World Ecology*, *27*(6), 496–514. DOI: 10.1080/13504509.2020.1764407

Saratun, M. (2016). Performance management to enhance employee engagement for corporate sustainability. *Asia-Pacific Journal of Business Administration*, *8*(1), 84–102. DOI: 10.1108/APJBA-07-2015-0064

Schaltegger, S., & Wagner, M. (2006). Integrative management of sustainability performance, measurement and reporting. *International Journal of Accounting. Auditing and Performance Evaluation, 3*(1), 1–19. DOI: 10.1504/IJAAPE.2006.010098

Shafi, R. M., Shahid, W., Osman, I., & Arifin, N. A. M. (2025). Mapping the Landscape: A Bibliometric Analysis of Global Trends in Sustainable Performance. *International Review of Management and Marketing, 15*(3), 228–245. DOI: 10.32479/irmm.18298

Soraya, B., Nurrochmah, A., & Hwihanus, H. (2024). Business Transformation Towards Sustainability: The Role of Green Accounting in Sustainability Management. *Journal of Environmental Economics and Sustainability, 1*(3), 1–8. DOI: 10.47134/jees.v1i3.342

Stauropoulou, A., & Sardianou, E. (2019, November). Understanding and measuring sustainability performance in the banking sector. [). IOP Publishing.]. *IOP Conference Series. Earth and Environmental Science, 362*(1), 012128. DOI: 10.1088/1755-1315/362/1/012128

Thomas, M., Maheswari, G. S., & Kavitha, M. (2025). PERFORMANCE MANAGEMENT AND SUSTAINABILITY GOALS. *Cuestiones de Fisioterapia, 54*(3), 1138–1156. DOI: 10.48047/4jfe3865

Van Eck, N. J., & Waltman, L. (2013). VOSviewer manual. *Leiden. Univeristeit Leiden, 1*(1), 1–53.

Wanday, J., & Ajour El Zein, S. (2022). Higher expected returns for investors in the energy sector in Europe using an ESG strategy. *Frontiers in Environmental Science, 10*, 1031827. DOI: 10.3389/fenvs.2022.1031827

Xu, X., Liu, Z., Liu, W., Pei, C., Wu, X., & Nie, Z. (2024). A sustainable development benchmarking framework for energy companies based on topic mining and knowledge graph: The case of oil and gas industry. *Renewable & Sustainable Energy Reviews, 196*, 114350. DOI: 10.1016/j.rser.2024.114350

Yang, Y., Xia, S., Huang, P., & Qian, J. (2024). Energy transition: Connotations, mechanisms and effects. *Energy Strategy Reviews, 52*, 101320. DOI: 10.1016/j.esr.2024.101320

Zhan, S. (2023). ESG and corporate performance: A review. In *SHS web of conferences* (Vol. 169, p. 01064). EDP Sciences.

Zharfpeykan, R., & Akroyd, C. (2022). Factors influencing the integration of sustainability indicators into a company's performance management system. *Journal of Cleaner Production, 331*, 129988. DOI: 10.1016/j.jclepro.2021.129988

Zou, Y., & Wang, M. (2024). Does environmental regulation improve energy transition performance in China? *Environmental Impact Assessment Review*, *104*, 107335. DOI: 10.1016/j.eiar.2023.107335

Chapter 4

Decentralized Energy Systems and Just Transition:
Digital Governance and Socio-Economic Impacts in the Energy Shift

A. Ayyoob
https://orcid.org/0009-0003-2839-3598
University of Calicut, India

Aparna Sajeev
https://orcid.org/0000-0001-5642-5460
University of Calicut, India

ABSTRACT

This study operates at the nexus between decentralised energy systems and just energy transition, studying how digital technologies such as blockchain and IoT transform energy governance with socio-economic inequalities in mind. While there are technological advances, integration still has not taken place between decentralised models and inclusive policy frameworks. Using a qualitative thematic analysis of recent literature (2019–2025), the study identifies how digital tools increase efficiency and transparency yet require regulation that can respond to these challenges. The findings present opportunities, from empowering communities to P2P trading, but also risks, including the emergence of regional inequalities and

DOI: 10.4018/979-8-3373-5302-9.ch004

labour displacement. The study puts forward integrative governance models that stretch the enabling potential of digital innovation toward justice-aligned transition frameworks, therefore providing critical knowledge for policy and practice.

INTRODUCTION AND SIGNIFICANCE

Prior to the global energy transition, a twin imperative was in place: decentralisation on the one hand, and socio-economic inclusion on the other. As climate change worsens and fossil-fuel dependence is less and less justifiable, energy systems around the world undergo fundamental transformation. The centrally organised and hierarchical models of energy generation and distribution are being challenged by decentralised, digitised infrastructures that facilitate localised production, management, and consumption. At the core of this energy paradigm shift are decentralised energy systems, which are defined by peer-to-peer (P2P) energy trading and community microgrids that reduce dependence on traditional utilities and offer increased technical efficiency with potential for further resilience and equity (Edem 2024). These systems give consumers the power to "presume," that is to generate, consume, and trade electricity through local networks. Paradoxically, Presumption depends neither on pure technological readiness nor on hospitality from classical regulators but rather on the availability of robust, adaptive, and inclusive governance frameworks that are able to handle their complexity.

Key to enabling these decentralised models are the various digital technologies such as blockchain, AI, IoT, and big data analytics. The possibilities provided by these new technologies include real-time condition monitoring, secure transactions, prediction of demand, and automated energy allocation, turning decentralised energy into something not only possible but also efficient and transparent (Onukwulu et al., 2023; Bañales, 2020). Despite the efficiencies of the inclusion of these technologies, their presence poses new challenges in issues of governance, regulation, and ethics. Existing energy policies are still tied to a centralised infrastructure and hence often lag in addressing the new challenges wrought by innovation and the less-digitised forms of governance that result (Brisbois, 2022). For example, issues such as data privacy, algorithmic biases, and platform control have been insufficiently addressed both in policy and practice. Simultaneously, however, energy transition is not only a technical exercise but also a deeply socio-economic one. The low-carbon transition has far-reaching implications for labour market changes, regional economies, and social equity. An economic collapse is a potential threat to fossil fuel-heavy regions such as Mpumalanga in South Africa, with jobs loss, and infrastructural obsolescence if the shift is not managed inclusively (Nel et al., 2023). Just transitions, therefore, require active measures-such as worker retraining and upskilling, investment in affected communities, and participatory policy-making that guarantees leaving no one behind (Xie et al., 2023; Bray et al., 2022). Nevertheless, while discussions concerning just transition have grown, there continues to be a disconnect between how socio-economic issues are incorporated and the latest digital governance innovations in decentralised systems.

This disconnect forms the core of the research problem. Whilst decentralised energy technologies and just transition frameworks seem to be gaining momentum, very few studies try to systematically investigate their interface and intersection. One might even argue that the literature addresses either the technical potential of decentralised systems or the socio-economic imperatives of transition but rarely both in an integrated manner. For example, Edem (2024) and Umar et al. (2024) give very thorough technical accounts of community microgrids and P2P trading while providing little consideration for labour or equity impacts. In contrast, works such as Bray et al. (2022) and Garcia-Casals et al. (2019) will bring social inclusion and justice into the discussion yet will often ignore the governance implications of digital infrastructures. This fragmented approach leaves a critical gap in theorising how to design digital governance systems to serve both sociotechnical utility and socioeconomic justice. This study seeks to fill the gap by exploring how decentralised energy systems supported by digital technologies can serve a just transition. In this line, the research is motivated and guided by three main questions: (1) How do digital technologies such as blockchain, IoT, and AI influence governance in decentralised energy systems? (2) What regulatory and ethical issues arise from the digitalisation of decentralised energy governance? (3) How can socio-economic inclusion be integrated into policy frameworks that govern the adoption of decentralising technologies?

The research stands very relevant in its interdisciplinary perspective. On a theoretical level, it contributes to new developing areas at the intersection of energy justice and digital innovation, whereby technologies are, therefore, not neutral tools but rather embedded within political and socio-economic contexts influencing their outcomes (Brisbois, 2022). On a practical level, the study provides some answers for policymakers, regulators, and community leaders on how to design governance systems that are not only technologically advanced but also socially inclusive and equitable.

Following the four main research goals, the first is to explore the digital-enabling scenario for decentralized energy systems such as P2P trading and microgrids; second, to understand the role that digital infrastructure plays in ensuring transparency, efficiency, and accountability in governance; third, to assess the socio-economic implications of these systems-one of the aspects being labour reskilling, regional development, and justice; and the last being to outline integrative policy approaches that align decentralized innovations with the objectives of a just and inclusive transition. This combining between these fields is aimed at shedding light on the nexus of digital governance and social justice in energy transitions. Only by understanding how these two are working together can envision and bring to reality energy futures that are sustainable and equitable.

REVIEW OF LITERATURE

The transition toward decentralized energy systems has emerged as one of the critical components in the worldwide-induced push toward sustainable energy infrastructure. Decentralization theoretically offers increased participation, greater flexibility, and efficiency in energy management, especially when it is aided by digital technologies. Brisbois (2022), Vasconcelos et al. (2022), and Bañales (2020) observe the process by means of which these digital inventions, such as blockchain, IoT, and AI, are reshaping the governance and operational workings of the contemporary energy systems.

The principal changes in decentralized systems have to do with towards participatory governance models. Brisbois (2022) claims that decentralized governance for energy should be multi-level and inclusive, allowing for a very broad range of stakeholders to engage and affect decision-making processes. Borrowing from governance approaches in water and forestry, adaptability and cooperation are highlighted as keys to the management of common resources. Akin to these are the words of Nolden (2019), who affirms that although decentralized models promote community ownership and innovations, structural barriers, such as outdated regulation and infrastructural restrictions, counteract their effects.

Among some digital technologies mentioned in this regard are blockchain and distributed ledgers for enhancing transparency, accountability, and efficiency in energy markets. According to Vasconcelos et al. (2022), blockchain applications support peer-to-peer (P2P) energy trading, enhance carbon emission traceability, and provide trust to consumers. Smart contracts on blockchains have largely automated such transactions, reducing their dependence on intermediaries, thereby offering more cost savings and operational efficiency (Aravind et al., 2024; Din & Su, 2024). Suthar and Pindoriya (2020) advance this argument further, by illustrating how blockchain enables prosumers to sell their excess energy directly, thereby democratizing energy access.

On the other hand, the IoT facilitates real-time monitoring, demand-side management, and microgrid optimization in energy systems. Sithole (2024) notes that IoT technologies offer load balancing and dynamic control mechanisms that are crucial to renewable energy integration. However, the massive data collection intrinsic to IoT systems would increase the risk of data privacy, thus making secure communications protocols and decentralized data control extra considerations (Kavitha et al., 2023). According to Benavides et al. (2023) and Basavanna et al. (2024), the use of IoT should improve local autonomy and system responsiveness, particularly in off-grid and rural settings.

Beyond IoT and blockchain, AI in its widest sense plays a central role in the optimization of decentral power generation systems. Artificial intelligence methods forecast energy demand with a high level of accuracy, provide dynamic pricing, and enhance the distribution network's efficacy (Zülfikaroğlu, 2024; Gopal et al., 2024). According to Alsaigh et al. (2022), AI operating on large datasets in real time contributes to operational efficiency and supports the intermittent renewable integration. However, concerns about bias, including decision transparency, remain. Jain and Mitra (2024) point out that biased data inputs could result in unfair pricing systems that discriminate against low-income users, while Hayawi et al. (2024) call for transparency and auditability of AI systems to build public trust on compliance.

Aligning conventional modes of governance to decentralized technologies continues to be a major challenge in the policy and regulatory sphere. Ahmad (2024) brings up the issues of voter apathy and whale manipulation in decentralized decision-making platforms, while Salami (2021) argues that the absence of Know Your Customer (KYC) in DeFi-based energy systems. Jurisdictional fragmentation makes enforcement and compliance all the more challenging, especially when digital energy platforms operate across national or regional boundaries (Ahmed, 2022). In view of these circumstances, scholars like Doan et al. (2023) advocate for dynamic pricing models and regulatory clarity surrounding crypto-assets. Ahmed (2022) sees the potential in using regulatory technologies (Reg-Tech) to automate compliance and enhance supervision over decentralized markets.

However, even with such advancements, there still remain some unsolved problems. With regard to integration, for instance, while blockchain and interoperating systems through IoT applications will benefit from enhanced efficiency in the system, they are often confronted with barriers in terms of scaling, interoperability, and regulatory support (Gururaja et al., 2024; Varshney et al., 2024). Moreover, very little has been done to dissect the ethical and social ramifications of algorithmic governance, especially in the instances of digital illiteracy. From another perspective, applications of community-based governance in energy systems are still coming into being and often get stuck in challenges related to coordination, equity, and trust (Brisbois, 2022; Ahmad, 2024).

Research Gap

While Technological, policy-related, and governance-related factors on decentralized energy systems are very well studied, some gaps remain in the literature. Often, blockchain, IoT, and AI are studied in isolation; their intersections are overlooked when it comes to participatory governance and transparency (see Vasconcelos et al., 2022; Aravind et al., 2024; Zülfikaroğlu, 2024). There is less empirical evidence on digital inclusion, algorithmic bias, and participation at the grassroots level (see Brisbois, 2022; Ahmad, 2024). Moreover, regulatory frameworks need their pace with innovation matters of compliance, consumer rights, and integration into DeFi remain under-researched (Salami, 2021; Ahmed, 2022). To fill these gaps, the study develops an integrated framework for adaptive and inclusive digital energy governance.

METHODOLOGY

Generically, the field of research followed here is qualitative and grounded on thematic analysis, with an aim to analyze the role of blockchain, the Internet of Things (IoT), and Artificial Intelligence (AI) in the formation of governance structures in decentralized energy systems. Thematic analysis fits this study perfectly, as it helps identify and interpret recurring patterns in the literature that provide insight into how digital technologies influence transparency, participation, and decision-making in energy governance. The whole study is about building a conceptual understanding concerning how these technologies act as facilitators for sustainable transitions while considering regulatory and policy concerns.

Secondary data are used in the research, which involve data culled from publication sources, such as peer-reviewed journals, conference proceedings, and highly recognized reports, published between 2019 and 2024. This selects sources with purposeful sampling and ensures they speak directly to digital technologies and decentralized energy governance. Search terms included "blockchain in energy governance," "IoT in microgrid management," "AI in energy pricing," and "peer-to-peer energy trading." Particularly relevant studies include those of Aravind et al. (2024), Din & Su (2024), Zülfikaroğlu (2024), and Jain & Mitra (2024). Grey literature was excluded to maintain academic rigor; focus was on clarity of methodology and empirical studies that are recent in date.

The analysis follows Braun and Clarke's (2006) six-phase framework. It began with familiarisation, which consisted of detailed reading and annotation of selected texts. Codes were then generated for data segments relevant to blockchain-enabled transactions, IoT-driven monitoring, and regulatory adaptation and were then organised into broader categories. The four overarching themes that emerged were Digital Governance within Decentralized Energy Systems, Blockchain and Smart Contracts, IoT and Real-time Monitoring, and Algorithmic Governance and the Adaptation of Regulations. Sub-themes such as transparency, data privacy, and community participation were streamlined for conceptual clarity. The final stage saw the entire process bound into a coherent narrative where the literature is interwoven to increase theoretical heft.

No ethical concerns in terms of human participants can be raised since the study is built on publicly available secondary data. Intellectual integrity has been maintained in that sources have been correctly cited, preserving the author's original intent of the said source. The limitations, however, exist in that thematic analysis is interpretative and is influenced invariably by the unique perspective of the researcher; secondly, a situation arises where challenges to real-world implementation are rarely given a voice owing to thematic analysis being overly dependent on published literature. Future research avenues include conducting case analyses and stakeholder interviews to augment the findings. This methodology provides a structured way to probe the interface between digital technology and governance, and thus provides an input for further research, regulatory design, and innovation toward sustainable energy.

THEMATIC ANALYSIS

Digital Governance in Decentralised Energy Systems

Digital governance in decentralised energy systems, thus, signifies a transformative change in energy production and how it is managed and consumed. Featuring transparency, stakeholder participation, and system optimisation facilitated by digital technologies, the energy systems become ever more fragmented. Given DERs disparagingly sit at the very centre of the concept behind Distributed Energy Resources (DER), alongside the timelines of installation, it is only fitting that governance structures are also constructed to reflect this decentralisation (Brisbois, 2022).

At the heart of this metamorphosis lies the creation of participatory governance models. Decentralised energy systems inherently involve multiple actors comprising households, local communities, micro-grid operators, and private energy providers. To deal with this diversity effectively, governance models must provide for multi-layered and participative decision-making across these layers, enabling stakeholders to influence decisions affecting the technical integrity of the grid itself (Brisbois, 2022). Not only is the participatory model inherently more democratic, but it also ensures that energy policies and operations are contextually appropriate, drawing on practices from other decentralised resource domains.

Digital technologies are considered to be the potential enabler of this participation framework. The distributed ledger systems, i.e. the blockchain, IoT systems, and the AI systems, are making the way for the digital transformation. For example, since energy transactions are recorded in an immutable ledger, a blockchain-based system ensures complete transparency in energy traceability across flows and carbon footprints, as well as in the discharge of contractual responsibilities, unlike before (Vasconcelos et al., 2022; Aravind et al., 2024). These tools have also been called upon as a solution to facilitate improved trust and accountability in an energy market that is, by nature, decentralised (Varshney et al., 2024).

Besides these, such digital technologies provide the Axis for integrating distributed energy resources (DER) such as solar PV, wind turbines, and battery storage systems outright into the grid. IoT devices gather real-time data and allow remote control of energy systems for optimal load balancing and improved distribution network hosting capacities (Bánales, 2020; Benavides et al., 2023). On the other hand, AI is in use for purposes of dynamic pricing, forecasting loads, and allocating resources more efficiently (Zülfikaroglu, 2024; Gopal et al., 2024).

Despite such advancements, some structural and regulatory barriers deter the full-scale implementation of decentralised systems. Energy regulations developed for centralised systems often fail to appreciate the intricacies of decentralised and community-based governance (Nolden, 2019). Further challenges are posed by digital-era complications concerning data security, algorithmic bias, and platform standardisation (Jain & Mitra, 2024; Kavitha et al., 2023; Hayawi et al., 2024).

The scholars and practitioners have shared a growing consensus that a hybrid model that exploits the robustness of centralised systems with the flexibility of decentralised governance would best pave the way forward (Ahmad, 2024). Regulatory adaptation will thus be required, including the formulation of frameworks validating P2P energy trading, dynamic pricing, and decentralised financial instruments, so as to build the leverage for this integration (Doan et al., 2023; Salami, 2021).

Consequently, a digital governance held promise to create decentralised energy systems that were more inclusive, transparent, and efficient. However, the realization of the above promise requires coordinated policy reforms, technological innovation, and sustained stakeholder engagement. The digitalisation of energy governance must be complemented by an ethical perspective and social safeguards to allow the benefits to be fairly distributed, and the risks to be well mitigated (Alsaigh et al., 2022; Basavanna et al., 2024).

Blockchain and Smart Contracts in Energy Transactions

Blockchains and smart contracts are altering the functioning of energy transactions largely within the peer-to-peer (P2P) energy trading ecosystem. The technologies confer unique abilities onto the operations of decentralised energy markets: transparency, traceability, automation, and decentralisation-all of which are inarguably necessary. The effects of these tools go beyond mere operational enhancements and amount to repositioning fundamentally how energy is commodified, exchanged, and governed.

Transparency and traceability are two of the most potent features possessed by blockchains. Using decentralised ledger technology, the records for all energy transactions are tamper-proof and fully transparent, thus nullifying any concern about data tampering and information asymmetry (Aravind et al., 2024; Varshney et al., 2024). This is especially vital in decentralised settings since the absence of a central authority can mean mechanisms for verifying the truthfulness of transactions must be kept extremely strong. Energy consumers and prosumers are able to verify with blockchain where their electricity came from, track the carbon intensity of their consumption, and hold the other party in the transaction accountable (Gururaja et al., 2024).

Smart contracts take things a step further by actually automating the execution of energy transactions. These self-executing contracts really let go of the manual process where real-time billing and payments are done directly through a third party, with intermediations being eliminated if payment triggers exist under the contract (Din & Su, 2024). Have an Energy Billing and Energy Payment smart contract from pilots that illustrate how automation can reduce transaction latency and operational costs and yet increase trust among the users (Aravind et al., 2024; Din & Su, 2024).

Another important set of values that blockchain is capable of offering is in reducing intermediations. Traditional energy transactions involve utilities and third-party service providers acting as intermediaries and creating inefficiencies and additional costs. In contrast, blockchain disrupts by enabling direct P2P transactions, where prosumers can set their own prices and sell surplus energy to their peers (Suthar & Pindoriya, 2020). It is a decentralisation that improves the efficiency of the marketplace, cuts down on costs, and provides people with access to electrical energy (Varshney et al., 2024; Gururaja et al., 2024).

Blockchain also supports the promotion of renewables through easier trading of green energy certificates for small-scale producers into energy markets without bureaucratic impediments. By lowering barriers to market entry for residential users, blockchain could speed up the proliferation of renewable technologies and thus climate objectives (Suthar & Pindoriya, 2020; Gururaja et al., 2024).

Yet, challenges are present. Scalability issues pertaining to blockchain, transaction throughput, energy consumption (especially for PoW-style consensus mechanisms), and issues arising from regulatory uncertainty stand to mar the large-scale adoption of blockchain (Din & Su, 2024). Moreover, for integration between blockchain platforms and IoT sensors as well as AI algorithms to work well, many interoperable standards and very strong cybersecurity protocols need to be put in place (Gururaja et al., 2024).

In short, the blockchain and smart contracts are by far the farthest leap through energy governance, to disperse transparent, ad-hoc decentralised transactions. Further successful integration into decentralised energy systems will require continued innovation, supportive regulation, and comprehensive stakeholder education.

IoT and Real-Time Energy Monitoring

The Internet of Things (IoT) has come up and evolved in providing energy management at-the-edge and changing paradigms in the generation, monitoring, and consumption of energy systems, basically in decentralized scenarios. IoT devices ensure the gathering of real-time data and have control applications that enormously advance areas such as demand-side management, grid stability, and microgrid optimization. From a technological standpoint, however, the IoT-grown decentralization challenges enable localized energy autonomy but raise security concerns in data privacy.

IoT for Demand-Side Management and Grid Stability. Technologies with real-time monitoring systems supply continuous data on energy production and use, thereby giving minute-level insights into energy profiles of residential or industrial customers. Such data then bind energy systems to inculcate consumption behavior into dynamic response, bettering energy conservation and cutting costs (Sithole, 2024). The automated control systems may oversee distribution profiles in real-time in response to demand variations, being thus the opens against balancing load, especially in microgrids (Benavides et al., 2023). IoT sensors also help the integration of renewable energy sources into the grid by monitoring the variable generation peaks, adjusting storage, or distribution accordingly (Basavanna et al., 2024). By doing so, they stabilize the grid and increase the hosting capacity of distributed energy resources.

Improving Microgrid Optimization. Being often community-managed or localized systems, IoT benefits microgrids greatly. Remote monitoring and diagnosis, predictive maintenance, or load forecasting allow systems to be either efficiently balanced or faults detected (Benavides et al., 2023). IoT-connected energy assets such as smart meters, solar panels, and battery storage systems can be integrated to implement demand response strategies in an automated fashion to stabilize supply and reduce operational inefficiencies (Sithole, 2024). In essence, these technologies are opening the doors to energy decentralization, wherein smaller communities or entities can manage energy resources with a high degree of independence.

Data Privacy and Decentralized Control. While these operational advantages accrue, the rise of IoT systems stands to create serious concerns with respect to data security and privacy. The enormous amounts of data generated—energy consumption patterns, appliance uses, and behavior information—can be exploited if not protected (Kavitha et al., 2023). As real-time data is more and more used in decentralized energy systems, the user privacy dilemma arises how to protect user privacy while letting enough data be accessed by the system for optimization. Some potential solutions for safeguarding users' data within IoT systems that Kavitha et al. (2023) have put forward include providing encryption protocols, decentralized authentication infrastructure, and secure firmware.

Furthermore, the IoT offers a great level of decentralization by enabling end-users to participate in energy governance. Consumers decide via smart home systems or community dashboards about when to consume or even sell electricity. Such decentralization improves system resilience and adaptability, facilitating quick responses to local energy demands in the absence of a centralized command center (Basavanna et al., 2024). Nonetheless, this would require a governance framework rethinking that includes technical literacy, inclusiveness, and equitable access to infrastructure.

In conclusion, the IoT-enabled decentralized energy system has become a transformative role in enhancing operational efficiency, enabling microgrid autonomy, and supporting participatory governance. Balancing these benefits brings with those increasing risks related to data privacy, cybersecurity, and equitable access to technology. A strong technical and regulatory framework will be necessary to fully harness the applications of the IoT tools toward energy transitions (Kavitha et al., 2023; Basavanna et al., 2024; Benavides et al., 2023; Sithole, 2024).

Algorithmic Governance and Artificial Intelligence in Energy Systems

AI is fast emerging as the backbone of decentralized energy systems. They provide efficiencies never seen before: from predictive analytics, dynamic pricing, and optimization of operations to energy forecasting. With every advance on technological grounds, there arise ethical questions, including ones of bias in the algorithms, lack of transparency of decisions, and accountability of decisions.

AI for Forecasting, Dynamic Pricing, and Distribution. AI is generally considered appropriate for demand forecasting so that system operators may anticipate energy needs and respond accordingly. Improved allocation of resources and less energy wastage are among the advantages offered by these AI predictive tools. Zülfikaroğlu (2024) emphasizes how an AI algorithm from smart meters, weather forecasts, and IoT devices to anticipate electricity consumption trends and automate responses. The capacity of AI to forccast makes dynamic pricing models more precise, ensuring that real-time prices reflect supply and demand variations.

Distribution optimization is achieved through the identification of inefficiency or potential faults in the grid by analyzing large and complex datasets (Gopal et al., 2024). For example, in cases of outages, voltage drops, or heavy loads, the AI-enabled system could dynamically reroute electricity flows, cut down on transmission losses, and balance loads much better than manual systems. Eventually, this stabilization of the grid allows intermittent renewables such as solar and wind to be integrated better (Alsaigh et al., 2022).

Ethical Concerns: Bias, Opacity, and Accountability. However, in spite of all good reasons to embrace AI, it brings the risk of algorithmic bias. Jain and Mitra (2024) piggyback on arguments that claim AI models trained on biased or incomplete data may produce unfair pricing schemes or an inequitable energy distribution, especially if it hits the poor or rural communities disproportionately. These biases become terrible in decentralized systems where communities deeply depend on algorithmic outputs for the management of their local energy.

Opacity is another vital issue. Many AI systems operate as "black boxes," offering little impedance to transparency in decision-making processes. The lack of responsibility will, therefore, erode the trust of stakeholders and impede the auditing and contesting of system errors (Hayawi et al., 2024). According to Alsaigh et al. (2022), given that AI is becoming deeply embedded in energy governance, ensuring the interpretability and traceability of algorithms cuts through paramount priorities on the regulatory agenda.

Balancing Efficiency with Ethical Governance. At the heart of unlocking AI potential is ethical governance, which must permeate the design and application of AI systems in energy governance. This entails the transparent design of algorithms, feedback mechanisms, and proper representation of diverse stakeholders in algorithmic design. Furthermore, regulators have the responsibility of creating a regulatory framework covering data ethics, AI accountability, and algorithmic fairness (Hayawi et al., 2024; Jain & Mitra, 2024).

In brief, AI is a great power in decentralized energy systems, capable of transforming energy management systems with forecasting, dynamic pricing, and grid optimization. But if not ethically governed, the benefits of AI might be outweighed by risks of bias, opacity, and inequity. A balanced approach foregrounding transparency, fairness, and stakeholder participation lay a foundation for these tools to enshrine sustainable and just energy transitions (Zülfikaroğlu, 2024; Gopal et al., 2024; Alsaigh et al., 2022).

Regulatory Adaptation to Decentralised Participation

With the evolution of distributed energy systems, they are increasingly questioning the adequacy of traditional regulatory frameworks. New governance regimes, such as peer-to-peer trading, dynamic pricing, and blockchain platforms, require regulators to adopt models that are sufficiently flexible and inclusive to address regulatory complexity across jurisdictions. This sub-theme looks at the problems of reconciling existing regulations with decentralized governance and emerging policy innovations to legitimize decentralized participation.

Challenges of Aligning Traditional Regulatory Models. A major problem in traditional decentralized governance is associated with voter apathy and manipulation. Ahmad (2024) stresses that lack of participation in token-based or community voting systems or dominance by "whales" (users with substantial control) can distort outcomes, thereby compromising the principles of democratic energy governance. In addition, the lack of KYC protocols, especially in decentralized finance applications with energy systems, undermines market integrity and consumer protection (Salami, 2021).

Another aspect about these regulations' enforcement is jurisdictional fragmentation. Since DeFi platforms and energy trading systems operate across multiple legal boundaries, it becomes challenging for regulators to enforce a single set of policies or ensure compliance (Ahmed, 2022). For example, a P2P energy-trading platform established on blockchain may have participants from different states or even countries and create legal uncertainty around licensing, taxation, and consumer rights.

Emerging Policies for Decentralised Participation. Dynamic pricing schemes have been developed and piloted to enhance market responsiveness and fairness to counterbalance all these challenges. According to Doan et al. (2023), dynamic pricing in decentralized P2P energy markets increases not just efficiency but also equity because it allows users to buy and sell electricity on the basis of real-time market conditions.

Besides, there is an increasing pursuit of clarity in crypto-asset regulations. Ahmed (2022) underlines that an operative platform deploying tokens or cryptocurrencies for energy trading must have clear-cut legal definitions and compliance obligations. Otherwise, any innovation could be crushed under the weight of legal uncertainty.

Another promising direction is that of Reg-Tech. Ahmed (2022) outlines how Reg-Tech can automate compliance monitoring and reporting to enable regulators to keep pace with rapidly evolving decentralized platforms. These tools can ensure P2P and DeFi-based energy systems proclaim compliance with consumer protection laws, anti-money laundering standards, and cybersecurity regulations.

Balancing Innovation and Oversight. The new policies are attempting to legitimize decentralized involvement; however, innovation is standing in the way of overregulation or poorly designed regulation. It is important to maintain a balance between flexibility and accountability. The policymakers must be in conversation with the developers, communities, and consumers so that frameworks can be co-created to foster innovation while protecting public interests.

In essence, it is both necessary and complex to alter the regulatory regime for the support of decentralized energy governance. Consideration must be given to redefining legal structures to meet technological innovation with equity, safety, and public confidence. Decentralized energy approaches, therefore, depend not just on technological advancement but also on regulation that is flexible and anticipatory (Ahmad, 2024; Salami, 2021; Doan et al., 2023).

Socio-Economic Dimensions of a Just Energy Transition

Socioeconomic just energy transition looks at a series of intersecting parameters that affect communities, labor markets, and equity amidst fossil fuel phase-out and adoption of renewable energies. The process of change is not merely technical but deeply embedded in social structures, thus requiring inclusive pathways for justice. Highly fossil-fuel-dependent regions, such as Mpumalanga, are at risk as they face employment loss, shrinking public revenues, and socio-economic destabilization (Henry et al., 2020; Cha & Pastor, 2022). At the same time, cleaner energy can boost public health and revenue streams (Henry et al., 2020).

There is one core issue at stake: equity; marginalized communities are those who most bear the brunt of environmental and social dangers." There should be participatory governance, community empowerment, and transparent policy-making in this process procedure (Cha & Pastor, 2022; Nurhidayah et al., 2024). If left unaddressed, disparities in accessing electricity and benefits from the transition would deepen further (Purnomo et al., 2024).

Policy design and legislative frameworks influence the nature of outcomes as well. For example, in Indonesia, efforts at very least towards just transition are challenged by very little stakeholder engagement and lack of transparency (Nurhidayah et al., 2024). Regulatory reforms and ending fossil fuel subsidies are a sine qua non, complemented by inclusive policies to ensure that all affected groups can benefit from ongoing renewable energy developments (Purnomo et al., 2024). The literature indicates an all-encompassing approach based on equity and combining social entrepreneurship with environmental justice and corporate social responsibility (Ayyoob & Sajeev, 2024a; 2025).

Re-Skilling Labor and Transforming the Workforce

Restructuring labor markets stands as a critical element in a just transition within the energy domain. As the energy sector shifts to renewables, new skills are required that are targeted particularly toward in digital technologies such as AI and IoT (Arcelay et al., 2021). Here is another factor. Outside the 1990s in developing nations, the current workforce has insufficient training in the cited skill areas and a growing divide exists between current capabilities and future needs (Datta et al., 2024). Micro-credentialing programs have appeared as a necessary avenue to empower workers to gather initial knowledge useful in their work (Datta et al., 2024).

Yet challenges still prevail: limited local opportunities, demographic inequalities, and unequal access to reskilling programs continue to worsen conditions, especially in disenfranchised localities (Briggs et al., 2022). There is also the risk of creating stranded skill sets where such educational institutions do not sufficiently respond to evolving market demands in their curricula (Vakulchuk & Overland, 2024).

Hence, reforms in education are necessary. The present curricula are inclined towards fossil fuel; hence the need to forge flexible, forward-looking curricula that evolve alongside technological development (Gorski et al., 2023). Social entrepreneurship could thus facilitate adaptation and inclusion in education and employment by creating new models to facilitate workforce transitions with environmental and social value as a priority (Ayyoob & Sajeev, 2024d).

Regional Inequities and Fossil Fuel Dependency

Regions such as South Africa's Mpumalanga bring to mind the socio-economic fragility that arises from fossil fuel dependence. The region employs over 100,000 workers in coal-related sectors, leaving its economy vulnerable to changes in environmental regulations and global energy markets (Nel et al., 2023). Such threats to livelihoods and service delivery arising from the coal phase-out could trigger social disturbances.

These vulnerabilities need to be tackled through tailored economic diversification strategies. The bundling of investments in renewable energy with international support and powerful policy mechanisms would ensure just outcomes (Nel et al., 2023). Programs for developing skills are an essential aspect of transitioning workers into new sectors (Clarke et al., 2024). However, these initiatives should be region-specific and address their socioeconomic conditions.

To build resilience over time, more general economic interventions supporting SMEs, investment in infrastructure, and innovation will be needed. Without a diversified economy, fossil fuel-dependent areas remain vulnerable to external shocks. A pulp-and-paper approach to energy risks ignoring the multifaceted nature of local economies.

A further transformation can be driven by CSR and sustainable commercial practices (Ayyoob and Sajeev, 2025). When innovation and CSR come together, companies help in economic diversification while pursuing environmental and social goals. At the regional level, policy should focus on securing cross-sector partnerships and inclusive policies for the purpose of equitable development.

Energy Access and Equity for Marginalized Communities

In designing a transition toward clean energy, increased and inclusive access must be maintained as the foremost point of consideration for largely marginalized communities. Energy-poverty situations and digital exclusions often intersect, creating multiple layers of disadvantage. The COVID-19 pandemic further deepened these inequalities as lower-income and minority groups faced energy and internet insecurities (Chen et al., 2024).

Injustices are persistent in history. Scheduled Castes and Tribes in India, for instance, are less able to get and access reliable energy, besides caste-based inequalities already entrenched (Pelz et al., 2020). Further, technological barriers-Having such installations with a high initial cost-impede the bereft population from access (Jiao & Tan, 2024).

Participation in equity-based financing and policy-making is required. Community engagement determines the success of energy justice strategies (Ikevuje et al., 2023). Local stakeholders can be involved in planning so that these energy systems are more responsive and equitable.

Decentralized solutions like off-grid solar systems hold promise as viable avenues for improving access and affordability in underserved areas (Jiao & Tan, 2024). New financing options may include targeted subsidies, alongside inclusive planning. The synergy between ESG principles and SDG targets (Ayyoob & Sajeev, 2024c) provides a framework for aligning corporate practices with inclusive energy goals.

Theoretically, these interventions are promising; however, lot of impediments still exist. There is a need for sustained financing, policy consistency, and regulatory harmonization to guarantee that equitable advantages from the expansion of energy provision are delivered and really endure. Without these safeguards, the marginalised will remain excluded in this transition.

Community Participation and Environmental Justice

Community participation is seen as a fundamental aspect of environmental justice and a just energy transition. Community microgrids have emerged as a platform for energy democracy. These decentralized systems enable local residents to own, operate, and profit from an energy resource, thereby giving them increased agency and equity (Eklund et al., 2023).

Social capital—trust, cooperation, and collective efficacy—deserves consideration in the utility of such systems. Strong social ties foster better governance and operation, which in turn encourages sustainable energy (Eklund et al., 2023). Microgrids can also be designed to prioritize low-income households, reducing energy bills when it comes to costs and improving reliability (Niraula et al., 2024).

Inclusive decision-making processes are budgetary. Participatory governance can hear local voices and thus means of systems well adapted to community needs (Gill-Wiehl et al., 2022). Nevertheless, barriers still exist. Households that are the most vulnerable may lack the money and capacity to be engaged in energy planning (Hanke et al., 2022).

To take on challenges at these levels, inclusive policies promoting community capacity-building must be created. Moreover, and best suited to this phase, efforts to integrate diversity, equity, and inclusion (Ayyoob & Sajeev, 2024b) must gain more attention. Government and civil society should stand together to ensure that the marginalized are not blocked from the opportunities provided by clean energy transitions.

To conclude, community microgrids are the embodiment of opportunity where environmental justice can enter the field of energy policy. Success depends on frameworks that warrant public incorporation, sharing of benefits on a just basis, and guarantees. A just transition is more than a technical challenge; it is a matter of rethinking the governance models through which all communities are engaged in the contemplation of a sustainable energy future.

DISCUSSION

The results of this study highlight digital technologies—with blockchain, IoT, and AI as examples—as agents of transformation in governance structures within decentralized energy systems. From the thematic analysis of current literature, several major insights emerged, highlighting the opportunities for and challenges in the application of these technologies to enable participatory, transparent, and effective energy management.

The first major theme emphasized the fact that digital governance enables participation and transparency in decision-making processes in decentralized energy models. Contrasting with those decisions taken by the so-called 'few' in centralized systems, digital governance frameworks in decentralized systems allow a broad range of stakeholders, including prosumers, local communities, and small energy entities, to engage actively (Brisbois, 2022). Technologies like blockchain or AI are not only tools for operation; instead, they are facilitating governance models that favor inclusivity, transparency, and accountability (Vasconcelos et al., 2022; Bañales, 2020).

Nolden in 2019 expressed that structural impediments such as regulatory lag, infrastructural gaps, and uneven literacy in digital matters obstruct the realisation of such a vision. In addition, if any digital governance tends to increase participation in energy democracy, it might conversely increase inequalities when a community is denied access to such technology or where disadvantaged groups are not able to utilize the platform adequately.

Applying blockchain technology to peer-to-peer energy trading is an example of trustless automated systems in which the need for centralized intermediaries is being rapidly diminished (Suthar and Pindoriya, 2020; Aravind et al., 2024). They ensure transparency and full traceability; smart contracts enable automatic payments, billing, and other operations based on predefined conditions, thus increasing overall operational efficiency (Din and Su, 2024).

These features considerably reduce transaction costs while increasing trust among users, thus having a very high possibility of developing decentralized energy markets. Nonetheless, scalability, energy consumption of blockchain networks, and the existence of strong regulatory frameworks remain critical issues (Varshney et al., 2024; Gururaja et al., 2024). Without clear-cut laws, since the legal status of smart contracts and digital assets remains dubious, applications of blockchain in energy trading could not gain universal legitimacy in the very near future.

Traditionally, however, IoT provides real-time visibility into energy production, distribution, and consumption; therefore, it is vital for microgrid optimization and demand-side management (Benavides et al., 2023; Sithole, 2024). IoT, by decentralizing control and enabling data-driven decision-making processes, enables local energy resiliency, especially in remote or off-grid locations (Basavanna et al., 2024).

On the flipside, there are certain issues around data privacy, cybersecurity, and surveillance that are increasingly imperiling the IoT due to its very functional need for always-on data exchange and connectivity (Kavitha et al., 2023). Hence, misuse of the energy consumption data can infringe privacy, discriminate, or profile behavior. So one should keep in mind that while IoT does appear to assist decentralization, its governance structure must also enforce strong ethical and regulatory mechanisms to prevent unintended consequences.

AI solutions bring tremendous potential in demand forecasting, dynamic pricing, and distribution network optimization within decentralized energy systems (Zülfikaroğlu, 2024; Gopal et al., 2024). These applications enhance grid reliability and economic efficiency and support nuclease-rich data-intense environments.

Governance challenges are activities introduced by their use. Algorithmic opacity and bias endanger fairness and accountability (Jain & Mitra, 2024). Biased algorithms can price vulnerable populations unequally or could deny them reliable access to energy. In addition, a lack of explainability of most AI systems erodes trust from the people and democratic oversight (Hayawi et al., 2024; Alsaigh et al., 2022). Thus, ethical AI design and governance, meaning transparent auditing and inclusive data processes, become paramount to ensuring fair outcomes in decentralized energy transitions.

Regulatory frameworks must be adapted to the particularities of decentralized digital energy. The traditional top-down models are often not able to accommodate the distributed, trans-national, and community-led nature of decentralized energy, thus yielding regulatory refrains or misalignment (Ahmad, 2024; Ahmed, 2022). As Salami (2021) points out, the absence of know-your-customer (KYC) standards in decentralized platforms could expose systems to fraud, manipulation, and financial crime.

Excess regulations might limit the growth of new ideas as well as community participation. Emerging practices in leveraging dynamic pricing models (Doan et al., 2023) and Reg-Tech for automated compliance (Ahmed, 2022) signal hybrid approaches where innovation is also put under the limelight. Hence, adaptive technologies must be technology-neutral and co-created alongside stakeholders to allow for market efficiency and social equity.

In the end, the integrations of digital technologies with decentralized energy systems mark a paradigm change in energy governance. While the technologies under study-blockchain, IoT, and AI-allow for transparency, efficiency, and community engagement, they otherwise create an ethical, technical, and regulatory mess that has to be proactively dealt with. One crucial takeaway is multi-level governance frameworks that are participatory, adaptable, and rooted in ethics. To make sure decentralized digital energy systems remain not only technically feasible but, even more so, socially just and environmentally viable, capacity building, inclusive policy-making, and continuous revisit to regulatory formulation need to go hand in hand with such governance frameworks.

LIMITATIONS AND SCOPE FOR FURTHER STUDIES

While this study offers deep insights into the subject of digital governance in decentralized energy systems, it is, however, limited by its use of secondary data via thematic analysis of extant literature, hence lacking contemporaneous real-time empirical evidence mainly from pilot studies or live decentralized energy platforms. Since developments also evolved quickly in blockchain, AI, and IoT, hence findings can become outdated with the emergence of further new innovations. A detailed comparative study between various geographic and regulatory jurisdictions was not attempted, somewhat thereby limiting generalizability to other countries. Future research could involve empirical case studies focusing on user experience with digital literacy and governance models embedded within real decentralized energy settings. Further, longitudinal studies could cast light on the sustainability and evenness considerations of the facilities concerned. Comparative studies on centralized versus decentralized digital governance regimes across various jurisdictions would, at the very least, be of considerable interest to policymakers regarding styles of ethical AI, cyber security, and participatory regulation.

IMPLICATIONS

The findings of this study reveal certain theoretical, practical, and policy implications. Theoretically, it feeds into the evolving discourse on digital governance in sustainable energy transitions, thereby putting weight on current models of energy decentralization with higher technology integration, especially with blockchain, IoT, and AI. It underlines the point that governance is not just institutional-anymore-but increasingly technological and has brought into question issues of algorithmic control, digital participation, and automated decision-making.

Practically, it presents a step-wise guideline for energy practitioners and technology players. It underscores the modus operandi by which smart contracts, AI-generated forecasting, and IoT-enabled demand-side operations might potentially transform energy trading and operations. The insights generated by this study could be channelled into designing energy platforms that are more secure, transparent, and inclusive. Care must be taken in the deployment of these technologies to consider ethical implications concerning data privacy, algorithmic fairness, and digital inequality.

The research also reveals that policies need to set adaptive frameworks for energy. Energy laws generally are lagging behind technological advancement, leading to governance vacuums. Integrating Reg-Tech, introducing legal clarity for crypto-assets, and instituting guidelines for the ethical use of AI may be proposed in this study to ensure strong oversight without stifling innovation. Encouraging multistakeholder engagement—including the local community, prosumers, and private tech companies—is important in building trust and an inclusive environment. Finally, governments should work on capacity building and promoting digital literacy so everyone can cooperate in the decentralized energy ecosystem.

The implications of the study, in essence, call for the balancing of innovation with accountability, decentralization with inclusivity, and efficiency with ethics in the continuously evolving field of digital energy governance.

CONCLUSION

This study looked at the emerging landscape of digital governance within decentralized energy systems, with consideration to blockchain, IoT, AI, and the regulatory dynamics that go along with them. Through a thematic analysis of the contemporary literature, it brings to light how digital technologies are transforming energy transactions and operations and with that the governance frameworks that underlie them. The key findings show that blockchain brings in potential transparency and trust in peer-to-peer energy trading, the power of IoT to enable real-time monitoring and microgrid optimization, and the growing influence of AI in forecasting and operational efficiency.

While they promise participation, efficiency, and decentralization within themselves, they pose new challenges, from data privacy, issues of algorithmic bias, to regulatory uncertainty. Engaging the ethical and governance dilemmas posed by algorithmic decision-making and decentralized control mechanisms requires earnest deliberation and proactive policymaking. And, the research finds regulation-related adaptation lagging behind technological progress, thereby limiting and providing opportunities for inclusive and secure energy markets.

One central finding is that digital governance must be designed with a human-centric and participative ethos. Technologies, therefore, should not maximize system efficiency with a poison pill tagging users but should empower users, protect rights, and democratize control. Balancing this way calls for collaborative setups where policy makers, technologists, and energy providers, along with their end-users, solve problems hand in hand.

Lastly, the study contends that integrating digital technologies successfully in decentralized energy systems revolves not only around the technology itself but also about governance, ethics, and inclusiveness. Future energy transitions must emphasize not only sustainability and efficiency, but also fairness, transparency, and digital equity—ensuring no stakeholder is left behind in the quest for a smarter, decentralized energy future.

REFERENCES

Ahmad, M. (2024). Community-Driven Finance Unveiling the Dynamics of DeFi Governance. *Advances in Finance, Accounting, and Economics*, •••, 33–48. DOI: 10.4018/979-8-3693-3346-4.ch003

Alsaigh, R., Mehmood, R., & Katib, I. (2022). AI Explainability and Governance in Smart Energy Systems: A Review. *ArXiv (Cornell University)*. https://doi.org// arxiv.2211.00069DOI: 10.48550

Aravind, A. R., Santhi, G. B., Patil, S. T., P, S., Sharma, G., Dhamone, J., & Ragu Nathan, S. (2024). Blockchain Technology in Energy Markets: Enabling Peer-to-Peer Energy Trading. *E3S Web of Conferences, 591*, 06002. DOI: 10.1051/ e3sconf/202459106002

Arcelay, I., Goti, A., Oyarbide-Zubillaga, A., Akyazi, T., Alberdi, E., & Garcia-Bringas, P. (2021). Definition of the Future Skills Needs of Job Profiles in the Renewable Energy Sector. *Energies*, *14*(9), 2609. https://www.mdpi.com/1996-1073/ 14/9/2609. DOI: 10.3390/en14092609

Ayyoob, A., & Sajeev, A. (2024a). Corporate Sustainability: The Role of Consumer Skepticism, Transparency, and Marketing in Greenwashing Practices. *Corporate Governance Insight*, *6*(2), 51–68. DOI: 10.58426/cgi.v6.i2.2024.51-68

Ayyoob, A., & Sajeev, A. (2024b). Diversity, Equity, and Inclusion: Employees Perspectives on Sustainable Business Practices. *Routledge EBooks*, 348–367. DOI: 10.4324/9781003487180-27

Ayyoob, A., & Sajeev, A. (2024c). Navigating Sustainability: Assessing the Imperative of ESG Considerations in Achieving SDGs. *ESG Frameworks for Sustainable Business Practices*, 53–84. DOI: 10.4018/979-8-3693-3880-3.ch003

Ayyoob, A., & Sajeev, A. (2024d). Role of Social Entrepreneurship in Sustainable Development. *Advances in Business Strategy and Competitive Advantage Book Series*, 193–224. DOI: 10.4018/979-8-3693-4338-8.ch008

Ayyoob, A., & Sajeev, A. (2025). Resilient and Sustainable Organizations. IGI Global EBooks, 35–58. DOI: 10.4018/979-8-3373-0608-7.ch003

Bañales, S. (2020). The enabling impact of digital technologies on distributed energy resources integration. *Journal of Renewable and Sustainable Energy*, *12*(4), 045301. DOI: 10.1063/5.0009282

Basavanna, Chavan, G., YM, D., & Verma, N. A. (2024). Smart Grid Using IOT. *International Journal of Advanced Research in Science Communication and Technology*, 580–587. DOI: 10.48175/IJARSCT-22769

Benavides, D., Arévalo, P., Ortega, A. C., Sánchez-Sutil, F., & Villa-Ávila, E. (2023). *Energy Management Model for a Remote Microgrid Based on Demand-Side Energy Control.* https://doi.org/DOI: 10.20944/preprints202311.1125.v1

Bray, R., Mejía Montero, A., & Ford, R. (2022). Skills deployment for a "just" net zero energy transition. *Environmental Innovation and Societal Transitions*, *42*, 395–410. DOI: 10.1016/j.eist.2022.02.002

Briggs, C., Atherton, A., Gill, J., Langdon, R., Rutovitz, J., & Nagrath, K. (2022). Building a "Fair and Fast" energy transition? Renewable energy employment, skill shortages and social licence in regional areas. *Renewable and Sustainable Energy Transition, 2*, 100039. https://www.sciencedirect.com/science/article/pii/S2667095X2200023X

Brisbois, M. C. (2022). *Governing Decentralized Electricity: Taking a Participatory Turn*. 325–346. DOI: 10.1002/9783527831425.ch15

Cha, J. M., & Pastor, M. (2022). Just transition: Framing, organizing, and power-building for decarbonization. *Energy Research & Social Science*, *90*, 102588. DOI: 10.1016/j.erss.2022.102588

Chen, C., Chen, W.-A., Greig, J., & Robinson, C. (2024). When concentrated disadvantage happens: Exploring the nexus of energy and internet insecurities among vulnerable households. *iScience*, *27*(6), 109895. DOI: 10.1016/j.isci.2024.109895 PMID: 38827408

Clarke, L., Curtis, M., Eisenberg, A., Grubert, E., Haggerty, J. H., James, A., Jensen, N., Kaufman, N., Krause, E., Raimi, D., Tingley, D., & Weber, J. (2024). *A research agenda for economic resilience in fossil fuel–dependent communities*. Environmental Research Energy., DOI: 10.1088/2753-3751/ad6d70

Datta, A., Coates, S., Rossiter, A., & Krishnamoorti, R. (2024). Reskilling and Upskilling for Decarbonization: Analyzing Micro-Credential Programs for Energy Workforce Development. *The Journal of Continuing Higher Education*, •••, 1–20. DOI: 10.1080/07377363.2024.2377777

Din, J., & Su, H. (2024). Blockchain-Enabled Smart Grids for Optimized Electrical Billing and Peer-to-Peer Energy Trading. *Energies*, *17*(22), 5744. DOI: 10.3390/en17225744

Doan, H. T. Bao, Kim, D., & Kim, H. (2023). Fully Decentralized Peer-to-Peer Community Grid with Dynamic and Congestion Pricing. *ArXiv (Cornell University)*. https://doi.org//arxiv.2308.04717DOI: 10.48550

Edem, K. (2024). Economic impact of digital twins on renewable energy investments. *Engineering Science & Technology Journal, 5*(7), 2232–2247. DOI: 10.51594/estj.v5i7.1318

Edem, K. (2024). Peer-to-peer energy trading: Innovations, regulatory challenges, and the future of decentralized energy systems. *World Journal of Advanced Research and Reviews, 24*(2), 172–186. DOI: 10.30574/wjarr.2024.24.2.3324

Eklund, M., Khalilpour, K., Voinov, A., & Hossain, M. J. (2023). Understanding the community in community microgrids: A conceptual framework for better decision-making. *Energy Research & Social Science, 104*, 103260. DOI: 10.1016/j.erss.2023.103260

Garcia-Casals, X., Ferroukhi, R., & Parajuli, B. (2019). Measuring the socio-economic footprint of the energy transition. *Energy Transitions, 3*(1-2), 105–118. DOI: 10.1007/s41825-019-00018-6

Gill-Wiehl, A., Miles, S., Wu, J., & Kammen, D. M. (2022). Beyond customer acquisition: A comprehensive review of community participation in mini grid projects. *Renewable & Sustainable Energy Reviews, 153*, 111778. DOI: 10.1016/j.rser.2021.111778

Gopal, J. N., Muthukaruppasamy, S., Rajagopal, V., Thomas, & Parimalarsundar, E. (2024). Artificial Intelligence Governance and Comprehensibility in Renewable Energy Systems. *Chapman and Hall/CRC EBooks*, 73–88. DOI: 10.1201/9781003457176-5

Gorski, A.-T., Gligorea, I., Gorski, H., & Oancea, R. (2023). Navigating the Disruptive Challenges and Opportunities of Digital Transformation in the Labour Market: Upskilling and Reskilling for the Fourth Industrial Revolution. *International Conference Knowledge Based Organization, 29*(3), 23–29. DOI: 10.2478/kbo-2023-0071

Gururaja, H. S., Hebbar, A., Poojary, A. S., Bharadwaj, A., & Rakshitha, B. R. (2024). Decentralized Energy Trading for Grids Using Blockchain for Sustainable Smart Cities. *Deleted Journal, 2*(02), 134–141. DOI: 10.47392/IRJAEH.2024.0025

Hanke, F., Guyet, R., & Feenstra, M. (2022). Energy communities' social role in a just energy transition. *Elsevier EBooks*, 195–208. DOI: 10.1016/B978-0-323-91135-1.00027-4

Hayawi, K., Shahriar, S., & Al-Ali, A. R. (2024). *Ethical Considerations in AI Applications in Smart Grid*. DOI: 10.20944/preprints202412.1883.v1

Henry, M. S., Bazilian, M. D., & Markuson, C. (2020). Just transitions: Histories and futures in a post-COVID world. *Energy Research & Social Science*, *68*, 101668. DOI: 10.1016/j.erss.2020.101668 PMID: 32839696

Ikevuje, A. H., Kwakye, J. M., Ekechukwu, D. E., Ogundipe, O. B., & Esiri, A. E.Augusta Heavens IkevujeJephta Mensah KwakyeDarlington Eze EkechukwuOlorunshogo Benjamin OgundipeAndrew Emuobosa Esiri. (2023). Energy justice: Ensuring equitable access to clean energy in underprivileged communities. *Magna Scientia Advanced Research and Reviews*, *8*(2), 211–218. DOI: 10.30574/msarr.2023.8.2.0097

Jain, V., & Mitra, A. (2024). Ethical Implications of Artificial Intelligence in Promoting Sustainable Clean Energy Solutions. *Practice, Progress, and Proficiency in Sustainability*, •••, 96–118. DOI: 10.4018/979-8-3693-6567-0.ch006

Jiao, L., & Tan, Y. (2024). Social Equity and Renewable Energy Access in Developing Countries. *MEDAAD*, *2024*, 27–33. DOI: 10.70470/MEDAAD/2024/005

Kavitha, C. R., Varalatchoumy, M., Mithuna, H. R., Bharathi, K., Geethalakshmi, N. M., & Boopathi, S. (2023). Energy Monitoring and Control in the Smart Grid. *Advances in Bioinformatics and Biomedical Engineering Book Series*, 290–316. DOI: 10.4018/978-1-6684-6577-6.ch014

Nel, E., Marais, L., & Mqotyana, Z. (2023). The regional implications of just transition in the world's most coal-dependent economy: The case of Mpumalanga, South Africa. *Frontiers in Sustainable Cities*, *4*, 1059312. Advance online publication. DOI: 10.3389/frsc.2022.1059312

Niraula, D., Wang, H., Wang, Z., & Wang, Y. (2024)... *Community-Based Microgrid Planning and Operation for Fostering Energy Justice*, *1–5*, 1–5. Advance online publication. DOI: 10.1109/NAPS61145.2024.10741738

Nolden, C. (2019). Transaction cost analysis of digital innovation governance in the UK energy market. *The Journal of Energy Markets*. DOI: 10.21314/JEM.2019.194

Nurhidayah, L., Alam, S., Utomo, N. A., & Suntoro, A. (2024). Indonesia's Just Energy Transition: The Societal Implications of Policy and Legislation on Renewable Energy. *Climate Law*, *14*(1), 36–66. DOI: 10.1163/18786561-bja10047

Onukwulu, C., Odochi, M., & Eyo-Udo, L. (2023). Decentralized energy supply chain networks using blockchain and IoT. *International Journal of Scholarly Research in Multidisciplinary Studies*, *2*(2), 066–085. DOI: 10.56781/ijsrms.2023.2.2.0055

Pelz, S., Chindarkar, N., & Urpelainen, J. (2020). Energy Access for Marginalized Communities: Evidence from Rural North India, 2015-2018. SSRN *Electronic Journal.*

Purnomo, V. D., Hamid, A., & Tedjosaputro, L. (2024). The Influence of Economic Law on a Just Energy Transition. *International Journal of Business and Applied Economics, 3*(6), 1077–1090. DOI: 10.55927/ijbae.v3i6.11950

Salami, I. (2021). Challenges and Approaches to Regulating Decentralized Finance. *AJIL Unbound, 115*, 425–429. DOI: 10.1017/aju.2021.66

Sithole, K. D.Khumbulani Derrick Sithole. (2024). Employing Internet of Things (IoT) devices for Monitoring and Controlling Energy Management Systems-A Review. *Journal of Electrical Systems, 20*(11s), 753–760. DOI: 10.52783/jes.7273

Suthar, S., & Pindoriya, N. M. (2020). *Blockchain and Smart Contract based Decentralized Energy Trading Platform.* DOI: 10.1109/NPSC49263.2020.9331883

Umar, A., Kumar, D., Ghose, T., Alghamdi, H., & Abdelaziz, A. Y. (2024). Decentralized Community Energy Management: Enhancing Demand Response through Smart Contracts in a Blockchain Network. *IEEE Access : Practical Innovations, Open Solutions, 12*, 80781–80798. DOI: 10.1109/ACCESS.2024.3409706

Vakulchuk, R., & Overland, I. (2024). The failure to decarbonize the global energy education system: Carbon lock-in and stranded skill sets. *Energy Research & Social Science, 110*, 103446–103446. DOI: 10.1016/j.erss.2024.103446

Varshney, N., Sreetharan, V., Vennila, C., Ahila, R., Mane, D. T., Kshirsagar, K., & Pragathi, B. (2024). Blockchain Technology in Energy Management Systems: Enhancing Security and Transparency. *E3S Web of Conferences, 591*, 01007–01007. DOI: 10.1051/e3sconf/202459101007

Vasconcelos, P. N., Silva, G. S., De Souza, A. C. Z., & Fuly, B. I. (2022, December 1). *Decentralized Governance and Energy Carbon Footprint: Opportunity for Re-Empowering People?* IEEE Xplore. DOI: 10.1109/IHTC56573.2022.9998383

Xie, J. J., Martin, M., Rogelj, J., & Staffell, I. (2023). Decarbonizing the US power system presents diverse challenges and opportunities in the changing employment landscape. *Research Square (Research Square).* DOI: 10.21203/rs.3.rs-2402526/v1

Zülfikaroğlu, S. (2024). AI-Driven Strategic Management and Decision Making for Energy Sector. *Next Frontier for Life Sciences and AI, 8*(1), 91. DOI: 10.62802/q7rkdb54

Chapter 5
Exploring Cyprus' Sustainable Development Through Energy Transition and Energy Hubs

Iordanis Katemliadis
https://orcid.org/0000-0003-2164-9856
University of Lancashire, Cyprus

Sotiris Apostolopoulos
https://orcid.org/0000-0002-1065-891X
Neapolis University, Pafos, Cyprus

Dimitrios Salampasis
Swinburne University of Technology, Australia

Amvrosios Prodromou
https://orcid.org/0009-0008-4787-6292
Neapolis University, Pafos, Cyprus

Nicholas Theodorakopoulos
https://orcid.org/0000-0002-2262-0054
Neapolis University, Pafos, Cyprus

ABSTRACT

Cyprus is considered an EU Member State that is isolated from the rest of Europe in terms of energy, and its energy mix continues to be dependent on imports of pe-

DOI: 10.4018/979-8-3373-5302-9.ch005

troleum and fuel oil. Simultaneously, it faces challenges in achieving its sustainable development goals, a fact that places the country low on the ranking of the European Sustainable Development Report 2025. However, the prevailing climatic conditions, combined with Cyprus's geostrategic position in the Eastern Mediterranean, provide Cyprus with the energy advantage to become a green energy producer. The current study examines Cyprus' sustainable development through energy transition and energy hubs. Specifically, it explores Cypriot energy planning and strategy and examines whether there is potential to transform Cyprus into a climate-neutral green country by 2050 and, at the same time, into an energy hub for Europe. Employing a multi-level approach, this study examines and evaluates the policies implemented for the energy transition and energy hubs whilst drawing policy recommendations.

INTRODUCTION

Cyprus is an EU Member State, with its economy heavily dependent on fuel oil (Tsangas et al., 2018; Pilavachi et al., 2009; Koroneos et al., 2005). Cyprus' electricity generation relies on the use of petroleum and fuel oil, as the country does not possess its own primary energy sources (Republic of Cyprus, 2021). In this light, its electricity generation system is directly dependent on imported fuels (Cyprus Electricity Authority, 2022). In 2018, Cyprus drew up a plan for electricity production from renewable energy sources (Republic of Cyprus, 2018). Since then, the country has been steadily increasing the share of renewable energy sources in its energy mix with the aim of reducing its dependence on oil and fuel oil, thereby strengthening its economy and geopolitical position (Tsangas & Zorpas, 2023; Taliotis et al., 2020a; Ripple et al., 2020).

The ambitious task of energy transition in Cyprus has been undertaken by the Cyprus Energy Regulatory Authority, which has set itself the goal of promoting the use of renewable energy sources and protecting the environment (CERA, 2022). Its energy transition strategy envisages the development of renewable energy sources to 30% by 2030, 75% by 2040 and 100% by 2050 (Poullikkas, 2024; Republic of Cyprus, 2020; Taliotis et al., 2020a). The goal is for Cyprus' energy mix in 2050 to be greenhouse gas emission-free, digitalized, interconnected, flexible, and decentralized (Poullikkas, 2020). Through electrical interconnections, it intends to completely remove its energy isolation and become an electricity exporting power (Venizelou & Poullikkas, 2025; Poullikkas, 2020). The climatic conditions prevailing in Cyprus are favourable for the development of extensive and sustainable solar and wind energy (Kalogirou, 2003).

At the same time, Cyprus, in cooperation with several other neighboring countries, intends to become a hub for the transmission of electricity to and from EU countries, as well as to and from the Middle East and Africa, thereby creating an energy security network (Sotiriou, 2025; Tsoukala, 2022; Wu et al., 2021). The agreements for the Israel-Cyprus-Greece electricity interconnection pipeline, named 'EuroAsia Interconnector', and the Egypt-Cyprus-Greece pipeline, named 'EuroAfrica Interconnector', will make Cyprus, if implemented, an energy hub due to its strategic position in the Eastern Mediterranean (Sotiriou, 2025; Pepe, 2024; Tsoukala, 2022; Mitchell, 2021).

Cyprus is facing considerable challenges when it comes to sustainable development and is currently ranked low in the European Sustainable Development Report of 2025 (32nd out of 34 countries). The country has managed to show some steady progress in various areas of the Sustainable Development Goals (SDGs) such as poverty reduction, but challenges remain in water management, waste recycling and CO2 emissions (Lafortune, et al., 2024). Especially the energy grid in Cyprus is supporting industries like the hospitality and tourism that is an energy intensive industry. The efforts of local stakeholders to move into a more sustainable future are well documented (Prodromou & Savvides, 2025) as well as their involvement in this effort (Katemliadis & Markatos, 2021).

This research examines Cyprus' sustainable development through energy transition and energy hubs. The research question posed is whether Cyprus has the potential, based on its energy planning and energy strategy, to become a climate-neutral green country by 2050 and, at the same time, to become an energy hub for Europe. Through a multi-level approach to sustainable development in Cyprus, this research holistically assesses the energy transition and energy hubs and concludes with policy recommendations.

LITERATURE REVIEW

The energy transition and its impact has attracted considerable research interest (Lamnatou et al., 2024; Yang et al., 2024; Rial, 2024; Psillaki et al., 2023). More specifically, this interest revolves around the transition areas from coal-fired to green power generation (Tranoulidis et al., 2022; Batas-Bjelić et al., 2015). Moreover, interest is also directed towards policies and financial instruments for energy transition and sustainable development of island regions (Icaza et al., 2024; Pacheco et al., 2022; Katsaprakakis et al., 2022). However, there exist additional island states such as for instance Cyprus and Malta whose specificities make them highly dependent on fuel oil and petroleum and where research interest in island states is more limited (Pinczynski et al., 2024; Bugeja et al., 2023; Kotzebue & Weissenbacher, 2020). These Mediterranean island states are energy dependent and energy isolated with their citizens also bearing a large financial burden for fossil fuel emissions (Taliotis et al., 2020b; Zorpas et al., 2017; Koroneos et al., 2005). Notably, EU island Member States are continuously increasing their participation in green energy sources in order to decouple from fossil fuels utilising the support of the EU (Tsangas & Zorpas, 2023; Taliotis et al., 2020a; Ripple et al., 2020). Nevertheless, the transition to sustainable modes of production and consumption has attracted research interest (Katemliadis & Markatos, 2021; Tsangas et al., 2018; Boukas & Chourides, 2016; Mavris, 2014). According to the international literature, Cyprus has the potential to become a model green country (Poullikkas, 2021) through the development of renewable energy installations and the installation of hydrogen systems for electricity generation (Zhang et al., 2020; Logan et al., 2020; Nikolaidis & Poullikkas, 2017). It also holds great prospects for the development of natural gas networks as has secured an exemption from the EU legal framework regarding gas from third countries (Taliotis et al., 2017; Tsakiris, 2017; Mirasgedis et al., 2004). Moreover, Cyprus has upgraded its strategic cooperation in the energy sector through bilateral and trilateral agreements (Venizelou & Poullikkas, 2025; Tsoukala, 2022) in order to gain energy security and reliability by reducing its energy dependencies (Marketos, 2021; Taliotis et al., 2017; Tsakiris, 2017). Venizelou & Poullikkas (2025) argue that Cyprus has great potential to change its energy mix, acquiring innovative renewable energy systems and simultaneously developing energy storage systems. Similarly, Poullikkas' (2021) research indicated that through proper and effective planning Cyprus has the potential from being an energy dependent state to become a green energy producer country to meet its needs and at the same time exporting electricity.

EU FIGURES THAT AFFECT MEMBER STATES

The European economy is at crossroads and is facing unprecedented challenges. The de-industrialisation that has started taking place back in 2022-2023 (Weiss et al., 2024), the start of the war in Ukraine and the consequent rise of energy prices constitutes some recent developments that are leading towards a profound effect on the continent. At the same time, the long-term commitment of the European Commission towards the Green Deal (GD), that aims to reduce gas emissions by at least 55% by 2030 is putting extra pressure to countries, businesses and citizens to adjust to the new realities (European Commission, 2019). There are reports that the demand for electricity will rise even though we have witnessed a drop in 2022 and 2023 (3.1% and 3.4% respectively) due to the pandemic (Ember.org 2024). Governments and system operators predict a 7% increase year per year until 20230 while McKinsey & Company in its recent predictions for the same period predicts a 2% increase (McKinsey & Company, 2024). The increase in power consumption will come mostly from demand generated by the use of electric vehicles, data centers, air conditioning, and 5G networks (International Energy Agency, 2025).

At the heart of the above discussion lies the need for energy grids to become more efficient, expand and interconnect to improve their security and functionality. The recent events in Spain and Portugal and France which they experienced widespread power outages is a bold reminder of this need (Brezar & Skopeliti, 2025). This becomes even more important if we think that our economies are increasingly reliant on good reliable power quality. Consequently, our digitalized economy depends on it.

Europe, in order to significantly boost its resilience, aims to improve cross-border electricity interconnections and electricity trade by creating a regulatory framework and a synchronised system of interconnected subregional grids (Pepe, 2024). The aim is to create cross-border electricity interconnections through which countries will be able to foster energy security, promote renewable energy integration, and potentially lower costs. This effort comes with considerable geopolitical risks, especially in this part of the world and the current fluidity that prevails. Furthermore, the financial costs are considerable and are estimated by the European Commission at €396 billion per year from 2021 to 2030 and €520-575 billion per year in the subsequent decades until 2050 (EPRS, 2023).

The energy transition from fossil-based energy systems to renewable energy sources meets the goal of reducing Greenhouse gas emissions that that will help combat climate change. This energy transition, as part of the green transition though, needs to take place in a sustainable way that will achieve its goal, but it will also allow the economy to grow and promote the well-being of European citizens. As a result, there is a strategic need to align the extensively increased businesses with sustainable development and quality of life (Makris, et al., 2025).

In the context of the European Green Deal (European Commission, 2019), 'Green Transition' is defined broadly by the European Commission as embracing long-term transformations of societal systems towards more sustainable modes of production and consumption (European Commission, 2020). Green transition is expected to substantially improve energy security, provided that the investments and infrastructure needed are aligned to treat new threats created by integrating renewables into the power grid (Kim et al., 2025).

Green Deal (GD) is an umbrella term and a strategy that the EU has deployed since 2021 and has a timeline until 2050. The aim is to reduce emissions by 55% by 2030 and become climate neutral by 2050. To support this strategic goal, EC will dedicate one third of the €1.8 trillion investment from the NextGenerationEU Recovery Plan (European Commission, 2019). The Green Deal (GD) consists of eight key policy objectives and one of them is named: building and renovating in an energy and resource-efficient way. This is the objective linked to Energy transitions and transformation.

The Figure below illustrates the overall change in net electricity generation over the period 2012-2022 according to Eurostat data (2024).

Figure 1. Overall change in net electricity generation, 2012-2022 (%, based on GWh) (Eurostat (online data code nrg_ind_peh))

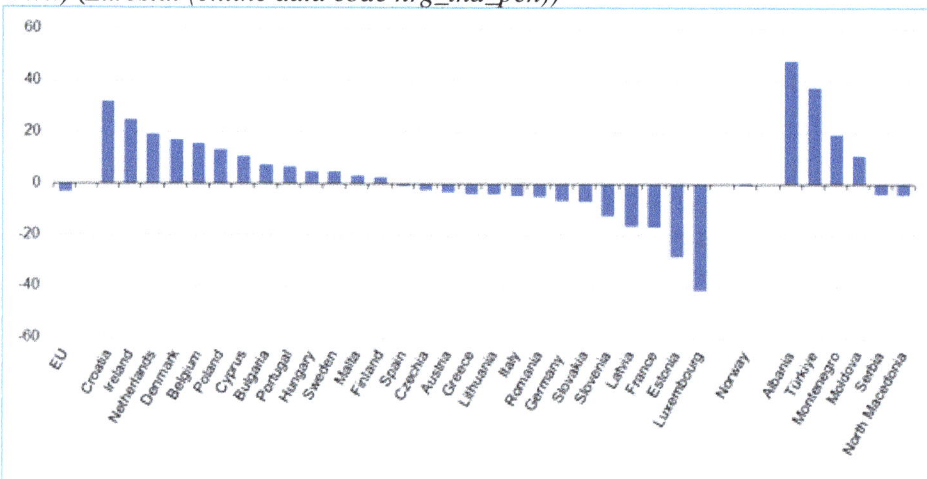

THE ENERGY SPECIFICITY OF CYPRUS

Cyprus, just like Malta, are two island countries in the Mediterranean that are members of the EU and their energy transition is also financed through EU funds (Venizelou & Poullikkas, 2025; Kotzebue & Weissenbacher, 2020). Cyprus, as an island country is experiencing some specific particularities in its energy sector that have an impact in the country's energy policies. The country is energy isolated from the rest of mainland Europe leading to a direct dependence on oil and fuel oil imports for its electricity production (Marketos, 2021; Taliotis et al., 2017; Tsakiris, 2017). However, this element carries three major advantages that must highlighted and exploited. The first advantage concerns the possibilities for highlighting and exploiting RES and in particular the exploitation of solar and wind energy (Agathokleous & Kalogirou, 2021; Taliotis et al., 2020a; Katsoufis, 2017). The second advantage concerns the potential natural gas and oil reserves that the country possesses, which by investing in them can become a strong energy power in the sensitive eastern Mediterranean region. Cyprus has in recent years granted ten hydrocarbon exploration licenses, leading to the discovering of three areas with significant quantities of natural gas (WEC, 2023). The third advantage concerns energy hubs and the passage of natural gas pipelines through Cypriot territory (Sotiriou, 2025; Pep, 2024; Tsoukala, 2022; Mitchell, 2021). The "EuroAsiaInterconnector" electricity interconnection hub that will connect Israel with Cyprus and Greece and the "EuroAfricaInterconnector" electricity interconnection that will connect Egypt with Cyprus and Greece will further consolidate the energy geostrategic position of Cyprus (Tsoukala, 2022). Additionally, the 2,000-kilometer EastMed submarine natural gas pipeline that has been designed in order to connect the sources of the Eastern Mediterranean with Greece via Cyprus, if implemented, will turn Cyprus into a strong energy power (Kolokasidou & Van Son, 2022; Tzogopoulos, 2022; Stratakis & Pelagidis, 2020; Troulis, 2019).

Positive prospects are emerging for Cyprus to transition from a carbon economy to a hydrogen economy by 2050 so that Cyprus can become a green energy country (Poullikkas, 2021; Panayiotou & Kalogirou, 2009). Cyprus has a high energy consumption in its transport sector system, thus; hydrogen will become an attractive solution and at the same time the development of RES can produce green hydrogen (Afanasiev et al., 2025). However, the current electricity production in Cyprus is mainly carried out by the Dhekelia and Vassilikos power plants using imported oil and fuel oil (Nikolaidis & Poullikkas, 2022; Ministry of Finance, 2022a). According to Poullikkas (2020) the existing energy model of Cyprus remains inefficient, significantly burdens the environment and does not respond to the planning of Cyprus's sustainable development. In order to reverse its negative energy model, in 2020 the country formulated its strategies, objectives and action plans that will lead it to be free of greenhouse gas emissions by 2050, and published the first National Strategic Plan for Energy and Climate (WEC, 2023). Through this extrovert plan, policies to support RES as well as circularity policies are being implemented. It should be noted that Cyprus has also progressed in the development of a liquefied natural gas terminal that will contribute to its energy security and its diversification from imported oil and fuel oil (WEC, 2023). At the same time, Cyprus is promoting energy storage systems that will maintain the stability of the system that is necessary in the case of RES (Therapontos et al., 2022). The goal is for energy storage to reach 160MW/4h through the Battery Energy Storage System (BESS) by 2030, and after 2033 another 80MW/8h pumping station to be installed (WEC, 2023).

Cyprus' subsidy system operates on two tracks. Firstly, subsidies are provided to support renewable energy technologies and, secondly, subsidies are provided for oil prices so that the system is not affected by fluctuations in oil prices. Subsidies particularly support photovoltaic systems, wind farms, and biomass/biogas plants. According to the final update of Cyprus' integrated national energy and climate plan for 2021-2030 towards wind farms, subsidies will be reduced when projects exceed a production threshold (Republic of Cyprus, 2024).

The prevailing climatic conditions in Cyprus, on the one hand, favour the development of RES, but on the other hand, the extreme heat wave increases electricity consumption in order to meet the needs for space cooling and seawater desalination (Ministry of Agriculture, Rural Development and Environment 2024; Giannakopoulos et al., 2016). It should be noted that Cyprus is one of the very first countries in the world to use over 90% solar water heaters in households and over 50% in hotels (Ministry of Agriculture, Rural Development and Environment 2024).

Based on the World Energy Cycle (WEC Trilemma) in 2023 Cyprus received a total score of 39.7 in energy security, 92.3 in energy equality and 68.0 in environmental sustainability and was ranked with an overall score of 65.8 in 40th place (WEC, 2023). Comparing this progress to previous years, it is found that in energy security Cyprus has achieved a score that ranks it among the countries with the highest improvement at a global level (WEC, 2023).

Table 1 below outlines the trends in greenhouse gas emissions for the period 2020-2022 as well as the rate of change from 1990 to 2022 (Ministry of Agriculture, Rural Development and Environment, 2024). It should be noted that in 2021 the percentage of greenhouse gas emissions that amounted to 37% of total emissions was due to electricity generation using oil and fuel oil (Ministry of Finance, 2022a).

Table 1. Greenhouse gas emissions by gas for the period 2020–2022

	2020	**2021**	**2022**	**Change from 1990 to 2022 (%)**
CO2 emissions without LULUCF	6925.65	7081.45	7118.39	53.23
CO2 emissions with LULUCF	6625.28	6836.73	6817.66	51.76
CH4 emissions without LULUCF	1063.16	1084.03	1077.52	38.77
CH4 emissions with LULUCF	1063.58	1089.51	1077.79	38.80
N2O emissions without LULUCF	172.74	175.28	174.84	20.01
N2O emissions with LULUCF	174.15	179.20	175.91	20.71
HFCs	334.51	359.20	383.73	100.00
PFCs	NO	NO	NO	0.00
Unspecified mix of HFCs and PFCs	NO	NO	NO	0.00
SF6	18.74	17.58	17.84	553.28
NF3	NO	NO	NO	0.00
Total (without LULUCF)	8514.80	8717.55	8772.32	57.48
Total (with LULUCF)	8216.26	8482.22	8472.93	56.41
Total (without LULUCF, with indirect)	8523.51	8725.57	8778.50	57.38
Total (with LULUCF, with indirect)	8224.97	8490.24	8479.11	56.30

Source: Ministry of Agriculture, Rural Development and Environment (2024). Rural Development and Environment

CYPRUS' NATIONAL ENERGY AND CLIMATE PLAN

Introductory Remarks

Cyprus is actively pursuing a comprehensive energy transition strategy in order to achieve sustainable development. This transition is largely guided by the country's National Energy and Climate Plan (NECP). This transition is neither easy nor straightforward it requires both internal and external reforms, so as to reduce energy consumption and foster the adoption of renewable energy and leverage the country's strategic geographic position so as to become a regional energy hub (Republic of Cyprus, (2020). The National Energy and Climate Plan, which was submitted in its final updated form in December 2024, provides a comprehensive and forward-looking outline of the Cyprus' strategy towards climate neutrality by 2050. The plan addresses five key dimensions: a) decarbonisation, b) energy efficiency, c) energy security, d) internal energy market and e) research, competitiveness and innovation (European Parliament, 2025; Republic of Cyprus, 2024; Republic of Cyprus, 2020).

Cyprus' Current Energy Landscape and the Imperative for Transition

Cyprus' isolated energy system and the country's overwhelming dependency on imported fossil fuels have been considered as pivotal contributors to the country's significant energy challenges. More than 90% of the country's internal energy consumption has been derived from petroleum products, with only a small fraction deriving from renewable sources of energy (Republic of Cyprus, 2020). This high dependence has been making the country particularly vulnerable to global energy price fluctuations and supply chain disruptions. This vulnerability is further underscored and perpetuated by the recent geopolitical events such as the war in Ukraine and the prolonged high-imported inflation (Republic of Cyprus, 2024).

It should be noted that Cyprus' net greenhouse gas emissions in 2023 decreased by 5.6% across the EU as a whole compared to 2005.While total emissions decreased by 4.7% during the period between 2005-2023, emissions from sectors covered by the Effort-Sharing legislation increased by 7.9% within the same period, slightly exceeding emissions from EU Emissions Trading System (EU ETS) sectors, which experienced a 14.9% reduction (European Parliament, 2025). The country's National Recovery and Resiliencc Plan (RRP), includes a chapter on REPowerEU and is dedicating 45% of its financial contribution (over €0.2 billion in loans and €1.02 billion in grants) towards green transition with a primary focus on transport and energy initiatives (European Parliament, 2025).

The Key Pillars of Cyprus' Energy Transition Strategy

As mentioned above, the Cyprus' Energy Transition Strategy has identified the following five critical pillars as pivotal enablers towards achieving a sustainable energy future:

Decarbonisation

Cyprus is strongly committed to reduce greenhouse gas emissions across all sectors of the economy, by placing strong emphasis on Renewable Energy Sources (RES). The revised strategy is setting a new national contribution target for RES in gross final energy consumption of 31%-34% by 2030 (Ministry of Energy, Commerce and Industry, 2023). Notably, in 2023, final energy consumption in Cyprus reached 20.7%, while wind energy remained stable (European Parliament, 2025). If Cyprus follows its plans for 2050 without compromise, it could become a green energy exporter.

However, in order to achieve this goal, its strategy should include:

i. Further promotion of renewable energy sources. Over the past 15 years, the Government of Cyprus has developed various support schemes and incentives aiming at boosting RES penetration, particularly in electricity generation (Republic of Cyprus, 2020).
ii. Increased electrification of the economy, especially within road transport, heating and cooling so as to displace the dependence and utilization of fossil fuels (Republic of Cyprus, 2024).
iii. Exploitation of bioenergy and waste-to-energy so as to contribute to the energy mix and effectively manage waste (Republic of Cyprus, 2024).

Energy Efficiency of Buildings

The energy efficiency of buildings is of strategic importance for Cyprus, which is why it is included in strategic plans for energy and climate. It is worth noting that final energy consumption is projected to reach 1.80 Mtoe in 2030 and national primary energy consumption is expected to reach 2.30 Mtoe by 2030. (Ministry of Energy, Commerce and Industry, 2023; Republic of Cyprus, 2024).

The measures and policies implemented in Cyprus include the following:

i. Building Renovations: Measures include incentives for building insulation, replacement of building windows, environmentally friendly heating and cooling systems, smart meters and efficient lighting. The Long-Term Building

Renovation Strategy, which was adopted in July 2024, aims at tripling the rate of comprehensive energy upgrades in buildings compared to the current rate of around 1000 per year. Last but not least, a new digital one-stop-shop platform dedicated to building renovations in currently under development (Republic of Cyprus, 2024).

ii. Public Sector buildings: Advancing energy upgrades across public buildings, including hospitals and schools (Republic of Cyprus, 2024).
iii. Transport Sector: Measures include the promotion of public and non-motorized transport, together with, the electrification of vehicles (Republic of Cyprus, 2024). Targeted investment interventions within public transport infrastructure such as bus lanes, bus stops and potentially tram systems in Nicosia, are foreseen (Republic of Cyprus, 2024, 2020).
iv. Smart Metering Infrastructure: Installation of smart metering systems in order to enhance energy management (Republic of Cyprus, 2024).

Energy Security

Given the country's historical energy isolation, a primary objective becomes the active enhancement of energy security. This particular strategy focuses on the diversification of energy sources and supply routes, together with, increasing the flexibility and fluidity of the national energy system (Republic of Cyprus, 2024, 2020).

Key measures and interventions include:

i. Introduction of Natural Gas (NG): This measure involves importing Liquefied Natural Gas (LNG) through the CyprusGas2EU project (Cyprus LNG Import Terminal) (Final Update of Cyprus' Integrated Energy and Climate Plan 2021-2030, 2024, p.8). The target for LNG availability for electricity consumption was initially planned for late 2021, however, it is now anticipated by the second quarter of 2026 (Final Update of Cyprus' Integrated Energy and Climate Plan 2021-2030, 2024, p.274). The project has received substantial funding from the "Connecting Europe Facility (CEF)" (Republic of Cyprus, 2020).
ii. Exploitation of Domestic Hydrocarbon Deposits: The utilization of the discovered hydrocarbon deposits within the Cyprus' Exclusive Economic Zone (EEZ), e.g. the Aphrodite gas field, is considered as a major contributing factor to the energy supply flexibility. The production from the Aphrodite field is expected to commence in 2025 and last for 18 years (Republic of Cyprus, 2020).
iii. Enhancing System Flexibility: This intervention includes increased RES penetration, together with, the adoption of demand response mechanisms and energy storage solutions. A plan for hybrid RES systems with storage is under preparation. While a 130 MW (1040 MWh) pumped-hydro storage unit was considered

cost-competitive for energy arbitrage and operational reserve purposes, its feasibility is subject to water availability in the country. The insufficient existing storage capacity (160 MW) highlights the pressing need for more incentives for storage systems to foster and enable greater RES penetration (Republic of Cyprus, 2024, 2020).

Internal Energy Market

The country's strategy for its internal energy market puts strong focus on integration with the wider EU network and the fostering of competition.

The strategy includes the following major initiatives and interventions:

i. Electricity Interconnection: A crucial and pivotal project is the promotion of the Great Sea Interconnector (previously known as EuroAsia Interconnector). This project concerns the establishment of a cross-border electrical interconnector between Greece, Israel and Cyprus (European Parliament, 2025). With this project, Cyprus aspires to put an end to its ongoing energy isolation, diversify resources and contribute to the decarbonization by enabling electricity imports largely from renewable sources. If and when this project is completed, this interconnector could transform Cyprus into a net exporter of electricity, predominantly from the installation of solar PV (Republic of Cyprus, 2020). This project has secured significant funding support from the CEF (Republic of Cyprus, 2024).

ii. Natural Gas Pipeline Infrastructure: This intervention involves the development of an internal natural gas pipeline (Republic of Cyprus, 2024).

iii. Investments in the Transmission System: The projects described and outlines in the Ten-Year Network Development Plan for the period 2024-2033 aim at upgrading and ensuring the safe operation of the electricity transmission system, with an estimated budget of €450 million (Republic of Cyprus, 2024).

iv. Competitive Electricity Market: Ensuring the promotion of necessary and critical regulatory frameworks and projects so as to enable a competitive electricity market (Republic of Cyprus, 2024).

Research, Innovation, and Competitiveness

Research, innovation and competitiveness are considered as pivotal and crucial enablers for Cyprus' energy transition. The national goals, as set out in "Innovate Cyprus", is to increase investments in Research and Innovation to 1.5% of GPD by 2030. Public funding in Research and Innovation is expected to generate positive results, especially when combined with measures and interventions to support entrepreneurial ventures and startups (Republic of Cyprus, 2024).

In the context of Research and Innovation the following measures and interventions are being considered:

i. Energy Efficiency and RES: This particular intervention revolves around the development of new or optimized RES technologies. These technologies put strong emphasis on solar energy and innovative solutions for energy saving and energy efficiency in buildings (Republic of Cyprus, 2020).
ii. Smart Grids and Energy Management: The focused and targeted utilization of advanced digital technologies incl. Artificial Intelligence and Internet of Things aims at providing better and more optimal monitoring of energy production, consumption, together with integrating RES in smart grids (Republic of Cyprus, 2024).
iii. Circular Economy and Water Management: Research endeavours within these critical areas are a priority, recognizing their interlinkage and interrelationships with energy and climate objectives (Republic of Cyprus, 2024, 2020).
iv. Collaborative Research Initiatives: Incentivizing and fostering partnerships between education institutions, Non-Governmental Organizations and government ministries (Republic of Cyprus, 2020).

Cyprus can also leverage and utilize the expertise developed within local universities and research centers such as the Cyprus Institute, the Cyprus University of Technology (CUT), the University of Cyprus (incl. the FOSS Research Centre for Sustainable Energy and KIOS Research Centre) and the Frederick University. All these institutions actively and significantly contribute to the various and interconnected aspects of energy and climate research. These Research and Innovation endeavours/projects receive financial support from EU funding programs like the Horizon 2020, LIFE and the European Structural and Investment Funds.

Cyprus as an Energy Hub

An integral and critical component in the country's sustainable development vision is its aspiration to become an energy hub within the Eastern Mediterranean region. Such vision involved leveraging the country's geographical location and the fostering of regional cooperation. This bold vision includes a number of foundational elements and components.

i. Regional Energy Infrastructure: The proposed EastMed Pipeline (the project that connects natural gas field in Israel to Greece, Cyprus and potentially Italy) and the Great Sea Interconnector are pivotal projects underpinning this bold ambition. The purpose and broader vision behind these projects are the diversification of energy sources and routes for the entire region, the strengthening of supply security and the promotion of market integration with the European Union (Republic of Cyprus, 2020).

ii. Export Potential: Significant investments in solar PV, together with, the full realization of electrical interconnections could contribute to Cyprus becoming a net exporter of electricity. This strategic transition from an isolated, import-dependent system to a potential exporter highlights and further strengthens the narrative behind the transformative potential renewable energy and regional integration (Republic of Cyprus, 2024).

iii. The Role of Natural Gas: For Cyprus, the long-term goal is decarbonization. At the same time natural gas is perceived as a transitional fuel, able to provide back-up for renewables, while enhancing energy security. Exploiting domestic gas reserves and importing LNG are also critical elements of this strategy. However, it is important to mention that the compatibility of fossil gas investments with the 2050 climate neutrality targets seeks further research and evaluation. All in all, there are calls for adopting a strategy to gradually phase out fossil gas use and its underlying infrastructure (Republic of Cyprus, 2024).

Benefits and Impacts of the Energy Transition

Considering the above-mentioned analysis and the presentation of Cyprus' Energy Transition Strategy, it is important to highlight a number of benefits and impacts that must be taken into account.

i. Economic Benefits: The energy transition is projected to have a positive macroeconomic impact, contributing to economic growth, increased gross value added, together with, creation of new jobs. Moreover, this transition is anticipated to lead to reduced volatility in energy prices (Republic of Cyprus, 2024). Overall, the cumulative additional investments required over the period 2020-2030 are estimated at over €1.4 billion. These investments are considered as feasible for the Cypriot economy (Final Update of Cyprus' Integrated Energy and Climate Plan 2020, p.261). Moreover, these investments are expected to yield net economic benefits to the society exceeding €500 million by 2030, alongside a 0.3% increase in national GDP (Republic of Cyprus, 2020).

ii. Employment: The transition towards renewable energy and energy efficiency sectors is expected to create a significant number of new jobs, particularly in the short to medium term. These jobs have the strong potential to offset job losses within traditional fossil fuel sectors due to this transition (Republic of Cyprus, 2024).

iii. Environmental Benefits: These additional policies and measures remain the preferred scenario towards achieving all the energy and environmental criteria within the EU's energy strategy. These measures and policies are projected to lead to reduced GHG emissions, improved quality of air, and consequently, fewer public health issues, including a reduction in pollution-related premature deaths (Republic of Cyprus, 2020).
iv. Social Equity: despite the substantial investments required, the implementation of these additional policies and measures are not expected to have a significant effect on energy affordability or social equity, especially for vulnerable households. One of the major reasons is the design of the policies so as to avoid disproportionate effects on low-income households (Republic of Cyprus, 2020). Energy poverty is a critical, complex and multidimensional phenomenon that required effort in terms of defining its context and developing an actionable mitigation plan (Republic of Cyprus, 2024).

Challenges and Future Outlook

Despite the fact that Cyprus has made significant and pivotal steps within its energy transition planning, there are still several pressing challenges and areas for continuous development and intervention:

i. Investment Mobilization: The targets and objectives that have been set are quite bold and ambitious and do require substantial private and public investments.
ii. Interconnection Dependence: The success of the increasing the penetration of RES technologies significantly depends on the realization of the Great Sea Interconnector project. Without this project, substantial investments in energy storage will be required so as to effectively meet the RES targets (Republic of Cyprus, 2020).
iii. Water Scarcity: Climate change is strongly and persistently affecting the sourcing of water in Cyprus. The prevalence of periodic droughts in the country may have a serious effect on the viability of large-scale pumped-hydro storage projects. Such negative prospects call for immediate exploration of alternative storage technologies and options (Republic of Cyprus, 2020).
iv. Research and Innovation: Continued investments in Research and Innovation projects and initiatives are vital for accelerating the deployment of new low-carbon technologies and adapting them to the specific needs and conditions of the country (Republic of Cyprus, 2020).

v. Public Awareness, Education and Training: The civic society must actively participate in the energy transition. To that respect, ongoing efforts are required in terms of raising public awareness via education, training and campaigns (Republic of Cyprus, 2020).
vi. Policy Evaluation: RES systems are still quite new and novel; therefore, a thorough evaluation of their environmental impact is required. In the same context, ongoing assessments of the alternative fuel adoption in transport infrastructure is also required (Republic of Cyprus, 2020).
vii. Cross-Border Harmonization of Assumptions: achieving accurate energy balance projections and trade potential requires agreement at a cross-jurisdictional level with the neighbouring counties regarding the underlying assumptions for large infrastructure projects (e.g. EuroAsia, Interconnector, EastMed pipeline) (Republic of Cyprus, 2020).

Concluding Remarks

The Republic of Cyprus has put together a very visionary and bold sustainable development strategy. This strategy is intrinsically linked to the country's ambitious energy transition strategy. Cyprus aims at transforming its energy landscape by focusing on five key pillars, namely; a) decarbonization, b) energy efficiency, c) energy security, d) internal energy market and e) research, innovation and competitiveness. Furthermore, strategic regional cooperation through cross-border projects incl. the Great Sea Interconnector and the EastMed Pipeline aims at re-positioning Cyprus in the international and Easter Mediterranean regional energy area not anymore as an energy-isolated island, but as a vital energy hub. The approach is quite multifaceted, complex and challenging, however, if executed correctly, it has the potential to yield environmental benefits, improved public health, significant economic growth and creation of new jobs. Overall, Cyprus' energy transition strategy has the potential to pave the way for a more resilient and sustainable future.

THE TERRITORIAL JUST TRANSITION PLAN OF CYPRUS

The dependence of Cypriot electricity production on fossil fuels is among the highest among EU countries (Ministry of Agriculture, Rural Development and Environment 2024; Kythreotou & Mesmeris, 2023; Cyprus Energy Agency, 2019). Over 8% of the country's GDP is allocated to the purchase of imported fossil fuels, although in recent years there have been investments in RES that are constantly changing the country's energy mix (Ministry of Agriculture, Rural Development and Environment, 2024; Ministry of Finance, 2022a; Zorpas et al., 2017). Mainly solar energy and secondarily wind energy are continuously showing increasing trends, contributing to the decarbonization of the Cypriot economy (Ministry of Finance, 2022a).

The Territorial Just Transition Plan of the Republic of Cyprus has stepped-in to address the consequences of the energy transition while creating sustainable development investments for the affected areas while also supporting workers (Republic of Cyprus, 2021, 2020, 2018; Cyprus Human Resources Development Authority, 2018). In order to identify the needs of the green economy of the Republic of Cyprus, a specific plan to identify the needs in knowledge and skills in green professions was developed in 2018 by the Cyprus Human Resources Development Authority (Cyprus Human Resources Development Authority, 2018). Through this plan, specific employment strategies as well as education and training strategies were developed. Based on the Territorial Just Transition Plan, the first power plant to be decarbonized by 2032 is Dhekelia. A similar path is planned to be followed after 2032 for the Vassilikos power plant.

The energy transition has an impact on small and medium-sized enterprises, which account for 99% of Cypriot enterprises and employ 60% of the workforce (Ministry of Finance, 2022a). These enterprises are expected to be supported through the Just Transition Fund in order to ensure their competitiveness as they bear the costs of the transition to a green economy (Georgiou et al., 2024; Kentas, 2023; Taliotis et al., 2020b). Among the sectors affected by the energy transition are power plants, businesses with a large energy footprint, the desalination sector since the cost of water production affects agriculture and tourism, as well as the motor vehicle and fuel trade sector due to the spread of electromobility (Ministry of Finance, 2022a). Further, a significant number of job losses and a possible reduction in economic activity are expected in these sectors. In power generation and related sectors, over 210 jobs are expected to be lost, and in Cypriot businesses, directly or indirectly, approximately 35,000 jobs (Ministry of Finance, 2022a). For every 10 jobs lost in electricity generation, 13.5 jobs are affected in the sectors that supply the production of the Electricity Authority of Cyprus (Cyprus Electricity Authority; Republic of Cyprus, 2021). In order to effectively address the problem with jobs losses, the Republic of Cyprus has proceeded towards upgrading the knowledge and skills of workers through education and training programs so that they can join the green economy (Republic of Cyprus, 2021, 2018; Cyprus Energy Agency, 2019; Cyprus Human Resources Development Authority, 2018). At the same time, the plan foresees the creation of green jobs in various sectors such as RES installations, transportation, installation, maintenance of networks and energy storage technologies as well as in various other sectors of the green economy (Psarros et al., 2020; Cyprus Electricity Authority, 2022; Republic of Cyprus, 2021, 2020).

The Partnership Agreement of the Single Operational Programme for Cyprus 2021-2027

The Partnership Agreement (PA) implements the programming and utilization of the resources received by Cyprus from the EU Cohesion Policy Funds for the period 2021-2027. At the same time, it ensures synergy and coordination between the Cohesion Policy Funds and the European Maritime, Fisheries and Aquaculture Fund (EMFAF). The Cohesion Policy Funds include: the European Regional Development Fund (ERDF), the European Social Fund+ (ESF+), the Cohesion Fund and the Just Transition Fund. Among the objectives of the PA is the goal for a greener and more resilient Europe with low carbon emissions towards a zero-emission economy. This objective is implemented through the promotion of just transition, green investments, circularity, climate change mitigation as well as sustainable urban mobility (Makris et al., 2024; Ministry of Finance, 2022b). The EU has paid exceptional attention to this objective by directing 45.8% of ERDF resources, 80% of Cohesion Fund resources and 80% of EMFAF resources to this objective, which collectively amount to 38.5% of the FTA. On this basis, Cyprus is committed to progressing in the implementation of nine national strategies (Ministry of Finance, 2022b). These strategies focused on the following: Prioritize green energy production by supporting RES; implement climate change adaptation policies; support the circular economy; support sustainable water management given that climate change exacerbates the problem of water scarcity; protect biodiversity; promote zero-emission public passenger transport systems by prioritizing sustainable urban mobility; ensure that the fishing sector is environmentally, economically and socially sustainable; support the aquaculture sector; and to strengthen international governance for sustainable management of the seas.

The Just Transition Fund

The main financial tool to support the energy transition of Cyprus, as well as all EU member states, is the Just Transition Fund (Apostolopoulos et al., 2025; Dimitrakopoulos et al., 2024). The Just Transition Fund constitutes a financial instrument integrated into the EU's cohesion policy (Kyriazi & Miró, 2023; Leppänen & Liefferink, 2022; Moesker & Pesch, 2022; Cameron et al., 2020). Through the resources of the Just Transition Fund in Republic of Cyprus, both public and private investments are developed in order to support through additional resources and practices the just green transition, economic and social sustainable development. In particular, the Just Transition Fund of Cyprus directs resources to the transmission and distribution of energy while simultaneously supporting energy storage systems in order to further enhance the penetration of RES in the energy mix. Businesses should be supported to adopt environmentally friendly technologies for the use of RES with the aim of reducing their energy footprint (Ministry of Finance, 2022b). Both human capital and infrastructure should be upgraded and strengthened. The transition to a green economy requires upgrading the accumulated knowledge and skills, which is why the resources of the Just Transition Fund are oriented in this direction (Cyprus Human Resources Development Authority, 2018). Moreover, a characteristic feature of Cyprus is the small number of students who choose to study technical and scientific professions related to the green economy. In order to overcome this weakness, the "Green Technical School" is being established through the Just Transition Fund, which aims to attract, train, motivate, develop, and therefore, equip the country's economy with qualified young people in green-related technologies (European Commission, 2022a). Furthermore, 14% of the Just Transition Fund resources are directed to the modernisation of the labour market with an emphasis to be given towards the vulnerable people as well as young people and women (Kentas, 2023; McCauley et al., 2023; European Commission, 2022b; Joi a & Dobrotă, 2022).

THE GEOSTRATEGIC POSITION OF CYPRUS IN RELATION TO ENERGY HUBS

The geostrategic position of the island of Cyprus which is located in the Eastern Mediterranean makes it an energy hub which will serve future strategic energy interconnections between Asia and Europe and between Africa and Europe (Venizelou & Poullikkas, 2025). Should this strategic plan, which is currently under implementation, signal the end of the energy isolation of the Republic of Cyprus, enhance energy security as well as network interoperability, and at the same time, it will significantly reduce dependence on fossil fuels (Tsoukala, 2022). The Israel-Cyprus-Greece electricity pipeline called "EuroAsiaInterconnector" and the Egypt-Cyprus-Greece electricity pipeline called "EuroAfricaInterconnector" are creating a new balance in the energy landscape of the Eastern Mediterranean, at the core of which Cyprus is located (Sotiriou, 2025; Tsoukala, 2022; Mitchell, 2021). In addition, these energy interconnections will change the existing balances and will lead to significant geopolitical implications as the relations between the states in the region are turbulent and intensely competitive (Koktsidis et al., 2021). Therefore, these developments lead to continuous revisions to thc project implementation schedules.

The Israel-Cyprus-Greece Electricity Pipeline

The Israel-Cyprus-Greece electricity pipeline called "EuroAsiaInterconnector" is a 2,000 MW high-voltage direct current multi-terminal interconnection, which is characterised by bidirectional capabilities, that is compatible with the EU Green Deal (Tsoukala, 2022).

The "EuroAsiaInterconnector" power pipeline, which in the literature is also known as the "Great Sea Interconnector", when implemented, will connect Chantera in Israel with Kofinou in Cyprus and Korakia in Crete; thus, bridging Asia with Europe in terms of energy (Official Journal of the European Union, 2022). From Crete, an existing power line will connect this massive project to mainland Greece, and from there, to the pan-European power grid. This constitutes a priority project for the EU, it is included in the list of major projects for this purpose, and is politically and financially supported by the EU (Official Journal of the European Union, 2022). In January 2022, the EU funded the project with 657 million euros through the Connecting Europe Facility (IPTO, 2023; European Commission, 2022c). In addition, to the previous amount, the project has also been financed with the amount of 100 million euros from the Recovery and Resilience Fund (IPTO, 2023). The "EuroAsiaInterconnector" project is a priority for the Republic of Cyprus since it safeguards its energy security, reduces its overall energy costs, develops the country to become a major energy hub and creates economic and social benefits for the entire population of the island (Tsoukala, 2022).

The Egypt, Cyprus, Greece Electricity Pipeline

Through its abundant sunshine and prevailing political stability, Egypt has the potential to permanently supply Europe with green energy. In this direction, through transnational collaborations, the implementation of the "EuroAfricaInterconnector" project was planned, which will transport electricity from Egypt via Cyprus to Greece and from there to the rest of Europe. This project will transform Cyprus into an energy hub and will remove it from its energy isolation. The final agreement for this project was signed between Cyprus, Egypt and Greece on 19 October 2021 (Republic of Cyprus Presidency 2021). This project, when implemented, shall enhance the security of energy supply by establishing the Republic of Cyprus as a key player in the development of the Eastern Mediterranean energy corridor (Republic of Cyprus Presidency 2021). The "EuroAfricaInterconnector" power pipeline will have a total length of 1,396 km and will reach up to 3,000 meters under the sea (Tsoukala, 2022).

Figure 2 and Figure 3 show the expected electricity interconnections in Cyprus. The first is the Israel, Cyprus, Greece 'EuroAsiaInterconnector' electricity pipeline and the second is the Egypt, Cyprus, Greece 'EuroAfricaInterconnector' electricity pipeline.

Figure 2. EuroAsiaInterconnector

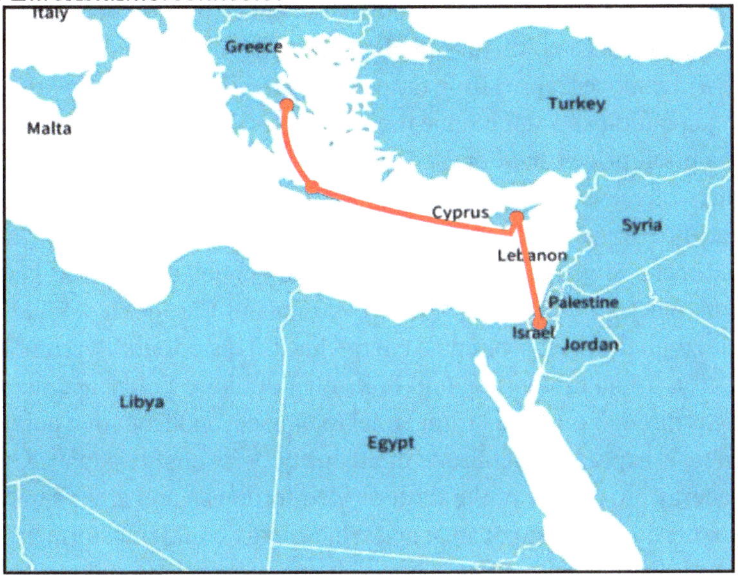

Figure 3. EuroAfricaInterconnector

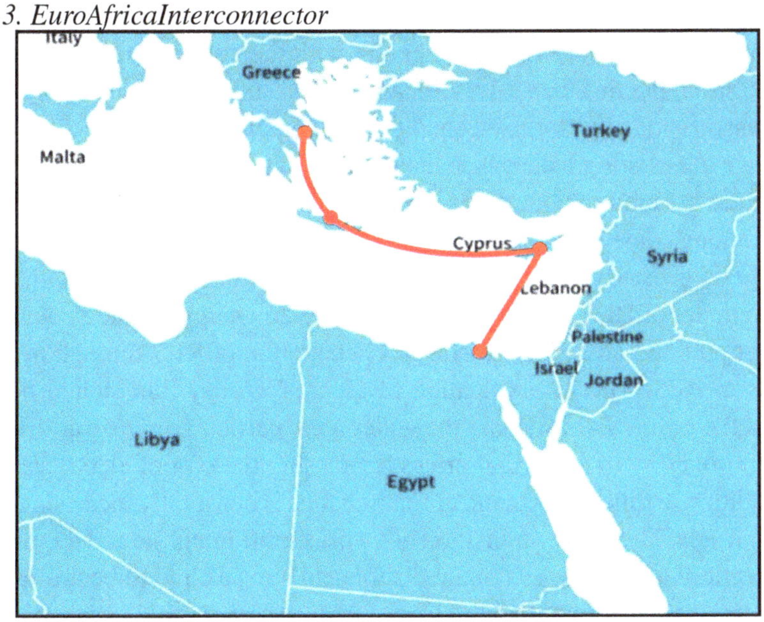

CONCLUSION AND POLICY RECOMMENDATIONS

The climatic conditions prevailing in Cyprus combined with its geostrategic position in the Eastern Mediterranean and its energy potential can provide Cyprus with a great energy advantage that could enable the country to become a producer of green energy. In fact a powerful stimulus exists, that if designed and implemented effectively will remove Cyprus from the polluting use of fuel oil and petroleum in its electricity generation and at the same time alleviate the vulnerability created by the fluctuations in oil prices and dependencies on oil producing countries. Moreover, Cyprus aims to achieve energy security and become a green energy country by 2050 free of greenhouse gas emissions. Simultaneously, it will cease to be an energy isolated and energy dependent country and will be connected due to its geostrategic position with electricity pipelines to Asia, Africa and Europe. It would be among the countries with the most prominent energy hubs. This possibility, combined with the exploration for natural gas and oil in its own offshore fields, and the transport of natural gas through pipelines from Israel to Greece, and from that point to many EU countries, is expected to enhance the security of its energy supply. Cyprus being at the energy crossroads of the Eastern Mediterranean and at the same time the potential for the development of solar and wind energy could transform the concrete country shortly into a major energy hub with major economic and social benefits.

In recent years, Cyprus may have started implementing energy transition plans with EU funding and signed transnational cooperation agreements with Greece, Israel and Egypt for electricity transmission, but these plans and the projects of the transnational agreements showcase, experience significant delays in their implementation. These shortcomings relate to dealing with the consequences of the energy transition, strengthening incentives for green investments and ensuring modern energy transmission and storage technologies. In addition, electricity interconnections change the existing balances in the region and have geopolitical implications. The latter are generating competitive reactions from countries within the area, as relations among some of the referred countries are highly turbulent.

Despite its geostrategic status and the existing climatic conditions that favour the development of RES, Cyprus continues to be an energy isolated country. This fact creates an imperative need for further penetration of RES through investment incentives in RES. The diversification of Cyprus' energy mix should be a high priority as the military conflicts in the wider area, particularly the war in Ukraine, have led to volatility in fossil fuel prices, increasing the cost of electricity generation. In addition to this deployment, climate change is constantly increasing Cyprus' electricity needs. The energy transition of Cyprus must integrate and align with EU policy directions and the targets of the EU for achieving a climate-neutral Europe in 2050 and therefore a climate-neutral Cyprus. Moreover, the EU resources directed towards Cyprus' energy transition and addressing its impacts on businesses and human resources are substantial and if utilised properly will bring positive results. Only from the EU cohesion policies on economic and social development and the Just Green Transition, Cyprus will be granted more than one billion euros in the period 2021-2027 (European Commission, 2022a). However, the development of RES alone cannot solve Cyprus' energy problem. Energy storage systems and energy mix systems are needed to ensure that the country's energy system remains stable. Venizelou & Poulikkas (2025) consider that the adoption of innovative RES systems require the simultaneous development of energy storage systems, sector coupling, smart grids, hydrogen development, digitalization as well as other advanced technologies. Power transmission lines with high temperature pipelines along with the digital system of transmission substations will serve the smart grids that the country needs. Rooftop solar PV systems for households, businesses and public buildings will require expansion to meet 100% of their energy needs via RES by 2050. Consequently, in order for Cyprus' energy transformation to be implemented, a specific framework of investment incentives will be necessary.

Cyprus is among the countries that can become a major hydrogen exporter by exploiting and investing in renewable energy sources. The energy strategy agreed with businesses, social partners and local stakeholders for the implementation of the energy targets should be implemented without any delays and shortfalls. Delays towards timeframes result to a negative effect, accumulating new costs and simultaneously creating confusion and uncertainty. The institutional and regulatory framework must be reliable, flexible and effective and serve the seamless energy transition and sustainable development. In order for the energy transition to be successful, it requires the consensus and support of society. Specifically, the consensus of local authorities, businesses, chambers of commerce, the scientific community and civil society.

A key element in energy transitions is the impact on the human resources working in power plants and the identification of new needs arising from the operation of green energy production (Papademetriou et al., 2025; Mezirow, 2018; Fotopoulos et al., 2015). In this regard, Cyprus has promptly proceeded with addressing this issue through a study on the detection of green knowledge and skills needs within its green economy (Cyprus Human Resources Development Authority, 2018). Moreover, electricity generation using fuel oil or oil does not employ a significant number of workers as required by the use of lignite (Sekar et al., 2022; Gunasundari, 2016). This has greatly facilitated the implementation of the knowledge and skills development programmes required by the green economy and the absorption of a share of workers into it. However, the problem in Cyprus is the impact of the energy transition on small and medium enterprises, which are the vast majority in Cyprus. Small and medium-sized enterprises in Cyprus will be affected by the energy transition as they will have to change their production and operating models. Specific financial instruments have been developed to support them accordingly. The European Regional Development Fund supports such enterprises in their financial diversification, digital transformation, and development of innovative and energy-saving technologies. Also, through the Just Transition Plan, the aforementioned businesses are taking advantage of a grant scheme for the use of renewable energy sources and energy-saving technologies in their production, as well as energy storage systems that will reduce their energy costs (Republic of Cyprus, 2024). Therefore, at this crucial point, a more specific intervention with financial support tools is required in order to help them to switch to sustainable ways of operating.

The EU has prioritized the implementation of energy efficiency in order to reduce the final energy consumption by improving the energy security of buildings and reducing energy-related costs. Cyprus has gained extensive experience in energy upgrades of residential and other buildings from 2018 to 2025 through national and European resources, which it should continue utilising in the future. However, it is necessary to accelerate the pace of energy renovation of buildings, invest in the Cypriot transport system, and implement an innovative green tax reform.

The transport plan provides for the gradual introduction of alternative fuels and green technologies. In Cyprus, the number of diesel-powered vehicles is decreasing and they are being replaced by hybrid and electric vehicles. By 2030, hybrid cars are expected to reach 209,000 and electric cars 42,090. Diesel fuel consumption is projected to fall from 349 million liters in 2022 to 275 million liters in 2030. At the core of the plan is the electrification of transport with the penetration of purely electric vehicles. To this end, smart vehicle charging systems and even vehicle-to-grid systems are planned (Republic of Cyprus, 2024). However, there has been a delay in the overall plan, and procedures need to be further accelerated.

REFERENCES

Afanasiev, A., Kandinskaia, O., Eminidou, S., & Sousanis, A. (2025). Hydrogen Valley in Cyprus: Insights and Strategies for Citizen Engagement. *Sustainability (Basel)*, *17*(2), 804. DOI: 10.3390/su17020804

Agathokleous, R. A., & Kalogirou, S. A. (2021). PV roofs as the first step towards 100% RES electricity production for Mediterranean islands: The case of Cyprus. *Smart Energy*, *4*, 100053. DOI: 10.1016/j.segy.2021.100053

Apostolopoulos, N., Makris, I., Apostolopoulos, S., & Liargovas, P. (2025). The Role of Rural Entrepreneurship in the Redevelopment of Rural Areas Previously Involved in Resource Extraction. In *Rural Entrepreneurship: Harvesting Ideas and Sowing New Seeds* (pp. 145-156). Emerald Publishing Limited. https://doi.org/DOI: 10.1108/S2040-72462025000020011

Batas-Bjelić, I., Rajaković, N., Ćosić, B., & Duić, N. (2015). A realistic EU vision of a lignite-based energy system in transition: Case study of Serbia. *Thermal Science*, *19*(2), 371–382. DOI: 10.2298/TSCI140613118B

Boukas, N., & Chourides, P. (2016). Niche tourism in Cyprus: Conceptualising the importance of social entrepreneurship for the sustainable development of islands. *International Journal of Leisure and Tourism Marketing*, *5*(1), 26–43. DOI: 10.1504/IJLTM.2016.077128

Brezar, A., & Skopeliti, C. (2025). *Portugal and parts of France hit by massive power outage.*

Bugeja, A., Azzopardi, B., & Loizou, E. (2023). A Comparative Analysis of Electric Mobility Operations in the Island States: A Case Study of Malta and Cyprus. *Journal of Energy: Energija, 72*(2.), 15-19. https://doi.org/DOI: 10.37798/2023722456

Cameron, A., Claeys, G., Midões, C., & Tagliapietra, S. (2020). *How good is the European Commission's Just Transition Fund proposal?* (No. 2020/04). Bruegel Policy Contribution.

CERA. (2022),Electricity, https://www.cera.org.cy/engb/ilektrismos/details/ilektrismos-adeiodotisi

Cyprus Electricity Authority. (2022). Electricity generation from Renewable Sources,(Greek),https://www.eac.com.cy/EL/EAC/Sustainability/Pages/Electricity Production.aspx

Cyprus Energy Agency, (2019). Public consultation on the National Energy and Climate Plan 2021-2030

Cyprus Human Resources Development Authority, (2018). Green skills needs identification in the Cyprus economy 2017-2027, Directorate of Research and Planning, (in Greek)

Dimitrakopoulos, P., Makris, I., & Apostolopoulos, S. (2024). The Impact of Major Energy Indexes on the Economic Growth of EU Countries, According to the Contribution of Industry to GDP. In *International Conference on Applied Economics* (pp. 59-74). Cham: Springer Nature Switzerland. https://doi.org/DOI: 10.1007/978-3-031-76658-9_4

Ember.org. (2024). "EU Electricity trends", https://ember-energy.org/latest-insights/european-electricity-review-2024/eu-electricity-trends/#electricity-demand

EPRS European Parliamentary Research Service. Briefing, (2023), https://www.europarl.europa.eu/RegData/etudes/BRIE/2023/754623/EPRS_BRI(2023754623_EN.pdf European Commission (2020), Technical support for implementing the European Green DealLuxembourg: Publications Office of the European Union, 2020 https://reform-support.ec.europa.eu/what-we-do/green-transition_en

European Commission. (2019). "The European Green deal", https://commission.europa.eu/strategy-and-policy/priorities-2019-2024/european-green-deal_en, accessed 9/7/2025

European Commission. (2022a). EU Cohesion Policy: more than €1 billion for Cyprus for economic and social development and a fair green transition in 2021-2027, Brussels, Press release, https://ec.europa.eu/commission/presscorner/api/files/document/print/en/ip_22_4326/IP_22_4326_EN.pdf

European Commission, (2022b). "Cohesion Open Data Platform".

European Commission. (2022c). Commission participates in launch of EuroAsia Electricity Interconnector, Commission participates in launch of EuroAsia Electricity Interconnector - European Commission European Parliament, (2025). Cyprus's climate action strategy, https://www.europarl.europa.eu/RegData/etudes/BRIE/2025/769495/EPRS_BRI(2025)769495_EN.pdf

Eurostat (2024). Electricity production, consumption and market overview https://ec.europa.eu/eurostat/statistics-explained/index.php?title=Electricity_production,_consumption_and_market_overview

Fotopoulos, N., Goulas, C., & Karra, V. (2015). Trade Union Education and Life Long Learning in Modern Greece. *World Journal of Educational Research (Los Angeles, Calif.)*, *2*(1), 15. Advance online publication. DOI: 10.22158/wjer.v2n1p15

Georgiou, A., Bednarova, L., & Kudelas, D. (2024). Climate Change Risk and Financial Performance in Cyprus: Management Perceptions. *Acta Montanistica Slovaca*, *29*(4), 920–932. Advance online publication. DOI: 10.46544/AMS.v29i4.11

Giannakopoulos, C., Psiloglou, B., Lemesios, G., Xevgenos, D., Papadaskalopoulou, C., Karali, A., Varotsos, K. V., Zachariou-Dodou, M., Moustakas, K., Ioannou, K., Petrakis, M., & Loizidou, M. (2016). Climate change impacts, vulnerability and adaptive capacity of the electrical energy sector in Cyprus. *Regional Environmental Change*, *16*(7), 1891–1904. DOI: 10.1007/s10113-015-0885-z

Gunasundari, K. (2016). A study on quality of worklife among mine workers with special reference to Neyveli Lignite Corporation. *International Journal of Advanced Research and Development*, *1*(7), 13–15.

Icaza, D., Vallejo-Ramirez, D., Guerrero Granda, C., & Marín, E. (2024). Challenges, roadmaps and smart energy transition towards 100% renewable energy markets in American Islands: A review. *Energies*, *17*(5), 1059. DOI: 10.3390/en17051059

International Energy Agency. (2025). Global Energy Review 2025, https://iea.blob.core.windows.net/assets/5b169aa1-bc88-4c96-b828-aaa50406ba80/GlobalEnergyReview2025.pdf

IPTO, (2023). IPTO is the new project promoter of the Greece-Cyprus-Israel Interconnection, IPTO is the new project promoter of the Greece-Cyprus-Israel Interconnection | IPTO

Joi a, D., & Dobrotă, C. E. (2022). Aspects of European funding for the energy transition-Just transition. *Ovidius Univ. Ann. Ser. Econ. Sci*, *22*(1), 289–297. DOI: 10.61801/OUAESS.2022.1.39

Kalogirou, S. A. (2003). The energy subsidisation policies of Cyprus and their effect on renewable energy systems economics. *Renewable Energy*, *28*(11), 1711–1728. DOI: 10.1016/S0960-1481(03)00062-4

Katemliadis, I., & Markatos, G. (2021). Stakeholders' involvement in sustainability planning and implementation: The case of Cyprus. *Worldwide Hospitality and Tourism Themes*, *13*(6), 709–718. DOI: 10.1108/WHATT-07-2021-0095

Katsaprakakis, D. A., Proka, A., Zafirakis, D., Damasiotis, M., Kotsampopoulos, P., Hatziargyriou, N., Dakanali, E., Arnaoutakis, G., & Xevgenos, D. (2022). Greek Islands' energy transition: From lighthouse projects to the emergence of energy communities. *Energies*, *15*(16), 5996. DOI: 10.3390/en15165996

Katsoufis, P. (2017). *Interconnection process of photovoltaic systems into the Electricity Authority of Cyprus's power grid using the method of self-production* (Master's thesis, Universitat Politècnica de Catalunya).

Kentas, G. (2023). Cyprus: Global Energy Crisis Is an Opportunity to Tackle Climate Change. In *Climate Change and the Future of Europe: Views from the Capitals* (pp. 19–22). Springer International Publishing., DOI: 10.1007/978-3-031-23328-9_5

Kim, J., Jaumotte, F., Panton, A. J., & Schwerhoff, G. (2025). Energy security and the green transition. *Energy Policy*, *198*, 114409. DOI: 10.1016/j.enpol.2024.114409

Koktsidis, P. I., Constantinou, C. M., & Joseph, J. (2021). Legal and Geopolitical Aspects of Energy Development in the Eastern Mediterranean and the EU's Strategic Involvement. In *The New Eastern Mediterranean Transformed: Emerging Issues and New Actors* (pp. 119–137). Springer International Publishing., DOI: 10.1007/978-3-030-70554-1_6

Kolokasidou, A., & Van Son, L. (2022). The EastMed Pipeline-How it is expected to impact the geostrategic position of Cyprus.

Koroneos, C., Fokaidis, P., & Moussiopoulos, N. (2005). Cyprus energy system and the use of renewable energy sources. *Energy*, *30*(10), 1889–1901. DOI: 10.1016/j.energy.2004.11.011

Kotzebue, J. R., & Weissenbacher, M. (2020). The EU's Clean Energy strategy for islands: A policy perspective on Malta's spatial governance in energy transition. *Energy Policy*, *139*, 111361. DOI: 10.1016/j.enpol.2020.111361

Kyriazi, A., & Miró, J. (2023). Towards a socially fair green transition in the EU? An analysis of the Just Transition Fund using the Multiple Streams Framework. *Comparative European Politics*, *21*(1), 112–132. DOI: 10.1057/s41295-022-00304-6

Kythreotou, N., & Mesmeris, T. (2023). 8th National Communication and 5th Biennial report under the UNFCCC of Cyprus. Department of Environment, Ministry of Agriculture, Rural 237 Development and Environment. Nicosia. Available online at https://unfccc.int/sites/default/files/resource/8th-NatCom-V1.4_submited301222.pdf

Lafortune, G., Fuller, G., Kloke-Lesch, A., Koundouri, P., & Riccaboni, A. (2024). Europe Sustainable Development Report 2023-2024: European Elections, Europe's Future and the Sustainable Development Goals.

Lamnatou, C., Cristofari, C., & Chemisana, D. (2024). Renewable energy sources as a catalyst for energy transition: Technological innovations and an example of the energy transition in France. *Renewable Energy*, *221*, 119600. DOI: 10.1016/j.renene.2023.119600

Leppänen, T., & Liefferink, D. (2022). Agenda-setting, policy formulation, and the EU institutional context: The case of the Just Transition Fund. *European Policy Analysis*, *8*(1), 51–67. DOI: 10.1002/epa2.1136

Logan, K. G., Nelson, J. D., McLellan, B. C., & Hastings, A. (2020). Electric and hydrogen rail: Potential contribution to net zero in the UK. *Transportation Research Part D, Transport and Environment*, *87*, 102523. DOI: 10.1016/j.trd.2020.102523

Makris, I., Apostolopoulos, S., & Dimitrakopoulos, P. (2024). Analysing supporting mechanisms, cultural shaping and incentives for circular economy: The case of local governments. *Interdisciplinary Environmental Review*, *23*(3), 218–235. DOI: 10.1504/IER.2024.139254

Makris, I., Apostolopoulos, S., Giannopoulos, V., Dimitrakopoulos, P., & Charalampakis, P. (2025). The Impact of Formal and Informal Institutional Elements on Land Mobility Within Rural Greece. *Sustainability (Basel)*, *17*(10), 4412. DOI: 10.3390/su17104412

Marketos, T. N. (2021). Energy Geopolitics Crossovers in Central & Eastern Mediterranean at the Prize of the Energy Union Policy. *Aspects of the Energy Union: Application and Effects of European Energy Policies in SE Europe and Eastern Mediterranean*, 47-71. https://doi.org/DOI: 10.1007/978-3-030-55981-6_3

Mavris, C. I. (2014). Social, cultural and environmental impacts of tourism in Cyprus: the need for sustainable development. *World Review of Entrepreneurship, Management and Sustainable Development 5, 10*(2-3), 197-215. https://doi.org/DOI: 10.1504/WREMSD.2014.060388

McCauley, D., Pettigrew, K. A., Todd, I., & Milchram, C. (2023). Leaders and laggards in the pursuit of an EU just transition. *Ecological Economics*, *205*, 107699. DOI: 10.1016/j.ecolecon.2022.107699

McKinsey & Company. (2024). "Global Energy perspective 2024" https://www.mckinsey.com/industries/energy-and-materials/our-insights/global-energy-perspective

Mezirow, J. (2018). Transformative learning theory. In *Contemporary theories of learning* (pp. 114–128). Routledge., DOI: 10.4324/9781315147277-8

Ministry of Agriculture. Rural Development and Environment, (2024). CYPRUS, National Greenhouse Gas Inventory 1990-2022, National Inventory Document under the United Nations Convention on Climate Change and the Paris Agreement, Ministry of Agriculture, Rural Development and Environment DepartmentofEnvironment, https://unfccc.int/sites/default/files/resource/CyprusNIR2024_vUNFCCC_v1.2.pdf

Ministry of Energy, Commerce and Industry, (2023). Revision of the National Energy and Climate Plan 2021-2030, (in Greek)

Ministry of Finance. (2022a). Just Transition Plan 2021-2027, Directorate General for Development, Ministry of Finance, (2022b). Partnership Agreement 2021-2027, Directorate General for Development, https://thalia.com.cy/wp-content/uploads/2022/07/%CE%A3%CE%95%CE%A3-2021-2027.pdf

Mirasgedis, S., Sarafidis, Y., Georgopoulou, E., Lalas, D. P., & Papastavros, C. (2004). Mitigation policies for energy related greenhouse gas emissions in Cyprus: The potential role of natural gas imports. *Energy Policy*, *32*(8), 1001–1011. DOI: 10.1016/S0301-4215(03)00048-X

Mitchell, G. (2021). Supercharged: The EuroAsia interconnector and Israel's pursuit of energy interdependence. *Mitvim Institute*, 1-8.

Moesker, K., & Pesch, U. (2022). The just transition fund–Did the European Union learn from Europe's past transition experiences? *Energy Research & Social Science*, *91*, 102750. DOI: 10.1016/j.erss.2022.102750

Nikolaidis, P., & Poullikkas, A. (2017). A comparative overview of hydrogen production processes. *Renewable & Sustainable Energy Reviews*, *67*, 597–611. DOI: 10.1016/j.rser.2016.09.044

Nikolaidis, P., & Poullikkas, A. (2022). A thorough emission-cost analysis of the gradual replacement of carbon-rich fuels with carbon-free energy carriers in modern power plants: The case of Cyprus. *Sustainability (Basel)*, *14*(17), 10800. DOI: 10.3390/su141710800

Official Journal of the European Union, (2022). COMMISSION DELEGATED REGULATION (EU) 2022/564 of 19 November 2021 amending Regulation (EU) No 347/2013 of the European Parliament and of the Council as regards the Union list of projects of common interest, L 109/14

Pacheco, A., Monteiro, J., Santos, J., Sequeira, C., & Nunes, J. (2022). Energy transition process and community engagement on geographic islands: The case of Culatra Island (Ria Formosa, Portugal). *Renewable Energy*, *184*, 700–711. DOI: 10.1016/j.renene.2021.11.115

Panayiotou, G., & Kalogirou, S. A. (2009). The introduction of Hydrogen Economy in Cyprus. In *Conference on the promotion of Distributed Renewable Energy Sources in the Mediterranean region*. https://doi.org/DOI: 10.1016/j.renene.2016.05.043

Papademetriou, C., Anastasiadou, S., Belias, D., & Ragazou, K. (2025). Integrating sustainability into human resource management: Building a greener workforce for the future. *Sustainability (Basel)*, *17*(3), 1113. DOI: 10.3390/su17031113

Pepe, J. M. (2024). Europe and the Emerging Geopolitics of Electricity Grids.

Pilavachi, P. A., Kalampalikas, N. G., Kakouris, M. K., Kakaras, E., & Giannakopoulos, D. (2009). The energy policy of the Republic of Cyprus. *Energy*, *34*(5), 547–554. DOI: 10.1016/j.energy.2008.08.007

Pinczynski, M., Kasperowicz, R., Azzopardi, B., Bilan, Y., & Štreimikienė, D. (2024). Malta's low carbon transition towards sustainability. *Sustainable Development (Bradford)*, *32*(5), 5120–5128. DOI: 10.1002/sd.2967

Poullikkas, A. (2020). Long-term Sustainable Energy Strategy: Cyprus' Energy Transition to Hydrogen Economy. https://doi.org/DOI: 10.31219/osf.io/js5hu

Poullikkas, A. (2021). Perspectives for the development of energy strategies—Challenges towards a hydrogen economy in Cyprus. *Green Energy Sustain*, *1*(0004), 0004. Advance online publication. DOI: 10.47248/HKOD902101020004

Poullikkas, A. (2024). Strategic objectives for Cyprus' energy transition, In Business news, https://inbusinessnews.reporter.com.cy/article/2024/9/19/796476/strategikoi-stokhoi-gia-tenenergeiake-metabase-tes-kuprou/

Prodromou, A., & Savvides, D. (2025). Cyprus Hotel Sustainable Development Through European and National Funding Schemes. In *Real-World Tools and Scenarios for Entrepreneurship Exploration* (pp. 459–496). IGI Global., DOI: 10.4018/979-8-3693-3100-2.ch017

Psarros, G. N., Papakonstantinou, A. G., & Papathanassiou, S. A. (2020, November). Integration of storage into large island power systems: The case of Cyprus. In *The 12th Mediterranean Conference on Power Generation, Transmission, Distribution and Energy Conversion (MEDPOWER 2020)* (Vol. 2020, pp. 270-275). IET. https://doi.org/DOI: 10.1049/icp.2021.1219

Psillaki, M., Apostolopoulos, N., Makris, I., Liargovas, P., Apostolopoulos, S., Dimitrakopoulos, P., & Sklias, G. (2023). Hospitals' energy efficiency in the perspective of saving resources and providing quality services through technological options: A systematic literature review. *Energies, 16*(2), 755. DOI: 10.3390/en16020755

Republic of Cyprus, (2018). Plan for the production of electricity from renewable energy sources with the final outcome of the integration of the projects within the competitive market, (2018)

Republic of Cyprus. (2020). Cyprus' Integrated National Energy and Climate Plan, Nicosia, Cyprus

Republic of Cyprus (2021). Territorial Just Transition Plan 2021-2027 (in Greek)

Republic of Cyprus. (2024). Final update of Cyprus' integrated energy and climate plan for 2021-2030, (in Greek), https://e-consultation.gov.cy/wp-content/uploads/2024/11/NECP_v.1.5.pdf

Republic of Cyprus Presidency. (2021). Joint Declaration of the 9th Greece-Cyprus-Egypt Trilateral Summit, in Athens, https://www.presidency.gov.cy/cypresidency/cypresidency.nsf/All/CB82614EC33A6A51C22587740031F9E0?OpenDocument

Rial, R. C. (2024). Biofuels versus climate change: Exploring potentials and challenges in the energy transition. *Renewable & Sustainable Energy Reviews, 196*, 114369. DOI: 10.1016/j.rser.2024.114369

Ripple, W. J., Wolf, C., Newsome, T. M., Barnard, P., & Moomaw, W. R. (2020). Corrigendum: World scientists' warning of a climate emergency. *Bioscience, 70*(1), 100–100. DOI: 10.1093/biosci/biz152

Sekar, A., Varghese, G. K., & Varma, R. (2022). Exposure to volatile organic compounds and associated health risk among workers in lignite mines. *International Journal of Environmental Science and Technology*, •••, 1–14. DOI: 10.1007/s13762-022-04056-4

Sotiriou, S. A. (2025). Connecting security with sustainable development in the Eastern Mediterranean and generating pay-offs for the European Union. *Mediterranean Politics, 30*(1), 208–217. DOI: 10.1080/13629395.2023.2279384

Stratakis, A. M., & Pelagidis, T. (2020). *The Competitive Advantage of the Forthcoming Mediterranean Energy Hub and its Implementation Alternatives: the EastMed Pipeline or an LNG Terminals Network?*.

Taliotis, C., Fylaktos, N., Partasides, G., Gardumi, F., Sridharan, V., Karmellos, M., & Papanicolas, C. N. (2020a). The effect of electric vehicle deployment on renewable electricity generation in an isolated grid system: The case study of Cyprus. *Frontiers in Energy Research, 8*, 205. DOI: 10.3389/fenrg.2020.00205

Taliotis, C., Giannakis, E., Karmellos, M., Fylaktos, N., & Zachariadis, T. (2020b). Estimating the economy-wide impacts of energy policies in Cyprus. *Energy Strategy Reviews, 29*, 100495. DOI: 10.1016/j.esr.2020.100495

Taliotis, C., Rogner, H., Ressl, S., Howells, M., & Gardumi, F. (2017). Natural gas in Cyprus: The need for consolidated planning. *Energy Policy*, *107*, 197–209. DOI: 10.1016/j.enpol.2017.04.047

Therapontos, P., Tapakis, R., Nikolaidis, A., & Aristidou, P. (2022). Increasing RES penetration in the Cyprus power system: current and future challenges. In *13th Mediterranean Conference on Power Generation, Transmission, Distribution and Energy Conversion (MEDPOWER 2022)* (Vol. 2022, pp. 313-318). IET.

Tranoulidis, A., Sotiropoulou, R. E. P., Bithas, K., & Tagaris, E. (2022). Decarbonization and transition to the post-lignite era: Analysis for a sustainable transition in the region of Western Macedonia. *Sustainability (Basel)*, *14*(16), 10173. DOI: 10.3390/su141610173

Troulis, M. (2019). The East Med pipeline project: Europe's fourth energy artery. *Civitas Gentium*, *7*(2), 153–168.

Tsakiris, T. (2017). Cyprus's natural gas strategy: Geopolitical and economic preconditions. *Mediterranean Quarterly*, *28*(1), 29–57. DOI: 10.1215/10474552-3882786

Tsangas, M., & Zorpas, A. A. (2023). Cyprus: Energy Policy. In *Encyclopedia of Mineral and Energy Policy* (pp. 205–212). Springer Berlin Heidelberg., DOI: 10.1007/978-3-662-47493-8_293

Tsangas, M., Zorpas, A. A., Jeguirim, M., & Limousy, L. (2018). Cyprus energy resources and their potential to increase sustainability. In *2018 9th International Renewable Energy Congress (IREC)* (pp. 1-7). IEEE. https://doi.org/DOI: 10.1109/IREC.2018.8362480

Tsoukala, O. (2022). Energy Diplomacy in the Eastern Mediterranean: The Cases of EuroAsia and EuroAfrica Interconnector Pipelines of Electricity Interconnection. *HAPSc Policy Briefs Series*, *3*(1). Advance online publication. DOI: 10.12681/hapscpbs.31005

Tzogopoulos, G. N. (2022). *EastMed Pipeline Could Be a Giant Step Towards Enhancing Regional Security*. Begin-Sadat Center for Strategic Studies.

Venizelou, V., & Poullikkas, A. (2025). Navigating the Evolution of Cyprus' Electricity Landscape: Drivers, Challenges and Future Prospects. *Energies (19961073), 18*(5). https://doi.org/DOI: 10.3390/en18051199

WEC. (2023). WEC Trilemma: Cyprus. World Energy Council, https://trilemma.worldenergy.org/#!/country-profile?country=Cyprus&year=2023

Weiss, A., Hernandez, D., & Gruenewald, D. (2024), "The electricity demand in Europe: growing or going? https://www.mckinsey.com/industries/electric-power-and-natural-gas/our-insights/electricity-demand-in-europe-growing-or-going, *accessed 9/7/2025*

Wu, C., Zhang, X. P., & Sterling, M. J. (2021). Economic analysis of power grid interconnections among Europe, North-East Asia, and North America with 100% renewable energy generation. *IEEE Open Access Journal of Power and Energy, 8*, 268–280. DOI: 10.1109/OAJPE.2021.3085776

Yang, G., Zhang, G., Cao, D., Zha, D., Gao, X., & Su, B. (2024). China's provincial-level sustainable energy transition requires accelerating renewable energy technological innovation. *Energy, 288*, 129672. DOI: 10.1016/j.energy.2023.129672

Zhang, G., Zhang, J., & Xie, T. (2020). A solution to renewable hydrogen economy for fuel cell buses–A case study for Zhangjiakou in North China. *International Journal of Hydrogen Energy, 45*(29), 14603–14613. DOI: 10.1016/j.ijhydene.2020.03.206 PMID: 32322133

Zorpas, A. A., Tsangas, M., Jeguirim, M., Limousy, L., & Pedreno, J. N. (2017). Evaluation of renewable energy sources (solar, wind, and biogas) established in Cyprus in the framework of sustainable development.

Chapter 6
Sustainable Development of Regions Affected by the Energy Transition:
The Case of Greece

Amvrosios Prodromou
https://orcid.org/0009-0008-4787-6292
Neapolis University, Pafos, Cyprus

Sotiris Apostolopoulos
https://orcid.org/0000-0002-1065-891X
Neapolis University, Pafos, Cyprus

Georgios Papadopoulos
https://orcid.org/0009-0006-8407-258X
University of the Peloponnese, Greece

Nikolaos Apostolopoulos
https://orcid.org/0000-0003-2810-5886
University of the Peloponnese, Greece

Nicholas Theodorakopoulos
https://orcid.org/0000-0002-2262-0054
Neapolis University, Pafos, Cyprus

ABSTRACT

The current study conducted across Greece's lignite regions – which are experiencing the consequences of the energy transition – and essentially aims to examine how the potential establishment of incentives could foster green entrepreneurship.

DOI: 10.4018/979-8-3373-5302-9.ch006

It has ultimately indicated the necessity of pursuing measures in order to prevent desertification. On top of utilising a Just Energy Transition plan of social legitimacy, building a concrete sustainable development model is required, and therefore, it must be implemented without any delays or deviations. Additionally, the energy crisis that erupted as a result of the war in Ukraine and the political decision to abandon the use of lignite in Greece by 2028, which is considered an unrealistic goal so far, have already put pressure on project implementation timeline. By conducting qualitative interviews with power plant workers and entrepreneurs located in lignite areas, this empirical research has specifically revealed that environmental goals should be supported by the revival of the local green economy.

INTRODUCTION

This book chapter mainly seeks to shed light on the effectiveness of incentives for the creation of sustainable business ecosystems within greek lignite areas where existing lignite-fired power plants are being impacted by the ongoing energy transition process. Whilst implementing the Green Deal, the European Commission has formulated its strategy for climate neutrality (Theodoropoulou et al., 2025; Apostolopoulos 2025a, 2025b). The EU's main objective focuses on reducing greenhouse gas emissions by 50% by 2030, and therefore, transform Europe into a climate neutral region by 2050 (Tranoulidis et al., 2024; Gema, 2023; Loewen, 2022). On this basis, mechanisms and financial programs have been activated in order to enable the EU to abandon the use of polluting fossil fuels in its electricity generation and switch to more sustainable Renewable Energy Sources (RES) in a fair way, without leaving anyone behind (Moesker & Pesch, 2022, European Commission, 2020). The Just Transition Mechanism was introduced with the main goal to substantially reduce the consequences of the areas affected by the transition, as well as to avoid their desertification (Abram et al., 2022; Filipović et al., 2022; Vanhercke & Verdun, 2021). The Just Transition Fund, which is a fundamental component of the Just Transition Mechanism, financially supports regions affected by the energy transition (Apostolopoulos et al., 2023; Strambo, 2020; Deal, 2020). In addition, the EU Member States, for their part, proceeded through consultation in order to create their own green energy transition plans (Brodny et al., 2024; Popescu et al., 2022; Rechsteiner, 2021). In terms of Greece, the country has in recent years shaped its policies in order to achieve the much-desired green transition and sustainable development by developing its own energy transition plan, and consequently, establish an environment that will deliver benefits for both its citizens and businesses (Wang et al., 2024a; Pegkas, 2024).

The country's strategy has focused on setting specific goals such as of eliminating lignite from electricity generation within a specific timeframe. More specifically, the strategy, and therefore, the country, is committed to reducing its lignite-based electricity generation by 80% by 2023 and to producing no energy from lignite by 2028 (Ministry of Environment and Energy, 2021). Despite the fact that the social stakeholders were positive in relation to the energy transition, they have disagreed on the tight timetable, considering that such time frameworks entailed risks and would mainly have socio-economic impacts in the areas using power plants lignite (Makris et al., 2024). As a result, it would be extremely difficult to achieve a green restart of local economies in such areas (Moesker & Pesch, 2022). Despite obvious practical difficulties, in 2024 Greece has reached a milestone, that of producing over 50% of its electricity from RES (Ember Energy, 2025). Greece's performance at an EU level has been exceptional as the country holds the third place in the production of energy from photovoltaics during the under-examination period, 2024 (Ember Energy, 2025). In 2023, the energy mix of production in the entire Greek territory derived from fossil fuels fell to 48.49% from 75% in 2013 (Ember Energy, 2025). As the use of fossil fuels in electricity generation decreased, electricity generation from RES increased, reaching an all-time high of 50.84% in 2023 (DEPEEP, 2024). The table below shows the energy mix from fossil fuels and from RES in 2023.

Table 1. Energy mix 2023

Fossil fuels					
Natural gas	**Oil**	**Lignite**	**Undetermined origin**		**TOTAL**
31.87%	**7.43%**	**9.16%**	**0.03%**		**48.49%**
Renewable Energy Sources					
Hydroelectric	Wind turbines	Solar	Biomass/ biogas	Unspecified origin	TOTAL
9.04%	23%	17.5%	1.30%	0.66%	50.84%

Source: DEPEEP, (2024)

Along with the environmental protection goal, Greece has also strategically focused on the green restart of local economies, especially those affected by the energy transition. However, the effectiveness and impact of such goals is hard to be measured using specific indicators, therefore, this research focuses towards this direction. More specifically, it examines the opinions, experiences and expectations of entrepreneurs in the affected areas as well as employees in power generation. In particular, it examines the energy transition with the degree of harmonization with and to environmental goals, with the performance of incentives for existing and new investments, with human resources as well as with the performance of domestic and European funding. A qualitative approach was carried out through the participation of fourteen individuals; seven entrepreneurs and seven employees in power generation.

LITERATURE REVIEW

The energy transition is an issue that has attracted research interest, as it is a key element in addressing climate change (Bashir et al., 2025; Balcilar et al., 2025; Sarabdeen et al., 2024; Rial, 2024). Similarly, there has been interest in the impacts of energy decarbonization (Goh et al., 2024; Oberthür et al., 2021; Papadis & Tsatsaronis, 2020; Inderwildi et al., 2020). Countries like Greece had based their electricity production on lignite extraction, creating major environmental issues for which there was research interest (Karasmanaki et al., 2020; Pavloudakis et al., 2020; Nanaki et al., 2016; Kolovos et al., 2002). However, according to international literature, the energy transition generates socioeconomic and environmental impacts in areas of electricity production using lignite (Pavloudakis et al., 2023; Tranoulidis et al., 2022; Ziouzios et al., 2021; Marinakis et al., 2020). In the case of Greece, the research has focused on various aspects of sustainable development of areas affected by the energy transition, such as the areas of electricity generation using lignite (Kaldellis et al., 2023; Kilintzis et al., 2023; Apostolopoulos, 2023; Pavloudakis et al., 2021). These includethe followinb: The assessment of the transition in relation to climate change mitigation, as well as the challenges created by the energy transition in Greece, including the obstacles and consequences of the transition (Fotis et al., 2025; Streimikiene et al., 2021; Ziouzios et al., 2021). Addressing the socioeconomic impacts of the energy transition through just transition governance (Pavloudakis et al., 2021). European funding to address the impacts of the energy transition and the sustainable development of lignite areas (Vlachou & Pantelias, 2022; Egli et al., 2022; Lypiridi, 2021; Marinakis et al., 2020). The impact of the Greek transition plan and its assessment (Zervas et al., 2021a). The restoration of the landscape from lignite extraction and the mitigation of resulting environmental problems that have been created (Pavloudakis et al., 2020). Investments in renewable energy sources and the integration of part of the former workforce that worked in lignite-based electricity plants (Makris et al., 2024; Apostolopoulos et al., 2023; Ioannidis et al., 2023; Ziouzios et al., 2021). However, what is absent from international literature on the case of the energy transition and the treatment of its consequences in Greece is a holistic approach based on four factors. The degree of harmonisation of lignite areas with environmental goals. The degree of efficiency of existing incentives for current enterprises as well as for attracting new enterprises to areas affected by the energy transition. The degree of utilization of human resources formerly employed in electricity generation using lignite following the upgrading of knowledge and skills through training programs. The degree of efficiency of domestic and European funding for the energy transition and addressing its consequences.

THE JUST TRANSITION DEVELOPMENT PLAN OF LIGNITE AREAS

The just transition development plan for Greece's lignite areas, which was completed after a public consultation foresees the withdrawal of all lignite electricity plants by 2028 (Ministry of Environment and Energy, 2020, 2021). Greece is one of the few EU member states that have opted for an immediate reduction in lignite dependence (Makris et al., 2024). This time-compressed energy transition has an impact on society, employment, enterprises and the local economy of lignite areas (Makris et al., 2024). The transition planning is supported by five principles (Ministry of Environment and Energy, 2020), namely; supporting labour intensive sectors in order to create employment opportunities in local communities; utilisation of natives advantages of the affected lignite areas; ensuring a rapid energy transition; promoting social and environmental sustainability; and, in the integration of modern technologies with simultaneous innovative interventions. The plan foresees the use of all available sources of financing, with emphasis on the Just Transition Fund, the InvestEU Special Scheme, the public sector loan facility and private capital (Theodoropoulou et al., 2025; Tranoulidis et al., 2024). In addition, it foresees the use of domestic and international financial institutions as well as of other European financing mechanisms such as the Partnership Agreement for Regional Development (ESPA) and the Infrastructure Fund (Ministry of Environment and Energy, 2020, 2021). For attracting investment in lignite areas, the plan provides for a set of incentives aimed at supporting existing enterprises, attracting new enterprises as well as natural persons affected by the transition (Kourkouridis et al., 2024; Dimitrakopoulos, et al., 2024; Pavloudakis et al., 2023; Tranoulidis et al., 2022). Attracting emblematic investments is a priority in the planning. Table 2 captures the form the incentives that have been included in the just transition development plan for Greece's lignite areas.

Table 2. Presents a package of incentives for new investments, for supporting existing enterprises as well as for supporting natural persons

Attracting new investments	Support for existing enterprises	Support for individuals
•Grant provision •Loans on favorable terms •Tax exemption •Tax relief •Exemption from fees •subsidy of insurance contributions •Guarantees	•Provision of grant for the transformation and modernization of production operations sustainability •Loans with favourable terms •Subside of loan obligations •Equity participation •Subsidisation of wage costs	•Income tax reduction •Enhancement of the foreseen benefits •Upgrading knowledge and skills through funded training programs •Housing loan subsidy

Furthermore, the design takes into account the needs of employees so that they can upgrade their knowledge and skills and integrate into the local green economy (Tranoulidis et al., 2022; Zervas et al., 2021b). At the core of the design is addressing the social impacts of the energy transition as well as enhancing social cohesion (Apostolopoulos et al., 2024). In addition, the spatial planning foresees in the just transition development plan for the lignite areas aims to address the environmental degradation of the landscape from lignite extraction. A series of landscape restoration projects as well as infrastructure development projects have been designed to be implemented (Government Committee SDAM, 2020; Ministry of Environment and Energy, 2020, 2021). Figure 1 illustrates power plants located in lignite regions.

Figure 1. Lignite regions with power plants in Greece (Ministry of Environment and Energy, 2020)

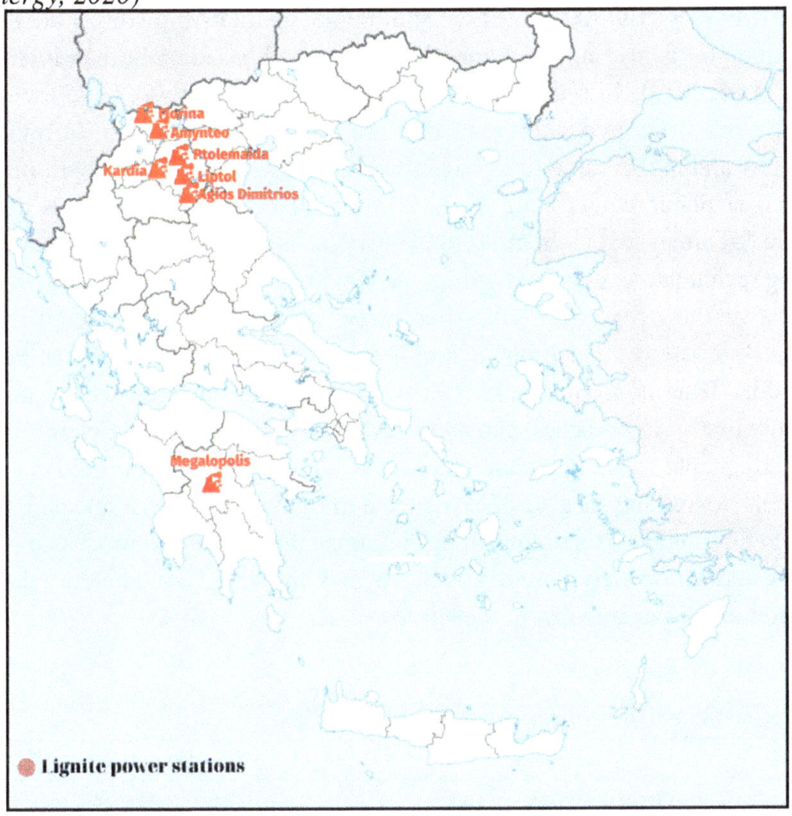

METHODOLOGY

A qualitative approach has been applied in order to detect, capture, and evaluate the views, experiences, and expectations of both, people working in lignite-fired power production and entrepreneurs operating in the environment in which the energy transition is taking place. Based on the existing international literature, the qualitative approach is appropriate and effective for detecting views and experiences within the environment in which the social, economic and environmental phenomenon is taking place (Ritchie et al., 2013; Hollway & Jefferson, 2012; Tuli, 2010; Denzin & Lincoln, 2011). The qualitative approach reliably clarifies complex phenomena in business environments with socio-economic dimensions (Javadian et al., 2020; Berglund, 2007; Neergaard & Ulhøi, 2007; Gartner & Birley, 2002). Also, based on international literature, the qualitative approach is considered effective for capturing, analyzing and interpreting opinions and experiences within social contexts (Willing, 2001; Fossey et al, 2002). At the same time, qualitative methods are suitable and accurate for analyzing behaviors that are formed based on human interaction. (Pathak et al, 2013). In order to collect the sample, the authors have focused on applying semi-structured interviews that are effective and reliable in qualitative approaches and at the same time have the ability to shed light in depth on social and business phenomena (Aung et al., 2021; Wallace et al., 2019; Hassink et al., 2016; Qu & Dumay 2011). In order to identify the appropriate sample, the targeted sampling technique was applied within the Regional Units of Arcadia, Kozani and Florina, as in these regions, lignite-fired power plants operate and are currently in the process of energy transition (Schmidt et al., 2020; Kral et al., 2010; Peterson et al., 2008; Watters & Biernacki, 1989). This sample selection technique, often used in qualitative approaches, can and does select cases that yield rich information in a small sample as its key characteristic is homogeneity (Patton, 2014). Seven individuals possessing long-term experience in lignite areas were selected, as well as seven entrepreneurs operating in lignite areas that are in the process of energy transition and whose activities are affected by it. Following, Table 1 shows the status and region of the participants in the survey.

Table 3. Characteristics of survey participants

R	Capacity	Area of activity
R1	Electricity generation worker	Regional Unit of Arcadia
R2	Electricity generation worker	Regional Unit of Kozani
R3	Electricity generation worker	Regional Unit of Florina
R4	Electricity generation worker	Regional Unit of Florina
R5	Electricity generation worker	Regional Unit of Arcadia
R6	Electricity generation worker	Regional Unit of Kozani
R7	Electricity generation worker	Regional Unit of Arcadia
R8	Entrepreneur	Regional Unit of Florina
R9	Entrepreneur	Regional Unit of Kozani
R10	Entrepreneur	Regional Unit of Arcadia
R11	Entrepreneur	Regional Unit of Florina
R12	Entrepreneur	Regional Unit of Arcadia
R13	Entrepreneur	Regional Unit of Kozani
R14	Entrepreneur	Regional Unit of Arcadia

Each one of the interviews lasted approximately 60 minutes and was conducted by telephone and online, taking into account that, based on international literature, telephone and online interviews provide equally reliable results as in-person interviews (Lo Iacono et al., 2016; Janghorban et al., 2014; Sullivan, 2012; Cachia & Millward, 2011). It should be noted that in order not to change and alter the existing individual characteristics, the sample size was based on the principles of adequacy and relevance (Morse & Field, 2013; Thompson, 1999). The collection of the sample began on the 3rd of May 2025 and was completed on the 30th of June 2025. The interviews were conducted in Greek, which is the native language of the research participants. To avoid ambiguities and misinterpretations during the data analysis, this was also carried out in Greek and only the references were translated into English, which is common in qualitative approaches in countries where their native language is not English (Van Nes et al., 2010; Maneesriwongul & Dixon, 2004; Temple & Young, 2004; Twinn, 1997). In the first phase, the interviews were examined separately using the inductive approach and then compared with each other (Creswell & Creswell, 2017; Goodrick & Rogers, 2015; Thomas, 2006). The emerging themes and subthemes are captured in Table 2. It should also be noted that the creation of the categories was determined in every detail and that the coding was both exclusive and independent (Makris et al, 2025; Krippendorff, 2018; Neuendorf, 2017; Smith et al., 1995).

Table 4. Thematic analysis

Theme	Sub-themes
Degree of alignment with environmental objectives	• Comprehensive sustainable transition plan • Environmental protection • Green restart of local economies • Sustainable business ecosystems
Incentive return for existing and new investments	• Incentives for green investments • Incentives for transforming existing businesses • Incentives for individuals • Integration into European and domestic support mechanisms
Human resources	• Creating quality jobs • Incentives for workforce retention • Incentives for hiring employees • Effective training programs
Performance of domestic and European financing	• Infrastructure projects and supporting projects of flagship investments • Projects in Renewable Energy Sources • Energy efficiency improvement and electromobility projects • Projects in the primary sector • Other projects in innovation and competitiveness

FINDINGS

The content of the research was then classified into four thematic sections. The first thematic section is concerned with the degree of alignment of the energy transition with environmental objectives. The second thematic section is concerned with the performance of incentives for existing and new investments. The third thematic section is concerned with human resources. Finally, the fourth thematic section is concerned with the performance of domestic and European funding in achieving the goal of the energy transition and addressing its impacts.

Degree of Alignment with Environmental Objectives

Four sub-themes emerged in this thematic section. The first is concerned with the integrated sustainable transition plan, the second with the protection of the environment, the third with the green restart of local economies, and the fourth with the sustainable business ecosystems.

The participants in the survey believe that there is a large gap between planning and implementation of the plan. They recognize that good planning brings quick results; however, they have expressed doubts over how much the plans in this specific case will yield. The compressed time for implementing the energy transition, as well as the energy crisis developed as a result of the war in Ukraine, are considered to have already impacted on the transition timelines. These elements create a deviation in the environmental goals of the plan.

> *"The plans may be good. However, everything is judged in practice. Complete de-reliance on lignite by 2028 is not an achievable goal. It will create dependencies on imported natural gas. Don't forget that we are already in an energy crisis as a result of the war in Ukraine." (R 2)*

> *"The energy crisis that arose with the war in Ukraine has already forced the government to revise the transition timetables. The deviations from the environmental goals that had been set are visible." (R 9)*

The participants in the survey believe that only when the transition plan is completed, lignite lands are restored and utilized, workers in the green economy find jobs, and businesses are supported in their transition, will the goal of environmental protection be achieved.

> *"Environmental protection in lignite areas is a difficult task. These areas have been burdened environmentally for many years. Plans to develop lignite areas*

are progressing very slowly. There are problems that will take many years to resolve." (R 7)

"Electricity generation using lignite has created major environmental problems with impacts on people and businesses. Addressing these consequences requires capital as well as strong incentives. Everything will be judged in practice." (R 12)

An important issue highlighted in the research is concerned with the restart of local economies that have been affected by the long-standing power generation using lignite and are now affected by the energy transition. This restart of local green economies is multifactorial and is still at the starting point. Green investments and strong incentives are highly required.

"Restarting the local economy is not an easy task that can be solved within a few years. Strong investments and employees with upgraded knowledge and skills are needed. I am not at all optimistic about the course of the restart so far." (R 1)

"Restarting the local economy affected by the energy transition, without providing thee appropriate green investments and financial incentives for existing businesses, will not be achieved. We hear many announcements from political officials, but little has been done to date." (R 14)

The participants in the survey believe that areas affected by the energy transition must acquire sustainable business ecosystems through state and European care and funding so that these areas can avoid abandonment and desertification.

"If the Megalopolis area does not acquire a sustainable business ecosystem, we are rapidly heading towards desertification. People will leave the area in search of better working and business conditions." (R 10)

"Only investments will give provide life to the region. If the working population does not immediately find work in the green economy, then internal migration will intensify." (R 5)

Incentive Return for Existing and New Investments

Four sub-themes emerged in this thematic section. The first is concerned with the incentives to attract green investments. The second is concerned with the transformation of existing businesses. The third is concerned with the incentives for individuals. Finally, the fourth is concerned with the integration of businesses into both, European and domestic support mechanisms.

The survey participants believe that creating a sustainable business ecosystem requires incentives for green investments in areas affected by the energy transition. When examining what kind of incentives, they mean, soft loans, grants, tax exemptions and reductions, social security contributions subsidies and fee exemptions emerged.

"In order to attract new investments in our region, which has been disproportionately affected by the energy transition, strong economic and fiscal incentives

are needed. Grants and tax breaks are required. A banking system that provides loans on favorable terms is also needed." (R 13)

An important issue that has strongly emerged during the research, is the support of existing businesses so that they can transform in order to move towards sustainability and withstand competition. By exploring what kind of incentives they mean, grants for the reform and modernization of existing businesses, loans on favorable terms, the subsidy of wage costs as well as the subsidy of their loan obligations emerged.

"Existing businesses are already experiencing the effects of the energy transition. Should strong incentives be not provided in order to enable transformation and modernisation of our businesses in the direction of the green economy, we will soon be locked down. We are in need of subsidies, favorable loans and a reduction in wage costs through subsidies." (R 11)

Furthermore, the support of individuals in order to deal with the impacts of the energy transition was also highlighted within this research. By exploring what kind of incentives the participants in the research mean, support for training programs to upgrade knowledge and skills, income tax equalization and housing loan subsidies were highlighted.

"Training programs are an important tool for upgrading the skills required for green energy production. However, this is not enough. Additional support is needed focusing on the reduction in income taxes as well as subsidizing our mortgage loans so that we can survive in the region." (R 3)

"The transformation and modernisation of local businesses and the green economy in general, requires specialized skills, which is why vocational training programs are essential." (R 9)

The research highlighted the need to support regions affected by the energy transition through European and domestic support mechanisms.

"Without the effective support of financial tools for businesses and human resources, the restart of the local economy will not be achieved. Our region is in need for financial resources for investment." (R 14)

Human Resources

Four sub-themes emerged in this thematic section. The first is concerned with the creation of quality jobs. The second is concerned with the incentives for retraining the workforce. The third is concerned with the incentives for hiring employees. Finally, the fourth is concerned with the results of training programs in the direction of evaluation and certification.

The participants in the survey believe that in areas affected by the energy transition, there should be investments accompanied by the creation of quality jobs.

"Positions in power generation are constantly decreasing. What will happen to this world without work? Investments are required, but they must be accompanied by the creation of green quality jobs. An energy transition without the support of human resources is doomed." (R 5)

"Local businesses, as they need to transform, require human resources. Towards this direction, there must be incentives in the direction of subsidizing wage costs." (R 12)

The survey highlighted the need for retraining the workforce so that it can seamlessly integrate into the green economy, which requires upgraded knowledge and skills on the part of employees. The survey participants believe that certified bodies should undertake the retraining of the workforce. However, they believe that the needs of the green economy in lignite areas must be firstly recorded and mapped.

"The workforce of workers in lignite-fired power generation is quite large. This workforce cannot be easily integrated into other jobs, unless it is retrained. Motivation is needed in this direction. Workers alone with their own resources cannot solve this problem. It is impossible for us to bear this burden." (R 1)

However, the entrepreneurs who participated in the survey also expressed the view that local businesses should be given incentives in order to hire more people. They believe that businesses have been affected by the energy transition to the extent that it is currently difficult for them to shoulder additional wage and insurance burdens on their own.

"Without state support, hiring in local businesses is almost impossible. The wage and insurance costs are high. Local businesses in the areas cannot afford the high cost as these businesses are mostly small and weak and cannot possibly bear it on their own." (R 8)

The participants in the survey expressed objections to the training programs developed in the past and how effective they were. They believe that all training programs should be certified and evaluated by external evaluators, and at the end provide certification to the trainee.

"My experience of the training programs in my home country is traumatic. Although those that concern workers in the power generation sector are in the same philosophy as those of the past, they will not offer anything new to the workers." (R 4)

"Training programs should be conducted by certified organizations. They should be continuously evaluated and at the end of the program, they should provide certification to the participants. I would say that these programs should be undertaken by the university in our region." (R 13)

Performance of Domestic and European Financing

Five sub-themes emerged in this thematic section. The first is concerned with the infrastructure projects as well as supporting projects for flagship investments. The second is concerned with projects in Renewable Energy Sources. The third is concerned with the energy efficiency and electromobility projects. The fourth is concerned with projects in the primary sector. Finally, the fifth is concerned with projects in innovation and competitiveness.

The participants in the survey consider it a positive element that in the development plan of their regions, important infrastructure projects as well as supporting projects of emblematic investments have been planned for implementation. Projects that are planned to be supported in their implementation by both European and domestic resources and mainly through the Just Development Transition Mechanism. However, they consider that these projects are delayed in terms of implementation, and consequently, this makes it difficult for the sustainable development of the regions affected by the energy transition.

"I recently participated in a conference on the sustainable development of the Megalopolis area where lignite mines operate. I cannot hide from you that I was impressed by the infrastructure projects and investments that are about to be made. However, the picture that we live and experience in the area is completely different. I hope the projects will be carried out and shall not remain on paper." (R 10)

The research revealed that projects in Renewable Energy Sources have progressed significantly in or near lignite areas. Large photovoltaic parks, mainly in lignite areas, have been created, which according to the participants in the research, contribute to the delignitisation of the areas, the creation of green jobs in energy and the utilization of existing electricity transmission networks. These projects are supported in addition to European and domestic financial instruments, by private capital and domestic and international financial institutions.

"What we see around us is that huge photovoltaic parks are being created. Some residents react to the image they see. I consider it positive because it helps in the delignitization of the area and in the creation of quality jobs. At the same time, the electricity transmission networks are also being utilized." (R 13)

In order for projects to improve energy efficiency and electromobility financed by the Just Development Transition Mechanism, the Green Fund and, domestic and international financial institutions, survey participants believe that they also have a positive impact on their region.

"Energy efficiency programs for homes and businesses to save energy have progressed considerably. Many homes and businesses have been funded in this direction. These programs continue leading to a positive impact." (R 6).

"From what we know about electric mobility in lignite regions, there are specialised incentives. Some are making use of these incentives and electric mobility infrastructure is constantly increasing. In our region, the creation of an electric mobility park is also planned that will develop products related to electric mobility. However, this project is still at the stage of issuing a building permit" (R 8).

In terms of the primary sector projects that are supported by the Rural Development Program, the Just Transition Mechanism, the Recovery Fund, as well as private and loan funds, the survey participants expressed concerns about whether they will be implemented. They believe that the primary sector in Greece is in a major crisis and that it will be difficult to implement the smart livestock farming units, hydroponics units and the production of protected designations of origin that are foreseen in the plans.

"In our area, the plan foresees the creation of an intelligent livestock and feed unit on an area of five acres. Local officials have told us that over two hundred jobs will be created. However, this project, like many others, is in the planning stages." (R 1).

"The primary sector in Greece is going through a major crisis. People are abandoning agricultural production and there is a shortage of labour projects. The planned agricultural and livestock activities in the lignite areas will hopefully be implemented. However, I am not optimistic that they will be implemented in this crisis in the primary sector." (R 5).

The participants in the survey believe that lignite regions need innovative and competitive projects that will give momentum to these regions and lead them to sustainable development.

"Without innovative and competitive projects, the lignite regions will be led to wither and abandonment. To achieve this, cooperation between the state machinery, the university, entrepreneurs and society itself is needed. The region of Western Macedonia has such potential and must develop it." (R 13).

DISCUSSION

The research highlighted the gap that exists in countries such as Greece between the planning of an initiative, an action, or a project and its completion within a specific timeframe and budget. This deployment may be due to internal weaknesses in planning, long-standing dysfunctions in services, as well as many external factors. With regard to the case of the energy transition and addressing its consequences, it has become apparent that there is concern in local communities about whether it will be implemented as planned. In particular, the compressed timeline for phasing out lignite-fired power generation, which must be completed by 2028, combined with the energy crisis that arose following the war in Ukraine, are causing shifts in the transition executing schedules as well as setbacks.

It also undermines with the environmental goals of the plan for a just transition in Greece's lignite regions. Towards overcoming these challenges without compromises or any delays and to successfully complete the energy transition, the concrete study has revealed that national and European support and funding for sustainable business ecosystems are needed so that these areas can avoid abandonment and desertification. Local economies need to be revitalized through robust incentives towards sustainable development. It is important not to deviate from the environmental objectives set out in the agreed plan on the just transition of lignite areas, a significant process which was established with the participation of social partners as well. Based on international literature, investment planning incorporates phases such as pre-investment, investment, and evaluation; however, the development of internal and external factors can create incompatibility with the overall design (Janeková et al., 2017; Hällgren & Wilson, 2007; Hällgren & Maaninen-Olsson, 2005). Usually, effective planning yields positive results, but the opposite can also occur if it is used or applied improperly in the field or if various obstacles ultimately disrupt it (Mizuno et al., 2000). In the case of the energy transition, certain studies indicate that the compressed timeframe for weaning off lignite use in electricity generation and the energy crisis that erupted with the war in Ukraine are creating pressure and obstacles to the faithful implementation of the program schedule (Makris et al., 2024; Panagiotopoulos & Roukanas, 2024; Kuzemko et al., 2022). International literature shows that the energy transition requires addressing its impacts through investment and the creation of quality jobs in order to revitalize local economies (Cappa et al., 2024; Hanson, 2023; Filipović et al., 2022; Leppänen & Liefferink, 2022). National and European resources can be utilised for this purpose, mainly in the field of renewable energy sources and sustainable business ecosystems generally (Adun et al., 2022; Moesker & Pesch, 2022). Additionally, international literature suggests that a successful energy transition requires a long-term strategy focused on sustainable development, financial instruments, and local revenue (Qadir et al., 2021; Manta et al., 2020; Cheung et al., 2019; Carley et al., 2018).

The present research has shown that creating a sustainable business ecosystem in regions affected by the energy transition requires incentives for green investments. These incentives can take the form of grants, loans with favourable terms, insurance contribution subsidies, fee exemptions, and tax incentive packages. In addition to attracting new investment projects, it has revealed that existing businesses also need support through incentives to transform and modernize in order to reduce their environmental footprint. These incentives may take the form of grants, wage subsidies, loans on favourable terms, and subsidies for their loan obligations. The research also showed the need to support individuals in dealing with the effects of the energy transition. This concrete support can take the form of training programs to upgrade knowledge and skills and enable them to seamlessly integrate into the local green economy. On top of that, it can be carried out through the form of income tax relief and mortgage subsidies. These findings are consistent with the findings of otherstudies in the international literature. The findings of these studies show that the sustainable development of lignite regions requires attracting investment and supporting existing businesses (Sophocleous & Prodromou, 2025; Liu et al., 2023; Li et al., 2023; Haldar, 2022; Sheng, 2020). In the light of this framework, support policies through financial tools and specific implementation timetables are highly required (Haldar, 2022; Sheng, 2020). Robust incentives are needed to attract investment, reform existing businesses, and develop a sustainable development plan (Filipović et al., 2022; Leppänen & Liefferink, 2022; Hrnčić et al., 2021). According to the international literature, the transition to a green economy requires human resources with upgraded knowledge and skills to meet the needs of local green economies (Prodromou et al., 2025; Papademetriou et al., 2025; Ziouzios et al., 2021; Marinakis et al., 2020; Alexander & Floyd, 2020; Pollin & Callaci, 2019).

Furthermore, the research highlighted the need to upgrade the knowledge and skills of human resources involved in lignite-based power generation. However, it is also necessary to upgrade the knowledge and skills of human resources coming from lignite-based power generation. This retraining and upskillng of the workforceshould be targeted at the needs of the local green economy and implemented by qualified training provider s. There should also be an evaluation of the programs, and participants should receive certification upon completion of the training. Local businesses need to be given incentives to hire retrained personnel. Comparing these findings with the findings of other recently published studies in international literature, it is observed a certain consistency. Studies conducted across the globe have revealed that the creation of quality jobs requires incentives to attract green investments and the restructuring of existing businesses (Tserkezos et al., 2025; Hanna et al., 2024; Wang et al., 2024b; Ram et al., 2022; Zervas et al., 2021a). The need to create quality jobs in areas affected by the energy transition has been highlighted by many researchers (Cappa et al., 2024; Curtis et al., 2024; Ioannidis et al., 2023; Cha, 2017). According to the international literature, this creates a need to upgrade the knowledge and skills of workers so that they can be absorbed into the local green economy (Pollin & Callaci, 2019; Vona 2019; Olson-Hazboun, 2018). Relevant research findings haveindicated that the energy transition will result in the loss of many jobs, which will need to be replaced within the local green economy (Bali Swain et al., 2022; Gambhir et al., 2018). However, it should be noted that a specific policy framework is required to govern and reinforce the entire learning process envisaged in the training programs (Zachmann et al., 2018). Upskilling programs must be implemented by certified bodies, being assessed, and issue certification to participants upon successful completion of the program (Ambu-Saidi et al., 2024; Nasirov et al., 2021; Kitivo et al., 2021).

The research pointed out the need of accelerating the implementation of infrastructure projects as well as the supporting projects of landmark investments that are planned to be carried out based on the plan for the just transition of lignite regions. Delays in the implementation of these projects hinder the sustainable development of regions affected by the energy transition. It is important that renewable energy projects have made significant progress in or near lignite regions because they are changing the country's energy mix and foster green jobs in the energy sector. However, research has indicated that investments in renewable energy sources must continue, making use of existing electricity transmission networks. Projects to improve energy efficiency and electrification should be intensified respectively, utilising resources from the Just Transition Mechanism, the Green Fund andnational and international financial institutions. Innovative and competitive projects will give momentum to sustainable development in regions affected by the energy transition. For projects in the primary sector, the research showed that they're really hard to be implemented, considering that the Greek primary sector is experiencing a major crisis and it doesn't easily have the capability to establish smart livestock farming units, hydroponic farms, and the certain protected designation of origin production as envisaged in the plans. However, these difficulties can be overcome if resources from the Rural Development Program, the Just Transition Mechanism, the Recovery Fund, and private and loan capital are leveraged. According to international literature, the reconstruction and reanimation of areas affected by the energy transition requires infrastructure projects, landscape restoration, the creation of an attractive business environment, and innovative interventions (Ragnitz et al., 2022; David & Schulte-Römer, 2021; Ziouzios et al., 2021; Spanidis et al., 2020; Deshaies, 2020). The creation of infrastructurebenefits existing businesses and attracts new investments (Zervas et al., 2021a; Pavloudakis et al., 2020). Based again on the international literature, rural areas involved in lignite mining have the potential to be reanimated by promoting smart agriculture, agricultural enterprises, and the use of renewable energy resources (Majeed et al., 2023; Chilvers et al., 2021; Bazilian et al., 2021; Marinakis et al., 2020). Research by Marinakis et al. (2020) on the lignite region of Megalopolis indicated that a diversified economy based on clean energy, smart agricultural production, entrepreneurship as well as research and innovation, and human resources leads to sustainable development.

CONCLUSIONS

The research revealed that the impact of the energy transition on lignite-producing regions is significant, and the revitalization of local economies in these areas is a complex, multifaceted process that is causing concern among residents. Building a sustainable development model that will heal the wounds of the past, will not create environmental problems, and will operate without exclusion is a necessity for lignite regions. In addition to the existence of a comprehensive plan for the fair development transition of lignite regions, which is the result of consultation and consensus, the effective implementation of this sustainable model without delays or deviations is also required. In this regard, certain factors must be at the core of the sustainable development of these regions. a) The implementation of the support plan for lignite regions must be oriented and harmonized with the environmental objectives that the restart of the local green economy should serve. b) Incentives to attract new investment and incentives to transform and modernize existing businesses must be supported at all stages of implementation by a flexible and effective institutional framework without delays in controls and the disbursement of funding to c) There can be no fair energy transition without addressing the consequences of the energy transition on human resources. The training and reintegration of human resources into the local green economy will maximize sustainable production potential. d) Uninterrupted allocation of domestic and European funding to infrastructure projects, flagship investments, landscape restoration, energy efficiency and electromobility, as well as renewable energy projects and projects in the primary sector. However, the current research highlighted the concerns and worries of employees and entrepreneurs in these regions about the very slow pace of implementation of the planned projects and initiatives for their region. The timetables for delignitisation must be realistic, and the compressed timeframe for complete independence from lignite for electricity generation by 2028 is causing concern about the impact this will have. Added to these concerns are those of the energy crisis with the war in Ukraine, which is putting pressure on the transition timeframe.

RESEARCH LIMITATIONS AND FUTURE RESEARCH PATHS

Like any research generally, the concrete study is subject to certain limitations. Future research could include government officials, local authorities, cultural bodies, civil society, and individual members of local communities from the areas affected by the energy transition in order to capture additional views and experiences.

REFERENCES

Abram, S., Atkins, E., Dietzel, A., Jenkins, K., Kiamba, L., Kirshner, J., Kreienkamp, J., Parkhill, K., Pegram, T., & Santos Ayllón, L. M. (2022). Just Transition: A whole-systems approach to decarbonisation. *Climate Policy*, *22*(8), 1033–1049. DOI: 10.1080/14693062.2022.2108365

Adun, H., Ishaku, H. P., Jazayeri, M., Dagbasi, M., Olusola, B., Okoye, T., & Dike, G. C. (2022). Decarbonization of EU energy sector: Techno-feasibility analysis of 100% renewables by 2050 in Cyprus. *Clean Technologies and Environmental Policy*, *24*(9), 2801–2824. DOI: 10.1007/s10098-022-02356-4

Alexander, S., & Floyd, J. (2020). The political economy of deep decarbonization: Tradable Energy Quotas for energy descent futures. *Energies*, *13*(17), 4304. DOI: 10.3390/en13174304

Ambu-Saidi, B., Fung, C. Y., Turner, K., & Lim, A. S. S. (2024). A critical review on training evaluation models: A search for future agenda. *Journal of Cognitive Sciences and Human Development.*, *10*(1). Advance online publication. DOI: 10.33736/jcshd.6336.2024

Apostolopoulos, N., Kakouris, A., Liargovas, P., Borisov, P., Radev, T., Apostolopoulos, S., Daskou, S., & Anastasopoulou, E. E. (2023). Just transition policies, power plant workers and green entrepreneurs in Greece, Cyprus and Bulgaria: Can education and retraining meet the challenge? *Sustainability (Basel)*, *15*(23), 16307. DOI: 10.3390/su152316307

Apostolopoulos, N., Liargovas, P., & Apostolopoulos, S. (2025a). European Transnational Cooperation Programmes Involving the Academic Community and Their Contributions to Sustainable Entrepreneurship. In *Real-World Tools and Scenarios for Entrepreneurship Exploration* (pp. 1–24). IGI Global., DOI: 10.4018/979-8-3693-3100-2.ch001

Apostolopoulos, N., Makris, I., Apostolopoulos, S., & Dimitrakopoulos, P. (2023). Resilience of rural micro-businesses in an adverse entrepreneurial environment: Adapting to the energy crisis. *Journal of Enterprising Communities: People and Places in the Global Economy*, *18*(5), 1023–1044. DOI: 10.1108/JEC-08-2023-0144

Apostolopoulos, N., Makris, I., Apostolopoulos, S., & Liargovas, P. (2025b). The Role of Rural Entrepreneurship in the Redevelopment of Rural Areas Previously Involved in Resource Extraction. In *Rural Entrepreneurship: Harvesting Ideas and Sowing New Seeds* (pp. 145-156). Emerald Publishing Limited. https://doi.org/DOI: 10.1108/S2040-72462025000020011

Aung, K. T., Razak, R. A., & Nazry, N. N. M. (2021). Establishing validity and reliability of semi-structured interview questionnaire in developing risk communication module: A pilot study. *Edunesia: Jurnal Ilmiah Pendidikan*, *2*(3), 600–606. DOI: 10.51276/edu.v2i3.177

Balcilar, M., Özkan, O., Usman, O., Saint Akadırı, S., & Zambrano-Monserrate, M. A. (2025). A global shift: How modern technologies are powering the energy transition in the face of climate change. *Journal of Environmental Management*, *384*, 125610. DOI: 10.1016/j.jenvman.2025.125610 PMID: 40315652

Bali Swain, R., Karimu, A., & Gråd, E. (2022). Sustainable development, renewable energy transformation and employment impact in the EU. *International Journal of Sustainable Development and World Ecology*, *29*(8), 695–708. DOI: 10.1080/13504509.2022.2078902

Bashir, M. F., Pata, U. K., & Shahzad, L. (2025). Linking climate change, energy transition and renewable energy investments to combat energy security risks: Evidence from top energy consuming economies. *Energy*, *314*, 134175. DOI: 10.1016/j.energy.2024.134175

Bazilian, M. D., Carley, S., Konisky, D., Zerriffi, H., Pai, S., & Handler, B. (2021). Expanding the scope of just transitions: Towards localized solutions and community-level dynamics. *Energy Research & Social Science*, *80*, 102245. DOI: 10.1016/j.erss.2021.102245

Berglund, H. (2007). Researching entrepreneurship as lived experience. In *Handbook of qualitative research methods in entrepreneurship*. Edward Elgar Publishing., DOI: 10.4337/9781847204387.00011

Brodny, J., Tutak, M., & Grebski, W. (2024). Empirical Assessment of the Efficiency of Poland's Energy Transition Process in the Context of Implementing the European Union's Energy Policy. *Energies*, *17*(11), 2689. DOI: 10.3390/en17112689

Cachia, M., & Millward, L. (2011). The telephone medium and semi-structured interviews: A complementary fit. *Qualitative Research in Organizations and Management*, *6*(3), 265–277. DOI: 10.1108/17465641111188420

Cappa, E., Lamperti, F., & Pallante, G. (2024). *Creating jobs out of the green: The employment effects of the energy transition* (No. 2024/21). LEM Working Paper Series.

Carley, S., Evans, T. P., & Konisky, D. M. (2018). Adaptation, culture, and the energy transition in American coal country. *Energy Research & Social Science*, *37*, 133–139. DOI: 10.1016/j.erss.2017.10.007

Cha, J. M. (2017). A just transition: Why transitioning workers into a new clean energy economy should be at the center of climate change policies. *Fordham Environmental Law Review*, *29*(2), 196–220.

Cheung, G., Davies, P. J., & Bassen, A. (2019). In the transition of energy systems: What lessons can be learnt from the German achievement? *Energy Policy*, *132*, 633–646. DOI: 10.1016/j.enpol.2019.05.056

Chilvers, J., Bellamy, R., Pallett, H., & Hargreaves, T. (2021). A systemic approach to mapping participation with low-carbon energy transitions. *Nature Energy*, *6*(3), 250–259. DOI: 10.1038/s41560-020-00762-w

Creswell, J. W., & Creswell, J. D. (2017). *Research design: Qualitative, quantitative, and mixed methods approaches*. Sage publications.

Curtis, E. M., O'Kane, L., & Park, R. J. (2024). Workers and the green-energy transition: Evidence from 300 million job transitions. *Environmental and Energy Policy and the Economy*, *5*(1), 127–161. DOI: 10.1086/727880

David, M., & Schulte-Römer, N. (2021). Phasing out and in: System transition through disassociation in the German energy transition–The case of light and coal. *Energy Research & Social Science*, *80*, 102204. DOI: 10.1016/j.erss.2021.102204

Deal, G. (2020). The Just Transition Mechanism: Making Sure No One Is Left Behind.

Denzin, N. K., & Lincoln, Y. S. (Eds.). (2011). *The Sage handbook of qualitative research*. Sage Publication.

DEPEEP, (2024). Energy mix 2023, file:///C:/Users/ A /Desktop/ ENERGY - MIX -2023.pdf

Deshaies, M. (2020). Metamorphosis of Mining Landscapes in the Lower Lusatian Lignite Basin (Germany): New uses and new image of a mining region. *Les Cahiers de la recherche architecturale urbaine et paysagère*, (7). https://doi.org/ DOI: 10.4000/craup.4018

Dimitrakopoulos, P., Makris, I., & Apostolopoulos, S. (2024). The Impact of Major Energy Indexes on the Economic Growth of EU Countries, According to the Contribution of Industry to GDP. In *International Conference on Applied Economics* (pp. 59-74). Cham: Springer Nature Switzerland. https://doi.org/DOI: 10.1007/978-3-031-76658-9_4

Egli, F., Polzin, F., Sanders, M., Schmidt, T., Serebriakova, A., & Steffen, B. (2022). Financing the energy transition: Four insights and avenues for future research. *Environmental Research Letters*, *17*(5), 051003. DOI: 10.1088/1748-9326/ac6ada

Ember Energy. (2025). Greece In Figures-data, https://github.com/GreeceInFigures/greeceinfigures-data/tree/main/energy

European Commission. (2020). A strong social Europe for just transitions. COM (2020) 14 Final, https://ec.europa.eu/commission/presscorner/detail/en/qanda_20_20

Filipović, S., Lior, N., & Radovanović, M. (2022). The green deal–just transition and sustainable development goals Nexus. *Renewable & Sustainable Energy Reviews*, *168*, 112759. DOI: 10.1016/j.rser.2022.112759

Fossey, E., Harvey, C., McDermott, F., & Davidson, L. (2002). Understanding and evaluating qualitative research. *The Australian and New Zealand Journal of Psychiatry*, *36*(6), 717–732. DOI: 10.1046/j.1440-1614.2002.01100.x PMID: 12406114

Fotis, G., Maris, T. I., & Mladenov, V. (2025). Risks, Obstacles and Challenges of the Electrical Energy Transition in Europe: Greece as a Case Study. *Sustainability (Basel)*, *17*(12), 5325. DOI: 10.3390/su17125325

Gambhir, A., Green, F., & Pearson, P. J. (2018). Towards a just and equitable low-carbon energy transition. *Grantham Institute Briefing Paper*, *26*. Advance online publication. DOI: 10.1080/14693062.2019.1657379

Gartner, W. B., & Birley, S. (2002). Introduction to the special issue on qualitative methods in entrepreneurship research. *Journal of Business Venturing*, *17*(5), 387–395. DOI: 10.1016/S0883-9026(01)00077-5

Gema, S. B. (2023). Financing the energy transition: The new paradigm for renewable energy investors. In *The Palgrave Handbook of Zero Carbon Energy Systems and Energy Transitions* (pp. 1–44). Springer International Publishing., DOI: 10.1007/978-3-030-74380-2_4-1

Goh, K. C., Kurniawan, T. A., Zainordin, N., Arifah, I. D. C., Abas, M. A., Masrom, M. A. N., Mohamed, S., Omar, R., Khoo, S. L., Gui, H. C., Toh, T. C., & Onn, C. W. (2024). Expediting decarbonization in energy, waste, and water sector through digitalization in sustainable smart cities (SSC): Case-studies in Malaysia and China based on Industry 5.0 paradigm. *Sustainable Cities and Society*, *117*, 105969. DOI: 10.1016/j.scs.2024.105969

Goodrick, D., & Rogers, P. J. (2015). Qualitative data analysis. *Handbook of practical program evaluation*, 561-595. https://doi.org/DOI: 10.1002/9781119171386.ch22

Government Committee SDAM. (2020). Just Transition Development Plan of Lignite Areas, (Greek), https://www.sdam.gr/sites/default/files/consultation/Master_Plan_Public_Consultation_ENG.pdf

Haldar, S. (2022). A landscape level analysis of entrepreneurship and sustainable energy transitions: Evidences from Gujarat, India. *Sustainable Development (Bradford)*, *30*(4), 489–502. DOI: 10.1002/sd.2244

Hällgren, M., & Maaninen-Olsson, E. (2005). Deviations, ambiguity and uncertainty in a project-intensive organization. *Project Management Journal*, *36*(3), 17–26. DOI: 10.1177/875697280503600303

Hällgren, M., & Wilson, T. L. (2007). Mini-muddling: Learning from project plan deviations. *Journal of Workplace Learning*, *19*(2), 92–107. DOI: 10.1108/13665620710728475

Hanna, R., Heptonstall, P., & Gross, R. (2024). Job creation in a low carbon transition to renewables and energy efficiency: A review of international evidence. *Sustainability Science*, *19*(1), 125–150. DOI: 10.1007/s11625-023-01440-y

Hanson, G. H. (2023). *Local labor market impacts of the energy transition: Prospects and policies* (No. w30871). National Bureau of Economic Research. https://doi.org/DOI: 10.3386/w30871

Hassink, J., Hulsink, W., & Grin, J. (2016). Entrepreneurship in agriculture and healthcare: Different entry strategies of care farmers. *Journal of Rural Studies*, *43*, 27–39. DOI: 10.1016/j.jrurstud.2015.11.013

Hollway, W., & Jefferson, T. (2012). Doing qualitative research differently: A psychosocial approach. https://doi.org/DOI: 10.4135/9781526402233

Hrnčić, B., Pfeifer, A., Jurić, F., Duić, N., Ivanović, V., & Vušanović, I. (2021). Different investment dynamics in energy transition towards a 100% renewable energy system. *Energy*, *237*, 121526. DOI: 10.1016/j.energy.2021.121526

Inderwildi, O., Zhang, C., Wang, X., & Kraft, M. (2020). The impact of intelligent cyber-physical systems on the decarbonization of energy. *Energy & Environmental Science*, *13*(3), 744–771. DOI: 10.1039/C9EE01919G

Ioannidis, F., Kosmidou, K., & Papanastasiou, D. (2023). Public awareness of renewable energy sources and Circular Economy in Greece. *Renewable Energy*, *206*, 1086–1096. DOI: 10.1016/j.renene.2023.02.084

Janeková, J., Fabianová, J., Onofrejová, D., Puskas, E., & Busa, M. (2017). Implementation of deviation analysis method in the utilisation phase of the investment project: A case study. *Polish. Journal of Management Studies*, *15*(1), 99–109. Advance online publication. DOI: 10.17512/pjms.2017.15.1.10

Janghorban, R., Roudsari, R. L., & Taghipour, A. (2014). Skype interviewing: The new generation of online synchronous interview in qualitative research. *International Journal of Qualitative Studies on Health and Well-being*, *9*(1), 24152. DOI: 10.3402/qhw.v9.24152 PMID: 24746247

Javadian, G., Dobratz, C., Gupta, A., Gupta, V. K., & Martin, J. A. (2020). Qualitative research in entrepreneurship studies: A state-of-science. *The Journal of Entrepreneurship*, *29*(2), 223–258. DOI: 10.1177/0971355720930564

Kaldellis, J. K., Boulogiorgou, D., Kondili, E. M., & Triantafyllou, A. G. (2023). Green Transition and Electricity Sector Decarbonization: The Case of West Macedonia. *Energies*, *16*(16), 5970. DOI: 10.3390/en16165970

Karasmanaki, E., Ioannou, K., Katsaounis, K., & Tsantopoulos, G. (2020). The attitude of the local community towards investments in lignite before transitioning to the post-lignite era: The case of Western Macedonia, Greece. *Resources Policy*, *68*, 101781. DOI: 10.1016/j.resourpol.2020.101781

Kilintzis, P., Samara, E., Topaloglou, L., Avlogiaris, G., & Kafetzopoulos, D. (2023). The role of Green Public Procurements in energy transition: The case of Western Macedonia. *Journal of Innovation and Entrepreneurship*, *12*(1), 87. DOI: 10.1186/s13731-023-00354-4

Kitivo, E. M., Kanga, A., & Mbae, J. (2021). Evaluation of training programmes: A review of selected models and approaches. *International Journal of Educational Theory and Practice*, *4*(1-4), 1–16.

Kolovos, N., Georgakopoulos, A., Filippidis, A., & Kavouridis, C. (2002). Environmental effects of lignite and intermediate steriles coexcavation in the southern lignite field mine of Ptolemais, northern Greece. *Energy Sources*, *24*(6), 561–573. DOI: 10.1080/00908310290086563

Kourkouridis, D., Salepaki, A., Kasselouris, G., & Baltas, E. (2024). Regional and Local Development in Energy Transition Areas: The Case of Lignite Phase-Out in the Peloponnese. *Sustainability (Basel)*, *16*(14), 6153. DOI: 10.3390/su16146153

Kral, A. H., Malekinejad, M., Vaudrey, J., Martinez, A. N., Lorvick, J., McFarland, W., & Raymond, H. F. (2010). Comparing respondent-driven sampling and targeted sampling methods of recruiting injection drug users in San Francisco. *Journal of Urban Health*, *87*(5), 839–850. DOI: 10.1007/s11524-010-9486-9 PMID: 20582573

Krippendorff, K. (2018). *Content analysis: An introduction to its methodology*. Sage publications., DOI: 10.4135/9781071878781

Kuzemko, C., Blondeel, M., Dupont, C., & Brisbois, M. C. (2022). Russia's war on Ukraine, European energy policy responses & implications for sustainable transformations. *Energy Research & Social Science*, *93*, 102842. DOI: 10.1016/j.erss.2022.102842

Leppänen, T., & Liefferink, D. (2022). Agenda-setting, policy formulation, and the EU institutional context: The case of the Just Transition Fund. *European Policy Analysis*, *8*(1), 51–67. DOI: 10.1002/epa2.1136

Li, B. (2023). The role of financial markets in the energy transition: An analysis of investment trends and opportunities in renewable energy and clean technology. *Environmental Science and Pollution Research International*, *30*(43), 97948–97964. DOI: 10.1007/s11356-023-29014-6 PMID: 37599346

Liu, W., Shen, Y., & Razzaq, A. (2023). How renewable energy investment, environmental regulations, and financial development derive renewable energy transition: Evidence from G7 countries. *Renewable Energy*, *206*, 1188–1197. DOI: 10.1016/j.renene.2023.02.017

Lo Iacono, V., Symonds, P., & Brown, D. H. (2016). Skype as a tool for qualitative research interviews. *Sociological Research Online*, *21*(2), 103–117. DOI: 10.5153/sro.3952

Loewen, B. (2022). Coal, green growth and crises: Exploring three European Union policy responses to regional energy transitions. *Energy Research & Social Science*, *93*, 102849. DOI: 10.1016/j.erss.2022.102849

Lypiridi, D. (2021). Just Transition Mechanism and Lignite Phase-Out in Greece: Challenges and Proposals for Policy Adoption. https://doi.org/DOI: 10.12681/hapscpbs.29493

Majeed, Y., Khan, M. U., Waseem, M., Zahid, U., Mahmood, F., Majeed, F., Sultan, M., & Raza, A. (2023). Renewable energy as an alternative source for energy management in agriculture. *Energy Reports*, *10*, 344–359. DOI: 10.1016/j.egyr.2023.06.032

Makris, I., Apostolopoulos, S., & Anastasopoulou, E. E. (2024). An Entrepreneurial Perspective on the Transition of Lignite Rural Areas to a New Regime within a Suffocating Timeframe. *Administrative Sciences*, *14*(4), 64. DOI: 10.3390/admsci14040064

Makris, I., Apostolopoulos, S., Giannopoulos, V., Dimitrakopoulos, P., & Charalampakis, P. (2025). The Impact of Formal and Informal Institutional Elements on Land Mobility Within Rural Greece. *Sustainability (Basel)*, *17*(10), 4412. DOI: 10.3390/su17104412

Maneesriwongul, W., & Dixon, J. K. (2004). Instrument translation process: A methods review. *Journal of Advanced Nursing*, *48*(2), 175–186. DOI: 10.1111/j.1365-2648.2004.03185.x PMID: 15369498

Manta, O., Gouliamos, K., Kong, J., Li, Z., Ha, N. M., Mohanty, R. P., Yang, H., Pu, R., & Yue, X. G. (2020). The architecture of financial networks and models of financial instruments according to the "Just transition mechanism" at the European level. *Journal of Risk and Financial Management*, *13*(10), 235. DOI: 10.3390/jrfm13100235

Marinakis, V., Flamos, A., Stamtsis, G., Georgizas, I., Maniatis, Y., & Doukas, H. (2020). The Efforts towards and Challenges of Greece's Post-Lignite Era: The Case of Megalopolis. *Sustainability (Basel)*, *12*(24), 10575. DOI: 10.3390/su122410575

Ministry of Environment and Energy. (2020). *Just Transition Development Plan of Lignite Areas, Just Development Transition Plan*; Brussels: European Commission. Availableonline: https://www.sdam.gr/sites/default/files/consultation/Master_Plan_Public_Consultation_ENG.pdf, (In Greek)

Ministry of Environment and Energy. (2021). Territorial Just Transition Territorial Plan for Megalopolis. Available online: https://www.sdam.gr/node/431, (In Greek)

Mizuno, O., Kikuno, T., Inagaki, K., Takagi, Y., & Sakamoto, K. (2000). Statistical analysis of deviation of actual cost from estimated cost using actual project data. *Information and Software Technology*, *42*(7), 465–473. DOI: 10.1016/S0950-5849(00)00092-6

Moesker, K., & Pesch, U. (2022). The just transition fund–Did the European Union learn from Europe's past transition experiences? *Energy Research & Social Science*, *91*, 102750. DOI: 10.1016/j.erss.2022.102750

Morse, J. M., & Field, P. A. (2013). *Nursing research: The application of qualitative approaches*. Springer.

Nanaki, E. A., Koroneos, C. J., & Xydis, G. A. (2016). Environmental impact assessment of electricity production from lignite. *Environmental Progress & Sustainable Energy*, *35*(6), 1868–1875. DOI: 10.1002/ep.12427

Nasirov, S., Girard, A., Peña, C., Salazar, F., & Simon, F. (2021). Expansion of renewable energy in Chile: Analysis of the effects on employment. *Energy*, *226*, 120410. DOI: 10.1016/j.energy.2021.120410

Neergaard, H., & Ulhøi, J. P. (Eds.). (2007). *Handbook of qualitative research methods in entrepreneurship*. Edward Elgar Publishing., DOI: 10.4337/9781847204387

Neuendorf, K. A. (2017). *The content analysis guidebook.* sage. https://doi.org/ DOI: 10.4135/9781071873045

Oberthür, S., Khandekar, G., & Wyns, T. (2021). Global governance for the decarbonization of energy-intensive industries: Great potential underexploited. *Earth System Governance*, *8*, 100072. DOI: 10.1016/j.esg.2020.100072

Olson-Hazboun, S. K. (2018). "Why are we being punished and they are being rewarded?" views on renewable energy in fossil fuels-based communities of the US west. *The Extractive Industries and Society*, *5*(3), 366–374. DOI: 10.1016/j.exis.2018.05.001

Panagiotopoulos, P. P., & Roukanas, S. A. (2024). Navigating the Greek Energy Crisis through a Multidimensional Approach: A Review Article. *Energies*, *17*(16), 3915. DOI: 10.3390/en17163915

Papademetriou, C., Ragazou, K., Belias, D., Anastasiadou, S., Garefalakis, A., & Prodromou, A. (2025). The Importance of Green Human Resources: Green Ambidexterity and Environmental Performance. In K. Ragazou, A. Garefalakis, C. Papademetriou, & A. Samara (Eds.), *Sustainability Through Green HRM and Performance Integration* (pp. 83-100). IGI Global Scientific Publishing. https://doi.org/DOI: 10.4018/979-8-3693-5981-5.ch004

Papadis, E., & Tsatsaronis, G. (2020). Challenges in the decarbonization of the energy sector. *Energy*, *205*, 118025. DOI: 10.1016/j.energy.2020.118025

Pathak, V., Jena, B., & Kalra, S. (2013). Qualitative research. *Perspectives in Clinical Research*, *4*(3), 192. DOI: 10.4103/2229-3485.115389 PMID: 24010063

Patton, M. Q. (2014). *Qualitative research & evaluation methods: Integrating theory and practice*. Sage publications.

Pavloudakis, F., Karlopoulos, E., & Roumpos, C. (2023). Just transition governance to avoid socio-economic impacts of lignite phase-out: The case of Western Macedonia, Greece. *The Extractive Industries and Society*, *14*, 101248. DOI: 10.1016/j.exis.2023.101248

Pavloudakis, F., Karlopoulos, E., Roumpos, C., & Koukouzas, N. (2021). Energy transition and the future of lignite mining in the region of Western Macedonia, Greece. In *Zlatibor: 10th International conference coal 2021* (pp. 13-16).

Pavloudakis, F., Roumpos, C., Karlopoulos, E., & Koukouzas, N. (2020). Sustainable rehabilitation of surface coal mining areas: The case of Greek lignite mines. *Energies*, *13*(15), 3995. DOI: 10.3390/en13153995

Pegkas, P. (2024). Energy consumption and human capital: Does human capital stimulate renewable energy? The case of Greece. *Journal of the Knowledge Economy, 15*(4), 1–28. DOI: 10.1007/s13132-024-01770-x

Peterson, J. A., Reisinger, H. S., Schwartz, R. P., Mitchell, S. G., Kelly, S. M., Brown, B. S., & Agar, M. H. (2008). Targeted sampling in drug abuse research: A review and case study. *Field Methods, 20*(2), 155–170. DOI: 10.1177/1525822X08314988

Pollin, R., & Callaci, B. (2019). The economics of just transition: A framework for supporting fossil fuel–dependent workers and communities in the United States. *Labor Studies Journal, 44*(2), 93–138. DOI: 10.1177/0160449X18787051

Popescu, C., Panait, M., Palazzo, M., & Siano, A. (2022). Energy transition in European Union—Challenges and opportunities. *Energy Transition: Economic, Social and Environmental Dimensions*, 289-312. https://doi.org/DOI: 10.1007/978-981-19-3540-4_11

Prodromou, A., Papademetriou, C., & Savvides, D. (2025). The Strategic Role of Green HRM in Hospitality Industry for Reaching Sustainable Development Goals. In K. Ragazou, A. Garefalakis, C. Papademetriou, & A. Samara (Eds.), *Sustainability Through Green HRM and Performance Integration* (pp. 1-38). IGI Global Scientific Publishing. https://doi.org/DOI: 10.4018/979-8-3693-5981-5.ch001

Qadir, S. A., Al-Motairi, H., Tahir, F., & Al-Fagih, L. (2021). Incentives and strategies for financing the renewable energy transition: A review. *Energy Reports, 7*, 3590–3606. DOI: 10.1016/j.egyr.2021.06.041

Qu, S. Q., & Dumay, J. (2011). The qualitative research interview. *Qualitative Research in Accounting & Management, 8*(3), 238–264. DOI: 10.1108/11766091111162070

Ragnitz, J., Markwardt, G., Schwartzkopff, J., Reitzenstein, A., Wehnert, T., Kurwan, J., & Beutel, J. (2022). Analysis of the historical structural change in the German lignite mining area of Lusatia (case study).

Ram, M., Osorio-Aravena, J. C., Aghahosseini, A., Bogdanov, D., & Breyer, C. (2022). Job creation during a climate compliant global energy transition across the power, heat, transport, and desalination sectors by 2050. *Energy, 238*, 121690. DOI: 10.1016/j.energy.2021.121690

Rechsteiner, R. (2021). German energy transition (Energiewende) and what politicians can learn for environmental and climate policy. *Clean Technologies and Environmental Policy, 23*(2), 305–342. DOI: 10.1007/s10098-020-01939-3 PMID: 33041746

Rial, R. C. (2024). Biofuels versus climate change: Exploring potentials and challenges in the energy transition. *Renewable & Sustainable Energy Reviews, 196*, 114369. DOI: 10.1016/j.rser.2024.114369

Ritchie, J., Ormston, R., McNaughton Nicholls, C., & Lewis, J. (2013). Qualitative research practice: A guide for social science students and researchers.

Sarabdeen, M., Elhaj, M., & Alofaysan, H. (2024). Exploring the influence of digital transformation on clean energy transition, climate change, and economic growth among selected oil-export countries through the panel ARDL approach. *Energies, 17*(2), 298. DOI: 10.3390/en17020298

Schmidt, D., Reyment, J., Webster, E., Kirby, S., & Lyle, D. (2020). Workplace-based health research training: A qualitative study of perceived needs in a rural setting. *Health Research Policy and Systems, 18*(1), 67. DOI: 10.1186/s12961-020-00580-2 PMID: 32539865

Sheng, C. (2020). Not just the state: The role of entrepreneurs in China's energy transition. *Energy Research & Social Science, 70*, 101814. DOI: 10.1016/j.erss.2020.101814

Smith, J. A., Langenhove, L. V., & Harre, R. (1995). Rethinking methods in psychology. https://doi.org/DOI: 10.4135/9781446221792

Spanidis, P. M., Roumpos, C., & Pavloudakis, F. (2020). A multi-criteria approach for the evaluation of low risk restoration projects in continuous surface lignite mines. *Energies, 13*(9), 2179. DOI: 10.3390/en13092179

Strambo, C. (2020). *Just transition and the geopolitics of decarbonization in the EU*. Mistra Geopolitics.

Streimikiene, D., Kyriakopoulos, G. L., Lekavicius, V., & Siksnelyte-Butkiene, I. (2021). Energy poverty and low carbon just energy transition: Comparative study in Lithuania and Greece. *Social Indicators Research, 158*(1), 319–371. DOI: 10.1007/s11205-021-02685-9 PMID: 33941997

Sullivan, J. R. (2012). Skype: An appropriate method of data collection for qualitative interviews? *The Hilltop Review, 6*(1), 10.

Temple, B., & Young, A. (2004). Qualitative research and translation dilemmas. *Qualitative Research, 4*(2), 161–178. DOI: 10.1177/1468794104044430

Theodoropoulou, S., Sabato, S., & Akgüç, M. (2025). National eco-social policies in the framework of EU just transition: The cases of Greece, Italy, Portugal and Spain. *Global Social Policy, 25*(1), 86–111. DOI: 10.1177/14680181241291122

Thomas, D. R. (2006). A general inductive approach for analyzing qualitative evaluation data. *The American Journal of Evaluation*, *27*(2), 237–246. DOI: 10.1177/1098214005283748

Thompson, C. (1999). Qualitative research into nurse decision making: Factors for consideration in theoretical sampling. *Qualitative Health Research*, *9*(6), 815–828. DOI: 10.1177/104973239900900609 PMID: 10662261

Tranoulidis, A., Sotiropoulou, R. E. P., Bithas, K., & Tagaris, E. (2022). Decarbonization and transition to the post-lignite era: Analysis for a sustainable transition in the region of Western Macedonia. *Sustainability (Basel)*, *14*(16), 10173. DOI: 10.3390/su141610173

Tranoulidis, A., Sotiropoulou, R. E. P., Bithas, K., & Tagaris, E. (2024). Lessons from European Union just transition toolkits: A regional investment framework for Greece and Germany. *Energy Research & Social Science*, *114*, 103603. DOI: 10.1016/j.erss.2024.103603

Tserkezos, D., Apostolopoulos, S., & Prodromou, A. (2025). Sustainable Entrepreneurship in Under-Resourced Rural Areas in Relation to The EU Rural Vision: The Greek Case. In Apostolopoulos, S., Paunovic, I., Macaulay, B., & Kakouris, A. (Eds.), *Entrepreneurial Opportunities in Disadvantaged Rural Communities* (pp. 1–22). IGI Global Scientific Publishing., DOI: 10.4018/979-8-3373-0214-0.ch001

Tuli, F. (2010). The basis of distinction between qualitative and quantitative research in social science: Reflection on ontological, epistemological and methodological perspectives. *Ethiopian journal of education and sciences, 6*(1): 97-108 https://doi .org/DOI: 10.4314/ejesc.v6i1.65384

Twinn, S. (1997). An exploratory study examining the influence of translation on the validity and reliability of qualitative data in nursing research. *Journal of Advanced Nursing*, *26*(2), 418–423. DOI: 10.1046/j.1365-2648.1997.1997026418.x PMID: 9292378

Van Nes, F., Abma, T., Jonsson, H., & Deeg, D. (2010). Language differences in qualitative research: Is meaning lost in translation? *European Journal of Ageing*, *7*(4), 313–316. DOI: 10.1007/s10433-010-0168-y PMID: 21212820

Vanhercke, B., & Verdun, A. (2021). The European semester as goldilocks: Macroeconomic policy coordination and the recovery and resilience facility. *Journal of Common Market Studies*, *60*(1), 204–223. DOI: 10.1111/jcms.13267

Vlachou, A., & Pantelias, G. (2022). Energy transitions: The case of Greece with a special focus on the role of the EU ETS. *Science and Society*, *86*(4), 516–545. DOI: 10.1521/siso.2022.86.4.516

Vona, F. (2019). Job losses and political acceptability of climate policies: Why the 'job-killing'argument is so persistent and how to overturn it. *Climate Policy*, *19*(4), 524–532. DOI: 10.1080/14693062.2018.1532871

Wallace, C., Farmer, J., & McCosker, A. (2019). Boundary spanning practices of community connectors for engaging 'hardly reached'people in health services. *Social Science & Medicine*, *232*, 366–373. DOI: 10.1016/j.socscimed.2019.05.034 PMID: 31132544

Wang, Q., Gontikaki, E., Stenzel, P., Louca, V., Küpper, F. C., & Spiller, M. (2024a). How to decarbonize Greece by comparing wind and PV energy: A land eligibility analysis. *Energies*, *17*(3), 567. DOI: 10.3390/en17030567

Wang, X., Ali, D. A., & Zhao, W. (2024b). Urban energy transition and quality of life: A comprehensive statistical study on the nexus of wind clean energy adoption, job creation, and livability. *Sustainable Cities and Society*, *108*, 105436. DOI: 10.1016/j.scs.2024.105436

Watters, J. K., & Biernacki, P. (1989). Targeted sampling: Options for the study of hidden populations. *Social Problems*, *36*(4), 416–430. DOI: 10.2307/800824

Zachmann, G., Fredriksson, G., & Claeys, G. (2018). The distributional effects of climate policies. *Bruegel. Blueprint Series, 28.*

Zervas, E., Vatikiotis, L., & Gareiou, Z. (2021a). Proposals for an environmental and social just transition for the post-lignite era in Western Macedonia, Greece. In *IOP Conference Series: Earth and Environmental Science* (Vol. 899, No. 1, p. 012049). IOP Publishing. https://doi.org/DOI: 10.1088/1755-1315/899/1/012049

Zervas, E., Vatikiotis, L., Gareiou, Z., Manika, S., & Herrero-Martin, R. (2021b). Assessment of the Greek national plan of energy and climate change—Critical remarks. *Sustainability (Basel)*, *13*(23), 13143. DOI: 10.3390/su132313143

Ziouzios, D., Karlopoulos, E., Fragkos, P., & Vrontisi, Z. (2021). Challenges and Opportunities of Coal Phase-Out in Western Macedonia. *Climate (Basel)*, *9*(7), 115. DOI: 10.3390/cli9070115

Chapter 7
Effects of Wind Farms Through the Forest Road Network on Grasslands in Greece

Aimilia Lempesi
https://orcid.org/0009-0002-8644-3867
Democritus University of Thrace, Greece

Christodoulos Daoutis
Democritus University of Thrace, Greece

Panos Dimitrakopoulos
https://orcid.org/0009-0000-8188-4255
University of the Peloponnese, Greece

ABSTRACT

Wind energy is central to the renewable transformation in Greece since this is among the main resources for fulfilling the National Energy and Climate Plan by achieving 70% renewable penetration by 2030. The majority of wind farms are situated within grassland habitats containing high cultural, ecological, and agricultural values often coincide with Natura 2000 protected areas. However, creation and operation of a wind farm could cause fragmentation accompanied by a long-term reduction in plant richness, soil degradation supporting invasive species establishment for a long-term effect on biodiversity, ecosystem function as well as landscape connectivity. This study discusses the impacts of wind energy development on Greek grassland ecosystems, observed through the forest road network that connect with the construction

DOI: 10.4018/979-8-3373-5302-9.ch007

of wind farms. Therefore, brings out a call for the involved infrastructure-related impacts in renewable energy planning and management to achieve climate goals without a loss in ecological integrity.

INTRODUCTION

Wind energy is considered one of the largest contributors to combatting climate change (Barthelmie & Pryor, 2021), and essential to Greece's transition to renewables (Wang et al., 2020). It is crucial to the NECP (National Energy and Climate Plan) which aims to achieve 70% of electricity generation from renewable sources by 2030 (Psarros & Papathanassiou, 2022).Wind farms, representing large-scale renewable energy, remain a crucial part of the shift to low carbon energy systems around the world (Veers et al., 2022). One of the main components of Greece's strategy to achieve the EU climate targets involve the construction of wind farms to reduce carbon emissions and energy diversification (Delagrammatikas & Roukanas, 2023). Greece has grown into a major player in the Mediterranean wind energy market by taking advantage of their wealth of wind resources particularly found in mountainous regions and islands (Soukissian et al., 2017; Chaskos et al., 2012).

Wind farms are often constructed in large areas of grassland, as the unobstructed area of land and steady winds are ideal for energy production (Rediske et al., 2021; Boccard, 2009). Greek grasslands serve an important role in supporting traditional livestock agriculture and have high cultural and ecological value (Psyllos et al., 2022; Hadjigeorgiou, 2011). There are different habitats exist, ranging from dry calcareous grasslands to mesic and semi-dry grasslands, to sub-alpine Nardus grasslands, and floodplain wet grasslands (Filis et al., 2024; Zotos et al., 2021). Wind farms in Greece pose a complicated environmental dynamic, overlap with the Natura 2000 protection areas, and grassland habitats (Xanthakis et al., 2022; Vlami et al., 2020). There is an increasing body of evidence that these dynamics could compromise ecological integrity (Zhao et al., 2025; Moustakas et al., 2023). Productivity and plant richness may decrease in grasslands, especially near proposed infrastructure, including access roads and turbine bases (Li et al., 2024; Diffendorfer et al., 2022).

The development of wind farm sites often requires the construction of entirely new access roads or the improvement of existing forest routes (Li et al., 2024). There are benefits and drawbacks to access roads, particularly regarding sensitivity in development to ecological factors in grassland areas (Auffret & Lindgren, 2020). Access roads are usually constructed for energy purposes, however, they often persist, in the long term, in the landscape (Lugo & Gucinski, 2000). When roads are built, they fragment habitat, degrade site and soil quality, increase the likelihood of invasive species establishment at a site, alter drainage systems, increase human access to the landscape, and expose landscapes to additional land use pressures (Kleinschroth et al., 2016; Mortensen et al., 2009). These factors, in turn, can have long-term and substantial effects on ecosystem function, biodiversity, and connectivity of landscapes (Arjmand et al., 2023; Kuklina et al., 2021). These effects tend to be undervalued in environmental impact assessments, overlooking the extent of their ecological consequences (Alamgir et al., 2017).

This research analyzes the ecological effects of wind energy development on grassland ecosystems in Greece through the forest road networks that accompany the construction of wind farms. Wind energy management needs to incorporate the impacts that relate to infrastructure for environmental impact assessments and land-use planning. It is considered for managers and decision makers to recognize that it is essential to balance renewable energy development priorities with conservation of biodiversity.

GRASSLANDS IN GREECE

Types of Natural Grasslands in Greece

There are three types of Grasslands in Greece (a. Mediterranean Pseudo-steppe, b. Calcareous and Sub-Mediterranean and c. Montane and Alpine). According to the deliverables of the CORINE program of the European Union (Copernicus Land Monitoring Service, 2021) the grassland areas (Figure 1) are larger than 1,000,000 ha.

Mediterranean Pseudo-Steppe Grasslands

The Greek natural grasslands are mostly dry Mediterranean formations found in water-limited climates and on shallow, nutrient-poor soils (Koukoura, 2007). One significant type is pseudo-steppe grassland (Natura 2000 code 6220) which is identified by annual grasses and herbs (Thero-Brachypodietea) that have adapted to hot summers and seasonal drought (Morabito et al., 2024). These grasslands, which develop in areas with high temperatures and noticeable summer dry spells lead the vegetation to completely desiccate. *Stipa capensis*, *Aira* and *Bromus* spp. are examples of drought-tolerant annuals (Janišová et al., 2010). There are plentiful in their flora during the spring and a range of Mediterranean forbs including species of *Tuberaria guttata* and *Centaurea* spp. (Korakis & Gerasimidis, 2010). The open structure of pseudo-steppes which is often semi-natural pasture with occasional dwarf shrubs (*Phlomis fruticosa*, *Sarcopoterium* sp.) has been maintained by extensive grazing on dry, infertile soils. Pseudo-steppe habitats have a high level of biodiversity (many endemic plants) and are given priority for conservation under EU law (Bruelheide et al., 2024).

Calcareous and Sub-Mediterranean Grasslands

The Festuco-Brometalia (Natura 2000 code 6210) are semi-natural dry grasslands found in Greece on calcareous substrates and in the foothills of mountains (Dengler et al., 2013). These incredibly varied communities which are home to many rare and endemic plants are located on the slopes and ridges of limestone or serpentinite plains (Bergmeier et al., 2009). *Festuca*, *Bromus*, and others are the most common perennial grasses and they are usually accompanied by a profusion of forbs such as geophytes and orchids (Janišová et al., 2010). In northern Greece the floral composition changes with elevation up to around 2000 meters above sea level (Janišová et al., 2010). A closely related habitat is the Eastern sub-Mediterranean dry grassland (62A0) which is located in the southern Balkans under less continental conditions, (Morabito et al., 2024). Another subtype of open vegetation that fosters drought-tolerant succulents and annuals in ledges and fissures is the rupicolous grassland (6110) a pioneer habitat on superficial calcareous soils (Janišová et al., 2010). Despite their high species diversity and conservation value calcareous grasslands require traditional low grazing intensity to preserve their ecological integrity (Dimopoulos et al., 2005).

Montane and Alpine Grasslands

Above the treeline, the high-altitude grasslands constitute another significant category. Perennial grasses and low herbs cover the 1,300 – 2,200 m altitude range of alpine and subalpine meadows (6170) on limestone massifs (Janišová et al., 2010). The *Nardus* grasslands (6230) which are species-rich swards generally found on alpine slopes or plateaus above 1,600 meters are dominated by *Nardus stricta* and other acid-tolerant grasses on siliceous substrates (Dalle Fratte & Cerabolini, 2023). High-altitude steppe-like communities with tussock grasses (*Stipa pennata*, *Festuca* spp.) are open in the Oro-Mediterranean grasslands (6280) are found between alpine zones and montane forests and on sunny slopes (Janišová et al., 2010). Although these alpine and montane grasslands usually survive longer than lowland grasslands climate change and the abandonment of traditional grazing are posing a growing threat to them (Sidiropoulou et al., 2022).

Figure 1. Grasslands of Greece

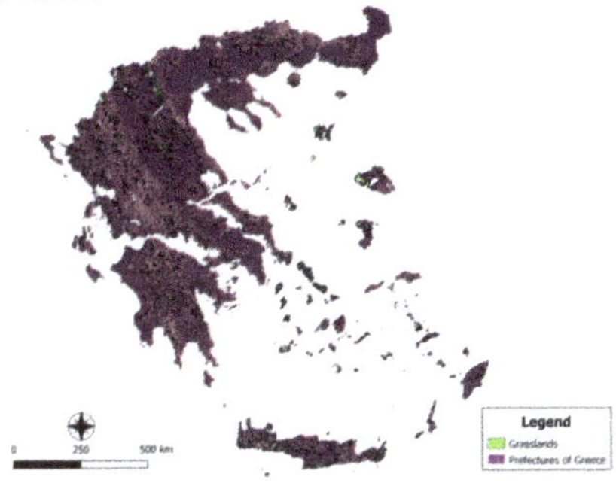

BIODIVERSITY OF NATURAL GRASSLANDS IN GREECE

Greek natural grasslands are distinguished by their significant endemism and high floristic diversity (Kokkoris et al., 2014). Annual grasses and forbs like *Stipa capensis*, *Bromus fasciculatus* and *Aira* species coexist with Mediterranean forbs (e. g. *Centaurea idaea* and *Sideritis syriaca*) in pseudo-steppe grasslands (Natura 2000 code 6220) and commonly occurring on Greek islands like Crete (Janišová et al., 2010). Springtime flora abundance is notable because drought-adapted annuals bloom at the same time (Hess et al., 2022).

Dry calcareous grasslands (Festuco-Brometalia 6210) consist of orchid species such as *Orchis* and *Ophrys* as well as of perennial grasses and geophytes which improves regional biodiversity (Janišová et al., 2010). Numerous of these locations are categorized as orchid-rich habitats. The rich in endemic or range restricted taxa sub-Mediterranean grasslands (62A0) and rupicolous pioneer communities (6110) contribute to floristic heterogeneity (Delipetrou et al., 2015). Alpine meadows (6170) and *Nardus stricta* swards (6230) are examples of high-elevation grasslands that sustain rare alpine endemics and cold-adapted species (Janišová et al., 2010).

Along with the diversity of plants these ecosystems also sustain a range of invertebrate communities especially pollinators that rely on the complex structural patterns and diverse flowering patterns of the grasslands (Minachilis et al., 2023). The survival of pollinator species such as butterflies and wild bees is essential for providing pollination services in neighboring agricultural areas according to recent research (Kougioumoutzis et al., 2022). Another key factor for biodiversity is that grasslands support a range of microbial communities that support soil health and nutrient cycling soil (Mencel et al., 2022).

According to the European Environment Agency (2020) Greek grasslands are frequently identified as priority habitats in European conservation policy frameworks and are thought to be crucial for preserving functional biodiversity in Mediterranean ecosystems. The importance of the Mediterranean Basins biodiversity cuts across national borders making it a global hotspot for conservation of biodiversity (Aurelle et al., 2022).

CONSERVATION STATUS OF GRASSLANDS IN GREECE

The majority of natural grasslands in Greece are rated as having an unfavorable conservation status despite their importance to biodiversity. Due to severe habitat degradation and deteriorating quality habitat types like 6220, 6210 and 62A0 are primarily classified as being in ''unfavorable–bad'' or ''unfavorable–inadequate'' condition (European Environment Agency, 2020). The main concerns are the abandonment of traditional grazing which results in secondary succession and afforestation and overgrazing especially in lowland and foothill regions which causes soil erosion and shrub encroachment (Chouvardas et al., 2022).

Another threat is the disruption of existing plant communities caused by changes in temperature and precipitation patterns particularly in montane and alpine grasslands (Berauer et al., 2019). These threats are made worse by habitat fragmentation brought on by infrastructure growth and agricultural intensification (Kallimanis et al., 2008).

Regarding protection, most of the natural grassland types are found in Natura 2000 sites and several of them (6210, 6230) have been named priority habitats by the EU Habitats Directives Annex I (Dimopoulos et al., 2006). Conservation strategies prioritize the maintenance or restoration of low grazing intensity regimes to prevent overgrazing and land abandonment, and when or if it is required are applied active management interventions (e.g. controlled burning and shrub clearance) (Chouvardas et al., 2022; Lempesi et al. 2017).

The main target for The European Union for the Biodiversity Strategy for 2030 aims for restoration projects to prevent and even reverse habitat degradation (European Environment Agency, 2020). However, to maintain traditional land-use practices that are essential for the maintenance of Greek grassland habitats, still require community involvement, strong monitoring frameworks and coordinated national management planning.

NATURA 2000 GRASSLANDS IN GREECE

Natural grasslands inside Greece's Natura 2000 sites include a variety of habitat types ranging from species-rich montane and alpine grasslands in the high mountains to semi-natural dry calcareous grasslands and pseudo-steppe annual grasslands in lowland Mediterranean zones (e.g. the Pindos range, Mt. Olympus) (Janišová et al., 2010). As reported by Copernicus Land Monitoring Service (2021) and Hellenic Data Service (2018), the Natura 2000 grasslands (Figure 2) are larger than 540,000 ha. These grasslands support a remarkable diversity of plant species including numerous rare and endemic taxa and offer significant ecosystem services making them hotspots for biodiversity with high ecological value (Zikos & Constantinidis, 2022). They do however face formidable conservation obstacles because of the extensive habitat degradation brought on by land-use changes and management problems (Sidiropoulou et al., 2022; Mallinis et al., 2020). Due to the fact that, traditional grazing has been discontinued in many places open grasslands have been fragmented and reduced in size by the encroachment of shrubs and forests (Oikonomou et al., 2023). It has been demonstrated that overstocking and intensive grazing which are frequently fueled by Common Agricultural Policy subsidies severely reduce vegetation and plant species at other locations having devastating consequences for grassland biodiversity (Kindvall et al., 2022; Fetzel et al., 2018). Consequently, despite being designated as Natura 2000 many grassland habitats in Greece continue to have unfavorable conservation status with the majority of assessments showing poor or bad conditions and continuous declines (Copernicus Land Monitoring Service, 2021).

Figure 2. Natura 2000 Grasslands in Greece

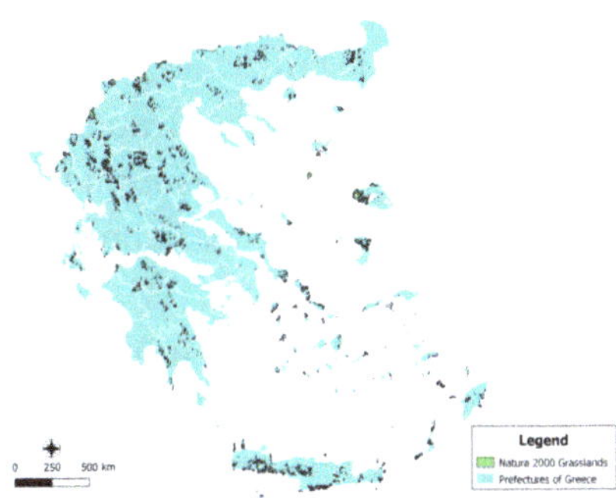

FOREST ROAD NETWORK IMPACTS

Advantages of Forest Roads for Wind Energy Development

Forest roads, in addition to contributing to easy access for the installation of wind turbines, also contribute to the efficiency of their operational maintenance (Wang et al., 2023). As operational costs are reduced in combination with increased reliability in the operation of wind farms due to improved access infrastructure (Gu et al., 2014). The easy access provided by these roads allows for regular surveys and timely repairs, as well as replacement of components, making long-term operational stability possible for wind energy projects (Kati et al., 2025). Forest roads also provide support for emergency intervention when equipment failure or during extreme weather conditions, hence minimizing both downtimes as well as financial losses (Ring et al., 2024). Also, these roads improve the lives of local people by opening new ways to the mountain land, helping them in farming, rural, and seasonal activities such as beekeeping and traditional livestock farming (Karamountzou & Vagiona, 2023). As a result, help economic development by supporting nature activities mainly on scenic routes used for outside activities (Guarino et al., 2020).

Forest roads play an especially important logistical function in accessing remote, mountainous grasslands where wind resources are the appropriate ones (Clifton et al., 2022). In the rugged landscapes typical of Greek upland grasslands found in Evvoia, Thrace, and the Aegean islands, fields usually require roads for turbine component transportation and construction machinery (Kati et al., 2025). In this way, they enable wind energy installations to access these sites, to help support national renewable energy targets, otherwise it would be unapproachable. Forest roads may also facilitate multipurpose land use by enhancing access fire prevention, silvicultural and grazing management, as well as ecotourism activities (Karamountzou & Vagiona, 2023). Road access is also significant in wildfire-prone regions since it eases observation and firefighting operations (Daoutis & Lempesi, 2023; Thompson et al., 2021). It is also very important to mention that access via the forest road network facilitates the monitoring of flora and fauna in such remote regions (Ring et al., 2024; Guarino et al., 2020).

The forest road network also has benefits from an economic point of view, as it has lower transport expenses and makes it possible to construct more efficient wind projects (Wang et al., 2023). They serve as access to areas of high winds for optimum yields, therefore maximizing the economic life of the wind farm (Kati et al., 2025; Gu et al., 2014). This directly and indirectly creates jobs during the construction and maintenance periods, raises local economies, which are consistent with regional development goals (Slattery et al., 2011). In addition, improved accessibility resulting from forest roads can improve emergency response times in cases of wildfires or other natural disasters on energy infrastructure and surrounding landscapes (Thompson et al., 2021). The forest roads are always seen as necessary infrastructural elements for the development of an onshore wind project within mountainous and semi-mountainous landscapes (Chaskos et al., 2012).

Disadvantages and Ecological Impacts

Forest roads are associated with long-term ecological degradation that goes far beyond the immediate impact of habitat loss (Trombulak & Frissell, 2000). When the forest and grassland landscapes are opened up for vehicle access, illegal activities such as grazing, logging, and dumping of waste increase the ecological disturbance (Keller & Berry, 2007). The road surface itself acts as an avenue for the spread of invasive species because vehicles transport seeds into old-growth grasslands (Mortensen et al., 2009). Forest road network also affects wildlife as animals move away from areas and find new habitats due to noise pollution and human presence near their habitat, especially birds in Natura 2000 grasslands (Konstantopoulos et al., 2020). Roads cut across the most significant ecological corridors and thus isolate populations of flora and fauna, which leads to a reduction in genetic diversity (Marsh et al., 2008). The cumulative fragmentation resulting from road expansion undermines the landscape-level ecological processes, hence conservation interventions (Mo et al., 2017).

Though roads present an immediate solution to logistics and economic resources, they introduce a significant amount of ecological disturbance across the grassland ecosystem (Phillips et al., 2020). The construction of roads within forests involves the removal of vegetation cover, excavation of slopes, and fragmentation of habitats, thus habitat loss, and finally, landscape fragmentation (Kati et al., 2025; Lempesi & Daoutis, 2024). In Natura 2000 grasslands in Greece, road networks greatly degrade their habitat values, especially in areas with high conservation values (Kati et al., 2025; Lawrence et al., 2021; Votsi et al., 2012). The expansion of roads in montane and subalpine grasslands increases the degree to which habitats are degraded by invasion processes and alters microhabitat conditions important for endemic flora and fauna (Guarino et al., 2020).

Forest roads compound the risks of soil erosion on steep grassland slopes by allowing for the easy destabilization of slopes and concentrating runoff along their surfaces (Yousefi et al., 2024). Roads may also shift hydrology patterns, nutrient cycles, and pollution resulting from dust as well as from the sediment (Kati et al., 2025). Vehicle traffic adds to exacerbating factors related to soil compaction through time, widening the road, hence increasing problems related to erosion and runoff (Alvis et al., 2022).

From a conservation standpoint, road development in Natura 2000 grasslands goes against the primary goal of sustaining or achieving a favorable conservation status of the habitats therein (Lawrence et al., 2021; Votsi et al., 2012). These are grasslands falling under priority annex I of the EU Habitats Directive, e.g. Calcareous dry grasslands (6210) and Mediterranean pseudo-steppes (6220), which are most prone to fragmentation by roads, reducing habitat quality as well as their connectivity (Meyer et al., 2020; Dimopoulos et al., 2006). In fact, biodiversity loss from this type of infrastructure-caused habitat fragmentation is a concern for scientists who consequently call for much more strictly enforced environmental assessment and spatial planning in strategic terms (Lawrence et al., 2021). Long-term ecological research indicates that the negative impacts associated with access roads on species composition, wildlife population, and landscape-scale ecosystem processes may long outlast actual road presence due to legacy effects even after restorative interventions have been made (Larson & Rew, 2022).

WIND ENERGY IN GREECE

National Targets for Wind Energy Development

Greece has gradually increased the participation of wind energy in its national targets. In the Renewable Energy Action Plan 2010, total installed capacity by 2020 was estimated at about 7.5 GW, where offshore capacity would constitute 300 MW but only about 3.6 GW has been realized (International Energy Agency, 2011). The target rephrased in the National Energy and Climate Plan (European Commission – NECP) of 2019 is approximately close to 7 GW by the year 2030. This target was even higher in the revision for the year 2024 with a prescription of about 8.9 GW onshore and around 1.9 GW of offshore capacities to be achieved by the year 2030 that will eventually reach some approximate values for onshore and offshore installations, that is 13 GW onshore and 11.8 GW offshore by 2050 (European Commission – NECP, 2025).

Investment Trends and Policy Frameworks

The reforms and support through regulation have notably improved the investment in wind energy in Greece (Dimitrakopoulos et al., 2025; Sendstad et al., 2022). Initially, this included grid access as well as the high feed-in tariffs (FiTs) being offered to investors (Doulos et al., 2019). However, there was a retroactive cut in FiT in 2014, which thus diminished investor confidence (Sendstad et al., 2022). Recently, Greece has further simplified licensing offshore wind turbines through competitive auctions and floating wind turbine pilot installations that were provided under Law 4964/2022 (Delagrammatikas & Roukanas, 2023). Offshore investment analysis has found large social benefits from reduced energy imports, local job creation, and carbon savings (Makris et al., 2022; Lavidas, 2019). Considering wind variability and local grid constraints, medium-sized investment models in the Aegean regions show satisfactory returns (Sakka et al., 2020; Konstantinidis et al., 2014).

Typical Locations and Regional Impacts

Greece wind farms often found in corridors of high potential winds, e.g. mountain ridges, island towers, and strong coastal zones (Pavlis, 2025; Palaiologou et al., 2011). Average hub height wind speeds in the Aegean islands and mountainous mainland reach between 8-10 m/s (Hadjipetrou & Kyriakidis, 2024). The greatest offshore resource establishes within the central and eastern parts of the Aegean Seas with power density estimates near 900 W/m^2 at 80 m and several zones are currently under plans for development (Soukissian et al., 2017). This has created local resistance where there is onshore concentration in places such as Evia and Thrace due to landscape transformation, biodiversity concerns, and tourism disruption (Pavlis, 2025; Moustakas et al., 2023). It is recommended the need for improved explicit stakeholder engagement and benefit-sharing frameworks be integrated into wind energy project planning to achieve social acceptance (Komninou et al., 2023).

Number of Wind Turbines in Greece

There were about 3,017 wind turbines (Figure 3) installed and running all over Greece, corresponding to a total installed capacity of approximately 5,351 MW by the end of 2024. Only in that year, 39 new turbines got connected, representing about 125.5 MW added (ELETAEN, 2025). It should be noted that offshore development is still at its planning stage since, there are no offshore machines running up to now (Soukissian et al., 2021). A national offshore framework would further installations in both the Aegean Sea and Ionian Sea (Delagrammatikas & Roukanas, 2023).

As regards the year 2025, Greece had about 5,507 MW per region, of wind energy capacity (Table 1), with 37 new turbines (152.2 MW) added in the first half of the year, as reported by ELETAEN until June. This extends the total number of turbines operating to about 3,045 which marks very robust growth year over year. Also, it has contracted or is under construction more than 1 GW that plus another 300 MW still on the way would bring total capacity up to 6.5 GW by 2026 (ELETAEN, 2024).

Figure 3. Wind turbines in Greece

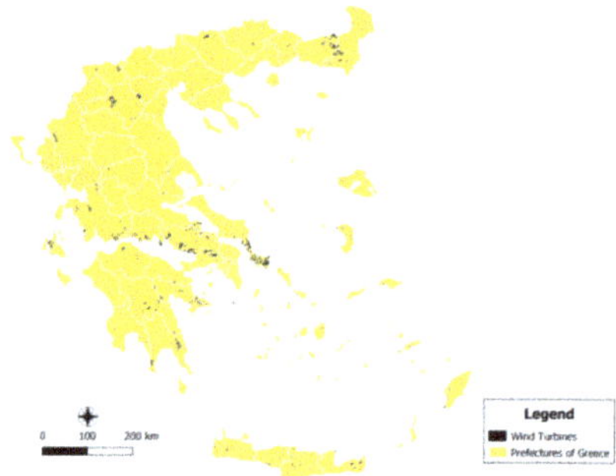

Table 1. Capacity (MW) per region of Greece for at the end of the first half of 2025

Region	Capacity (MW)
Central Greece	2,427
Peloponnese	709
Eastern Macedonia Thrace	535
Western Greece	417
Western Macedonia	414
Crete	203
Central Macedonia	158
Attica	187
Ionian Islands	122
Southern Aegean	112
Epirus	110
Thessaly	73
North Aegean	41

*There is a deviation of ±1 MW due to rounding
(ELETAEN, 2025)

Wind Turbines in Protected Natura Grasslands

The installation of wind turbines in natural grasslands and mountain habitats that fall within areas protected by the Natura 2000 network is a major problem for the development of wind energy in Greece (Kati et al., 2021). According to Karamountzou and Vagiona (2023), 27,19% of wind farms (Figure 4) are in places that are within the boundaries of the Natura 2000 protected areas. As demonstrated by Copernicus Land Monitoring Service (2021), ELETAEN (2025) and Hellenic Data Service (2018), they are contained in 18 Prefectures (Figure 5) of Greece. In particular, they are located in the Prefectures of Phocis, Aetolia-Acarnania, Heraklion, Evros, Serres, Chios, Achaea, Florina, Evia, Cephalonia, Rhodope, Lesvos, Cyclades, Lasithi, Magnesia, Arta, Dodecanese & Phthiotis. Wind farms mainly take place in mountainous grasslands of Evia and Thrace, as well as Cycladic islands, where the resources for both wind elements and conservation interests coincide. These areas are of particular importance, as distinguished by natural and cultural heritage (Vlami et al., 2020; Evans, 2012), but also for the large number of endemic species of flora and fauna (Kati et al., 2025). Installations like these have been known to cause habitat fragmentation effects on bird species and changes in the vegetation structure of grasslands (Kati et al., 2021). This increases the growing debate on where future wind developments are to be sited such that biodiversity impacts are minimized while still allowing renewable energy growth (Latinopoulos & Kechagia, 2015). The scientific community of environmental sciences wants greater exclusion zones and cumulative impact assessments included in the licensing processes to protect these ecologically important grassland habitats that exist within Natura 2000 areas.

Figure 4. Wind turbines in Natura 2000 grasslands in Greece

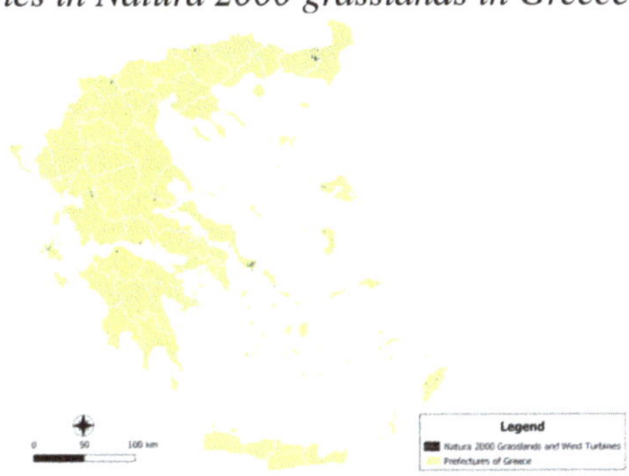

Figure 5. Prefectures with Natura 2000 grasslands and wind turbines in Greece

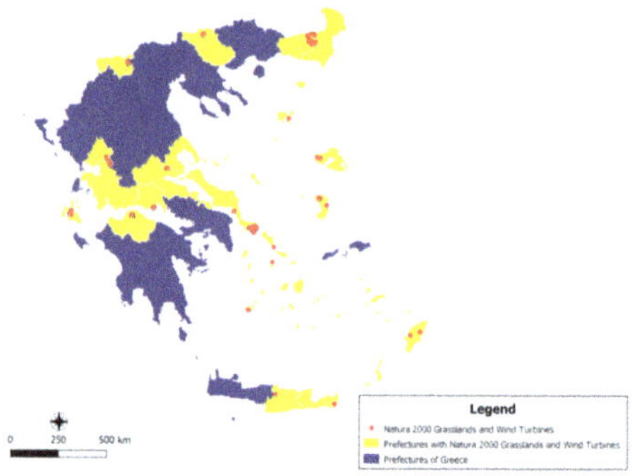

SOLUTION AND RECOMMENDATIONS

Sustainable wind energy development in Greece should begin with the induced impacts of infrastructure, especially access roads, into EIA (Environmental Impact Assessments) by changing from the current superficial accounting of only siting turbines to EIA reports that largely underestimate the ancillary infrastructures' ecological impacts. Thus, holistic cumulative impact-based EIAs are proposed as best practice, including consideration of the whole lifecycle of road networks together with long-term persistence in landscapes. High-resolution biodiversity mapping should already be available for use together with other spatial planning tools, both for recognizing high-value ecological zones such as sensitive grasslands and Natura 2000 sites and further providing alternative locations based on lower environmental risks. The application of ecological connectivity models to planning would prevent habitat fragmentation caused by expansion.

Where roads are unavoidable, should upgrade and or realign existing routes rather than create new ones from scratch widen road widths to further reduce the disturbance footprint use permeable surfaces to keep natural drainage patterns intact. Wildlife crossings, erosion control structures, and immediate restoration of roadside vegetation using native species will restore the negative impacts that might have been caused. Post-construction rehabilitation programs should be made a condition where disturbed areas have to be brought back actively into their ecological function rather than leaving them to recover passively, which propagates vulnerability due to invasiveness. This would not only conserve biodiversity but also reduce long-term maintenance costs for operators.

The management of renewable energy should be regarded as ecological stewardship. It should involve facilitated multi-stakeholder planning processes with energy developers and conservation authorities, local governments, and community representatives. This will enable shared decision-making and reconciliation of competing land-use priorities. Provision for the integration of local knowledge—traditional grazing and land management practice in wind farm siting to ensure cultural landscape value should be put into policy frameworks that can incentivize this integration through subsidies or certification schemes for biodiversity-sensitive projects. Such integrated action will make possible the achievement of Greece's ambitious climate targets while maintaining the ecological integrity of its grasslands.

CONCLUSIONS

The development of wind energy is central to Greece's efforts on its national and EU climate commitments. Wind farms in the grasslands raise issues of competing values toward government policy. They are significantly decarbonizing, and at the same time working for energy diversification. However, infrastructure development poses long-term ecological damages such as habitat fragmentation, soil degradation, hydrology changes, and other impacts relating to biodiversity losses—all because of an access road. These impacts are always minimized during environmental assessment procedures; a recipe for cultural and ecological values that belong to the Greek grasslands getting lost. Therefore, at this level, integrated planning is a requirement for an infrastructure support system towards a sustainable energy transition ensuring ecosystem integrity while attaining climate goals. That ideal state must ensure that renewable energy resource exploitation is balanced with biodiversity conservation.

REFERENCES

Alamgir, M., Campbell, M. J., Sloan, S., Goosem, M., Clements, G. R., Mahmoud, M. I., & Laurance, W. F. (2017). Economic, socio-political and environmental risks of road development in the tropics. *Current Biology*, *27*(20), R1130–R1140. DOI: 10.1016/j.cub.2017.08.067 PMID: 29065299

Alvis, A. D., Luce, C. H., & Istanbulluoglu, E. (2022). How does traffic affect erosion of unpaved forest roads? *Environmental Reviews*, *31*(1), 182–194. DOI: 10.1139/er-2022-0032

Arjmand, A., Kiadaliri, H., Kazemnezhad, F., & Eshagh Nimvari, M. (2023). Effects of forest roads on vegetation biodiversity and soil characteristics in Hyrcanian forests. *Nordic Journal of Botany*, *2023*(12), e04039. DOI: 10.1111/njb.04039

Auffret, A. G., & Lindgren, E. (2020). Roadside diversity in relation to age and surrounding source habitat: Evidence for long time lags in valuable green infrastructure. *Ecological Solutions and Evidence*, *1*(1), e12005. DOI: 10.1002/2688-8319.12005

Aurelle, D., Thomas, S., Albert, C., Bally, M., Bondeau, A., Boudouresque, C. F., Cahill, A. E., Carlotti, F., Chenuil, A., Cramer, W., Davi, H., De Jode, A., Ereskovsky, A., Farnet, A. M., Fernandez, C., Gauquelin, T., Mirleau, P., Monnet, A. C., Prévosto, B., & Fady, B. (2022). Biodiversity, climate change, and adaptation in the Mediterranean. *Ecosphere*, *13*(4), e3915. DOI: 10.1002/ecs2.3915

Barthelmie, R. J., & Pryor, S. C. (2021). Climate change mitigation potential of wind energy. *Climate (Basel)*, *9*(9), 136. DOI: 10.3390/cli9090136

Berauer, B. J., Wilfahrt, P. A., Arfin-Khan, M. A., Eibes, P., Von Heßberg, A., Ingrisch, J., Schloter, M., Schuchardt, M. A., & Jentsch, A. (2019). Low resistance of montane and alpine grasslands to abrupt changes in temperature and precipitation regimes. *Arctic, Antarctic, and Alpine Research*, *51*(1), 215–231. DOI: 10.1080/15230430.2019.1618116

Bergmeier, E., Konstantinou, M., Tsiripidis, I., & Sýkora, K. V. (2009). Plant communities on metalliferous soils in northern Greece. *Phytocoenologia*, *39*(4), 411–438. DOI: 10.1127/0340-269X/2009/0039-0411

Boccard, N. (2009). Capacity factor of wind power realized values vs. estimates. *Energy Policy*, *37*(7), 2679–2688. DOI: 10.1016/j.enpol.2009.02.046

Bruelheide, H., Jandt, U., Fernández, N., Marmol-Guijarro, A., Smets, B., Buchhorn, M., Giagnacovo, L., Milli, G., Jimenez-Alfaro, B., & Álvarez-Martínez, J. M. (2024). D5. 2 Past-to-present EBV modelled datasets and status indicator for selected terrestrial habitats in the Habitats Directive. *ARPHA* Preprints, *5*, e128158. DOI: 10.3897/arphapreprints.e128158

Chaskos, D. C., Bartzokas, A., & Pnevmatikos, J. D. (2012). Wind energy in NW Greece. In Helmis, C. G., & Nastos, P. T. (Eds.), *Advances in Meteorology, Climatology and Atmospheric Physics* (pp. 43–50). Springer., DOI: 10.1007/978-3-642-29172-2_7

Chouvardas, D., Karatassiou, M., Tsioras, P., Tsividis, I., & Palaiochorinos, S. (2022). Spatiotemporal changes (1945–2020) in a grazed landscape of northern Greece, in relation to socioeconomic changes. *Land (Basel), 11*(11), 1987. DOI: 10.3390/land11111987

Clifton, A., Barber, S., Stökl, A., Frank, H., & Karlsson, T. (2022). Research challenges and needs for the deployment of wind energy in atmospherically complex locations. *Wind Energy Science Discussions, 7*, 2231–2254. DOI: 10.5194/wes-7-2231-2022

Copernicus Land Monitoring Service. (2021). OBSERVER: Copernicus Land Monitoring Service. *The Copernicus Land Monitoring Service: Measuring the effectiveness of protecting. Natura, 2000*, •••. https://www.copernicus.eu/en/news/news/observer-copernicus-land-monitoring-service-measuring-effectiveness-protecting-natura

Dalle Fratte, M., & Cerabolini, B. E. (2023). Extending the interpretation of Natura 2000 habitat types beyond their definition can bias their conservation status assessment: An example with species-rich Nardus grasslands (6230*). *Ecological Indicators, 156*, 111113. DOI: 10.1016/j.ecolind.2023.111113

Daoutis, C., & Lempesi, A. (2023). The effect of the forest road network on grassland ecosystems with the contribution of the SWOT-AHP method. In: *E3S Web of Conferences, 4th International Conference on Environmental Design (ICED2023)*, 20-22 October 2023, Athens, Greece, 436, 09007, EDP Sciences. https://doi.org/ DOI: 10.1051/e3sconf/202343609007

Delagrammatikas, G., & Roukanas, S. (2023). Offshore wind farm in the Southeast Aegean Sea and energy security. *Energies, 16*(13), 5208. DOI: 10.3390/en16135208

Delipetrou, P., Sarika, M., Michopoulos, P., & Karetsos, G. (2015). *Mountain grassland composition, structure, and ecology in Mt. Oiti and Mt. Kallidromo. Deliverable A.5.1. for the project LIFE11 NAT/GR/2014 - ForOpenForests*. Hellenic Agricultural Organization Dimiter - Institute of Mediterranean Forest Ecosystems and Forest Products Technology, National and Kapodistrian University of Athens - HSPN.

Dengler, J., Bergmeier, E., Willner, W., & Chytrý, M. (2013). Towards a consistent classification of European grasslands. *Applied Vegetation Science*, *16*(3), 518–520. Advance online publication. DOI: 10.1111/avsc.12041

Diffendorfer, J. E., Vanderhoof, M. K., & Ancona, Z. H. (2022). Wind turbine wakes can impact down-wind vegetation greenness. *Environmental Research Letters*, *17*(10), 104025. DOI: 10.1088/1748-9326/ac8da9

Dimitrakopoulos, P., Makris, I., & Apostolopoulos, S. (2025). The Impact of Major Energy Indexes on the Economic Growth of EU Countries, According to the Contribution of Industry to GDP. In: Tsounis, N., Vlachvei, A. (eds) *Advances in Applied Macroeconomics*. ICOAE 2024. Springer Proceedings in Business and Economics. Springer, Cham, pp 59–74. https://doi.org/DOI: 10.1007/978-3-031-76658-9_4

Dimopoulos, P., Bergmeier, E., & Fischer, P. (2005). Monitoring and conservation status assessment of habitat types in Greece: Fundamentals and exemplary cases. *Annali di Botanica*, *5*, 7–20.

Dimopoulos, P., Bergmeier, E., & Fischer, P. (2006). Natura 2000 habitat types of Greece evaluated in the light of distribution, threat and responsibility. *Biology and Environment*, *106B*(3), 175–187. DOI: 10.1353/bae.2006.0015

Doulos, I. G., Katsoulakos, N. M., & Kaliampakos, D. C. (2019). The Electricity Market Structure in Greece and the Paradox of Renewable Energy Sources. *Journal of Power and Energy Engineering*, *7*(4), 1–26. DOI: 10.4236/jpee.2019.74001

European Commission. National energy and climate plans - NECP. (2025). In *Energy Union*. Retrieved July 29, 2025, from https://energy.ec.europa.eu/strategy/energy-union_en#national-energy-and-climate-plans

European Environment Agency. (2020). State of Nature in the EU – Results from reporting under the nature directives 2013–2018. EEA Report No DOI: 10/2020. 10.2800/088178

Evans, D. (2012). Building the European union's Natura 2000 network. *Nature Conservation*, *1*, 11–26. DOI: 10.3897/natureconservation.1.1808

Fetzel, T., Petridis, P., Noll, D., Singh, S. J., & Fischer-Kowalski, M. (2018). Reaching a socio-ecological tipping point: Overgrazing on the Greek island of Samothraki and the role of European agricultural policies. *Land Use Policy*, *76*, 21–28. DOI: 10.1016/j.landusepol.2018.04.042

Filis, E., Kyrkas, D., Mantzos, N., Dimitrakopoulos, P. G., Fotiadis, G., & Konstantinou, M. (2024). Grassland flora of ultramafic areas in Northern Pindus (Greece). *Ecological Research*, *39*(4), 531–542. DOI: 10.1111/1440-1703.12469

Gu, H., Wang, J., Lin, Q., & Gong, Q. (2014). Automatic contour-based road network design for optimized wind farm micrositing. *IEEE Transactions on Sustainable Energy*, *6*(1), 281–289. DOI: 10.1109/TSTE.2014.2369432

Guarino, R., Vranakis, M., Rodríguez-Rojo, M. P., Giuga, L., & Pasta, S. (2020). Grasslands and Shrublands of the Mediterranean Region. In Goldstein, M., & Della Sala, D. (Eds.), *Encyclopedia of the World's Biomes* (Vol. 2, pp. 638–655). Oxford Academic Press, Elsevier., DOI: 10.1016/B978-0-12-409548-9.12119-0

Hadjigeorgiou, I. (2011). Past, present and future of pastoralism in Greece. *Pastoralism: research, policy and practice, 1*, 1-22. https://doi.org/DOI: 10.1186/2041-7136-1-24

Hadjipetrou, S., & Kyriakidis, P. (2024). High-resolution wind speed estimates for the Eastern Mediterranean Basin: A statistical comparison against coastal meteorological observations. *Wind*, *4*(4), 311–341. DOI: 10.3390/wind4040016

Hellenic Data Service. (2018). *Natura 2000 Network and Protected Areas [Data set]*. https://data.hellenicdataservice.gr/dataset/289263ce-04be-4e88-941c-81ea79da7169

Hellenic Wind Energy Scientific Association (ELETAEN). (2024). *HWEA Status & Challenges for the Supply Chain for Offshore Wind in Greece* [PDF report]. Retrieved from https://eletaen.gr/wp-content/uploads/2024/09/2024-09-02-OW-Supply-Chain-Greece-ELETAEN-consolidated-report.pdf

Hellenic Wind Energy Scientific Association (ELETAEN). (2025). *HWEA Wind Energy Statistics* [PDF report]. Retrieved from https://eletaen.gr/wp-content/uploads/2025/07/2025-07-09-eletaen-statistic-s1-f-2025.pdf

Hess, M. C., Gómez-Ruiz, P. A., Morellato, L. P. C., & Buisson, E. (2022). Phenological patterns of herbaceous Mediterranean plant communities in spring. *Plant Ecology and Evolution*, *155*(2), 207–220. DOI: 10.5091/plecevo.86335

International Energy Agency. (2011). *Energy Policies of IEA Countries: Greece 2011 Review*. OECD/IEA. Retrieved from https://www.iea.org/reports/energy-policies-of-iea-countries-greece-2011-review

Janišová, M., Bystrica, B., Vrahnakis, M., Dengler, J., & Rūsiņa, S. (2010). In *The European Dry Grassland Group. Bulletin*, *6*, •••. https://edgg.org/publ/bulletin/Bulletin_EDGG_06.pdf

Kallimanis, A. S., Tsiafouli, M. A., Pantis, J. D., Mazaris, A. D., Matsinos, Y., & Sgardelis, S. P. (2008). Arable land and habitat diversity in Natura 2000 sites in Greece. *Journal of Biological Research (Thessaloniki)*, *9*, 55–66.

Karamountzou, S., & Vagiona, D. G. (2023). Suitability and sustainability assessment of existing onshore wind farms in Greece. *Sustainability (Basel), 15*(3), 2095. DOI: 10.3390/su15032095

Kati, V., Kassara, C., Vrontisi, Z., & Moustakas, A. (2021). The biodiversity-wind energy-land use nexus in a global biodiversity hotspot. *The Science of the Total Environment, 768*, 144471. DOI: 10.1016/j.scitotenv.2020.144471 PMID: 33454485

Kati, V., Spiliopoulou, K., Stefanidis, A., & Kassara, C. (2025). Sacrificing Wilderness for Renewables? Land Artificialization from Inadequate Spatial Planning of Wind Energy in Evvoia, Greece. *Land (Basel), 14*(6), 1296. DOI: 10.3390/land14061296

Keller, G. R., & Berry, J. (2007). Reduced Impact Logging Road Issues in Tropical Forests. *Transportation Research Record*, 1989-1(1), 98-106. DOI: 10.3141/1989-11

Kindvall, O., Franzén, M., Askling, J., Forsman, A., & Johansson, V. (2022). Subsidized Common Agricultural Policy grazing jeopardizes the protection of biodiversity and Natura 2000 targeted species. *Animal Conservation, 25*(5), 597–607. DOI: 10.1111/acv.12773

Kleinschroth, F., Healey, J. R., Sist, P., Mortier, F., & Gourlet-Fleury, S. (2016). How persistent are the impacts of logging roads on Central African forest vegetation? *Journal of Applied Ecology, 53*(4), 1127–1137. DOI: 10.1111/1365-2664.12661

Kokkoris, I., Dimitrellos, G., Kougioumoutzis, K., Laliotis, I., Georgiadis, T., & Tiniakou, A. (2014). The native flora of Mountain Panachaikon (Peloponnese, Greece): New records and diversity. *Journal of Biological Research (Thessaloniki), 21*(1), 9. DOI: 10.1186/2241-5793-21-9 PMID: 25984492

Komninou, M., Afouxenidis, A., Souliotis, N., & Karoulas, G. (2023). Windfarms, social acceptability, and the environmental divide in Greece. *Qeios, RZBUH5*. Advance online publication. DOI: 10.32388/RZBUH5

Konstantinidis, E. I., Kompolias, D. G., & Botsaris, P. N. (2014). Viability analysis of an offshore wind farm in North Aegean Sea, Greece. *Journal of Renewable and Sustainable Energy, 6*(2), 023116. Advance online publication. DOI: 10.1063/1.4871484

Konstantopoulos, K., Moustakas, A., & Vogiatzakis, I. N. (2020). A spatially explicit impact assessment of road characteristics, road-induced fragmentation and noise on birds species in Cyprus. *Biodiversity (Nepean), 21*(1), 61–71. DOI: 10.1080/14888386.2020.1736154

Korakis, G., & Gerasimidis, A. (2010). Vegetation and habitat types. In Catsadorakis, G., & Källander, H. (Eds.), *The Dadia–Lefkimi–Souli Forest National Park, Greece: Biodiversity, Management and Conservation* (pp. 85–93). WWF Greece.

Kougioumoutzis, K., Kaloveloni, A., & Petanidou, T. (2022). Assessing climate change impacts on Island bees: The Aegean Archipelago. *Biology (Basel), 11*(4), 552. DOI: 10.3390/biology11040552 PMID: 35453751

Koukoura, Z. (2007). Natural grasslands—a case study in Greece. In *Nutrient Cycling in Terrestrial Ecosystems* (pp. 257–269). Springer Berlin Heidelberg. DOI: 10.1007/978-3-540-68027-7_9

Kuklina, V., Bilichenko, I., Bogdanov, V., Kobylkin, D., Petrov, A. N., & Shiklomanov, N. (2021). Informal road networks and sustainability of Siberian boreal forest landscapes: Case study of the Vershina Khandy taiga. *Environmental Research Letters, 16*(11), 115001. DOI: 10.1088/1748-9326/ac22bd

Larson, C. D., & Rew, L. J. (2022). Restoration intensity shapes floristic recovery after forest road decommissioning. *Journal of Environmental Management, 319*, 115729. DOI: 10.1016/j.jenvman.2022.115729 PMID: 35853306

Latinopoulos, D., & Kechagia, K. (2015). A GIS-based multi-criteria evaluation for wind farm site selection. A regional scale application in Greece. *Renewable Energy, 78*, 550–560. DOI: 10.1016/j.renene.2015.01.041

Lavidas, G. (2019). Energy and socio-economic benefits from the development of wave energy in Greece. *Renewable Energy, 132*, 1290e1300. DOI: 10.1016/j.renene.2018.09.007

Lawrence, A., Friedrich, F., & Beierkuhnlein, C. (2021). Landscape fragmentation of the Natura 2000 network and its surrounding areas. *PLoS One, 16*(10), e0258615. DOI: 10.1371/journal.pone.0258615 PMID: 34673802

Lempesi, A., & Daoutis, C. (2024). The effect of trail paths in grassland ecosystems using a SWOT-AHP Hybrid method. In: *Proceedings of 24th International Multidisciplinary Scientific GeoConference SGEM 2024*, 29th June 2024 to 8th July 2024, Albena, Bulgaria, STEF92 Technology, ()DOI: 10.5593/sgem2024/3.1/s14.49

Lempesi, A., Eleftheriadou, A., Delivasi, Z., Psyllidou, A., Korakis, G., & Kyriazopoulos, A. P. (2017). Effects of grazing intensity on the regeneration of woody species in an oak woodland. *Notulae Botanicae Horti Agrobotanici Cluj-Napoca, 45*(2), 597–601. DOI: 10.15835/nbha45210908

Li, L., Ma, W., Duan, X., Wang, S., Wang, Q., Gu, H., & Wang, J. (2024). Effects of Wind Farm Construction on Soil Nutrients and Vegetation: A Case Study of Linxiang Wind Farm in Hunan Province. *Sustainability (Basel), 16*(15), 6350. DOI: 10.3390/su16156350

Lugo, A. E., & Gucinski, H. (2000). Function, effects, and management of forest roads. *Forest Ecology and Management, 133*(3), 249–262. DOI: 10.1016/S0378-1127(99)00237-6

Makris, I., Babalos, V., & Dimitrakopoulos, P. (2022). A study of the energy efficiency formula for the development of economic progress policies in Greece. *Energy Sources. Part B, Economics, Planning, and Policy, 17*(1), 2007178. DOI: 10.1080/15567249.2021.2007178

Mallinis, G., Chrysafis, I., Lempesi, A., Kyriazopoulos, A. P., Korakis, G., Kokkoris, I. P., & Dimopoulos, P. (2020). Spatiotemporal changes of forest line and canopy cover between 1945-2019 at selected Greek mountains. In: *Eighth International Conference on Remote Sensing and Geoinformation of the Environment (RSCy2020)*, 26 August 2020, Paphos, Cyprus, 11524, 236-248, 2020 Society of Photo-Optical Instrumentation Engineers (SPIE). https://doi.org/DOI: 10.1117/12.2570735

Marsh, D. M., Page, R. B., Hanlon, T. J., Corritone, R., Little, E. C., Seifert, D. E., & Cabe, P. R. (2008). Effects of roads on patterns of genetic differentiation in red-backed salamanders, Plethodon cinereus. *Conservation Genetics, 9*(3), 603–613. DOI: 10.1007/s10592-007-9377-0

Mencel, J., Mocek-Płóciniak, A., & Kryszak, A. (2022). Soil microbial community and enzymatic activity of grasslands under different use practices: A review. *Agronomy (Basel), 12*(5), 1136. DOI: 10.3390/agronomy12051136

Meyer, B. C., Kirsten, F., Sattler, D., & Heinrich, J. (2020). The Land Use-Land Degradation Nexus in Mediterranean Landscapes:–Drivers of Changes And Key Processes at Selected Natura 2000 Sites of Crete, Greece. *Journal of Environmental Geography, 13*(1-2), 13–24. DOI: 10.2478/jengeo-2020-0002

Minachilis, K., Kantsa, A., Devalez, J., Vujic, A., Pauly, A., & Petanidou, T. (2023). High species turnover and unique plant–pollinator interactions make a hyperdiverse mountain. *Journal of Animal Ecology, 92*(5), 1001–1015. DOI: 10.1111/1365-2656.13898 PMID: 36754546

Mo, W., Wang, Y., Zhang, Y., & Zhuang, D. (2017). Impacts of road network expansion on landscape ecological risk in a megacity, China: A case study of Beijing. *The Science of the Total Environment, 574*, 1000–1011. DOI: 10.1016/j.scitotenv.2016.09.048 PMID: 27668852

Morabito, A., Musarella, C. M., & Spampinato, G. (2024). Diversity and ecological assessment of grasslands habitat types: A case study in the Calabria Region (Southern Italy). *Land (Basel), 13*(6), 719. DOI: 10.3390/land13060719

Mortensen, D. A., Rauschert, E. S., Nord, A. N., & Jones, B. P. (2009). Forest roads facilitate the spread of invasive plants. *Invasive Plant Science and Management, 2*(3), 191–199. DOI: 10.1614/IPSM-08-125.1

Moustakas, A., Georgiakakis, P., Kret, E., & Kapsalis, E. (2023). Wind turbine power and land cover effects on cumulative bat deaths. *The Science of the Total Environment, 892*, 164536. DOI: 10.1016/j.scitotenv.2023.164536 PMID: 37268139

Oikonomou, D., Vrahnakis, M., Yiakoulaki, M., Xanthopoulos, G., & Kazoglou, Y. (2023). Grazing as a management tool in Mediterranean Pastures: A meta-analysis based on a literature review. *Land (Basel), 12*(7), 1290. DOI: 10.3390/land12071290

Palaiologou, P., Kalabokidis, K., Haralambopoulos, D., Feidas, H., & Polatidis, H. (2011). Wind characteristics and mapping for power production in the Island of Lesvos, Greece. *Computers & Geosciences, 37*(7), 962–972. DOI: 10.1016/j.cageo.2010.05.025

Pavlis, E. (2025). Large-Scale Wind Energy Farm Development and its Impacts on the Landscape: A Review of the Greek Case. *Journal of Landscape Ecology, 18*(3), 24–61. DOI: 10.2478/jlecol-2025-0019

Phillips, B. B., Bullock, J. M., Osborne, J. L., & Gaston, K. J. (2020). Ecosystem service provision by road verges. *Journal of Applied Ecology, 57*(3), 488–501. DOI: 10.1111/1365-2664.13556

Psarros, G. N., & Papathanassiou, S. A. (2022). Electricity storage requirements to support the transition towards high renewable penetration levels–Application to the Greek power system. *Journal of Energy Storage, 55*, 105748. DOI: 10.1016/j.est.2022.105748

Psyllos, G., Hadjigeorgiou, I., Dimitrakopoulos, P. G., & Kizos, T. (2022). Grazing land productivity, floral diversity, and management in a semi-arid Mediterranean landscape. *Sustainability (Basel), 14*(8), 4623. DOI: 10.3390/su14084623

Rediske, G., Burin, H. P., Rigo, P. D., Rosa, C. B., Michels, L., & Siluk, J. C. M. (2021). Wind power plant site selection: A systematic review. *Renewable & Sustainable Energy Reviews, 148*, 111293. DOI: 10.1016/j.rser.2021.111293

Ring, E., Wallgren, M., Marald, E., Westerfelt, P., Djupstrom, L., Davidsson, A., & Sonesson, J. (2024). Forest roads in Sweden-infrastructure with multiple uses and diverse impacts. *Silva Fennica, 58*(4). Advance online publication. DOI: 10.14214/sf.24021

Sakka, E. G., Bilionis, D. V., Vamvatsikos, D., & Gantes, C. J. (2020). Onshore wind farm siting prioritization based on investment profitability for Greece. *Renewable Energy, 146*, 2827–2839. DOI: 10.1016/j.renene.2019.08.020

Sendstad, L. H., Hagspiel, V., Mikkelsen, W. J., Ravndal, R., & Tveitstøl, M. (2022). The impact of subsidy retraction on European renewable energy investments. *Energy Policy, 160*, 112675. DOI: 10.1016/j.enpol.2021.112675

Sidiropoulou, A., Chouvardas, D., Mantzanas, K., Stefanidis, S., & Karatassiou, M. (2022). Impact of Transhumant Livestock Grazing Abandonment on Pseudo-Alpine Grasslands in Greece in the Context of Climatic Change. *Land (Basel), 11*(12), 2126. DOI: 10.3390/land11122126

Slattery, M. C., Lantz, E., & Johnson, B. L. (2011). State and local economic impacts from wind energy projects: Texas case study. *Energy Policy, 39*(12), 7930–7940. DOI: 10.1016/j.enpol.2011.09.047

Soukissian, T., Papadopoulos, A., Skrimizeas, P., Karathanasi, F., Axaopoulos, P., Avgoustoglou, E., Kyriakidou, H., Tsalis, C., Voudouri, A., Gofa, F., & Katsafados, P. (2017). Assessment of offshore wind power potential in the Aegean and Ionian Seas based on high-resolution hindcast model results. *AIMS Energy, 5*(2), 268–289. DOI: 10.3934/energy.2017.2.268

Soukissian, T. H., Karathanasi, F. E., & Zaragkas, D. K. (2021). Exploiting offshore wind and solar resources in the Mediterranean using ERA5 reanalysis data. *Energy Conversion and Management, 237*, 114092. DOI: 10.1016/j.enconman.2021.114092

Thompson, M. P., Gannon, B. M., & Caggiano, M. D. (2021). Forest roads and operational wildfire response planning. *Forests, 12*(2), 110. DOI: 10.3390/f12020110

Trombulak, S. C., & Frissell, C. A. (2000). Review of ecological effects of roads on terrestrial and aquatic communities. *Conservation Biology, 14*(1), 18–30. DOI: 10.1046/j.1523-1739.2000.99084.x

Veers, P., Dykes, K., Basu, S., Bianchini, A., Clifton, A., Green, P., Holttinen, H., Kitzing, L., Kosovic, B., Lundquist, J. K., Meyers, J., O'Malley, M., Shaw, W. J., & Straw, B. (2022). Grand Challenges: Wind energy research needs for a global energy transition. *Wind Energy Science Discussions, 7*(6), 2491–2505. DOI: 10.5194/wes-7-2491-2022

Vlami, V., Danek, J., Zogaris, S., Gallou, E., Kokkoris, I. P., Kehayias, G., & Dimopoulos, P. (2020). Residents' views on landscape and ecosystem services during a wind farm proposal in an island protected area. *Sustainability (Basel), 12*(6), 2442. DOI: 10.3390/su12062442

Votsi, N., Mazaris, A., Kallimanis, A., Zomeni, M., Vogiatzakis, I., Sgardelis, S., & Pantis, J. (2012). Road effects on habitat richness of the Greek Natura 2000 network. *Nature Conservation, 1*, 53–71. DOI: 10.3897/natureconservation.1.2086

Wang, Q., Gontikaki, E., Stenzel, P., Louca, V., Küpper, F. C., & Spiller, M. (2024). How to decarbonize Greece by comparing wind and PV energy: A land eligibility analysis. *Energies, 17*(3), 567. DOI: 10.3390/en17030567

Wang, Y. D., Yin, F. K., Shen, L., & Wu, C. Z. (2023). A Review of Road Design for Wind Farms in China. *Applied Sciences (Basel, Switzerland), 13*(7), 4075. DOI: 10.3390/app13074075

Xanthakis, M., Katsimanis, N., & Antonopoulos, N. (2022). Impact of a wind farm on the avifauna of a Mediterranean mountainous environment. *Environmental Sciences Proceedings*, *21*(1), 11. DOI: 10.3390/environsciproc2022021011

Yousefi, S., Emami, S. N., Nekoeimehr, M., Rahmati, O., Imaizumi, F., Gomez, C., & Valjarevic, A. (2024). A hot-spot analysis of forest roads based on soil erosion and sediment production. *Land (Basel), 13*(10), 1583. DOI: 10.3390/land13101583

Zhao, X., Li, F., Yuan, Y., Ari, G., Yan, Y., Zhang, Q., Olhnuud, A., & Liu, P. (2025). Wind farms reduce grassland plant community diversity and lead to plant community convergence. *BMC Ecology and Evolution*, *25*(1), 10. DOI: 10.1186/s12862-025-02350-6 PMID: 39815201

Zikos, A., & Constantinidis, T. (2022). Plant diversity of Mts. Oligirtos and Farmakas (NE Peloponnisos, Greece) with emphasis on their endemic flora. *Plants, 11*(19), 2649. DOI: 10.3390/plants11192649 PMID: 36235515

Zotos, A., Kosma, C., Triantafyllidis, V., Kakabouki, I., Kehayias, G., Roussis, I., Mavroeides, A., Tataridas, A., & Bilalis, D. (2021). Plant species diversity of the wet meadows under natural and anthropogenic interventions: The case of the Lakes Amvrakia and Ozeros (W. Greece). *Notulae Botanicae Horti Agrobotanici Cluj-Napoca*, *49*(3), 12435–12435. DOI: 10.15835/nbha49312435

Chapter 8
Towards a Sustainable Future:
Analyzing India's Path of Energy Transition

Sebak Kumar Jana
https://orcid.org/0000-0002-3532-4350
Vidyasagar University, India

Khushboo Singh
https://orcid.org/0009-0009-7268-2171
Vidyasagar University, India

Abstract

India's energy transition reflects a delicate balance between rapid economic expansion, increasing energy consumption, and the need for a cleaner environment. Even though coal still occupies a dominant position in the overall national energy portfolio, renewables, led by solar and wind, have made significant inroads as costs have dropped and policy support, along with technological advances, has strengthened. But that vast renewable potential has so far gone largely unused, slowing the path to decarbonization. The country's dual identity as both a leading coal consumer and a rapidly emerging renewable power highlights the complexity of its energy transition. It also calls for accelerating the deployment of renewables, expanding the grid infrastructure, and investing in green finance. As India proceeds along this trajectory, its decisions will not just shape domestic energy security but also have a large bearing on the success of global climate and sustainable development goals.

DOI: 10.4018/979-8-3373-5302-9.ch008

INTRODUCTION

The breadth and depth of challenges in 2022 for the world's electricity demand and generation patterns are not only acute, but incredibly nuanced, should we continue under business-as-usual (BAU). Worldwide electrification demand grew to a total of 28,510 TWh, indicating a dramatic increase in the reliance on electricity as an enabler of growth and decarbonisation. The largest economies like China (31% of global demand), the United States (15%) and European Union (10%) still led in consumption, with India although home to over 1.3 billion people responsible for just 6% (1,836 TWh) at very low per capita levels of consumption based on available data –1.3 MWh- demonstrating continuing inequality over access/affordability of energy between developed and developing economies as a whole (IEA, 2023a). On the supply side, there was no substantial change in the market with fossil fuels dominating electricity generation and reaching 61% of total generation, including 36% for coal and a 22% share for fossil gas. Clean sources accounted for less than 40 per cent collectively, with the key contributors being hydro- (15%) nuclear power (9.2%), wind (7.6%) and solar PV (4.5%) (IEA, 2023b). Despite the rapid expansion of wind (17% increase in 2022) and solar power (24%) this year, these growth rates are still below that required for a BAU trajectory, with coal and gas either increasing or flatlining rather than declining at a rate appropriate to meet net-zero alignment. The IEA's Stated Policies Scenario (STEPS) – one that includes current intentions, but not yet enhanced ambition – projects a 75% increase in electricity demand by 2050 and continued substantial reliance on fossil fuels, which pushes back the timing of deep decarbonization and increases climate risk (IEA, 2021). This has important policy implications — in the absence of an increased rate of deployment of renewables, lock-in impacts and stranded assets will reinforce carbon-dominated systems and undermine progress towards the goals set out in the Paris Agreement. For emerging economies, such as India, the challenge is two-fold – not only meeting the increasing need for power (which, according to The National Electricity Plan will be 2473.7 BU at 2031-32) but also reducing reliance on coal (the predominant source of power in the country). Policies like the Ministry of Power's Electricity (Rights of Consumers) Amendment Rules, 2023 – that are enabling time-of-day tariffs to better balance demand with solar generation – are a step in the direction for flexible, consumer-driven clean energy uptake. However, such measures could face limitations in terms of cost-effectiveness, technology diffusion, and institutional readiness under business-as-usual scenarios. This means that the global electricity transition under BAU is at risk of maintaining a two-tier world, dragging down emissions reductions and locking us into a high-carbon future – unless policy ambition increases significantly (Tripathy, 2025).

India's journey in energy transition is marked by the web of complications that impede a cleaner electric future. Key hurdle lies in infrastructural limitations, notably the sluggish roll-out of smart meters, which are critical for implementation of solar-hour ToD (time-of-day) tariffs. Although policies have been put in place, as of March 2025 only 10.8% of approved smart meters were installed, allowing for real-time monitoring and pricing of consumption across states, including Bihar and Madhya Pradesh, while even more developed states such as Gujarat and Maharashtra needed accelerated coverage to allow for effective tariff take-up (IEA, 2023). In the absence of a strong metering infrastructure, consumer awareness and behavioural changes are limited, particularly in rural and under-privileged households where information asymmetries and the cost of smart appliances act as obstacles (Luthra et al., 2015). But the structural problems run deeper than infrastructure alone: India continues to depend heavily on fossil fuels as a means of power. Coal and gas dominate the power sector, driving industrial growth and energy access and locking in a carbon-intensive economy. The International Energy Agency (IEA) Net Zero scenario demands annual coal generation declines in the region of 8% through 2030, a level nowhere to be seen in India's path, where new coal infrastructure is developed and commissioned (IEA, 2022). Further complicating this scenario is the stagnation of nuclear and hydropower plants that have traditionally supplied reliable, low-carbon electricity but are limited by high costs, regulatory barriers, and climate change-induced water stress (IRENA, 2023). Although intermittent renewable energies such as solar and wind have grown impressively, high growth rates are hampered by bottlenecks in the supply chain of strategic minerals, inflexibility of the electricity grid, and land use conflicts – particularly in densely populated areas (Gielen et al., 2021). Key to achieving that will take a huge amount of investment in transmission, storage, and demand-side management (DSM), areas in which the country remains behind. In addition, energy transition in India needs to do justice to concerns about equity, as per capita electricity consumption lags well behind global averages—1.3 MWh compared with 12 MWh in the U.S., indicating persistent access inequities (World Bank, 2023). Poor households are those who commonly use cheaper fossil-based energy, and clean forms of energy have a higher upfront cost, which makes equitable access for all to renewables an urgent challenge. The transition could even exacerbate regional and socio-economic inequalities without sufficient international funding, technology transfer and inclusive policies. So India's transition challenge is threefold: it needs to strike a balance between decarbonization and economic growth, modernize its infrastructure with inclusiveness and manage global supply chain and climate risks through coordinated policy action, strong investments, and international collaboration for a sustainable future.

LITERATURE REVIEW

India's energy transition has become a central locus of development, sustainability, and climate policy discourse—deserving at once of its ambitions and contradictions. Historical reviews establish that the push towards renewable energy dates back to the 1980s when concerns for fossil fuel vulnerability and climate change solicited investments in wind and solar, thereby India ranking fourth globally in installed renewable capacities, targeting 450–500 GW by 2030 (Lal et al., 2025). But the shift is highly uneven; even though urban households are making a faster transit towards modern fuels, rural ones predominantly rely on bio-mass because of sociocultural, technical and economic reasons- as can be seen in Bihar villages where clean energy use remains inefficient due to caste structures, remoteness factor and trust (Pachauri & Jiang, 2008; Goswami et al., 2017). These inequalities connect to broader justice discussions, for transition costs and benefits are unevenly spread, with rural and disadvantaged groups having to shoulder too much of the burden, a gap that is not properly recognised in planning (Haldar et al., 2023; Pandey &Sharma, 2021). This inequality is exacerbated by the structural challenges of coal-dependent India, where, due to coal's deep integration into employment, regional economies, and political power networks, there is paradoxical expansion of coal even in the wake of commitments to renewables (Roy& Schaffartzik 2021). Political economy studies also demonstrate that coal phase-out will be most detrimental to Eastern coal mining states, necessitating redistributive policies such as recycling carbon tax revenue in order to prevent regional inequality (Ordonez et al., 2023). At the same time, analyses in sectors underline that industries need to decouple growth from their carbon emissions through optimal least-cost pathways and non-dispatchable renewable integration, motivating innovative electricity storage forms such as battery or pumped hydro power plants for demand matching (Trudeau et al., 2011; Gulagi et al., 2017, 2018). AI and information management are both considered among the technologies that would enable optimization and efficiency of the grid; however, there has been fragmented policy support (Chawla et al., 2022). Clean energy potential also differs by states – Gujarat at the top and Uttar Pradesh at the bottom of the ideal zone, hinting at differentiated policy priority (Indrajayanthan & Mohanty, 2022). Water is also another dominant constraint as the power sector reinforces the nexus among resource systems (Prajapati et al., 2025). The legislative and policy frameworks, such as Saubhagya, Ujjwala, and rooftop solar initiatives, portray advancements in access but also lingering governance deficits (Jairaj & Kumar, 2019). Singh and Jana (2024b) investigated trends in India's energy security for the period 1970–2022, showing increasing reliance on imports of energy sources and calling for more investment in renewables to maintain the sustainability of long-term requirements. In addition, Singh and Jana (2024a) employed panel data to examine carbon emission determinants in the South Asian countries during the period 2000–2022 and concluded that greater levels of energy intensity, as well as renewable energy consumption lead to higher carbon intensities, and that income levels exert a negative impact on emissions while having a larger proportion of renewables in the total energy mix reduces emissions, thus calling for regional cooperation towards sustainable energy development. Jana and Singh (2022) evaluated the development pattern and factors of renewable energy in India, with large variations among different states; they revealed that the potential for renewable energy and state GDP might be important determining factors of RE development. Jana, Ghosh and Karmakar (2019) also examined the state-wise outlook of renewable energy in India and discovered large inter-state disparities with ANOVA showing noticeable variation among states' renewables development against potential, which suggests that there is ample scope for generating more power from renewable resources, especially solar.

All told, India has emerged as a global leader in renewable energy ambition, but the shift is far from predictable or homogeneous; instead, it is entangled with justice questions, technological capacity, socio-political formation, and ecological limitation. The consensus in the literature is that not only will India's transition to a low-carbon future remain contingent on renewables expansion, but it must also address equity issues, provide policy coherence, and integrate justice dimensions into transition plans for a sustainable and inclusive energy future.

The renewable energy sector in India is a vital support for India's power vis-à-vis the challenges of energy security, economic development, and climate change mitigation, yet an array of economic, technical, social, and policy barriers restrict its growth. Research indicates that Renewable Energy Technologies (RETs) are characterised by structural factors hindering their uptake in India despite strong policy initiatives and investment prospects. Luthra et al. (2015) highlight several barriers in terms of finance, technology, market and policy, applying the Analytical Hierarchy Process to show financial constraints, awareness deficiency and governance weakness as the predominant hindrances. Likewise, Sen and Ganguly (2017) remark that for sustainable development, renewable energy is indispensable, and India's dependence on fossil fuels continues to be an issue because of the inadequacies in infrastructure and policy. There are specific challenges that technologies face. Kulkarni and Anil (2020) identify the high 'upfront costs' and financing as key barriers for wind energy while Chakraborty et al. (2016) believe that even though solar industry is the cheapest sustainable energy after falling cost, they have to work on below elements such as awareness of consumers, R&D's weakness and lack of finance. Majid (2020) adds that although the demand for renewable energy has made India an attractive market with potential for generating jobs, there is relatively little progress, mainly due to inconsistent policies, investment risks and technology lapses. At the macro level, institutional and market barriers in terms of energy poverty and overambitious government targets have been identified as key obstacles that hinder efficient diffusion of RETs (Bhide and Monroy 2011; Reddy et al.2004). Further complexity is highlighted in sector-specific analyses. Govindan and Shankar (2016) for example, attribute capital costs being overly high is the major obstacle to offshore wind energy, but Pal et al. (2023) and the bottlenecks of these solar PV applications include affordability, regulatory barriers as well as long paybacks. For biofuels, Paul et al. (2025) note that despite their sustainability potential, unsuitable policies and implementation environments limit their widespread adoption. Financial constraints are still widespread, and Sarangi (2018) observed that the mobilization of green finance is impeded by high interest rates, short-term loans, and insufficient debt financing. They also expand to the related areas beyond energy technologies: Saha et al. (2021) noted a lacklustre response in terms of the green buildings uptake owing to disjointed policy frameworks, low public zeal, and indeterminate payback periods; meanwhile, Kumar et al. (2017) focus on weak waste-to-energy measures resulting from a lack of technical knowledge and inadequate structure for waste removal. Exogenous shocks aggravated some of these difficulties; Hoang et al. (2021) show how COVID-19 has destabilized global supply chains and financing, affecting the speed of India's transition, but also provide insights for sustainable forms of policy-making and resilience. Taken together, the review indicates that renewable energy uptake in India is hindered by a combination of financial, institutional, infrastructural, and sociocultural challenges. Addressing these will involve comprehensive policy reforms, risky financing schemes, enhanced R&D support and inclusive awareness generation to shift renewable energy from an add-on to a mainstream power source for sustainable development.

India's solar energy sector has become a linchpin in its own turn to renewables, encouraged by both the promise of plentiful sunshine and policy imperative to enhance energy security, cut carbon emissions, and upend the country's power mix. Early reviews focused that solar had potential to complement conventional resources, but its scaled-up development remained restrained due to high capital investment, grid interconnection, funding, and customers' approach (Kar et al., 2016). Researchers have been consistently advocating the requirement of solar energy to fulfill India's electricity scarcity, for environmental advancement and less reliance on imported fossil fuel (Sharma, Tiwari, & Sood, 2012). The launch of the Jawaharlal Nehru National Solar Mission (JNNSM) in 2010, a national action plan on climate change's strategy, was a key policy shift that sought to spearhead growth in solar capacity through grid-based as well as off-grid strategies as India's relief from debilitating poverty and scarcity issues, while simultaneously advancing it as a source of clean energy (Kapoor et al., 2014). The latter studies highlighted the impact of support policy, technology research and education on the development of solar energy in India, but also identified bottlenecks associated with rooftop installation and technology innovation (Raina & Sinha, 2019). Local experiences, such as in Rajasthan, showed that apart from climatic benefits, governance, infrastructure and enabling policy regimes are critical to attract mega-investments (Pandey et al., 2012), while state measures taken in Odisha exemplified the role of specific local policies for tapping solar potential (Sahu, 2016). Technical and resource assessments also highlighted elevated solar insolation levels in India, where most of the cities exhibited continual photovoltaic performance; however, there are barriers still to be addressed along the regulatory, financial and infrastructure axis (Irfan et al., 2020). Forecasting studies emphasized the necessity of improvements in solar radiation forecasts for better integration of variable solar power on the national grid (Mohanty, Patra, Sahoo & Mohanty, 2017) as reviews on solar photovoltaics reflected consistent policy support and industrial expansion in both solar parks and applications (Sahoo, 2016). Historical background showed that India is no newcomer to renewables, having supported multiple technologies since the time of its independence, yet solar has assumed importance due to global pressures of climate and the growth imperative posed by electricity in rural areas (Bhattacharya & Jana, 2009). Estimates for future point to a huge potential endowment in southern Asia that makes India an ideal solar playground, especially in states like Rajasthan and Gujarat through large-scale developments throughout the Thar Desert (Singh & Singh, 2016; Suman & Ahamad, 2018). Notwithstanding significant accomplishments— with capacity increasing from less than 10 MW in the early 2000s to thousands of MW within a decade (Yenneti, 2016)— solar advancement is being hampered by systemic challenges necessitating thoughtful policy allocation, innovative finance, strong infrastructure, and acceptance at the community level. In general, based on literature, India's solar transition is making tremendous progress and has potential, yet long-term sustainability can only be achieved if there is sustained policy coherence and support from institutions, as well as technological innovations to lead globally in the solar sphere.

While the renewable energy sector in India has seen rapid expansion, price and cost dynamics have been complex, determined by financing arrangements, policy instruments, and technological possibilities. Access to finance is one of several primary obstacles, where high-cost debt, short loan tenors, and floating-interest rates increase the price of renewable energy by close to 30% as compared to global benchmarks, emphasizing the requirement for low-cost, long-term credit instruments (Shrimali et al., 2013; Farooquee & Shrimali, 2016). Although programs, such as RECs, were formed to establish a market-based incentive for renewables adoption, they have had varying levels of success with low participation rates and violation of renewable purchase obligations in addition to less-than-efficient operation of REC price structures (Shrimali & Tirumalachetty 2013; Gupta & Purohit 2013; Shereef & Khaparde 2017). With these limitations in place, solar energy markets have expanded gradually through ambitious government goals, such as reaching 100 GW by 2022, albeit with problems around grid integration, high upfront capital expenditure and poor domestic manufacturing (Kar et al.,2016). Other than the expansion-planning analysis, some studies dealing with market interactions show that additional costs arising from system reliability due to intermittency of solar and wind power also cause congestion in interstate power flow, which has its impact on electricity price and dispatch. In addition, REC and electricity price relationships changed over time since the positively correlated pairs turned into negative when markets matured (Shrestha & Kakinaka, 2023), whilst empirical evidence indicates that electricity trading volumes and wholesale prices are dominant factors for the long-term solar REC prices (Irfan, 2021). Policy breakthroughs, such as interest-rate subsidies and currency hedging facilities, coupled with competitive auctions, have lowered the cost of finance and enhanced investor confidence in solar energy while making it more competitively priced (Farooquee & Shrimali, 2016; Thapar et al., 2016). Meanwhile, the ongoing discussion regarding least-cost renewable targets indicates that India's solar-focused pathway is not always the cheapest option and that optimized wind–solar mixtures can be more cost-effective in reducing CO2 emission levels at a competitive price (Deshmukh et al, 2021). Decentralized solar and hybrid solutions demonstrate some effectiveness in electrifying rural health centers and combating energy poverty on the microeconomic front, although their cost-efficiency is directly dependent on proximity to the national grid and incurred costs (Munuswamy et al., 2011). Comparative insights also demonstrate that, despite considerable progress, India remains behind countries such as China in terms of utilizing low-cost finance for rural renewables (Liming, 2009). After all, the solar story in India is an age-old paradox, that of enormous untapped potential versus financial barriers and structural inefficiencies, and market dislocation. RECs, hedging facilities, and renewable purchase obligations have all been developed to bridge these gaps, but more systemic changes across financing models, build-out of requisite infrastructure, and enforcement of policy are necessary for solar to be cost-competitive and deployed at scale in the Indian sector as a sustainable element of India's energy portfolio.

The discussion on energy transition in emerging markets has strongly broadened to note the interaction between technological innovation and sociopolitics, institutions, as well as finance. Academics emphasize that the shift to sustainable energy systems in the South is important for combating climate change and interacting with development (Cantarero, 2020). Historical context reminds us that windows open and close due to uneven trajectories over time in relation to health, literacy, welfare gains from partaking in energy shifts, and countless data gaps which limit association between energy transition and economic growth, causing further ambiguity as well as policy misalignments (Elias & Victor 2005). Clean energy pathways of developing countries are limited by infrastructural shortcomings, inefficiency in the power grid, and demographic pressure, but should be considered a specific technological solution to adopt sustainable development, even if it is an industrial development (Babayomi et al., 2022). The international climate policy and donor support are also important, and energy aid is sensitive to the renewable energy and electrification needs of recipients, especially after Kyoto (Kim, 2019). Yet structural financial barriers – in particular the high cost of capital – continue to be a key challenge, with lowering these through international climate finance and risk sharing facilities regarded as essential to scaling up renewable deployment (Calcaterra et al., 2024; Briera & Lefèvre, 2024). Cases such as the *Uttam Urja* initiative in India show that successful transitions to rural energy depend on tailor-made solutions, creative financing schemes and community involvement (Rehman et al., 2010). Similarly, the spatiotemporal decompositions of the rural energy upgrading in China shed some light on how urbanization, technological progress, and consumption patterns shape diverse regions with different resource endowment, suggesting a need for differentiated policies (Han et al., 2022). At the household level, there is empirical evidence from Vietnam suggesting the "energy stacking", including more than one form of energy use at the same time by households, could be determined by income, education and occupational diversity (Vo & Ho, 2024). The global discussions also emphasise that though financing and governance issues have been hanging around, developing economies have the chance to leapfrog to renewable futures by exploiting rich resources available to them and reforming institutional set-ups for investment influx (Ram et al., 2022). Monetary/Voucher incentives along with policy measures including location-based policies and feed-in tariffs are found to be effective in encouraging the adoption of solar power, as evidenced by Vietnam (Nguyen & Ponomarenko, 2025). In addition, a macroscopic examination shows that the effect of energy consumption on GDP is more significant in non-OECD countries, indicating that energy transition approaches should be designed according to economic development patterns in order for them to be sustainable (Komarova et al., 2022). On the whole, it seems from reviewing the literature that while developing countries have their peculiar structural, financial and governance challenges in the transition to clean energy, they also present a transformative potential for mismatches between scaling up clean energy and sustainable development can be realigned, provided that these policies incorporate local historical factors, technological advances, global finance cooperation and progressive participation.

RENEWABLE ENERGY POLICY IN INDIA

India has positioned itself as a global leader in renewable energy development by with aggressive policies and programmes striving to achieve the objectives of energy security, sustainable growth, and commitments made under the Paris Agreement and SDGs. So far, policy design and implementation across solar, wind, biomass, biofuels, and emerging technologies such as green hydrogen have been led by the Ministry of New and Renewable Energy (MNRE). India is constructing a well-laid-out renewable energy ecosystem by means of a mix of mega infrastructure projects, at the consumer level schemes and research and development (R&D) programs. This present article attempts to review the key renewable energy policies in India based on the source of energy, and focusing on international collaboration as well as domestic institutions.

Institutional and International Frameworks

India's renewables industry is favourable not only due to domestic policies, but also due to participation in the global market. TERI (The Energy and Resources Institute) was established in 1974 and has been at the forefront of research, public advocacy, and cutting-edge clean solutions, significantly impacting the challenges arising from climate change through adaptive measures (TERI, 2024). On the international front, India is an associate member of the International Energy Agency (IEA), which provides policy recommendations on achieving economic growth, energy security, and environmental protection goals (IEA, 2024). India also partners with the International Renewable Energy Agency (IRENA) for policy and technical assistance in the promotion of renewable energy penetration (IRENA, 2023). Others partnerships, such as REEEP and REN21, promote the sharing of experiences internationally (REN21, 2024). These collaborations ensure India plugs in to best practices while customising renewable energy policies for Indian realities.

Solar Energy Policy in India

Solar energy gets the most overarching policy-oriented attention in India with the country aiming to exploit its huge solar future. JNNSM (Jawaharlal Nehru National Solar Mission, launched in 2010) — part of the NAPCC — which initially set out to achieve 20 GW by 2022 and later revised the target to 100 GW, was scaled up further in the MNRE targets document (MNRE, 2023a). The mission supports rooftop (40 GW) as well as large-scale grid projects (60 GW) to make solar energy prices competitive with fossil fuels.

Another significant partnership is the International Solar Alliance (ISA) started by India and France at COP21 in 2015. ISA, now a compact body with more than 110 member states and is headquartered in Gurugram, concentrates on facilitating investments, exchange of good solar industry practices, and accelerating deployment of solar technologies in the tropical countries (ISA, 2024). At the national level, some mega policies are the Scheme for Development of Solar Parks and Ultra-Mega Solar Projects aimed at 40 GW capacity by providing all state-of-the-art infrastructure such as land, evacuation system and statutory clearances (MNRE, 2023b).

To encourage domestic module manufacturing, the Production-Linked Incentive (PLI) Scheme for HR Solar PV modules backs domestic module production and mitigates import dependence (MNRE 2023c). This is complemented by the CPSU Phase-II Scheme, which aims at 12,000 MW of solar projects for self-use or use by the government with VGF in the person country-made projects. At the domestic level, the PM-Surya Ghar: Muft Bijli Yojana, aims to install for rooftop solar, in one crore households giving up to 300 units of free electricity every month (PIB [Press Information Bureau], 2024). In agriculture, there is the PM-KUSUM Scheme, which encourages solar pumps and decentralized plants in order to lessen subsidy burdens, as well as transmission losses (MNRE, 2023d). Combined, these projects represent utility-scale, distributed generation, manufacturing, and agriculture aspects of solar power.

Wind Energy Policy in India

Wind power is the fourth biggest market in India and, on a global scale. National Offshore Wind Energy Policy (MNRE, 2015) announced in the year 2015 has designated National Institute of Wind Energy (NIWE) as the nodal agency for resource assessment and project facilitation. Viability gap funding is being provided for offshore wind projects, and the first of such projects is to be located in Gujarat and Tamil Nadu (MNRE, 2023e). Onshore wind power development has been supported by GBI (generation-based incentives), AD (accelerated depreciation), and competitive bidding; subsequently, there shall be over 40 GW of installed capacity by 2023 (Central Electricity Authority [CEA], 2023). Wind projects are not only expected to help diversify the renewable profile, but they also incorporate the latest technologies for higher efficiencies and hybrid projects – combining wind with solar generation.

Biomass and Bioenergy Policies

Biomass energy in India is used primarily for domestic cooking, rural and urban electricity production, and industrial thermal applications. The National Project on Biogas Development (NPBD) initiated in 1981–82 supplied rural families with biogas plants for cooking facilities, enhancement of sanitary conditions, and output of organic fertilizers (MNRE, 2023f). The Community and Institutional Biogas Plant Programme (CIBPP) focused on the poor by promoting community-model installations in a complementary manner.

National Programme on Improved Chulha (1986–87) aimed to minimize deforestation and health problems through the distribution of improved cookstoves with chimneys (MNRE, 2023g). Also, the Village Energy Security Programme (VESP) 2004; a scheme that is meant to cover remote and specimen cluster of villages for electrification with only one mini/micro- hydro project and promote village based products by exploiting renewable energy potential, including biomass-based electricity generation in off grid villages. Waste-to-energy, biomass briquettes and cogeneration projects that combine environmental protection with renewable energy supply have been focused on under the Bio-Energy Programme (2021–26) (MNRE, 2023h).

The National Policy on Biofuels (2009, revised in 2018) goal is to blend 20% of ethanol with petrol and 5% biodiesel with diesel, as well as promote non-edible oilseeds-based biofuels and advanced biofuels (MNRE, 2023i). This policy has financial incentives, minimum support prices for feedstock, and R&D support as its basis.

Green Hydrogen and Emerging Technologies

India's energy transition policies also apply to emerging areas. The National Green Hydrogen Mission (2023), with a budget allocation of 19,744 crore, would help in making India a green hydrogen hub for its production, utilization, and export (MNRE, 2023j). The task is accompanied by packages for electrolyser production (including incentive measures), pilot projects in the steel and transport sectors, as well as demand creation with compulsory purchase.

The Green Energy Corridors (GEC) programme seeks to enhance intra-state and inter-state renewable power evacuation grids and integrate variable renewables into the national grid (CEA, 2023). Complimentary missions also comprise the RE-RTD Programme (Renewable Energy Research & Technology Development programme) and Human Resource Development schemes which support invention, fellowships, and skill development (MNRE, 2023k).

The policy framing in India for renewable energy is one of the most holistic frameworks; covering infrastructure, manufacturing capacity, agriculture, R&D, and international partnerships. Solar policies (like JNNSM, PLI Scheme, PM-KUSUM) have pushed through rapid growth, while wind energy policies (including offshore) diversify the pool. All those biomass and biofuel programs for rural energy needs and waste management, and additional sundry benefiting plans like the Green Hydrogen Mission, are forward-looking. Participation in global forums such as the IBASA, IEA, and IRENA promotes international alignment and cooperation. Importantly, these policies not only help limit climate change and reduce dependence on fossil fuels, but also support livelihoods, enhance access to energy and encourage industrial development. In the future, it will be crucial for community engagement, financing, and grid dispatch to ensure that India's renewable energy path remains viable.

RENEWABLE ENERGY DEVELOPMENT IN INDIA

Domestic Energy Production in India

The changes in India's energy production within the country between 2000 and 2023 are part of Figure 1. The coal production tripled and extended from c.a. 5.5 million TJ to more than 16.1 million TJ, thereby reinforcing its excessive predominance in the energy mix of India . Only crude oil production fell 1.56 million TJ to 1.41 million TJ, and natural gas increased mildly from 966,000 TJ to about 1.25 million TJ. High rates of growth were also experienced in nuclear energy, which increased by almost threefold from 184,385 TJ to 522,949 TJ and hydropower generation, which doubled (from 268,063 TJ to 518,004 TJ). Renewable sources, such as wind and solar, have seen the most growth by far — increasing from 7,549 TJ in 2000 to 830,227 TJ in 2023. Biofuels and waste also had a large increase, rising from 5.26 million TJ to almost 8.97 million TJ. All in all, the numbers point to a dramatic increase in use of coal (which India has in abundance), and impressive buildout of solar, wind and other non-fossil energy sources – i.e., slowly but surely expanding away from a reliance on hydrocarbons.

Figure 1. Domestic energy production in India (TJ)

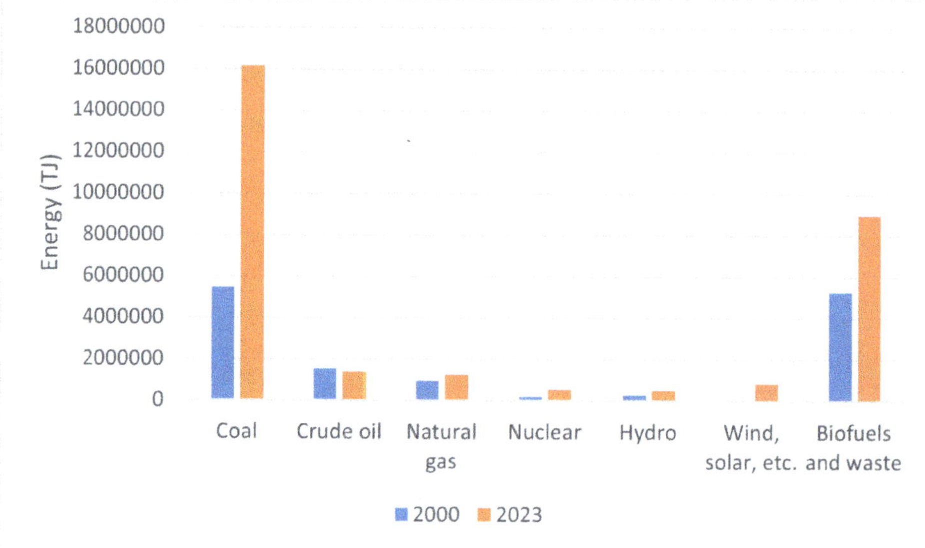

Source: IEA

Table 1 presents the trend of India's domestic energy mix during 2000–2023. Coal's share rose steadily from 39.9% in 2000 to the projected 54.5% in 2023, highlighting its increasing dominance of the energy mix. In comparison, the proportion of crude oil has steadily decreased from 11.4% in 1976 to 4.7%, indicating that our energy dependency issue is not a result of insufficient domestic production of petroleum or crude oil. Natural gas swung, reaching 9.2 percent in 2010 before falling to 4.2 percent by 2023. Nuclear grew a little, going from 1.3% in 2000 to 1.8% in 2023, and hydro was roughly flat, but a touch lower, going from 2.0% down to 1.7%. Meanwhile, renewable sources, including wind and solar — and despite a meager 0.1% share in 2000 — reached an impressive 2.8% by last year (2023). Meanwhile, biofuels and waste, at 38.4% of "renewables" in 2000, fell to 30.3% by 2023 (but are still a big part). Overall, the numbers show coal creeping ahead (in terms of absolute production) with renewables growing in increments and all the rest — oil, gas, hydro, and biofuels — either sliding or holding steady.

Table 1. Share of different sources in domestic energy production in India (TJ)

Year	Coal	Crude oil	Natural gas	Nuclear	Hydro	Wind, solar, etc.	Biofuels and waste	Total
2000	39.9	11.4	7.0	1.3	2.0	0.1	38.4	100.0
2005	43.3	10.0	6.9	1.2	2.5	0.2	36.0	100.0
2010	45.4	9.2	9.2	1.5	2.3	0.4	32.1	100.0
2015	50.0	7.9	5.0	1.8	2.2	0.9	32.1	100.0
2020	48.5	6.2	4.1	2.0	2.4	2.2	34.6	100.0
2023	54.5	4.7	4.2	1.8	1.7	2.8	30.3	100.0

Source: IEA

Installed power capacity in India from 1998 to 2024 by energy source presented in Table 2 displays growth statistics relating to the installed power capacity of India between 1998 and up to the year 2024 according to different sources of energy. The thermal power capacity, which is the mainstay of electricity generation, has grown from 64,005 MW in 1998 to 243217 MW in 2024, though it has trended towards diminishing proportions. Nuclear advanced at a slow but steady pace from 2,225 MW up to 8,180 MW over the same period. India also added substantial new large hydro capacity, but at a more gradual rate – 21,904 MW in 1998 to 46,928 MW in 2024 – due to the economic limits associated with its geography. The most spectacular growth was in the renewable energy sources (RES), which soared from under 1 GW in 1998 to nearly 191 GW in 2024, and made a substantial contribution to the energy mix of India. Thus, total installed capacity increased more than four times (from around 89,000 MW in 1998 to nearly 442,000 MW in 2024), illustrating India's continued progression toward a diversified and cleaner energy system.

Table 2. Electricity from renewable energy sources in India

Year	Thermal (MW)	Nuclear (MW)	Large Hydro	RES* (MW)	Total (MW)
March, 1998	64005	2225	21904	968	89102
March, 2000	70193	2680	23857	1155	97885
March, 2002	74429	2720	26269	1628	105046
March, 2004	77969	2720	29507	2488	112684
March, 2006	82410	3360	32326	6191	124287
March, 2008	91907	4120	35909	11125	143061
March, 2010	102454	4560	36863	15521	159398
March, 2012	131603	4780	38990	24504	199877

continued on following page

Table 2. Continued

Year	Thermal (MW)	Nuclear (MW)	Large Hydro	RES* (MW)	Total (MW)
March, 2014	168255	4780	40531	34988	248554
March, 2016	210675	5780	42783	45924	305163
March, 2018	222907	6780	45293	69022	344002
March, 2020	230600	6780	45699	87028	370106
March, 2022	236109	6780	46723	109885	399497
March, 2024	243217	8180	46928	190573	441970

Source: CEA Report (Various years 2015- 2024)

This picture of electricity generation (from different energy sources) in India for the period 2015-2025, given below in Table 3 presents a mixture of not breaking away from dependence on conventional and food-for-power sources, but witnessing a spectacular ascendence of renewables. Coal, predictably and against all odds, would prevail as the most reliable fuel for power generation over the years – growing from 835,291 MU in 2015 to 1,331,867 MU in 2025 - with intermittent declines like that seen in FY20-21 too bad for it to erode its overall dominance within the energy mix. However, oil and natural gas generation fell from 42,651 MU in 2015 to 32,023 MU by the year 2025, showing lower dependence on this fossil fuel because of meager domestic reserves as well as an inclination towards cleaner options. Nuclear-based capacity also rose over the period, from 36,102 MU in 2015 to 56,681 MU by 2025, indicating India's move towards diversification of base load capacity through atomic power. Big hydro power generation, however, saw ups and downs – it reached a peak of 162,099 MU in 2023 but then declined in the year after and again to partially recover by 2025 (148,634 MU), reflecting largely the monsoon-dependent nature of hydro and availability of water. Renewable power plants, on the other hand, chief among them solar and wind, have performed impressively. Solar energy in particular experienced an explosive uptick from 4,599 MU in 2015 to 144,150 MU by 2025 - becoming one of the leading fast-growing sectors driving India's clean energy growth story. Wind energy also rose significantly, from 33,768 MU in 2015 to reach 83,347 MU by 2025, indicative of the increased wind capacity in states such as Tamil Nadu, Gujarat, and Maharashtra. Small-hydro projects were also a tame contributor, maintaining between 7692 MU (in 2018) and 11,568 MU (in 2025) through the years, which reflected their weak but consistent role in localized energy supply. Biomass-based power generation had fallen to 15,292 MU in 2015 and was expected to increase only marginally to 15,944 MU by 2025, showing a supporting act as cover with its own set of environmental problems. On balance, the table suggests that although coal remains the dominant source of electricity generation in India, such rapid growth in solar and wind points to a structural change in its power system toward renewables. Nuclear and hydro are still of significance, but as relatively smaller donors, whereas oil and gas are losing traction. This trend shows India's twin challenge of managing fast-paced growth in electricity demand while protecting the environment and making the energy sector environmentally sustainable, as renewables are increasingly being promoted as a means to meet a significant share of future energy security.

Table 3. Power generation from different energy resources in India in Million Units (MU)

Year	Coal	Oil and Natural Gas	Nuclear Power	Large Hydro	Solar Power	Wind Power	Small-Hydro	Biomass-Power
31.03.2015	835291	42651	36102	129244	4599	33768	8060	15292
30.03.2016	895340	47673	37414	121377	7448	33029	8355	16949
30.03.2017	944022	49495	37916	122378	13499	46004	7673	14372
31.10.2018	986591	50556	38346	126123	25871	52666	7692	15611
31.03.2019	1022265	50049	37813	134894	39268	62036	8703	16752
31.03.2020	994197	48641	46472	155769	50131	64646	9451	14108
31.03.2021	981443	51168	43029	150300	60402	60150	10258	16437
31.03.2022	1078581	36230	47112	151627	73484	68640	10464	18325
31.03.2023	1182096	24115	45861	162099	102014	71814	11171	18554
31.03.2024	1294852	31697	47937	134054	115975	83385	9485	16989
31.03.2025	1331867	32023	56681	148634	144150	83347	11568	15944

Source: Energy statistics of India and CEA Report (Various years 2015- 2024)

Table 4 gives the installed capacity of various RESs in India during the 16-year period (2009–2024) from which it is evident that its RES-profile has changed phenomenally over these years. Biomass generation has grown to date from 1,752 MW in 2009 to 10,356 MW in 2024 largely coming from what uses agricultural residues and other organics for power production. Even though still marginal in size compared to other sources, waste-to-energy increased from a minuscule 59 MW of generating capacity in 2009 to 586 MW in 2024, reflecting growing concern with urban waste and energy generation. Wind energy, being one of the first commercially implemented renewable technologies in India, saw a steady rise from 10,242 MW in 2009 to 45,882 MW in 2024 with substantial contributions coming from states like Tamil Nadu, followed by Gujarat and Maharashtra. Small hydro also grew modestly from 2,181 MW in 2009 to 4,958 MW in 2024, with geographical and environmental constraints. The largest increase came from solar power, which rose from a meager 2 MW of capacity in 2009 to a whopping 81,761 MW in 2024 and it became the fastest-growing intermittent source of power in the country's energy mix. This growth is a result of ambitious government programmes like the National Solar Mission, solar parks and policy support around the feed-in tariffs which generated massive investment. Large hydro power, historically a leading renewable source of energy witnessed no significant addition from 36,075 MW in 2009 to just 46,928 MW in 2024 indicating saturation of viable project locations and increasing anxiety over environmental and displacement issues. The bigger picture clearly indicates a change in the renewable energy investments landscape of India and solar plus wind have been dominating capacity addition complemented by significantly smaller contributions from biomass, waste-to-energy, small hydro etc. That old hydro number is still a very significant portion, but it has less relative weight these days as new technologies have surged. Solar & wind had between them together, become over 127,000 MW of capacity by 2024, thereby completely overshadowing all other contenders and proving that India was serious about moving to cleaner and more sustainable energy. This shift reflects the effectiveness of policy and that country's response to the increasing demand for energy, climate pledges, and the quest for energy security. The data therefore underscores that India has made impressive strides in diversifying its energy mix, getting away from fossil fuels, and further establishing itself as a leader in renewable energy deployment across the world.

Table 4. Installed Capacity of different renewable energy sources of India (In MW) from 2009-2024

Year	Biomass energy	Waste to energy	Wind Power	Small Hydro	Solar Energy	Large Hydro power
31.03.2009	1752	59	10242	2181	2	36075
31.03.2010	2200	65	11807	2735	10	36164
31.03.2011	2665	72	14156	3043	35	36633
31.03.2012	3135	90	17353	3395	941	38235
30.03.2013	3601	96	19047	3632	1686	38694
30.03.2014	4014	107	21132	3804	2631	39631
31.03.2015	4419	115	23500	4055	3743	40318
30.03.2016	4832	115	26862	4273	6705	41729
30.03.2017	8182	114	32275	4380	12231	43164
31.10.2018	9408	138	34982	4507	24312	44144
31.03.2019	9778	138	35622	4575	28181	44237
31.03.2020	9875	148	37739	4665	34628	44492
31.03.2021	10146	388	39243	4769	41191	45047

continued on following page

Table 4. Continued

Year	Biomass energy	Waste to energy	Wind Power	Small Hydro	Solar Energy	Large Hydro power
31.03.2022	10206	477	40353	4849	53952	45560
31.03.2023	10248	554	42629	4901	66728	45688
31.03.2024	10356	586	45882	4958	81761	46928

Source: Energy statistics of India and CEA Report (Various years)

Growth contributors for various energy sources between 2008 and 2023 are in Table 5, which demonstrates the trend of increasing shares due to the renewable energy mix. Regarding non-renewable, solar power showed the largest annual increase as it exceeded 62.71%, which indicated great policies and investments in PV technologies. Waste-to-energy (14.86%) and biomass & wind power (12.69% and 9.83%, respectively) posted robust growth, while small hydro increased by a moderate 4.57%. On the other hand, large hydro grew by a meagre 1.87%, reflecting capacity addition saturation. Nuclear power grew at a slow pace (3.95%), and the traditional sources such as coal (7.15%) and natural gas (3.37%) also increased on a moderate level. Take, for example, Diesel-based generation, it showed a negative growth of -6.97% indicating its waning nature in favour – limitations related to cost and environment. In nominal terms, non-conventional sources added 15.66% more total capacity if we do not count large hydro and 9.18% if we do, compared to just 6.47% for conventional sources overall. This led to the total installed capacity in India increasing by 7.49% annually, with CEA reporting an overall growth of 7.35%, highlighting a substantial buildup of renewable power capacity as part of the country's power mix during this year.

Table 5. Growth trend of different energy sources in India for the period 2008-2023

Different sources of energy	Growth Rate
Biomass Power	12.69% ***
Waste to Energy	14.86% ***
Wind Power	9.83% ***
Small Hydro	4.57% ***
Large Hydro	1.87% ***
Solar power	62.71% ***
Nuclear Energy	3.95% ***
Total capacity generation from non-conventional sources (Without large hydro)	15.66% ***

continued on following page

Table 5. Continued

Different sources of energy	Growth Rate
Total capacity generation from non-conventional sources (With large hydro)	9.18% ***
Coal	7.15% ***
Natural Gas	3.37% ***
Diesel	(-6.97% ***)
Capacity generation from conventional sources	6.47% ***
Total installed capacity	7.49% ***
Total All India Installed Electricity Generating Capacity	7.35% ***

*** indicates level of significance at 1%

Source: Authors' estimation based on Energy statistics of India and CEA Report (Various years)

Figure 2 gives the present scenario of renewable and hydro energy in India as on 31st March, 2024 in regard to the potential capacity of these sources, their installed capacity and percentage of potential being harnessed. Regarding the sources, wind power has highest potential of 11,63,856 MW, and installed capacity is 45,882.21 reflecting its utilization for only 55% stands at a meager level on a practical basis of mere 3.94%, leaving an untapped vast potential source. Solar, with a potential of 7,48,990 MW, second only to Wind, has an installed capacity of 81,760.79 MW, which is about 10.92% harnessing, depicting a lukewarm progress with considerable scope is still present for exploiting this potential commercially. Power generation from large hydropower, with a potential of 1,33,410 MW, has been realized to the extent of 46,928.17 MW (35.18% utilization), which is the highest among all sources and the most effectively tapped source in one way. Biomass energy and small hydro are slightly behind with 24.50% and 23.46% exploitation ratio, respectively, whereas waste-to-energy, having a lesser potential of 2,556 MW, harnessed only 585.8 MW (22.92%). The study indicates that in the case of India, development of biomass, small hydro, and large hydro has made significant progress, but potential assessments show solar and wind resources as being grossly underutilized. This presents a challenge and an opportunity for policy actions, infrastructure developments, and technological innovations to ensure that clean energy generation is maximized while future energy needs are served in a sustainable fashion.

Figure 2. Percentage of harness from potential capacity from different sources of power in India in 2024

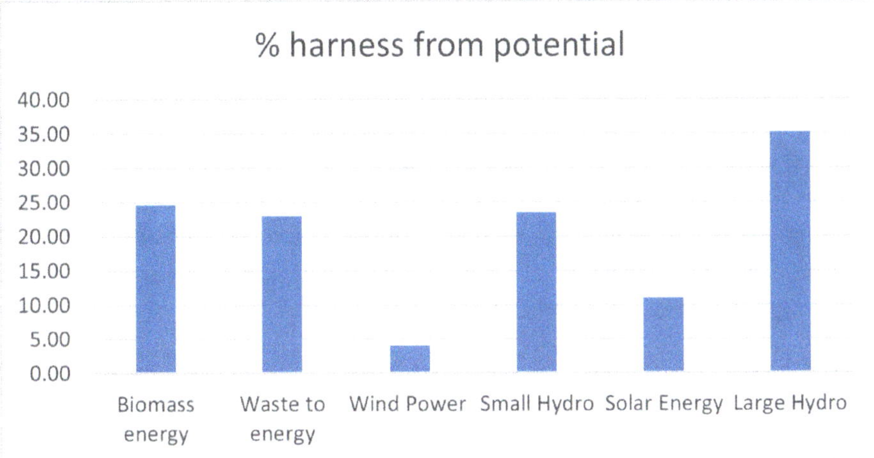

Source: Energy Statistics of India, 2025

The world and Indian energy consumption status over various sources from the year of 1990 to 2024 is shown in Table 6, which indicates an increase of share (%) (India) against the world total for global energy requirements. Worldwide, cumulative energy use soared from 95,354.33 TWh in 1990 to 175,271.75TWh in 2024 to assume substantial scale on the back of industrialisation, population growth, and overall technological advances. Total consumption in India increased even more rapidly, from 2,288.41 TWh in 1990 to 11,276.18 TWh in 2024: together with the increase of its share on the world level (from 2.40% to 6.43%), this suggests the emergence of this country as a major consumer of energy resources. In individual sources, the fastest-growing worldwide sources are wind and solar power, which have gone from 10.09 TWh and 1.08 TWh in 1990 to 6,124.47 TWh and 5,150.57 TWh in 2024 respectively. India's uptake has surged, with wind energy climbing from virtually nothing (0.08 TWh) in 1990 to 198.89 TWh in 2024 and increasing its global share to 3.25%, while solar has grown from essentially zero a quarter-century ago to reach 333.55 TWh by the same cut-off point during the reference period, or around 6.48% of world total electricity generation, showcasing India's rising lead on solar expansion. Even classic renewables like hydro grew, the Indian hydro generation rising from 184.44TWh to 381.78 TWh, but its share increased just a smidgen from 3.08% to 3.52%. Nuclear and biomass are smaller percentages in India, at 1.94% and 3.48%, respectively, with shares for both increasing to 2024, but suggest diversification in the Indian energy centre. Fossil fuels, however, still dominate. Coal, oil and natural gas account for the major portion of global energy consumption; while specifically in India, much of that climate-damaging form increased dramatically from 1,276.03 TWh in 1990 to its marked presence at 6,379.83 TWh in 2024 as it contributions jumped rose from a mere 4.92% to today's towering-at-13.91%, forming a statement on continued reliance by India on coal-based power for powering the nation and industry. Oil and natural gas consumption expanded too, with India's share in world oil increasing from 1.83% to 5.48%, and that in natural gas stretching from 0.60% to 1.70%. The analysis demonstrates two key trends: the widespread uptake of RE, in both the world and India (in particular wind and solar), albeit that fossil fuels, especially coal, continue to be predominant in the Indian energy mix, raising sustainability concerns. India's increasing share of the global energy pie is not only the result of economic growth (and rising energy intensity), but also a sign that India needs additional, more robust policies to reconcile its need for energy security, economic development, and its climate commitments.

Table 6. Energy consumption (TWh) of the World (source-wise) from 1990-2024

	World		**India**		**% share of India in the world**	
Year	1990	2024	1990	2024	1990	2024
Wind Energy	10.09	6124.47	0.08	198.89	0.83	3.25
Hydro Power	5996.17	10860.75	184.44	381.78	3.08	3.52
Solar Energy	1.08	5150.57	0.00	333.55	0.00	6.48
Nuclear	5557.21	6871.84	17.76	133.42	0.32	1.94
Biomass Energy	106.57	1366.88	0.00	47.63	0.00	3.48
Other RE	366.33	2476.36	0.12	69.53	0.03	2.81
Coal	25928.91	45850.54	1276.03	6379.83	4.92	13.91
Oil	37906.82	55292.08	694.06	3028.74	1.83	5.48
Natural Gas	19481.15	41278.27	115.93	702.82	0.60	1.70
Total	95354.33	175271.75	2288.41	11276.18	2.40	6.43

Source: Energy statistics of India, CEA Report (Various years) and Statistical Review of World Energy, 2025

Contributions of various sources in electricity generation by sources in India and the world for 2024 have been presented in Figure 4. There are significant differences between India's situation and global average. Coal globally contributes 34% to electricity production, while in India, it is an overwhelming 48%, reflecting India's continued reliance on coal and the transition away from it that other countries have made. In contrast, 23% of the world's electricity is generated from natural gas, but only 6% in India, mainly due to the availability of smaller gas reserves and infrastructure hurdles. There is an interesting pattern when it comes to renewables: that is, while its share elsewhere in the world is 17%, in India it stands at a much higher 33%—thus reflecting the fact that India has actually been very aggressive on solar, wind, biomass, and other such cleaner sources of power. Hydroelectricity contributes to 14% of generation globally and that in India is also not too different at 11%, indicating usage is relatively balanced but under-utilised in case of India. Just 2% of India's electricity is supplied by nuclear, which meets 9% of global needs, and much more could be if the country had greater installed capacity or was prepared to adopt better policy. The "other" category makes up 3% of it globally, but holds little sway in India. Overall, it is clear that India has a long way to go toward decarbonization and that we should not be overly blithe about the lesson of this comparison for other developing countries. While India remains coal-centric, its transition to renewables is in fast gear—at least in parts of the electricity sector—and in doing so is emerging as anything but a laggard on the global-scale green transition stakes (as appears gas and nuclear under-employed).

Figure 3. Contribution of different energy sources in electricity generation in India and in the World for 2024

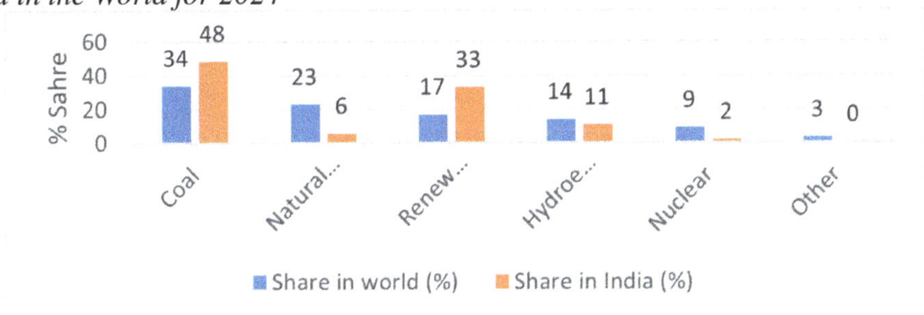

Source: CEA Report, 2024 and Statistical Review of World Energy, 2025

Table 7 indicates renewable energy production trends for the top five countries (the US, China, Germany, India and Japan) in exajoules from the years 2014 to 2024, illustrating how wide-ranging growth rates and energy output figures vary from one another. China takes a clear lead in the world, leaping from 5.12 exajoules in 2014 to 13.9 by 2024 and reflecting its massive investment in solar, wind, and hydro with the steepest growth of all countries. Back in the U.S., the second-largest producer of bioenergy, still sees solid but relatively slower growth: a more modest transition from 4.34 exajoules to 6.65 over the same decade-long period, indicating its slow adoption of cleaner technologies even as it sticks to fossil fuels. Germany is traditionally one of the renewables front runners, but exhibits a slower growth pattern, from 1.06 exajoules to 1.49 over the decade and plateau post-2020 which indicates market saturation/maturity with scaling up constraints. India's path is striking, as it starts from a lowly 0.88 exajoules in 2014 but steadily climbs to reach 1.77 exajoules by 2024, ahead of Germany and indicating the policy-driven drive for solar and wind power it's experiencing under its national renewable energy targets. Japan, too, hasn't shown any sign of faltering, growing from 0.73 in 2014 to 1.29 in 2024, albeit at a much slower pace than India. The broader picture in the data is of a global trend towards increasing concentration, as the gap between China and 'the rest' continues to grow, with its output in 2024 more than twice that of the US, though India is forecast to register the fastest relative growth rate among developing economies. This has been the case on the backdrop of different national priorities, technological capabilities, and policy environment, where China and India are evidently leading renewable energy producers, US is making some progress, Germany is following recent stagnation, and Japan is gradually progressing as collectively they form a future landscape for global renewables.

Table 7. Renewable energy production (in Exajoules) trend of the top 5 countries in the world from 2014 to 2024

Year	US	China	Germany	India	Japan
2014	4.34	5.12	1.06	0.88	0.73
2015	4.44	5.51	1.16	0.84	0.84
2016	4.80	6.04	1.17	0.88	0.79
2017	5.10	6.68	1.27	0.95	0.84
2018	5.18	7.42	1.30	1.10	0.9
2019	5.26	8.21	1.36	1.24	0.92
2020	5.34	8.96	1.43	1.29	1.01
2021	5.62	10.14	1.34	1.38	1.1
2022	6.04	11.06	1.43	1.59	1.14
2023	6.24	11.99	1.49	1.63	1.21
2024	6.65	13.90	1.49	1.77	1.29

Source: Statistical Review of World Energy, 2025

Figure 4 shows the decreasing trend of LCOE in solar power across top five solar-producing countries – China, USA, India, Germany, and Japan, along with the world average from 2010 to 2023. The LCOE (in $/MWh) is the average revenue per MWh required to recover all the lifetime costs of generating electricity, including the cost for capital, plant operation and maintenance, fuel, and financing. Solar in 2010 was not as cheap; LCOEs were between $0.24/kWh (USA) to $0.49/kWh (Japan), world average being $0.46/kWh. The LCOE has decreased systematically and sharply across all countries with technological progress, large-scale deployment, module price falling down and beneficial policies. By 2023, LCOE plummeted with China ($0.0356/kWh) as the cheapest, followed by India ($0.0481/kWh) and Germany

($0.0626/kWh), while Japan ($0.1101/kWh) was the most expensive among the five countries. The mean centralised and all-in LCOE decreased to USD 0.0442/kWh, or close to 90% less than in 2010. All in all, the data unequivocally underscores how solar power continues to grow more and more competitive throughout the world — in fact, it's already one of the cheapest sources of energy as we make our global pivot toward sustainable energy.

Figure 4. Trend of LCOE (USD/Kwh) of solar power in top 5 countries of the world in 2023

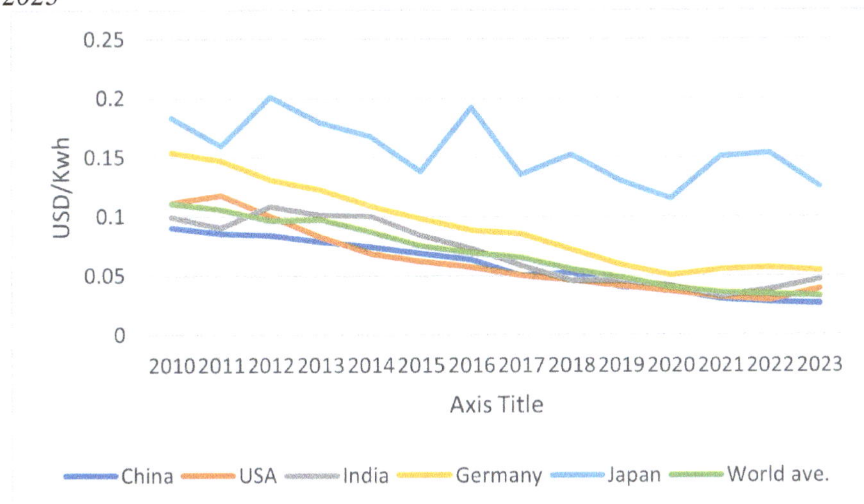

Source: IRENA

India's energy transition needs to be seen in the context of global energy dynamics, with renewables on a steep growth path, but fossil fuels – especially coal – still dominant. Wind and solar comprised 12% of global electricity in 2023; the two technologies had crossed the one-terawatt capacity threshold within the past two years. But even with these victories, the UN global stock take makes clear that progress has been far too slow to get anywhere near the 1.5^{0}C climate goal: coal and gas consumption have gone up by 22% and 37%, respectively, since 2010. And while the US and European Union are increasingly retiring coal plants, thanks to favorable policies in declining prices for cleaner alternatives amid stable electricity demand, Asia is a mixed picture. Electricity use in the area has tripled since the beginning of this century, ensuring coal remains a long-term spine of growth. China alone added 27 gigawatts of new coal capacity in 2022, countering declines in coal use around the world, while India remains dependent on coal to power industrialization and growing household demand. India, however, is also grandiosely pursuing one of the world's most aggressive renewable agendas – aiming for 500GW of non-fossil capacity by 2030 – a clear sign that its energy transition has to be taken seriously. In contrast to OECD countries, where the growth in renewables undercuts the dominance of fossil fuels, India has to invest heavily just to keep up with rising energy consumption and improve its grid – although falling prices for solar and wind make it a better deal than ever.

At the same time, India's role in the global shift to a new energy paradigm is critical and serves as a microcosm of the larger transitions required of developing economies looking to balance growth with sustainability. At home, the steep falls in the costs of renewables, plus advances in battery storage technologies, raised prospects that clean energy could compete with coal for battery baseload and peak load power. This parallels what is happening also in advanced economies, but is complicated by India's high rate of demand growth and its infrastructural bottlenecks that make integration of renewables into the grid at scale more challenging. Further, India's continued coal dependence is driven not just by economics but also by restrictions on financing and policy adaptation, emphasizing the need for international collaboration. That India needs to speed up closing the financing gaps, upgrade its grid systems, and effectively tap technology transfers to align with global climate goals would have a large appendix. India lags behind OECD countries such as the U.S. and Germany in renewable penetration, but its trajectory of growth is among the highest globally and has seen it as a petri dish for decarbonization efforts around the world. The fate of India's transition will matter enormously, if it succeeds, in rapidly scaling low-cost renewables while phasing down coal we are witnessing not only would the optimal conditions for domestic energy security be met, but also a monumental contribution to the global fight against climate change, tipping Asia's coal-fixated power landscape toward a cleaner future.

CONCLUSION

In conclusion, India's energy transition is a complex picture of fast economic growth humming with impressive amounts of energy consumption, and the need to move quickly down a different path for long-term sustainability. On the other hand, coal has reinforced its position as the dominant source of energy domestically speaking, with a production in 2023 that exceeds half of total output (this fact underlines the CPU dependence from fossil fuels). Meanwhile, oil and gas have also stagnated or declined in terms of their relative share, highlighting domestic reserve constraints. But the dramatic increase is in renewable energy sources, and – specifically – in the quantum growth of solar and wind power capacity on the back of policy support, price fall, and technological progress. There have also been modest but steady contributions from nuclear and hydro, with biofuels and waste as well contributing significantly although those contributions have been falling in relative terms. Combined, these trends represent a slow transition in India's energy mix — one that still relies heavily on coal in the immediate future but increasingly will rest on renewables to be sustained over the long term.

At the same time, the data reveals both progress and unmet potential. Installed renewable capacity has increased almost two-hundredfold since the late 1990s and solar is now overtaking all other sources in terms of adding new supply, while wind has solidified its place as a major provider. However, the exploitation of available potential is very small, especially for solar and wind, which are among the smallest contributors to the energy system despite having a large resource base. It is thus on two levels that India's dual role stands out: as a country that remains more coal-dependent than the rest of the world, and one that has now surpassed many advanced economies in renewable growth rates, establishing itself as a key player in the global energy transition. Next, how India treads its path will rely on the acceleration of infrastructure investment ó in grid connection, climate finance, and global technology collaboration. By making use of these opportunities, India can meet not just its own energy security requirements but also become a global leader in shaping sustainable energy futures.

REFERENCES

Babayomi, O. O., Dahoro, D. A., & Zhang, Z. (2022). Affordable clean energy transition in developing countries: Pathways and technologies. *iScience*, *25*(5).

Bhattacharya, S. C., & Jana, C. (2009). Renewable energy in India: Historical developments and prospects. *Energy*, *34*(8), 981–991.

Bhide, A., & Monroy, C. R. (2011).. . *Renewable & Sustainable Energy Reviews*, *15*(2), 1057–1066.

Briera, T., & Lefèvre, J. (2024). Reducing the cost of capital through international climate finance to accelerate the renewable energy transition in developing countries. *Energy Policy*, *188*, 114104.

Calcaterra, M., Aleluia Reis, L., Fragkos, P., Briera, T., de Boer, H. S., Egli, F., & Tavoni, M. (2024). Reducing the cost of capital to finance the energy transition in developing countries. *Nature Energy*, *9*(10), 1241–1251.

Cantarero, M. M. V. (2020). Of renewable energy, energy democracy, and sustainable development: A roadmap to accelerate the energy transition in developing countries. *Energy Research & Social Science*, *70*, 101716.

Central Electricity Authority. (2023). *Report on optimal generation capacity mix for 2030*. Government of India.

Central Electricity Authority. (2024). *Annual Report 2024*. New Delhi: Central Electricity Authority. https://cea.adgstaging.in/annual-report/?lang=en

Chakraborty, S., Sadhu, P. K., & Goswami, U. (2016).. . *Problemy Ekorozwoju*, *11*(2), 75–80.

Govindan, K., & Shankar, M. (2016).. . *International Journal of Energy Sector Management*, *10*(2), 266–282.

Chawla, Y., Shimpo, F., & Sokołowski, M. M. (2022). *Artificial intelligence and information management in the energy transition of India: Lessons from the global IT heart*. Digital Policy. *Regulation & Governance*, *24*(1), 17–29.

Deshmukh, R., Phadke, A., & Callaway, D. S. (2021)... *Proceedings of the National Academy of Sciences of the United States of America*, *118*(13), e2008128118.

Energy Institute. (2025). *Statistical Review of World Energy 2025*. London: Energy Institute. https://www.energyinst.org/statistical-review/home

Elias, R. J., & Victor, D. G. (2005). Energy transitions in developing countries: a review of concepts and literature. Program on Energy and Sustainable Development, Working Paper. Stanford: Stanford University.

Farooquee, A. A., & Shrimali, G. (2016).. . *Energy Policy*, *95*, 518–528.

Gielen, D., Boshell, F., Saygin, D., Bazilian, M. D., Wagner, N., & Gorini, R. (2021). The role of renewable energy in the global energy transformation. *Energy Strategy Reviews*, *24*, 38–50.

Goswami, A., Bandyopadhyay, K. R., & Kumar, A. (2017). Exploring the nature of rural energy transition in India: Insights from case studies of eight villages in Bihar. *International Journal of Energy Sector Management*, *11*(3), 463–479.

Gulagi, A., Bogdanov, D., & Breyer, C. (2017). The demand for storage technologies in energy transition pathways towards 100% renewable energy for India. *Energy Procedia*, *135*, 37–50.

Gulagi, A., Bogdanov, D., & Breyer, C. (2018). The role of storage technologies in energy transition pathways towards achieving a fully sustainable energy system for India. *Journal of Energy Storage*, *17*, 525–539.

Gupta, S. K., & Purohit, P. (2013).. . *Renewable & Sustainable Energy Reviews*, *22*, 380–392.

Haldar, S., Peddibhotla, A., & Bazaz, A. (2023). Analysing intersections of justice with energy transitions in India-A systematic literature review. *Energy Research & Social Science*, *98*, 103010.

Han, J., Zhang, L., & Li, Y. (2022). Spatiotemporal analysis of rural energy transition and upgrading in developing countries: The case of China. *Applied Energy*, *307*, 118225.

Hoang, A. T., Nižetić, S., Olcer, A. I., Ong, H. C., Chen, W. H., Chong, C. T., & Nguyen, X. P. (2021).. . *Energy Policy*, *154*, 112322.

Indrajayanthan, V., & Mohanty, N. K. (2022). Assessment of clean energy transition potential in major power-producing states of India using multi-criteria decision analysis. *Sustainability*, *14*(3), 1166.

International Energy Agency (IEA). (2021). *World Energy Outlook 2021*. IEA.

International Energy Agency (IEA). (2022). *World Energy Outlook 2022*. IEA.
International Energy Agency (IEA). (2023). *Tracking Clean Energy Progress 2023*. IEA.

International Energy Agency (IEA). (2023a). *Electricity Market Report 2023*. IEA.

International Energy Agency (IEA). (2023b). *Global Energy Review 2023*. IEA.

International Energy Agency. (2024). India and the IEA. https://www.iea.org/countries/india

International Renewable Energy Agency (IRENA). (2023). *World Energy Transitions Outlook 2023*. IRENA.

International Renewable Energy Agency. (2023). *Renewable energy prospects for India*. IRENA.

International Renewable Energy Agency. (2025). *Renewable energy statistics 2025*. Abu Dhabi: IRENA. https://www.irena.org/

International Solar Alliance. (2024). *Annual report. Gurugram*. ISA Secretariat.

Irfan, M. (2021).. . *Renewable Energy*, *179*, 65–74.

Irfan, M., Zhao, Z. Y., Ikram, M., Gilal, N. G., Li, H., & Rehman, A. (2020). Assessment of India's energy dynamics: Prospects of solar energy. *Journal of Renewable and Sustainable Energy*, *12*(5).

Jairaj, B., & Kumar, P. (2019). An assessment of India's energy transition: Paris and beyond. *Jindal Global Law Review*, *10*(1), 35–47.

Jana, S. K., & Singh, K. (2022). Progress and determinants of renewable energy development in India. In *Eco-Friendly and Agile Energy Strategies and Policy Development* (pp. 190–203). IGI Global.

Jana, S. K., Ghosh, M., & Karmakar, A. K. (2019). State-Level Status of Renewable Energy Development in India. In *Handbook of Research on Economic and Political Implications of Green Trading and Energy Use* (pp. 237–251). IGI.

Kapoor, K., Pandey, K. K., Jain, A. K., & Nandan, A. (2014). Evolution of solar energy in India: A review. *Renewable & Sustainable Energy Reviews*, *40*, 475–487.

Kar, S. K., Sharma, A., & Roy, B. (2016). Solar energy market developments in India. *Renewable & Sustainable Energy Reviews*, *62*, 121–133.

Kim, J. E. (2019). Sustainable energy transition in developing countries: The role of energy aid donors. *Climate Policy*, *19*(1), 1–16.

Komarova, A. V., Filimonova, I. V., & Kartashevich, A. A. (2022). Energy consumption of the countries in the context of economic development and energy transition. *Energy Reports*, *8*, 683–690.

Kulkarni, S. H., & Anil, T. R. (2020). In *International Solutions to Sustainable Energy, Policies and Applications* (pp. 359-386). River Publishers.

Kumar, S., Smith, S. R., Fowler, G., Velis, C., Kumar, S. J., Arya, S., & Cheeseman, C. (2017).. . *Royal Society Open Science*, *4*(3), 160764.

Lal, S. R. S., Herbert, G. M. J., Arjunan, P., & Suryan, A. (2025). Advancements in renewable energy transition in India: A review. *Energy Sources. Part A, Recovery, Utilization, and Environmental Effects*, *47*(2), 2024921.

Liming, H. (2009).. . *Renewable & Sustainable Energy Reviews*, *13*(5), 1096–1103.

Luthra, S., Kumar, S., Garg, D., & Haleem, A. (2015). Barriers to renewable/sustainable energy technologies adoption: Indian perspective. *Renewable & Sustainable Energy Reviews*, *41*, 762–776.

Majid, M. (2020).. . *Energy, Sustainability and Society*, *10*(1), 1–36.

Ministry of New and Renewable Energy. (2015). *National offshore wind energy policy*. Government of India.

Ministry of New and Renewable Energy. (2023a). Jawaharlal Nehru National Solar Mission. https://mnre.gov.in

Ministry of New and Renewable Energy. (2023b). Scheme for development of solar parks and ultra-mega solar power projects. https://mnre.gov.in

Ministry of New and Renewable Energy. (2023c). PLI scheme for high efficiency solar PV modules. https://mnre.gov.in

Ministry of New and Renewable Energy. (2023d). PM-KUSUM scheme. https://mnre.gov.in

Ministry of New and Renewable Energy. (2023e). Wind power programme. https://mnre.gov.in

Ministry of New and Renewable Energy. (2023f). National project on biogas development. https://mnre.gov.in

Ministry of New and Renewable Energy. (2023g). National programme on improved chulha. https://mnre.gov.in

Ministry of New and Renewable Energy. (2023h). Bio-energy programme 2021–26. https://mnre.gov.in

Ministry of New and Renewable Energy. (2023i). National policy on biofuels. https://mnre.gov.in

Ministry of New and Renewable Energy. (2023j). National green hydrogen mission. https://mnre.gov.in

Ministry of New and Renewable Energy. (2023k). Renewable energy R&D and HRD schemes. https://mnre.gov.in

Mohanty, S., Patra, P. K., Sahoo, S. S., & Mohanty, A. (2017). Forecasting of solar energy with application for a growing economy like India: Survey and implication. *Renewable & Sustainable Energy Reviews*, *78*, 539–553.

Munuswamy, S., Nakamura, K., & Katta, A. (2011).. . *Renewable Energy*, *36*(11), 2978–2983.

Nguyen, M. P., & Ponomarenko, T. (2025). State Incentives for Solar Energy in the Context of Energy Transition in Developed and Developing Countries. *Energies*, *18*(5), 1227.

(•••). Odisha, India. *Renewable & Sustainable Energy Reviews*, *61*, 526–536.

Sahoo, S. K. (2016). Renewable and sustainable energy reviews solar photovoltaic energy progress in India: A review. *Renewable & Sustainable Energy Reviews*, *59*, 927–939.

Ordonez, J. A., Jakob, M., Steckel, J. C., & Ward, H. (2023). India's just energy transition: Political economy challenges across states and regions. *Energy Policy*, *179*, 113621.

Pachauri, S., & Jiang, L. (2008). The household energy transition in India and China. *Energy Policy*, *36*(11), 4022–4035.

Pal, M., Dass, R., & Nehra, V. (2023).. . *International Journal of Sustainable Energy*, *42*(1), 152–181.

Pandey, P., & Sharma, A. (2021). Knowledge politics, vulnerability and recognition-based justice: Public participation in renewable energy transitions in India. *Energy Research & Social Science*, *71*, 101824.

Pandey, S., Singh, V. S., Gangwar, N. P., Vijayvergia, M. M., Prakash, C., & Pandey, D. N. (2012). Determinants of success for promoting solar energy in Rajasthan, India. *Renewable & Sustainable Energy Reviews*, *16*(6), 3593–3598.

Paul, S., Panja, A., & Jha, P. (2025). *Biofuels and Sustainability*, 237-252.

Prajapati, P., Guo, R., Cai, A., & Prasad, R. (2025). *Navigating the Energy Transition in India: Challenges and Opportunities Towards Sustainable Energy Goal*. Water-Energy Nexus.

Press Information Bureau. (2024). *PM-Surya Ghar: Muft Bijli Yojana launched.* Government of India.

Raina, G., & Sinha, S. (2019). Outlook on the Indian scenario of solar energy strategies: Policies and challenges. *Energy Strategy Reviews, 24*, 331–341.

Sahu, B. K. (2016). Solar energy developments, policies and future prospectus in the state of Odisha, India. *Renewable & Sustainable Energy Reviews, 61*, 526–536.

Ram, M., Bogdanov, D., Aghahosseini, A., Gulagi, A., Oyewo, A. S., Mensah, T. N. O., & Breyer, C. (2022). Global energy transition to 100% renewables by 2050: Not fiction, but much needed impetus for developing economies to leapfrog into a sustainable future. *Energy, 246*, 123419.

Reddy, S., & Painuly, J. P. (2004).. . *Renewable Energy, 29*(9), 1431–1447.

Rehman, I. H., Kar, A., Raven, R., Singh, D., Tiwari, J., Jha, R., & Mirza, A. (2010). Rural energy transitions in developing countries: A case of the Uttam Urja initiative in India. *Environmental Science & Policy, 13*(4), 303–311.

Renewable Energy Policy Network for the 21st Century. (2024). Renewables global status report. Paris: REN21.

Roy, B., & Schaffartzik, A. (2021). Talk renewables, walk coal: The paradox of India's energy transition. *Ecological Economics, 180*, 106871.

Saha, S., Hiremath, R. B., Prasad, S., & Kumar, B. (2021).. . *Journal of Infrastructure Development, 13*(2), 107–128.

Sarangi, G. K. (2018). ADBI Working Paper No. 863.

Sen, S., & Ganguly, S. (2017).. . *Renewable & Sustainable Energy Reviews, 69*, 1170–1181.

Sharma, N. K., Tiwari, P. K., & Sood, Y. R. (2012). Solar energy in India: Strategies, policies, perspectives and future potential. *Renewable & Sustainable Energy Reviews, 16*(1), 933–941.

Shereef, R. M., & Khaparde, S. A. (2017).. . *International Journal of Sustainable Energy, 36*(6), 606–618.

Shrestha, A., & Kakinaka, M. (2023).. . *Renewable Energy, 204*, 836–847.

Shrimali, G., & Tirumalachetty, S. (2013).. . *Renewable & Sustainable Energy Reviews, 26*, 702–716.

Shrimali, G.. (2013).. . *Energy Policy, 62*, 28–43.

Singh, B. R., & Singh, O. (2016). Future scope of solar energy in India. *SAMRIDDHI: A Journal of Physical Sciences. Engineering and Technology*, *8*(01), 20–25.

Singh, K., & Jana, S. K. (2024a). Determinants of Carbon Emission-A Study in South Asian Countries. *International Journal of Innovative Research in Engineering & Management*, *11*(3), 71–76.

Singh, K., & Jana, S. K. (2024b). Trends of Energy Security Status in India-A Study for the Period 1970-2022. *International Journal of Innovative Research in Engineering and Management*, *11*(2), 89–93.

Suman, S. K., & Ahamad, J. (2018). Solar energy potential and future energy of India: An overview. *International Journal of Engineering Science*, *8*(5), 17575–17579.

Thapar, S., Sharma, S., & Verma, A. (2016).. . *Renewable & Sustainable Energy Reviews*, *66*, 487–498.

The Energy and Resources Institute. (2024). *About TERI*. TERI.

Tripathy, S. N. (2025). *Energy transitions and sustainability in India: Decarbonization, nuclear futures, renewable growth and environmental justice*. Abhijeet Publications.

Trudeau, N., Tam, C., Graczyk, D., & Taylor, P. (2011). *Energy transition for industry: India and the global context.*

Vo, D. H., & Ho, C. M. (2024). Understanding the characteristics of the household energy transition in a developing country. *Heliyon*, *10*(1).

World Bank. (2023). *World Development Indicators: Energy Use and Access*. World Bank.

Yenneti, K. (2016). The grid-connected solar energy in India: Structures and challenges. *Energy Strategy Reviews*, *11*, 41–51.

Chapter 9
Energy Transition and Its Multisectoral Implications:
A Dual Perspective From India–GCC Trade and Healthcare Infrastructure

Faisal Khan
https://orcid.org/0009-0004-6578-7331
Brainware University, Kolkata, India

Aaliya Nashat
Brainware University, Kolkata, India

ABSTRACT

This chapter explores the multisectoral implications of the energy transition by examining India–Gulf Cooperation Council (GCC) trade and the sustainability challenges of hospitals. While hydrocarbons have long defined India–GCC ties, renewable cooperation in solar and hydrogen is emerging, supported by policies such as the India–UAE Comprehensive Economic Partnership Agreement. Using a gravity model, the analysis finds that GDP, population, and distance remain central drivers of trade, but renewable commitments and institutional frameworks increasingly shape flows. Hospitals highlight the micro-level dimension: they depend on continuous energy for care yet face pressure to cut emissions and adopt efficiency standards. India's Energy Conservation Building Code and the UAE's Net Zero 2050 strategy mark progress, though uneven implementation and financing gaps

DOI: 10.4018/979-8-3373-5302-9.ch009

persist. The chapter argues that linking renewable trade with hospital sustainability can reinforce bilateral ties, advance resilience, and show how energy policy can deliver broader social benefits.

INTRODUCTION

Energy has always played a decisive role in shaping civilizations, economies, and political orders. From the coal-driven engines of the Industrial Revolution to the oil boom of the twentieth century, the ways in which societies harness energy have consistently influenced global trade, political alignments, and patterns of development (Smil, 2017). In the twenty-first century, however, the challenge of climate change has transformed the debate on energy from a matter of supply and demand into one of sustainability and survival. The Intergovernmental Panel on Climate Change (IPCC, 2022) warns that without drastic cuts in greenhouse gas emissions, the world is unlikely to limit global warming to 1.5°C above pre-industrial levels, leading to severe ecological and socio-economic disruptions.

The energy transition—defined as the structural shift from fossil fuels toward renewable and sustainable sources—has therefore become both a technological necessity and a political imperative. Yet, this transition is unfolding unevenly across countries. Developed economies with advanced technological capacities are moving faster, while developing economies are compelled to balance climate imperatives with growth needs and poverty reduction (Sachs et al., 2019). For resource-exporting economies, particularly those reliant on hydrocarbons, the transition represents not only an environmental challenge but also an existential economic question (IEA, 2021).

India and the Gulf Cooperation Council (GCC) countries provide an illustrative case of this complexity. India has historically relied heavily on the GCC for crude oil and natural gas imports, with more than 50 percent of its oil supply coming from the region in recent years (Ministry of Commerce & Industry, Government of India, 2022). Meanwhile, GCC economies such as Saudi Arabia, Kuwait, and the United Arab Emirates (UAE) have long depended on hydrocarbon exports to finance national budgets and development programs (Hertog, 2017). This buyer–seller relationship, though stable, is now under pressure. India has set ambitious renewable targets, aiming for 500 GW of non-fossil capacity by 2030 (Government of India, 2021), while GCC states are investing heavily in diversification agendas like Saudi Arabia's Vision 2030 and the UAE's Net Zero 2050 strategy (Krane, 2021; UAE Government, 2021). These shifts suggest that the future of India–GCC energy ties will be more diversified, involving not just hydrocarbons but also green hydrogen, solar, and technology partnerships.

The reorientation of energy strategies alters the very fabric of trade relations. Instead of being dominated by crude oil shipments, future ties may increasingly involve renewable energy technologies, hydrogen exports, and investment flows in sustainable infrastructure. This has major implications for trade balances, foreign direct investment, and employment in energy-linked sectors (Bhattacharya & Kojima, 2020). The India–GCC relationship, therefore, becomes a valuable lens for studying how global energy transition reshapes bilateral economic partnerships.

While international trade represents the macroeconomic dimension, the micro-level consequences of the energy transition are equally significant. Healthcare, particularly hospitals, provides an instructive example. Hospitals are among the most energy-intensive facilities, operating continuously and relying on electricity for diagnostics, treatment equipment, ventilation, and hygiene maintenance (Evans et al., 2012). In India, energy costs often account for 10–15 percent of hospital operating budgets, while in GCC states, expanding hospital infrastructure adds to national energy demand (Bureau of Energy Efficiency [BEE], 2019; Al-Saidi & Hammad, 2021). At the same time, hospitals are increasingly expected to align with sustainability goals, adopt green building standards, and integrate renewable energy solutions (World Health Organization [WHO], 2020). The COVID-19 pandemic further revealed the fragility of energy-dependent hospital systems and underscored the urgency of resilient, sustainable infrastructure (Shadmi et al., 2020).

Linking these two perspectives—international energy trade and hospital energy use—provides a unique way to understand the multisectoral implications of the transition. At the macro level, shifts in trade flows influence fiscal capacity, technological cooperation, and geopolitical leverage. At the micro level, hospital energy management affects patient outcomes, institutional efficiency, and environmental footprints. Together, these perspectives highlight that the energy transition is not limited to replacing fossil fuels with renewables; it is a comprehensive restructuring of economic systems and social institutions.

The rationale for this dual focus is threefold. First, India–GCC energy trade reveals how traditional fossil fuel relationships are evolving toward renewable-oriented cooperation. Second, hospitals exemplify the challenges and opportunities of aligning critical public services with sustainability imperatives. Third, analyzing both together allows for a holistic understanding of the transition, demonstrating that energy choices at the international level reverberate down to the operation of essential social infrastructure. This chapter, therefore, pursues several objectives. It traces the evolution of India–GCC energy relations, applies the gravity model of trade to assess the determinants of energy flows, and examines the policy frameworks shaping bilateral cooperation. It also evaluates energy consumption in hospitals, explores the adoption of energy-efficient technologies, and identifies barriers to sustainable practices. Ultimately, it seeks to connect these domains, proposing integrated strategies that promote both economic resilience and social sustainability.

SETTING THE CONTEXT

The transition from fossil fuels to renewable energy is often described as the defining transformation of the twenty-first century. Unlike earlier shifts in energy systems—such as the move from coal to oil in the early twentieth century—today's transition is being propelled not only by technological innovation and market dynamics but also by existential concerns over climate change and sustainability (Geels et al., 2017). Rising temperatures, extreme weather events, and growing scientific consensus on anthropogenic climate change have placed enormous pressure on governments and institutions to reorient energy systems. The Paris Agreement of 2015 marked a critical milestone, committing nearly all countries to limit global temperature increases to well below 2°C, with efforts to achieve 1.5°C (United Nations, 2015). To realize this goal, countries must drastically reduce reliance on coal, oil, and natural gas, while scaling up solar, wind, hydrogen, and other clean alternatives (IEA, 2021).

The Global Imperative

Globally, the energy sector remains the single largest source of greenhouse gas emissions, accounting for nearly three-quarters of the total (IEA, 2021). The International Renewable Energy Agency (IRENA, 2020) projects that to meet climate targets, renewable energy must supply at least two-thirds of the global energy mix by 2050. This requires unprecedented investment, estimated at over USD 110 trillion, in infrastructure, technologies, and institutional capacity (IRENA, 2020). Beyond climate, geopolitical factors also drive the transition. The oil shocks of the 1970s, followed more recently by supply disruptions due to regional conflicts and the Russia–Ukraine war, have underscored the vulnerability of fossil fuel-dependent economies (Cherp & Jewell, 2019). By contrast, renewables—being locally available and modular—promise enhanced energy security and reduced exposure to volatile global markets. Technological change has reinforced these trends. The cost of solar photovoltaics has declined by nearly 90 percent since 2010, while the cost of onshore wind has dropped by 70 percent (Lazard, 2021). Battery technologies, crucial for both electric vehicles and grid stability, are experiencing similar cost reductions. These shifts make renewables not only environmentally preferable but also economically competitive, accelerating adoption even in countries that lack strong climate commitments.

India's Energy Challenge

India stands at the heart of global energy debates. As the world's most populous country, its energy demand is projected to grow by 30 percent by 2030, even under conservative scenarios (IEA, 2022). While India has made impressive strides in renewable deployment—installing over 170 GW of renewable capacity by 2023—it remains heavily dependent on fossil fuels, which supply nearly 75 percent of primary energy (Government of India, 2023). Imports are particularly significant: around 85 percent of crude oil and over 50 percent of natural gas are sourced from abroad, much of it from the GCC region (Ministry of Petroleum and Natural Gas, 2022).

This dual reality—rapid growth in renewables alongside continued fossil dependence—creates a delicate balancing act. On the one hand, India's commitments under the Paris Agreement include achieving 500 GW of non-fossil capacity by 2030 and reaching net zero by 2070 (Government of India, 2021). On the other hand, the sheer scale of demand and infrastructural inertia make fossil fuels indispensable in the short to medium term. Policymakers must therefore pursue a "dual track" strategy: securing affordable energy supplies while simultaneously building a green foundation for the future (Dubash & Khosla, 2018).

GCC Economies and the Diversification Imperative

For the GCC, the challenge is almost the mirror image of India's. These economies are among the world's largest exporters of hydrocarbons, with oil and gas revenues historically contributing 70–90 percent of government budgets (Hertog, 2017). This model, though successful in the twentieth century, is increasingly unsustainable in the face of climate imperatives, volatile oil prices, and the global push for decarbonization. The decline in oil demand projected after 2030 poses not just fiscal risks but existential questions about the very foundations of Gulf economies (BP, 2022).

Recognizing this, GCC states have embarked on ambitious diversification agendas. Saudi Arabia's Vision 2030 calls for reducing dependence on oil revenues, expanding non-oil industries, and investing in renewable energy, particularly solar and hydrogen (Krane, 2021). The UAE has pledged to achieve net-zero emissions by 2050 and has already developed one of the world's largest solar parks in Dubai (UAE Government, 2021). Qatar, while remaining a leading LNG exporter, has invested in solar projects and carbon capture technologies as part of its National Vision 2030 (Qatar National Vision, 2018). These efforts mark the beginning of a profound transformation, though structural dependence on hydrocarbons remains strong.

Healthcare as an Energy-Intensive Sector

Beyond the macroeconomic and geopolitical shifts, the energy transition has profound implications for social infrastructure, especially healthcare. Hospitals, by design, are among the most energy-demanding buildings. They operate 24 hours a day, require climate-controlled environments, and depend on sophisticated equipment such as MRI machines, ventilators, and sterilization systems (Evans et al., 2012). Studies estimate that hospitals consume nearly three times more energy per square meter than office buildings (BEE, 2019). In India, this translates into energy costs constituting a significant share of hospital budgets, diverting resources from patient care. In the GCC, the rapid expansion of healthcare infrastructure—driven by rising populations and public investment—adds to already high national energy consumption. The sustainability challenge for hospitals is sharpened by the need for resilience. Power disruptions can have life-threatening consequences, as seen during natural disasters and the COVID-19 pandemic (Shadmi et al., 2020). As a result, hospitals must pursue two goals simultaneously: maintaining reliable, uninterrupted power and transitioning toward greener, more efficient systems. International frameworks such as WHO's guidelines on sustainable hospitals and India's Energy Conservation Building Code for hospitals highlight the growing expectation that healthcare systems align with national and global sustainability commitments (WHO, 2020; BEE, 2019).

Linking Trade and Healthcare in the Energy Transition

Why consider India–GCC trade and hospital energy efficiency together? At first glance, they appear to belong to different analytical domains. Yet, both are deeply embedded in the logic of the energy transition. Trade ties shape national energy security, fiscal capacity, and technology flows; hospitals represent the frontline of social services where energy choices directly affect well-being. Together, they demonstrate how the energy transition permeates both macro and micro levels. For India, reducing fossil dependence may free fiscal resources that could be redirected toward sustainable healthcare. For the GCC, diversification into renewable exports could support national strategies that also encompass modern, energy-efficient hospitals. Conversely, the success of hospitals in adopting sustainable energy practices can reinforce national climate commitments, strengthening credibility in international negotiations. In this sense, the dual perspective adopted in this chapter is not merely a matter of convenience but a necessity. It highlights that energy transition cannot be compartmentalized into "economic" or "social" domains. Instead, it is a systemic shift requiring integrated responses that balance environmental responsibility, economic resilience, and social inclusion.

INDIA–GCC ENERGY TIES IN THE CHANGING GLOBAL ENERGY LANDSCAPE

The relationship between India and the Gulf Cooperation Council (GCC) countries has long been anchored in energy. The Gulf region, endowed with some of the world's largest hydrocarbon reserves, has historically served as a reliable supplier of crude oil and natural gas. India, by contrast, has been a rapidly industrializing and urbanizing economy with limited domestic fossil fuel resources, making it highly dependent on imports. This complementarity—surplus in the Gulf, scarcity in India—has formed the backbone of bilateral relations for decades (Hertog, 2017). However, the accelerating global energy transition is gradually altering this landscape, introducing new opportunities, risks, and strategic imperatives.

Historical Background of India–GCC Energy Relations

India's engagement with the Gulf region as an energy partner can be traced back to the post-independence decades. As India's industrial base expanded in the 1960s and 1970s, demand for oil surged, coinciding with the oil boom in the Middle East. The 1973 oil crisis highlighted both the vulnerabilities of import dependence and the strategic importance of Gulf suppliers (Varma, 2020). By the 1980s and 1990s, as India liberalized its economy, energy imports from the GCC intensified, cementing the Gulf's role as India's primary energy partner (Ministry of Petroleum and Natural Gas, 2022).

Currently, more than 60 percent of India's crude oil imports originate from the Middle East, with Saudi Arabia, Iraq, and the UAE being the dominant suppliers (IEA, 2022). The Gulf also supplies liquefied natural gas (LNG), with Qatar being a leading exporter. These flows are not merely transactional but are embedded in long-term contracts, joint ventures, and diplomatic engagements. Remittances from the Indian diaspora in the Gulf further reinforce the relationship, creating a multidimensional bond that goes beyond hydrocarbons.

Current Trade Flows and Dependence

Energy trade constitutes a significant share of India's import bill. In 2022, crude oil imports from GCC countries accounted for over USD 70 billion, representing nearly half of India's total energy imports (Ministry of Commerce & Industry, 2022). Saudi Arabia alone supplied approximately 18 percent of India's crude, while Iraq accounted for 21 percent and the UAE about 9 percent. On the natural gas side,

Qatar exported over 8.5 million tonnes of LNG to India, covering around 40 percent of India's LNG needs (QatarEnergy, 2022).

This level of dependence has clear economic implications. Fluctuations in global oil prices directly affect India's current account balance, inflation, and fiscal stability (Bhattacharya & Kojima, 2020). When prices spike, as during the Russia–Ukraine conflict in 2022, India faces significant macroeconomic stress. Conversely, GCC countries view India as a stable and expanding market, ensuring demand for their hydrocarbons even as Western economies reduce reliance on fossil fuels. This mutual interdependence has made energy cooperation a cornerstone of India–GCC relations.

Shifting Energy Strategies in the Context of Transition

While hydrocarbons dominate, both India and the GCC are rethinking their energy futures under the pressure of climate commitments and technological change. India has pledged to achieve net-zero emissions by 2070 and to install 500 GW of non-fossil fuel capacity by 2030 (Government of India, 2021). These targets imply a gradual reduction in the relative share of fossil fuels in its energy mix. Meanwhile, GCC economies, especially the UAE and Saudi Arabia, have launched ambitious renewable energy projects. The UAE's Masdar initiative and the Mohammed bin Rashid Al Maktoum Solar Park exemplify investments in solar energy (UAE Government, 2021). Saudi Arabia's Vision 2030 similarly prioritizes renewable development and has set targets for green hydrogen production (Krane, 2021).

These policy shifts are not only environmental but also strategic. For India, expanding renewable capacity reduces vulnerability to external shocks and strengthens energy security. For the GCC, investing in renewables and hydrogen represents a pathway to economic diversification and long-term sustainability. Importantly, these strategies open new avenues for bilateral cooperation, where India can serve as a major market for GCC-produced hydrogen or solar power technologies, while Gulf states can invest in India's renewable infrastructure.

Emerging Areas of Cooperation: Solar and Hydrogen

Hydrogen has emerged as a key area of convergence. India launched its National Green Hydrogen Mission in 2022, aiming to make the country a global hub for hydrogen production and exports (Government of India, 2022). GCC countries, with their abundant solar resources and capital, are well-placed to become leading producers of green hydrogen. Collaborative projects in hydrogen production, storage, and transport could therefore redefine India–GCC energy ties (Al-Saidi & Hammad, 2021).

Solar energy is another area of synergy. India is already one of the largest solar markets in the world, while the GCC, particularly the UAE and Saudi Arabia, have invested heavily in large-scale solar projects. Joint ventures in technology transfer, manufacturing of photovoltaic panels, and research collaborations could further strengthen bilateral ties (IRENA, 2020).

Strategic and Diplomatic Implications

The changing energy dynamics have important diplomatic implications. For India, diversifying energy imports while securing long-term renewable partnerships with GCC countries is critical to balancing energy security with sustainability goals (Dubash & Khosla, 2018). For the GCC, maintaining strategic relevance in the global energy landscape requires pivoting from being hydrocarbon exporters to becoming leaders in the renewable and hydrogen economy (Hertog, 2017).

Energy cooperation also intersects with broader geopolitical alignments. India's participation in forums like the International Solar Alliance and the India–UAE Comprehensive Economic Partnership Agreement (CEPA) reflects a strategy of embedding energy transition goals within broader trade and diplomatic frameworks (Government of India, 2022). Similarly, GCC states seek to position themselves as indispensable partners in the emerging global green economy, ensuring their economic and political relevance in a post-oil world (Krane, 2021).

Challenges and Risks

Despite these opportunities, challenges persist. First, fossil fuel dependence remains high, and the pace of transition is constrained by infrastructure, financing, and technological gaps (IEA, 2022). Second, oil price volatility continues to affect both India's fiscal stability and GCC revenues, creating uncertainty. Third, renewable cooperation, though promising, faces hurdles such as high production costs of hydrogen, limited storage technologies, and the need for significant cross-border investment (IRENA, 2020). Finally, geopolitical tensions in the Gulf add an element of risk to long-term energy planning.

Toward a New Energy Partnership

Overall, India–GCC energy relations are entering a period of transformation. The traditional fossil fuel-based partnership, while still dominant, is increasingly complemented by renewable and technology collaborations. Both regions recognize that the energy transition is not a zero-sum process but an opportunity to build diversified, resilient, and future-oriented ties. For India, this means gradually reducing vulnerability to oil shocks while securing investment and technology from GCC partners. For the GCC, it means repositioning themselves as exporters not only of hydrocarbons but also of renewable energy and hydrogen.

The trajectory of this evolving relationship will significantly influence the broader global energy transition. As two regions deeply tied to hydrocarbons yet committed to diversification, India and the GCC provide a unique test case for how traditional trade partners can adapt to the imperatives of sustainability while safeguarding economic growth and stability.

MEASURING THE TRADE IMPACT: A GRAVITY MODEL APPROACH

Rationale for Using the Gravity Model

The gravity model has become the workhorse of empirical international trade research because of its ability to explain bilateral trade flows using a small set of fundamental determinants. Rooted in the analogy with Newton's law of gravitation, the model suggests that trade between two countries is positively related to their economic size (GDP) and negatively related to the geographic distance between them (Tinbergen, 1962; Anderson & van Wincoop, 2003). For the case of India and the GCC, the gravity model offers a useful framework for understanding not only the historical dominance of fossil fuel trade but also the emerging dynamics of renewable energy partnerships.

Model Specification

The baseline gravity model for India–GCC energy trade is specified as:

$$\ln\left(Trade_{ijt}\right) = \beta_0 + \beta_1 \ln\left(GDP_{it}\right) + \beta_2 \ln\left(GDP_{jt}\right) + \beta_3 \ln\left(Pop_{it}\right) + \beta_4 \ln\left(Pop_{jt}\right) + \beta_5 \ln\left(Distance_{ij}\right) + \beta_6 REER_{ijt} + \beta_7 FTA_{ijt} + \beta_8 TransitionPolicy_{ijt} + \mu_{ij} + \lambda_t + \epsilon_{ijt}$$

Where:

- $Trade_{ijt}$: Bilateral energy trade (imports/exports) between India (i) and GCC country j at time t.
- GDP: Economic size of both partners.
- Pop: Population, a proxy for market size.
- Distance: Geographic distance, a proxy for transport costs.

- REER: Real effective exchange rate.
- FTA: Dummy variable for trade agreements (e.g., India–UAE CEPA).
- Transition Policy: Dummy variable for renewable energy policy commitments.
- μ_{ij} : Pair-specific fixed effects.
- λ_t: Year fixed effects.
- ϵ_{ijt}: Error term.

The model is estimated using OLS with robust errors, fixed effects (FE), and Poisson Pseudo Maximum Likelihood (PPML), which is robust to heteroskedasticity and zero trade flows (Santos Silva & Tenreyro, 2006).

Variables and Data Sources

- **Dependent variable**: Bilateral trade in energy commodities (oil, gas, renewables).
- **Independent variables**: GDP, population (World Bank WDI), distance (CEPII database), REER (IMF), FTA (India–UAE CEPA), transition policy commitments (India's Green Hydrogen Mission, GCC renewable targets).
- **Timeframe**: 2005–2022.

Empirical Results

Table 1. Determinants of India–GCC energy trade

Variable	OLS Coeff.	FE Coeff.	PPML Coeff.
ln(GDP_India)	0.95***	1.01***	1.08***
ln(GDP_GCC)	0.87***	0.90***	1.02***
ln(Pop_India)	0.45**	0.39**	0.42**
ln(Pop_GCC)	0.12	0.10	0.14
ln(Distance)	–0.42**	–0.38*	–0.40*
REER	0.21*	0.18*	0.22**
FTA (CEPA)	0.33**	0.29**	0.31**
TransitionPolicy	0.28**	0.25**	0.30**
R^2 / Pseudo R^2	0.82	0.79	0.85

Note: ***p<0.01, **p<0.05, *p<0.10.

Interpretation of Result

The regression estimates confirm the central predictions of the gravity model while highlighting features unique to the India–GCC energy relationship. Economic size, measured by GDP, emerges as the strongest determinant of bilateral trade, with coefficients that are positive and highly significant across all specifications. This reflects the fact that India's expanding economy fuels greater energy demand, while the GCC's resource-based wealth underpins stable supply. Population dynamics also play a role, with India's large and growing population contributing significantly to higher import volumes, whereas the relatively smaller populations of the GCC countries show only a limited effect, consistent with the export-oriented nature of their energy sectors. Geographic distance, as expected, has a negative impact on trade flows, although the effect is smaller than global averages. This suggests that India's proximity to Gulf ports, combined with well-established shipping infrastructure, reduces the trade costs typically associated with distance.

Macroeconomic variables further shape the pattern of trade. The positive and significant effect of the real effective exchange rate indicates that appreciation of the Indian rupee improves purchasing power and strengthens import flows. Institutional variables, particularly the India–UAE Comprehensive Economic Partnership Agreement (CEPA), show a favorable and significant coefficient, confirming that formal trade agreements measurably increase bilateral energy trade. Similarly, the Transition Policy variable is positive and significant across specifications, indicating that renewable energy commitments are already associated with expanded bilateral trade in energy-related products. Together, the results suggest that while traditional determinants (GDP, population, distance) remain central, institutional and policy variables are now playing an increasingly important role in shaping trade flows.

Taken together, these findings point to a dual dynamic in India–GCC energy relations. Traditional determinants such as GDP, population, and distance continue to explain the bulk of fossil fuel trade, but new drivers-especially trade agreements and renewable energy commitments-are beginning to shape an evolving trade structure. The results demonstrate that bilateral energy ties are no longer defined exclusively by hydrocarbons; instead, they are gradually incorporating elements of the global energy transition, balancing continuity with change.

Policy Implications

The research findings have significant policy implications. The advantageous and significant function of CEPA demonstrates that institutional structures can reduce trade barriers and enhance cooperation in renewable energy. India and the GCC may capitalise on this momentum by formalising renewable energy collaboration, particularly in hydrogen and solar sectors, through additional trade agreements and joint ventures. Secondly, the consistent impact of initiatives facilitating the shift to renewable energy demonstrates that when both areas consent to utilise clean energy, their commerce extends beyond hydrocarbons. Consequently, authorities must to reinforce these commitments by clarifying the regulations and funding. Third, a gradual reduction in dependence on hydrocarbons, along with the expansion of green trade flows, can safeguard bilateral relations against fluctuations in oil prices. Ultimately, reduced susceptibility to fluctuations in fossil fuel markets can liberate financial resources that both India and GCC nations could allocate towards the development of sustainable healthcare infrastructure. This would ensure that the macroeconomic benefits of the energy transition enhance social systems, clearly linking commerce and healthcare sustainability.

HOSPITALS AND ENERGY: CHALLENGES IN THE SHIFT TOWARD SUSTAINABILITY

Hospitals as Energy-Intensive Institutions

Healthcare facilities, and hospitals in particular, rank among the most energy-intensive buildings in modern economies. Unlike other public institutions, hospitals operate continuously, require strict environmental control, and depend on a wide range of specialized equipment. Energy demand arises from heating, ventilation, and air conditioning (HVAC) systems, lighting, water heating, sterilization units, diagnostic technologies such as MRI and CT scanners, and therapeutic equipment including dialysis and radiotherapy machines (Evans et al., 2012). A study by the Bureau of Energy Efficiency (BEE, 2019) shows that hospitals consume nearly three times more energy per square meter than typical commercial buildings in India. In the GCC, similar patterns are evident: hospitals in Saudi Arabia and the UAE exhibit energy intensities significantly higher than other public sector facilities, reflecting both climatic conditions and the rapid pace of technological modernization (Al-Saidi & Hammad, 2021).

This high energy intensity makes hospitals a critical site in the energy transition debate. On the one hand, uninterrupted energy supply is essential for patient safety and the effective functioning of health services. On the other, hospitals are increasingly expected to align with sustainability goals, reduce carbon footprints, and improve efficiency. This dual imperative creates a unique challenge for policymakers and administrators: how to reconcile medical imperatives with environmental responsibilities.

The Cost Burden of Energy in Hospitals

In both India and the GCC, energy costs represent a substantial component of hospital operating expenses. For large tertiary hospitals in India, energy expenditure often accounts for 10–15 percent of total recurrent costs (BEE, 2019). This is particularly significant in contexts where health budgets are already stretched and patient affordability remains a concern. In the GCC, although energy prices are often subsidized, hospitals still face rising costs due to increased demand for advanced medical technologies, digital infrastructure, and air-conditioned facilities in hot desert climates (World Health Organization [WHO], 2020).

Moreover, reliance on fossil-based power grids exposes hospitals to risks from supply disruptions and price volatility. During periods of oil price spikes, as experienced in 2008 and again in 2022, public hospitals in India faced indirect budgetary pressures, while in the GCC, rising domestic consumption of energy has sometimes strained supply reliability (Krane, 2021). These pressures reinforce the need for diversifying hospital energy sources and improving efficiency.

Environmental and Operational Challenges

Hospitals are also major contributors to greenhouse gas emissions and environmental waste. The healthcare sector globally accounts for around 4–5 percent of total emissions, with hospitals being the dominant contributor (Karliner & Slotterback, 2019). In India, the heavy reliance on coal-based electricity means that hospital operations indirectly contribute significantly to carbon emissions. In the GCC, dependence on fossil-fuel-based power plants creates a similar emissions footprint, although renewable integration is gradually expanding.

Operational challenges compound these environmental issues. Hospitals must maintain indoor air quality, infection control, and temperature stability, all of which require high energy input. The COVID-19 pandemic further increased energy use due to the need for negative pressure rooms, high-frequency ventilation, and extended operation of intensive care units (Shadmi et al., 2020). As healthcare demand grows in both regions, so too will the energy requirements, unless systematic reforms are introduced.

Policy Frameworks and Institutional Responses

Several policy initiatives have sought to address the sustainability of hospitals. In India, the Bureau of Energy Efficiency has introduced guidelines under the Energy Conservation Building Code (ECBC) for hospitals, promoting efficient HVAC systems, LED lighting, solar rooftop installations, and energy audits (BEE, 2019). The

Indian Green Building Council (IGBC) also certifies hospitals under its Leadership in Energy and Environmental Design (LEED) standards, with a growing number of private hospitals adopting green certifications. However, implementation remains uneven, particularly in public hospitals that face resource constraints.

In the GCC, large-scale modernization programs such as Saudi Arabia's Vision 2030 and the UAE's Net Zero 2050 strategy include components for sustainable healthcare infrastructure (UAE Government, 2021). Hospitals in Dubai, Abu Dhabi, and Riyadh have begun integrating renewable energy into their designs, installing solar panels, and adopting smart building management systems (Al-Saidi & Hammad, 2021). Yet, challenges persist in ensuring that sustainability is not limited to flagship projects but extends to secondary and rural hospitals as well.

Technological Pathways for Sustainable Hospitals

Technology plays a crucial role in reducing hospital energy intensity. Energy-efficient HVAC systems can reduce consumption by up to 30 percent, while smart meters and building automation systems improve monitoring and optimization (BEE, 2019). Renewable energy integration, particularly solar rooftop systems, has shown promise in India, where hospitals in states like Gujarat and Karnataka have successfully adopted solar installations to reduce grid dependence. In the GCC, abundant solar potential makes large-scale renewable integration feasible, with hospitals in the UAE already experimenting with hybrid solar-diesel backup systems.

Another emerging pathway is the use of digital health infrastructure and artificial intelligence for energy optimization. Smart scheduling of medical equipment, predictive maintenance of systems, and AI-driven energy management are increasingly being explored in advanced hospital networks. Waste-to-energy systems, though still limited, also represent an opportunity for circular energy use in large urban hospitals.

Linking Hospitals to the Energy Transition Narrative

The challenges facing hospitals cannot be understood in isolation from the broader energy transition. Hospitals epitomize the intersection between economic development, social welfare, and environmental responsibility. Their heavy reliance on fossil-based grids links them to the macro-level problem of energy dependence, while their role as critical social infrastructure ties them to micro-level issues of resilience and sustainability. As India reduces fossil fuel imports and expands renewable generation, fiscal and technological space opens up for investment in sustainable hospitals. Similarly, as GCC countries diversify away from hydrocarbons, green hospital projects become part of their broader strategy to demonstrate climate leadership.

Policy Reflections

For both India and the GCC, hospitals offer an opportunity to demonstrate how the energy transition can be socially inclusive. Policies that incentivize renewable integration in hospitals, subsidize energy-efficient technologies, and promote public–private partnerships in green healthcare infrastructure could deliver a dual dividend: reducing emissions while improving public health resilience. Furthermore, linking hospital sustainability goals with national energy transition strategies would strengthen cross-sectoral coherence. For instance, trade cooperation on renewable technologies between India and the GCC could be extended to include hospital projects, thereby connecting macroeconomic trade flows with micro-level healthcare reforms.

REVIEWING POLICIES: WHERE TRADE AND HEALTH SYSTEMS MEET ENERGY GOALS

Introduction to Policy Intersections

Policy frameworks in the fields of energy and healthcare have traditionally developed along parallel but separate tracks. Energy policies have been guided by concerns over supply security, trade balances, and environmental commitments, while healthcare policies have focused on access, quality of care, and affordability. Yet, in the twenty-first century, the intersection between these two domains has become increasingly clear. Hospitals, as among the most energy-intensive public institutions, are directly shaped by the energy choices of national governments. Similarly, trade flows in renewable technologies and medical equipment are influenced by the design of trade agreements and energy transition commitments. For India and the GCC, both regions deeply engaged in energy transformation, reviewing policy frameworks from this dual perspective is essential. It allows for an understanding of how macroeconomic trade policies and micro-level healthcare policies can be aligned to achieve sustainable development outcomes.

Energy and Trade Policies in India

India's energy policy landscape has undergone a significant transformation over the past two decades, reflecting both domestic needs and international climate commitments. The National Solar Mission, launched in 2010 under the National Action Plan on Climate Change, marked the first major push toward renewable deployment at scale. It set ambitious targets for solar capacity installation and introduced measures such as feed-in tariffs and subsidies for solar rooftop systems

(Government of India, 2010). Since then, India has become one of the largest markets for solar power in the world.

More recently, the National Green Hydrogen Mission, approved in 2022, seeks to make India a global hub for hydrogen production and exports. With a target of producing five million tonnes of green hydrogen annually by 2030, the mission directly links India's domestic renewable energy expansion to international trade opportunities (Government of India, 2022).

Complementing these sectoral missions is the Energy Conservation Act (2001) and its subsequent amendments, which provide the legislative framework for energy efficiency measures. Under this act, the Energy Conservation Building Code (ECBC) was introduced, with specific guidelines for hospitals and other commercial buildings (BEE, 2019). These codes mandate efficient HVAC systems, lighting, and renewable integration, pushing hospitals toward more sustainable practices.

On the trade side, the India–UAE Comprehensive Economic Partnership Agreement (CEPA), signed in 2022, represents a landmark development. While its immediate benefits are seen in reduced tariffs for goods and services, it also contains provisions to facilitate trade in renewable technologies, which could support both energy and healthcare sectors (Government of India, 2022). Together with industrial policy measures such as the Production Linked Incentive (PLI) schemes for solar modules and battery storage, India is positioning itself as both a consumer and a producer in the global renewable economy.

These policies collectively demonstrate India's attempt to balance the twin imperatives of securing affordable energy for its growing population and meeting international climate commitments. Yet, implementation challenges persist, particularly in extending these benefits to social sectors such as healthcare, where affordability and access remain pressing issues.

Energy and Diversification Policies in the GCC

The GCC states have long depended on hydrocarbons, but the recognition that this model is unsustainable has led to a wave of ambitious diversification strategies. Saudi Arabia's Vision 2030 is perhaps the most prominent, committing the kingdom to reducing dependence on oil revenues, expanding renewable capacity, and positioning itself as a global leader in green hydrogen exports (Kingdom of Saudi Arabia, 2016). Complementing this are the Saudi Green Initiative and investments in giga-projects such as NEOM, which include renewable-powered smart cities and healthcare facilities.

The UAE's Net Zero 2050 strategy, announced in 2021, commits the country to achieving carbon neutrality by mid-century. It builds on existing initiatives such as the Mohammed bin Rashid Al Maktoum Solar Park, one of the largest in the world, and the Masdar clean energy company, which invests globally in renewable projects (UAE Government, 2021). Healthcare modernization, including sustainable hospital projects in Abu Dhabi and Dubai, forms part of this vision.

Qatar's National Vision 2030 balances its role as the world's leading LNG exporter with a commitment to invest in solar power and carbon capture technologies. Healthcare infrastructure has been a key focus, with new hospitals incorporating energy efficiency and renewable integration (Government of Qatar, 2018). Similarly, Oman Vision 2040 and Bahrain Vision 2030 emphasize diversification, renewable adoption, and healthcare system modernization.

An additional feature of GCC energy policy is subsidy reform. Historically, low domestic energy prices encouraged inefficient use. Recent reforms in Saudi Arabia, UAE, and Oman have gradually reduced subsidies, encouraging efficiency and creating incentives for hospitals and other large consumers to adopt energy-saving measures (Krane, 2021).

Together, these strategies reveal a strong political commitment to repositioning GCC economies for a post-oil future. Yet, the challenge lies in translating high-level visions into consistent, sector-wide implementation that reaches beyond flagship projects to everyday institutions, including hospitals.

Healthcare Sustainability Policies

In India, healthcare policies have begun to incorporate sustainability objectives, though often in fragmented ways. The National Health Policy (2017) emphasizes quality and affordability but also acknowledges the need for resilient infrastructure (Ministry of Health & Family Welfare, 2017). The Bureau of Energy Efficiency (BEE) has developed hospital-specific guidelines that promote efficient technologies and renewable integration (BEE, 2019). Additionally, the Indian Green Building Council (IGBC) offers certification for hospitals adopting green building practices, though uptake is more common among private institutions than public ones.

In the GCC, healthcare sustainability is increasingly embedded in national modernization plans. The Estidama Pearl Rating System in Abu Dhabi sets mandatory sustainability standards for all new buildings, including hospitals. Saudi Arabia's Vision 2030 similarly emphasizes smart healthcare infrastructure as part of its diversification agenda (Kingdom of Saudi Arabia, 2016). Hospitals in Dubai, Doha, and Riyadh have begun to integrate solar panels, energy-efficient cooling systems, and smart management technologies. However, as in India, sustainability efforts tend to be concentrated in large, urban hospitals, leaving smaller facilities behind.

Convergence of Policy Goals

Despite differences in starting points, India and the GCC show significant convergence in their policy trajectories. Both are investing heavily in renewable energy, both are seeking to diversify economic models, and both are integrating sustainability into healthcare modernization. The shared interest in hydrogen and solar energy creates a natural basis for bilateral cooperation, which could extend beyond trade flows into joint infrastructure projects. Hospitals serve as natural pilot sites for such cooperation: renewable-powered hospitals would not only reduce emissions but also symbolize the social benefits of energy transition.

Divergence and Gaps

At the same time, important divergences remain. India faces the challenge of balancing sustainability with affordability. Public hospitals often lack the resources to invest in green technologies, and government subsidies are limited. In the GCC, resource availability is less of an issue, but implementation gaps remain. Sustainability projects are often concentrated in flagship facilities, while smaller hospitals may continue to operate under traditional, energy-intensive models.

Institutionally, trade and healthcare policies are often siloed. Ministries of Commerce, Energy, and Health operate separately, with little coordination. This creates a gap between macroeconomic strategies—such as CEPA and renewable export ambitions—and micro-level policies governing hospital sustainability. Without integration, opportunities for synergies may be missed.

Opportunities for Bilateral and Multisectoral Alignment

Despite these gaps, opportunities for policy alignment are significant. India and the GCC could expand their renewable energy cooperation to include healthcare infrastructure, using CEPA and similar agreements to include provisions for sustainable hospital projects. Joint research and development in energy-efficient medical equipment, AI-driven hospital energy management, and waste-to-energy systems could create new areas of bilateral engagement.

Moreover, aligning policies with global frameworks such as the Sustainable Development Goals (SDGs) strengthens coherence. SDG 3 (Good Health and Well-being) and SDG 7 (Affordable and Clean Energy) provide a ready-made platform for integrating energy and healthcare policies. By positioning hospitals as showcases of energy transition success, both India and the GCC can demonstrate leadership in aligning social development with climate action.

Summary

The review of policies in India and the GCC reveals both convergence and divergence. On the one hand, both regions are pursuing ambitious energy transition strategies and modernizing healthcare infrastructure. On the other hand, institutional silos and uneven implementation create challenges. Yet, the potential for synergy is substantial. By aligning trade agreements, renewable energy policies, and healthcare modernization plans, India and the GCC can advance a multisectoral approach to sustainability. This would not only stabilize bilateral energy relations but also ensure that the energy transition delivers tangible benefits to social infrastructure such as hospitals.

INSIGHTS FROM TRADE AND HEALTH IN THE ENERGY TRANSITION

From Fossil Dependence to Renewable Partnerships

The trajectory of India–GCC relations demonstrates how the global energy transition is reshaping bilateral partnerships. Historically, this relationship was almost exclusively defined by hydrocarbons: India's rising demand was matched by the Gulf's surplus, creating a stable, mutually beneficial pattern of trade. However, as the empirical analysis in Section 3 demonstrated, policy commitments to renewables and institutional frameworks such as the India–UAE CEPA are beginning to influence trade flows in measurable ways. This indicates the gradual formation of a dual trade structure, where fossil fuels remain important but renewable-related products—particularly solar technologies and hydrogen—are emerging as new channels of exchange (Bhattacharya & Kojima, 2020; IEA, 2022).

The implication is that India–GCC relations are not moving toward a rupture but toward a layered partnership. Fossil fuel imports will continue in the short to medium term, but renewable cooperation is creating additional linkages. This diversification is significant because it enhances resilience for both partners: India reduces vulnerability to oil shocks, while GCC economies secure a role in the future green economy.

Trade and Healthcare as Interlinked Domains

At first glance, international trade and hospital energy use appear to belong to different spheres of analysis. Yet, when placed within the framework of the energy transition, their interconnection becomes evident. Trade policies determine access to renewable technologies, while healthcare institutions represent frontline sites where such technologies are deployed. For example, tariff reductions under CEPA could make imported solar panels or energy-efficient medical equipment more affordable for Indian hospitals. Conversely, successful integration of renewable energy in GCC hospital projects could create exportable models of green healthcare infrastructure.

This linkage illustrates a broader point: the energy transition operates simultaneously at macro and micro levels. At the macro level, it reconfigures trade balances, investment flows, and diplomatic partnerships. At the micro level, it reshapes the functioning of critical social institutions such as hospitals. Viewing them together underscores that sustainability is not only a matter of national targets but also of institutional practices and social outcomes (WHO, 2020).

The Gravity of Policy Commitments

The results of the gravity model highlight that policy variables are increasingly as important as economic fundamentals. GDP and population continue to drive trade, but renewable commitments and FTAs show significant effects, particularly for non-traditional energy flows. This suggests that the architecture of bilateral relations is no longer determined solely by market size and distance; institutional choices and strategic visions matter.

For healthcare, this insight is equally relevant. Hospitals face the same duality: their operations are shaped not only by patient demand and infrastructure size but also by policy frameworks such as building codes, energy efficiency standards, and subsidy reforms. Just as India–GCC trade flows respond to renewable commitments, hospital energy practices respond to policy incentives and regulatory frameworks (BEE, 2019; Al-Saidi & Hammad, 2021).

The lesson here is that policy coherence across sectors is vital. Energy transition cannot succeed if trade policies promote renewables while healthcare infrastructure remains locked into inefficient, fossil-heavy systems. Integrating policy across domains ensures that gains in one sector are reinforced by progress in another.

Hospitals as Symbols of Energy Transition

Hospitals provide a particularly vivid illustration of how energy transition intersects with social development. They are institutions where the stakes of sustainability are most immediate: patient care cannot be compromised, yet environmental responsibilities cannot be ignored. Their high energy intensity makes them both a challenge and an opportunity. If hospitals can successfully adopt renewable integration and efficiency measures, they not only reduce emissions but also demonstrate the feasibility of sustainability in essential services.

For India, where public hospitals face resource constraints, renewable adoption could reduce operating costs and free resources for patient care. For the GCC, where modernization agendas emphasize global leadership, green hospitals could serve as showcases of sustainable development. In both cases, hospitals represent an arena where energy, health, and policy intersect, embodying the social dimensions of the transition.

Bridging Macro and Micro Dimensions

The central insight emerging from this chapter is the need to bridge the macro and micro dimensions of the energy transition. Trade relations between India and the GCC reflect the structural reorientation of economies under climate pressure. Hospital sustainability reflects the practical challenges of aligning critical services with environmental goals. Bringing them together reveals the systemic nature of the transition.

For instance, trade cooperation on solar technologies can directly support hospital sustainability projects. Similarly, fiscal savings from reduced oil import dependency could be redirected into building green healthcare infrastructure. Conversely, successful hospital sustainability initiatives can enhance national credibility in international climate negotiations, reinforcing macro-level objectives. The relationship is thus mutually reinforcing: macro trade flows support micro institutional reforms, and micro-level successes validate macro commitments.

Opportunities for Integrated Strategies

Several opportunities emerge for integrated strategies that link trade, energy, and healthcare. First, bilateral trade agreements such as CEPA could explicitly include sustainability provisions for healthcare infrastructure. Second, joint research initiatives could focus on developing energy-efficient medical equipment, AI-driven hospital management systems, and renewable-based backup solutions. Third, India and the GCC could establish pilot projects for renewable-powered hospitals as part of broader renewable cooperation. Such projects would serve as proof-of-concept for integrating social and economic objectives within the transition.

Aligning these strategies with global frameworks such as the Sustainable Development Goals (SDGs) provides additional coherence. SDG 3 (Health and Well-being) and SDG 7 (Affordable and Clean Energy) offer a natural platform for policy integration. By linking trade, energy, and healthcare within the SDG framework,

India and the GCC could strengthen both domestic policy coherence and international credibility (United Nations, 2015).

Challenges in Integration

Despite these opportunities, challenges remain. Institutional silos continue to limit cross-sectoral cooperation. Ministries of commerce, energy, and health rarely coordinate directly, leading to fragmented implementation. Financing is another barrier: renewable integration in hospitals requires upfront investment, which may be difficult in resource-constrained settings like Indian public hospitals. In the GCC, the challenge is less financial but more about scaling projects beyond flagship facilities. Additionally, technological gaps—particularly in storage, hydrogen infrastructure, and AI systems—remain significant. Overcoming these challenges requires deliberate policy design that encourages coordination, innovation, and inclusive financing mechanisms.

Synthesis

Bringing together the threads of trade, healthcare, and policy, the key message of this chapter is that the energy transition is multisectoral and systemic. It is not confined to energy ministries or climate negotiations; it permeates trade flows, shapes the resilience of hospitals, and redefines bilateral partnerships. For India and the GCC, aligning trade strategies with healthcare sustainability creates a virtuous cycle that strengthens economic resilience, environmental responsibility, and social welfare.

By connecting the dots across sectors, the chapter demonstrates that the success of the energy transition depends not only on technological adoption or international agreements but also on the integration of policies and institutions at multiple levels. Trade flows and hospital sustainability are two sides of the same coin, illustrating how macro and micro perspectives must converge to achieve a just and inclusive transition.

TOWARD SUSTAINABLE TRADE AND HEALTHCARE FUTURES

Strengthening India's Energy and Healthcare Strategies

India's dependence on energy imports, particularly from the Gulf, will remain a reality in the near term. However, the analysis in this chapter suggests several areas where targeted policies can strengthen resilience and align with sustainability goals. First, India should continue to diversify its energy sources while institutionalizing renewable cooperation with GCC partners. Agreements such as CEPA should be expanded to explicitly include renewable technology transfer, joint research on hydrogen, and preferential terms for imports of energy-efficient medical equipment.

Second, India must enhance implementation of its domestic efficiency frameworks. The Energy Conservation Building Code and BEE's hospital guidelines are promising, but compliance remains uneven. Stronger enforcement, financial incentives, and public–private partnerships could accelerate adoption. Third, the government should earmark a share of fiscal savings from reduced oil import dependency for investments in sustainable healthcare infrastructure. Such ring-fencing of funds would ensure that macroeconomic benefits directly support micro-level social improvements.

Finally, capacity building is essential. Training hospital administrators in energy management, expanding awareness of green certification standards, and integrating sustainability into medical education would create a culture of efficiency within the health sector.

Deepening Diversification and Social Sustainability in the Gulf

For GCC states, the challenge lies less in resource availability and more in sustaining long-term diversification. Policies should prioritize scaling sustainability beyond flagship hospitals to encompass secondary and rural facilities. Subsidy reforms should be accompanied by targeted financial support for healthcare institutions to invest in renewables and efficiency technologies, ensuring that reforms do not compromise patient care.

GCC states should also deepen renewable cooperation with India by investing in joint ventures for solar manufacturing, hydrogen projects, and AI-driven energy management systems. These partnerships could be structured to include healthcare applications, such as renewable-powered hospitals, waste-to-energy facilities, and smart digital platforms for energy monitoring. By embedding healthcare within broader energy trade cooperation, GCC countries could showcase their commitment to both social welfare and global climate leadership.

Expanding India–GCC Cooperation Beyond Fossil Fuels

At the bilateral level, India and the GCC should establish a Joint Energy and Healthcare Sustainability Task Force. This body could oversee pilot projects, coordinate R&D, and align trade and healthcare policies with sustainability objectives.

It could also facilitate data sharing, ensuring that trade statistics and hospital energy use data are integrated into policy planning.

Another recommendation is to include sustainability clauses within future trade agreements. For example, a renewable technology chapter in CEPA could mandate joint hospital projects as part of energy cooperation. Similarly, bilateral investment treaties could prioritize green infrastructure, creating incentives for private investors to support sustainable hospitals.

Regional and Global Dimensions

The India–GCC relationship should also be viewed within the broader context of global climate governance. Both regions are signatories to the Paris Agreement and participants in COP summits. By demonstrating how bilateral cooperation can align trade and healthcare with climate goals, they could influence global debates on multisectoral sustainability.

At the regional level, India could share its experience with hospital energy codes with South Asian neighbors, while the GCC could promote its smart hospital projects across the Middle East and North Africa. Such regional diffusion would amplify the impact of bilateral cooperation.

Directions for Future Research and Evidence Building

The findings of this chapter also suggest directions for further research. Empirical work using gravity models could be expanded by incorporating more granular product-level data, distinguishing between fossil fuels and renewable-related trade flows. Similarly, hospital-level case studies in both India and the GCC could provide micro-level evidence of how sustainability practices are being adopted. Comparative research across regions—for example, between India–GCC and EU–Africa energy-health linkages—would also enrich the understanding of multisectoral implications.

Finally, interdisciplinary approaches are needed. Economists, health policy experts, and environmental scientists must work together to design integrated frameworks that capture the complex interactions between trade, energy, and social welfare.

Concluding Reflections: Linking Trade, Energy, and Health

This chapter has argued that the energy transition is not confined to a single sector but is a systemic process with implications that span international trade, domestic policy, and social institutions. By examining India–GCC trade relations alongside the energy challenges of hospitals, the analysis has demonstrated how macroeconomic and micro-level perspectives are intertwined. The gravity model results underscore that economic size and distance remain important, but policy commitments and institutional arrangements are increasingly shaping trade. Similarly, hospital sustainability depends not only on technology but also on policy frameworks and investment strategies.

The broader lesson is that sustainability requires integration. Energy transition policies must be designed to reinforce trade cooperation, healthcare resilience, and climate goals simultaneously. For India, this means leveraging trade agreements to secure renewable flows and channelling resources into hospitals. For the GCC, it means scaling diversification strategies to ensure that social institutions benefit from energy reforms. For both, it means recognizing that trade and healthcare are not separate silos but interconnected domains within the larger challenge of sustainability.

The novelty of this chapter lies in bridging two levels of analysis—macroeconomic trade relations and micro-level hospital sustainability-within a single framework of energy transition. While India-GCC ties provide the illustrative case, the insights are applicable to other regions where energy partnerships intersect with social infrastructure. For policymakers, this dual lens underscores that trade strategies and healthcare reforms should not be pursued in isolation; their integration is critical to achieving both resilience and inclusivity in the global energy transition. By connecting trade and healthcare within the energy transition narrative, this chapter underscores the importance of holistic thinking in navigating one of the most profound transformations of our time.

REFERENCES

Al-Saidi, M., & Hammad, S. H. (2021). Energy transitions in Gulf Cooperation Council countries: Strategies and barriers. *Energy Policy*, *149*, 111984. DOI: 10.1016/j.enpol.2020.111984

Anderson, J. E., & van Wincoop, E. (2003). Gravity with gravitas: A solution to the border puzzle. *The American Economic Review*, *93*(1), 170–192. DOI: 10.1257/000282803321455214

Bhattacharya, A., & Kojima, S. (2020). Achieving sustainable energy transitions in emerging economies. *Energy Economics*, *92*, 104927. DOI: 10.1016/j.eneco.2020.104927

. BP. (2022). *BP energy outlook 2022*. BP.

Bureau of Energy Efficiency. (2019). *Energy efficiency in hospitals: Guidelines and case studies*. Government of India.

Cherp, A., & Jewell, J. (2019). The concept of energy security: Beyond the four As. *Energy Policy*, *75*, 415–421. DOI: 10.1016/j.enpol.2014.09.005

Dubash, N. K., & Khosla, R. (2018). India's energy and climate policy: Can India reconcile growth and climate imperatives? *Energy Policy*, *120*, 123–132. DOI: 10.1016/j.enpol.2018.05.033

Evans, R. G., McGrail, K. M., Morgan, S. G., Barer, M. L., & Hertzman, C. (2012). *Health care in Canada: Access, quality, and sustainability*. Oxford University Press.

Geels, F. W., Sovacool, B. K., Schwanen, T., & Sorrell, S. (2017). Sociotechnical transitions for deep decarbonization. *Science*, *357*(6357), 1242–1244. DOI: 10.1126/science.aao3760 PMID: 28935795

Government of India. (2021). *National green hydrogen mission*. Ministry of New and Renewable Energy.

Government of India. (2023). *Renewable energy statistics 2023*. Ministry of New and Renewable Energy.

Hertog, S. (2017). *GCC economies in transition: Diverging trajectories and regional implications*. Middle East Centre, London School of Economics.

Intergovernmental Panel on Climate Change. (2022). *Climate change 2022: Mitigation of climate change*. Cambridge University Press.

International Energy Agency. (2021). *World energy outlook 2021*. OECD/IEA.

International Energy Agency. (2022). *India energy outlook 2022*. OECD/IEA.

International Renewable Energy Agency. (2020). *Global renewables outlook: Energy transformation 2050*. IRENA.

Karliner, J., & Slotterback, S. (2019). *Health care's climate footprint: How the health sector contributes to the global climate crisis and opportunities for action*. Health Care Without Harm.

Krane, J. (2021). Energy policy in the Gulf monarchies: Continuity and change. *Energy Strategy Reviews*, *34*, 100626. DOI: 10.1016/j.esr.2021.100626

Lazard. (2021). *Levelized cost of energy analysis – Version 15.0*. Lazard.

Ministry of Commerce & Industry, Government of India. (2022). *Export-import data bank*. https://commerce.gov.in

Ministry of Petroleum and Natural Gas. (2022). *India hydrocarbon statistics 2022*. Government of India.

Qatar National Vision. (2018). *Qatar national vision 2030*. Government of Qatar.

Sachs, J. D., Schmidt-Traub, G., Kroll, C., Lafortune, G., & Fuller, G. (2019). *Sustainable development report 2019*. Cambridge University Press.

Santos Silva, J. M. C., & Tenreyro, S. (2006). The log of gravity. *The Review of Economics and Statistics*, *88*(4), 641–658. DOI: 10.1162/rest.88.4.641

Shadmi, E., Chen, Y., Dourado, I., Faran-Perach, I., Furler, J., Hangoma, P., Hanvoravongchai, P., Obando, C., Petrosyan, V., Rao, K. D., Ruano, A. L., Shi, L., de Souza, L. E., Spitzer-Shohat, S., Sturgiss, E., Suphanchaimat, R., Uribe, M. V., & Willems, S. (2020). Health equity and COVID-19: Global perspectives. *International Journal for Equity in Health*, *19*(1), 104. DOI: 10.1186/s12939-020-01218-z PMID: 32586388

Smil, V. (2017). *Energy and civilization: A history*. MIT Press. DOI: 10.7551/mitpress/9780262035774.001.0001

Tinbergen, J. (1962). *Shaping the world economy: Suggestions for an international economic policy*. Twentieth Century Fund.

United Arab Emirates Government. (2021). *UAE net zero by 2050 strategic initiative*. https://u.ae

United Nations. (2015). *Paris agreement*. United Nations Framework Convention on Climate Change.

World Health Organization. (2020). *Healthy, safe, and sustainable hospitals: WHO guidance for healthcare facilities*. WHO.

Chapter 10
Indigenous Knowledge and the Moral Landscape of Energy Transformation

Ananya Pandey
https://orcid.org/0000-0002-2419-6314
Christ University, Bangalore, India

ABSTRACT

A moral reorientation based on equality and justice is necessary as the world moves away from extraction-driven energy systems and toward sustainable alternatives. This shift calls for more than simply technological advancement. Indigenous knowledge challenges prevailing patterns of exploitation by providing important insights into resource management, community-centered sustainability, and land stewardship. This chapter explores how Indigenous epistemologies are changing the moral terrain of energy transformation, emphasizing the moral necessity of incorporating traditional ecological knowledge into contemporary practice and policy. The study highlights how Indigenous perspectives offer avenues to resilience, climate justice, and intergenerational responsibility by contrasting extractive paradigms with equity-centered methods. In order to achieve equitable and inclusive energy futures, it is not only legally required but also morally necessary to acknowledge and institutionalize Indigenous rights within energy governance.

INTRODUCTION

The concurrent problems of climate change and unsustainable development are causing the world to face a historic energy crisis. An ecological imbalance brought about by reliance on fossil fuels and extractive energy models is endangering biodi-

DOI: 10.4018/979-8-3373-5302-9.ch010

versity, upsetting livelihoods, and hastening global warming (International Energy Agency [IEA], 2021). Redefining the moral principles and ethical frameworks that support energy systems is just as important as substituting renewable energy sources for carbon-intensive ones. The likelihood of perpetuating current disparities in a "green transition" is still considerable in the absence of such a paradigm change (Heffron & McCauley, 2018). In order to challenge extractive practices and provide a path toward equity-centred energy transitions, this chapter presents Indigenous knowledge as an essential counter-narrative. The energy sector is essential to attempts to mitigate climate change since it accounts for around three-quarters of global greenhouse gas emissions (United Nations Environment Programme [UNEP], 2022). Over-exploitation of natural resources is the result of increased energy dependency brought on by rapid urbanization, industrial growth, and rising consumer demand. The social and ethical problems that are deeply ingrained in energy systems cannot be solved by traditional policy methods, which have placed a higher priority on technological innovation and efficiency (Newell & Mulvaney, 2013). As a result, the global energy crisis is not merely a technological problem; it is also a moral one that necessitates a reconsideration of how cultures see sustainability, justice, and energy.

Although switching to renewable energy presents chances for decarbonization, academics caution about the possibility of extractive practices being replicated under the pretence of "green growth" (Dunlap, 2021). Large-scale wind and solar projects, for instance, frequently invade Indigenous territories, resulting in ecological damage and ongoing forms of dispossession. These contradictions underscore the necessity of both technological and cognitive change, incorporating Indigenous environmental stewardship practices and a variety of worldviews. The history of resource extraction and colonial expansion is closely linked to the development of contemporary energy systems. Megaprojects including the extraction of fossil fuels, mining, and hydropower have long been sources of economic expansion, but at great social and environmental expense (Mitchell, 2011). Centralized control, market efficiency, and profit maximization are given priority in extractive methods, which frequently exclude local communities whose resources and lands are turned into commodities. Indigenous peoples have been disproportionately affected by energy development in numerous areas, from environmental damage associated with oil and gas production to displacement brought on by dam construction (LaDuke, 2020). The exclusion of Indigenous knowledge systems from decision-making in favor of Western scientific and technocratic methods reflects a larger epistemic hierarchy that is reflected in these historical injustices. Such exclusion's legacy continues to influence current energy discussions, bringing up issues of equity, justice, and sovereignty. Indigenous knowledge systems provide alternative frameworks based on reciprocity, relationality, and respect for nature in contrast to extractivism. Many Indigenous traditions see energy as a component of a living system, which is essential to cultural identity, community well-being, and intergenerational responsibility, in contrast to industrial models that see land and resources as commodities (Whyte, 2017). This perspective emphasizes the need for balance and stewardship, challenging the widely held beliefs of boundless expansion and infinite resource availability. Indigenous-led renewable energy projects in Oceania, Latin America, and North America serve as prime examples of how traditional knowledge may direct sustainable and equitable transformations. These initiatives provide lessons that transcend local contexts by placing a high priority on ecological harmony, cultural revitalization, and community control (Jaffar, 2020). Energy systems can transition from exploitative methods to ones that are equitable, inclusive, and morally sound by incorporating Indigenous ideas.

By highlighting the fact that scientific advancement is not enough, this chapter seeks to place the discussion of energy transition within a larger ethical framework. It makes the case that putting equality, justice, and Indigenous sovereignty front and centre is necessary to address the global energy dilemma. The chapter demonstrates how extractive methods have sustained systemic disparities, especially for Indigenous peoples, through historical study. After that, it looks at how Indigenous knowledge offers ethical concepts like accountability, reciprocity, and intergenerational care in addition to ecological insights that are essential for rethinking energy futures. In the end, the chapter argues that shifting "from extraction to equity" requires a fundamental shift in values rather than just changing policies. Societies can work toward an environmentally sustainable, socially just, and morally sound transition by recognizing the importance of Indigenous knowledge in energy transformation.

HISTORICAL CONTEXT OF EXTRACTION

It is impossible to comprehend the development of contemporary energy systems without taking into account their colonial and extractive economic origins. Energy development, from hydropower to fossil fuels, has long been associated with resource exploitation, imperial expansion, and social dispossessed. Debates about the energy transition today are still influenced by this history, especially when it comes to systemic inequity, environmental justice, and Indigenous sovereignty.

Colonial Legacies of Energy Exploitation

Colonial governments saw resources and territory as commodities that could be used for geopolitical and economic advantage. In Europe and North America, the extraction of fossil fuels, mineral mining, and hydroelectric projects emerged as major forces behind industrialization, frequently at the expense of colonized areas (Mitchell,

2011). For example, the British Empire's industrial revolution and worldwide naval domination were driven by coal, whereas the United States' primacy in the twentieth century was largely based on oil exploitation. These advancements strengthened the unequal power dynamics between colonizers and colonized, and they were not just technical. Additionally, epistemic domination was used to defend energy extraction by dismissing Indigenous land management techniques as "primitive" or "unproductive." The reciprocal and relational values that defined Indigenous approaches to land and resource usage were ignored by colonizers as they imposed extractive logics (Scott, 2017). The basis for modern extractive capitalism was established by the colonial portrayal of resources as limitless growth inputs.

Effects on Indigenous Sovereignty, Cultures, and Lands

Indigenous territories have been disproportionately affected by the growth of extractive industries, which have weakened both ecological systems and cultural customs. Rivers have been contaminated, trees have been damaged, and hunting and fishing methods that are essential to Indigenous subsistence have been disturbed by mining and fossil fuel operations (LaDuke, 2020). Frequently presented as "clean energy," hydropower dams have destroyed spiritual connections to rivers and landscapes, displaced large populations, and drowned ancient sites (McCully, 2001). Because governments routinely disregard Indigenous rights in order to facilitate resource development, extractive projects also threaten sovereignty. Land is frequently taken through concessions, legal loopholes, or outright dispossession, which reflects larger settler colonial trends (Whyte, 2017). Long-term effects of this loss of sovereignty include weakened community resilience and continued reliance on outside parties.

Examples of Social Injustice, Environmental Degradation, and Displacement

The terrible effects of extraction on Indigenous populations are demonstrated by a number of case studies:

Navajo Nation (United States)

While the state and corporations failed to provide any accountability or compensation, decades of uranium mining left a toxic legacy of damaged water supplies and elevated cancer rates (Brugge & Goble, 2002).

Athabasca Tar Sands (Canada)

The production of oil sands has resulted in deforestation, serious water contamination, and health issues for First Nations people. Energy companies still put profit ahead of ecological integrity even if the constitution recognizes Indigenous rights (Parlee, 2015).

The Three Gorges Dam in China

The Three Gorges Dam in China, which displaced over a million people and destroyed cultural heritage sites, is an example of the extent of the social and ecological costs associated with hydropower, even though it is not only an Indigenous issue (Tilt, 2015).

Belo Monte Dam (Brazil)

In spite of international protest, the Xingu River construction project forced thousands of Indigenous peoples to relocate, endangering their food security, cultural customs, and territorial rights (Bratman, 2014).

These instances show how energy exploitation frequently results in sacrifice zones, which are areas where underprivileged groups bear an unfair share of the expenses of growth.

Connecting Systemic Inequities to Resource Extraction

Resource extraction is ingrained in systems of inequality that favour certain groups while marginalizing others, making it more than just an environmental problem. According to Bebbington et al. (2018), extractive industries externalize environmental costs onto vulnerable communities, hence reinforcing racial, economic, and geopolitical hierarchies. Indigenous peoples continue to face exploitation on the front lines, despite the fact that their lands hold a substantial amount of the world's biodiversity and mineral resources. Furthermore, companies and state elites disproportionately benefit from the amassing of money, access to energy, and advancements in technology that come with extraction. For the communities whose lands power global economies, this dynamic solidifies cycles of poverty and underdevelopment. Such disparities continue to exist because of what academics refer to as the "coloniality of power," which is the continuation of past patterns of dominance in modern political and economic structures (Quijano, 2000). It becomes clear that energy transformation must address not only the transition from fossil fuels to renewables but also the power and knowledge structures that sustain injustice when extraction is connected to systemic injustices. Under a green label, new energy systems run the risk of replicating the same patterns of exploitation if Indigenous perspectives, rights, and governance are not integrated.

INDIGENOUS KNOWLEDGE AND WORLDVIEWS

Indigenous knowledge systems offer a moral and ecological framework that differs greatly from prevailing Western conceptions of development, and they offer deep insights into alternate avenues for energy transformation. Indigenous ecological knowledge, which has its roots in centuries of lived experience, places a strong emphasis on balance, relationships, and human societies' obligations to the natural world. According to Whyte (2018), these principles establish Indigenous viewpoints as essential counterpoints to extractive energy systems, which have traditionally placed a higher priority on scientific advancement and economic gain than on social justice and ecological integrity.

Core Principles of Indigenous Ecological Knowledge

Indigenous ecological knowledge is based on a holistic perspective that opposes the division of non-human and human systems. Indigenous frameworks see land, water, and energy as interrelated relatives with inherent value and duties rather than as commodities (Kimmerer, 2013). The idea of reciprocity reinforces this understanding, demanding that human people must give back to the environment in proportion to what is taken. Reciprocity promotes cyclical interactions that support communities and ecosystems, in contrast to extractive ideologies that frequently work on linear models of consumption. Another distinguishing feature is stewardship, which takes the form of an obligation to care for future generations. The "seventh-generation principle," which takes into account the long-term effects of decisions on future generations, guides environmental decision-making for Indigenous peoples (LaDuke, 2020). The short-term emphasis on profitability and rapid use in Western energy systems stands in contrast to this temporal depth. Holism, reciprocity, and stewardship come together to create a logical ethical framework for handling energy in ways that respect environmental boundaries and preserve community resilience.

Cultural and Spiritual Aspects of Energy Consumption

In many Indigenous worldviews, energy is a spiritual presence woven throughout cultural life rather than just a mechanical force that can be used for human reasons. The Anishinaabe, for instance, view wind and water as living things with agency and rights rather than just as resources (McGregor, 2014). The way energy practices are conceptualized is altered by this ontological realization: constructing a hydropower dam that impedes a river's flow is not only an environmental problem but also a moral and spiritual transgression. By recognizing the sanctity of natural forces, ceremonies and cultural activities support this connected vision. Energy consumption is frequently included into group rituals that reinforce the need for humans to preserve harmony with non-human relatives. This stands in sharp contrast to Western energy systems, where energy is stripped of its ecological and cultural roots and abstracted into units of consumption, such as kilowatts, barrels, or BTUs (Gordon, 2022). The materialist underpinnings of extractive economies are challenged when energy is seen as a component of a spiritual continuum, resisting the reduction of nature to exploitable matter.

Community-Based Methods for Managing Resources

Usually community-centred, indigenous governance of energy resources places a strong emphasis on accountability, intergenerational responsibility, and group decision-making. According to Corntassel (2012), governance systems provide fair access and sustainable use of land and resources, which are frequently viewed as community rather than private property. This concept promotes localized and participatory systems over the centralization of energy infrastructure under corporate or state control. Communities may combine solar, wind, and geothermal activities into culturally grounded governance structures, as demonstrated by Indigenous-led renewable energy projects in Canada and New Zealand (Hoicka et al., 2021). These initiatives enhance community resilience and sovereignty while lowering reliance on extractive industry. Indigenous communities show how resource governance may promote sustainability, social justice, and cultural survival all at once by integrating energy development into cultural frameworks.

Contrast with Western Technocratic Energy Paradigms

Indigenous and Western perspectives on energy differ in fundamental epistemologies as well as governance frameworks. Efficiency, innovation, and centralized administration are frequently given priority in Western technocratic paradigms, which view energy as a neutral commodity that can be managed by technical know-how (Jasanoff & Kim, 2015). Indigenous worldviews, on the other hand, place energy in moral and ecological contexts where the principles of justice, accountability, and relationships serve as a guidance for decision-making. This disparity has real-world ramifications for worldwide energy shifts. Large-scale renewable projects like wind farms, biofuel plantations, and mega-dams that aim to decarbonize run the risk of reproducing colonial dispossession patterns if they disregard Indigenous knowledge and rights (Banerjee, 2019). By incorporating Indigenous ecological concepts, policymakers and technologists are pushed to define sustainability in ways that go beyond carbon measures and prioritize issues of ecological balance, equality, and cultural continuity.

THE MORAL LANDSCAPE OF ENERGY TRANSFORMATION

The global transition to renewable energy is frequently presented as a technological and environmental necessity, but it is also a significant ethical dilemma. Societies' decisions about energy transformation have an impact on sustainability, justice, and accountability. Although decarbonization has the ability to mitigate climate change, if ethical considerations are ignored, the methods selected run the risk of replicating imbalances. By discussing ideas of energy justice, climate justice, and intergenerational equality, this part explores the moral implications of energy transitions. Indigenous viewpoints that reinterpret "energy security" as a relational and ethical obligation are also highlighted.

Ethical Challenges of Energy Transition

A number of moral conundrums are brought up by the search for renewable energy. Underlying power dynamics are revealed by decisions concerning the location of solar or wind farms, who pays for the infrastructure, and who benefits from them. Questions of consent, sovereignty, and distributive justice are raised by the frequent occurrence of large-scale renewable projects on Indigenous or marginalized areas (Sovacool, 2021). Furthermore, extractive logics that have substantial negative effects on the environment and society are reproduced when vital minerals are mined for clean technologies. The shift must therefore be viewed as a moral endeavor that influences global equity and justice rather than just a technical replacement of fossil fuels with renewable energy sources (Heffron & McCauley, 2017).

Concepts of Energy Justice, Climate Justice and Intergenerational Equity

Distributional justice (how benefits and burdens are shared), procedural justice (who has the authority to make decisions), and recognition justice (whose voices and worldviews are acknowledged) are the three interrelated dimensions that make up the framework for assessing energy systems (Jenkins et al., 2016). The disparities inherent in both reliance on fossil fuels and the shift to renewable energy sources are highlighted by these dimensions. By emphasizing the disproportionate effects of climate change on marginalized communities, especially Indigenous peoples, low-income groups, and countries in the Global South, climate justice broadens this concept (Schlosberg & Collins, 2014). Fairness and accountability in global climate governance are called into question by its insistence that people who contribute the least to greenhouse gas emissions frequently shoulder the greatest responsibilities. By requiring current generations to preserve ecological integrity for the sake of future generations, intergenerational fairness significantly expands the moral landscape. This idea is quite similar to Indigenous worldviews, which base judgments on a sense of duty to the unborn. Intergenerational fairness in energy transitions stresses sustainability as a moral duty rather than a choice for policy and questions short-term economic calculations.

Indigenous Perspective on Moral Responsibility

By placing moral responsibility within a relational ontology that acknowledges the agency of non-human entities, the earth, and the water, indigenous frameworks expand and enhance these justice paradigms. Energy is a sacred power that should be treated with humility and reciprocity, not just as a resource (Kimmerer, 2013). According to Whyte (2017), the moral obligation to care for the land is inextricably linked to the obligation to care for the community and the next generation in many Indigenous cultures. For example, the Haudenosaunee "Great Law of Peace" places a strong emphasis on integrating long-term stewardship into governance by taking into account how actions would affect the seventh generation to come (LaDuke, 2020). This notion demands accountability that goes beyond the here and now by redefining energy decisions as moral commitments rather than technical calculations. Indigenous groups make moral arguments based on obligations to their ancestors, descendants, and the land itself when opposing extractive projects, such as pipelines, hydroelectric dams, and lithium mines (McGregor, 2014).

Reframing Energy Security through Indigenous Lenses

"Energy security" is frequently defined in the mainstream discourse in terms of affordability, geopolitical stability, and dependable supply. However, this framing stems from paradigms that prioritize unbroken access to resources and are state-centric and consumption-driven. According to Indigenous perspectives, sovereignty, ecological integrity, and cultural survival are all inextricably linked to energy security. Diversifying supply sources alone won't provide true energy security; instead, energy practices must respect Indigenous rights and support living systems (Powys Whyte et al., 2019). Energy security can be reframed as energy sovereignty, when communities exercise control over energy production in accordance with cultural values and ecological care, as exemplified by Indigenous-led renewable projects in Canada and Aotearoa New Zealand (Hoicka et al., 2021). This strategy promotes decentralized, robust, and morally sound energy systems in opposition to prevailing models that link security to expansion and centralization.

FROM EXTRACTION TO EQUITY: PATHWAYS FOR CHANGE

Systemic change based on justice, sovereignty, and inclusivity is necessary to go from extractive to equitable energy systems; it goes beyond simple technology replacement. The worldwide push to decarbonize must avoid reproducing the exploitation and dispossession patterns that defined previous periods of resource expansion and fossil fuel use. Rather, Indigenous governance frameworks, leadership, and expertise should be at the forefront of reform initiatives. Examples of Indigenous-led renewable energy projects, shared governance tactics, and the critical role that sovereignty and rights play in creating just energy futures are all covered in this section.

Transitioning from Exploitative to Equitable System

Energy systems have historically been organized around exploitative principles that put state control, industrial expansion, and profit maximization ahead of ecological balance and the welfare of communities. Indigenous peoples were uprooted by these institutions, which also damaged ecosystems and consolidated wealth in the hands of corporations (Scott, 2020). Conversely, distributive, procedural fairness, guaranteeing equitable access to energy resources, meaningful participation in decision-making, and acknowledging different worldviews are all emphasized by equitable energy systems (Heffron & McCauley, 2017). Rethinking infrastructure ownership and size is necessary for a fair transition. Whether they are renewable or fossil fuel-based, large, centralized projects frequently reinforce inequity and dependency (Pandey & Goyal, 2025). Decentralized, community-centered, environmentally conscious initiatives that not only provide energy but also give communities social, political, and cultural power are given priority in equitable models. In order to achieve this paradigm shift, energy system governance must incorporate the fundamental Indigenous values of reciprocity, stewardship, and accountability (Kimmerer, 2013).

Indigenous-Led Renewable Energy Projects

Indigenous people are spearheading renewable energy projects globally, challenging the dominance of extractive practices. These initiatives show how energy growth may promote justice, sustainability, and cultural survival all at once. In addition to producing renewable energy, Indigenous nations in Canada have built wind, solar, and run-of-river hydro projects that boost the local economy and create jobs. One of British Columbia's biggest solar power projects, for example, is run by the T'Sou-ke Nation on Vancouver Island and is intended to complement their traditional principles of sustainability and self-determination (Sovacool et al., 2020). Similar to this, Māori communities in Aotearoa, New Zealand, have made investments in geothermal projects that combine the development of renewable energy with traditional stewardship practices, strengthening sovereignty and encouraging ecological responsibility (Harris, 2021). Microgrid projects in isolated Indigenous areas provide other examples of creativity. These technologies improve energy security, represent localized government structures, and lessen reliance on expensive and environmentally harmful diesel generators. Such initiatives represent a move toward justice in energy supply, governance, and ownership by integrating Indigenous cultural values and ecological principles.

Joint Governance and Co-Development of Energy Plans

Beyond individual initiatives, fundamental governance reforms are necessary for equitable transitions. Indigenous communities are guaranteed to be active decision-makers rather than passive stakeholders through shared governance and co-creation of energy policies. Energy planning can be shaped by cultural values, traditional knowledge, and community interests through co-management agreements between states, utilities, and Indigenous nations (Baker & De Silva, 2021). For instance, collaborations between energy providers and Aboriginal communities in Australia have demonstrated that shared decision-making promotes legitimacy, trust, and more sustainable results (Bargh, 2019). Destroying colonial power structures and acknowledging Indigenous rights to land and resources as the cornerstones of equitable participation are necessary for co-creation, which is more than just a formality. Without this understanding, shared governance runs the risk of become tokenistic and promoting structural injustices rather than addressing them.

Indigenous Rights and Sovereignty in Fair Transformation

The declaration of Indigenous sovereignty and rights is essential to the transition from extraction to equity. The right to free, prior, and informed consent (FPIC) in resource development is upheld by international legal documents such the United Nations Declaration on the Rights of Indigenous Peoples (UNDRIP) (United Nations, 2007). The ability of Indigenous people to decide the sources, ownership, and uses of energy in accordance with traditional and ecological values is known as energy sovereignty, and it extends beyond simple participation (Powys Whyte et al., 2019). Therefore, the goals of energy transformation are reframed by indigenous sovereignty from a limited emphasis on carbon reduction to a more comprehensive pursuit of justice, resilience, and decolonization. Additionally, energy transitions support intergenerational knowledge transfer, language preservation, and cultural renewal when Indigenous sovereignty is respected. These factors highlight the fact that energy justice involves more than just allocating kilowatts; it also entails empowering communities to prosper according to their own rules.

CHALLENGES AND BARRIERS

The incorporation of these worldviews into mainstream policy and practice is not without difficulties, despite the fact that Indigenous knowledge and leadership offer vital avenues toward equitable energy transformation. Conflicts between economic growth models and Indigenous self-determination, the risk of tokenism, the delicate balance between modernization and cultural preservation, and issues of technology transfer, capacity building, and ownership must all be addressed in order to make the shift from extractive to equitable systems. In order to stop well-meaning energy transitions from replicating colonial patterns under different guises, it is imperative that these issues be addressed honestly.

Risk of Tokenism or Superficial Inclusion

The risk of tokenism—the superficial inclusion of Indigenous knowledge without significant changes in power or governance—is one of the most urgent issues. While continuing to function within extractive frameworks, governments and companies frequently symbolically evoke Indigenous perspectives, for as by mentioning traditional values in sustainability reports (Parsons et al., 2016). By limiting Indigenous knowledge to cultural ornamentation rather than the basis for systemic change, such activities weaken its transformational potential. Power-sharing and decision-making authority are necessary for true inclusion; consultation alone is insufficient. According to the United Nations Declaration on the Rights of Indigenous Peoples (UNDRIP), free, prior, and informed consent (FPIC) must be regarded as a legally binding norm rather than a formality (United Nations, 2007). The inclusion of Indigenous voices runs the risk of replicating the very injustices that energy transitions aim to eliminate if they are not accompanied by true empowerment.

Modernization and Cultural Preservation in Balance

Maintaining Indigenous cultural practices while embracing contemporary energy technologies presents another challenge. Although renewable technologies like solar, wind, and microgrids present chances for energy sovereignty and sustainability, they may also bring with them consumerism, efficiency, and commodification ideals that are at odds with Indigenous worldviews (Whyte, 2017). For example, if solar installations are run using private ownership models instead of collective governance, they may unintentionally change community relations even while they may lessen reliance on fossil fuels. Indigenous leaders frequently emphasize that cultural extinction or absorption into prevailing paradigms shouldn't be the price of modernization (McGregor, 2014). Therefore, the difficulty lies in designing energy systems that combine technological advancement with cultural preservation, making sure that technology enhances Indigenous identities and customs rather than diminishes them.

Conflicts Between Indigenous Self-Determination and Economic Growth Models

The paradigms of economic growth that emphasize investment, market expansion, and green capitalism are frequently used to frame mainstream energy transitions. However, these models often clash with Indigenous peoples' desires for sovereignty, self-determination, and environmental care. Energy development is a matter of political autonomy and cultural survival for many Indigenous communities, in addition to being an economic opportunity (LaDuke, 2020). When governments support massive renewable energy projects on Indigenous territories without giving proper acknowledgment to their rights or governance frameworks, conflicts result. Even supposedly “green” projects like biofuel plantations or mega-dams have the potential to replicate ecological disruption and displacement patterns. These disputes underscore the shortcomings of growth-centered strategies and the demand for substitute success indicators based on ecological balance, cultural resilience, and well-being (Simpson, 2017).

Transfer of Technology, Development of Capacity, and Ownership Frameworks

Equitable energy transitions also depend on resolving issues with ownership, capacity building, and technology transfer. Many Indigenous communities encounter difficulties obtaining the financial resources and technical know-how required to plan, carry out, and manage renewable energy projects, especially those located in remote areas (Henderson, 2013). Projects run the risk of becoming dependent on outside parties, which would compromise sovereignty and long-term viability, unless intentional efforts are made to develop local capability. Structures of ownership are very important. Local communities rarely receive significant benefits from projects owned or controlled by external businesses, but community-owned and co-managed projects strengthen energy sovereignty (Sovacool et al., 2020). Transferring technology shouldn't just involve gear; it should also involve sharing technological know-how, financial literacy, and governance abilities in ways that respect Indigenous customs and interests.

POLICY AND LEGAL DIMENSIONS

Without addressing the legal and policy frameworks governing resource management, Indigenous rights, and climate action, it will be impossible to make the shift from extractive energy systems to equitable and sustainable alternatives. The degree to which Indigenous knowledge and sovereignty are integrated into energy transformation processes is influenced by national laws, international governance regimes, and community-based initiatives. The persistence of extractive logics in many legal systems is shown in the substantial implementation gaps that continue notwithstanding advancements in international recognition.

International Agreements

A number of international agreements offer normative underpinnings for integrating Indigenous rights into the energy transition. Indigenous communities' rights to self-determination, cultural preservation, and free, prior, and informed consent (FPIC) for projects on their lands are outlined in the United Nations Declaration on the Rights of Indigenous Peoples (UNDRIP) (United Nations, 2007). When it comes to energy transition, FPIC is essential to preventing the replication of colonial patterns of dispossession in renewable energy projects, like massive wind farms or hydropower developments. In a similar vein, ratifying governments are legally required to acknowledge Indigenous peoples' rights to land, consultation, and decision-making involvement under International Labour Organization (ILO) Convention No. 169 (ILO, 1989). Despite being ratified by a small number of nations, its provisions are frequently cited when promoting energy governance that is inclusive of Indigenous peoples (Rodríguez-Garavito, 2019). By acknowledging the importance of traditional knowledge in adaptation and mitigation plans, the Paris Agreement (UNFCCC, 2015) also subtly links Indigenous rights to international climate responsibilities. Since decarbonization is a key component of climate goals, Article 7.5 of the Agreement emphasizes the importance of respecting Indigenous knowledge systems in climate action, which logically applies to energy transitions (Magni, 2017). However, because these rules are not legally enforceable, their impact has been limited, and many Indigenous people are still not given meaningful involvement in the planning of renewable energy.

National-Level Recognition

The way that national legal systems handle Indigenous rights and land claims varies greatly. In certain situations, Indigenous power over natural resources has been strengthened by statutory frameworks or constitutional recognition. For instance, Māori co-governance of energy and environmental resources is made possible by the Treaty of Waitangi settlements in New Zealand (Mutu, 2019). Similarly, although there are still issues with balancing these rights with extractive industries, Canada's acknowledgment of Aboriginal title has created opportunities for collaborations led by Indigenous peoples in renewable energy (Papillon & Rodon, 2017; Joshi & Pandey, 2025). However, in the United States, energy projects on Indigenous territories frequently face legal and administrative obstacles that compromise autonomy, even if the federal government recognizes tribal sovereignty (Jorgensen, 2016). Legal protections are in place in other places, such Latin America and portions of Africa, but they are usually compromised by extractive policies that put an emphasis on economic growth and foreign investment. Large-scale mining and oil operations, for example, continue to uproot Indigenous communities in Ecuador and Bolivia despite the constitution's protection of their rights. This reflects conflicts between development-driven governance models and Buen Vivir (living well) worldviews (Gudynas, 2011).

Gaps in Implementation

Even in cases when legal frameworks are in place, they are frequently not applied consistently or sufficiently. For instance, instead of being a true process of negotiation and consent, FPIC is usually reduced to a formality (Colchester & Ferrari, 2007). Furthermore, a number of renewable energy initiatives marketed as sustainable have come under fire for sustaining "green colonialism," in which Indigenous territories are once more singled out for resource exploitation through the development of wind farms, solar farms, or biofuel (Whyte, 2017; Pandey & Joseph, 2025). This suggests that renewable energy transitions run the risk of replicating the inequities of fossil fuel economies in the absence of systemic change (Pandey, 2025).

Policy Pathways for Embedding Indigenous Knowledge

Policy approaches must actively incorporate Indigenous knowledge into energy planning and governance in order to remedy these deficiencies. First, in order to guarantee meaningful involvement, FPIC's legal requirements must be strengthened both domestically and internationally. Second, equity can be advanced through institutional co-governance mechanisms like shared administration of energy infrastructure or Indigenous advisory boards in national energy agencies. Third, incorporating Indigenous knowledge systems into technical planning provides a comprehensive viewpoint that is frequently absent from technocratic energy plans. Examples of these systems include land use assessments, biodiversity considerations, and long-term stewardship practices (McGregor, 2018). Lastly, reevaluating ownership and benefit-sharing structures is necessary to advance Indigenous sovereignty in energy transitions. Indigenous-led renewable energy initiatives, including community-owned wind farms in New Zealand and solar microgrids in Canada, are prime examples of how energy transformation may coexist with environmental care, local economic growth, and cultural values (Krupa, 2012). Incorporating these approaches into national energy strategy guarantees the legitimacy and durability of the transition while also improving equity.

CONCLUSION

In energy systems, the transition from extraction to equity is a historical reflection as well as a forward-looking outlook. Extractive modes of development have exploited natural resources and the populations that live closest to them for ages, putting short-term profits ahead of long-term sustainability. Indigenous lands and societies are deeply scarred by this legacy of social injustice, ecological degradation, and relocation, which has solidified injustices that still exist today. However, there is a chance to forge a new course—one that centres energy transformation around fairness, justice, and moral responsibility—as the globe confronts the pressing challenges of climate change and the shift to low-carbon economies. Indigenous knowledge is central to this new course. Indigenous worldviews, which are based on the ideas of reciprocity, stewardship, and interconnectivity, provide an alternative perspective to exploitative paradigms that consider the earth as nothing more than a resource to be exploited. Rather, they present energy as a relationship that needs to be cultivated with responsibility and care between people, the land, and future generations. Energy transitions can transcend technical solutions and adopt a comprehensive, cultural, ethical, and ecological vision of sustainability by putting Indigenous ecological wisdom front and centre. A basic ethical requirement is highlighted by the combination of the history of extraction and the promise of equity: energy transformation cannot reproduce the colonial and exploitative logics of the past. Despite being presented as sustainable, renewable energy initiatives run the risk of sustaining "green colonialism" if they ignore Indigenous knowledge and rights. Restructuring power dynamics, ownership structures, and decision-making procedures is necessary for equity; it goes beyond simply replacing fossil fuels with renewable energy sources like wind and solar. In addition to being consulted, indigenous communities must be given the authority to lead and co-create energy futures. In this setting, reclaiming Indigenous knowledge is an act of inventiveness and survival rather than nostalgia. It creates opportunities for environmentally restorative, culturally grounded, and community-led renewable initiatives. Examples of how knowledge systems can support contemporary technology to increase resilience include solar microgrids in isolated communities, wind farms run by Indigenous cooperatives, and hybrid energy systems with a traditional stewardship foundation. These instances show that, as long as it respects Indigenous peoples' sovereignty and voices, energy transition may be both technologically advanced and equitable. As this chapter has shown, the shift is not just technical but also deeply moral. It pushes society, business leaders, and legislators to prioritize justice, accountability, and respect for Indigenous sovereignty over efficiency and profitability. Humanity can both combat climate change and mend the wounds from its extractive history by integrating equality and Indigenous knowledge into energy transitions. Therefore, moving from extraction to equity is both morally required and a sign of a more sustainable, inclusive future.

REFERENCES

Baker, S., & De Silva, L. (2021). Co-governance of renewable energy projects with Indigenous communities. *Energy Research & Social Science*, *79*, 102160. DOI: 10.1016/j.erss.2021.102160

Banerjee, S. B. (2019). Decolonizing sustainability. *Organization Studies*, *40*(3), 437–456. DOI: 10.1177/0170840618818992

Bargh, M. (2019). Indigenous peoples and renewable energy: Rights and recognition. *Indigenous Policy Journal*, *30*(2), 1–16.

Bebbington, A., Bebbington, D. H., Sauls, L. A., Rogan, J., Agrawal, S., Gamboa, C., & Verdum, R. (2018). Resource extraction and infrastructure threaten forest cover and community rights. *Proceedings of the National Academy of Sciences of the United States of America*, *115*(52), 13164–13173. DOI: 10.1073/pnas.1812505115 PMID: 30509982

Bratman, E. (2014). Contradictions of green development: Human rights and environmental norms in light of Belo Monte Dam activism. *Journal of Latin American Studies*, *46*(2), 261–289. DOI: 10.1017/S0022216X14000042

Brugge, D., & Goble, R. (2002). The history of uranium mining and the Navajo people. *American Journal of Public Health*, *92*(9), 1410–1419. DOI: 10.2105/AJPH.92.9.1410 PMID: 12197966

Colchester, M., & Ferrari, M. F. (2007). *Making FPIC—free, prior and informed consent—work: Challenges and prospects for indigenous peoples*. Forest Peoples Programme.

Corntassel, J. (2012). Re-envisioning resurgence: Indigenous pathways to decolonization and sustainable self-determination. *Decolonization*, *1*(1), 86–101.

Dunlap, A. (2021). *Renewable energy: A critical introduction*. Routledge.

Gordon, J. (2022). *Energy cultures: Technology, justice, and everyday life*. Cambridge University Press.

Gudynas, E. (2011). Buen Vivir: Today's tomorrow. *Development*, *54*(4), 441–447. DOI: 10.1057/dev.2011.86

Harris, M. (2021). Māori approaches to energy: Geothermal development and Indigenous sovereignty. *Energy Policy*, *156*, 112403. DOI: 10.1016/j.enpol.2021.112403

Heffron, R. J., & McCauley, D. (2017). The concept of energy justice across the disciplines. *Energy Policy, 105*, 658–667. DOI: 10.1016/j.enpol.2017.03.018

Heffron, R. J., & McCauley, D. (2018). What is the 'just transition'? *Geoforum, 88*, 74–77. DOI: 10.1016/j.geoforum.2017.11.016

Henderson, C. (2013). *Aboriginal power: Clean energy and the future of Canada's First Peoples*. Rainforest Editions.

Hoicka, C. E., Savic, K., & Campney, A. (2021). Reconciliation through renewable energy? A survey of Indigenous communities, involvement, and peoples in Canada. *Energy Research & Social Science, 74*, 101897. DOI: 10.1016/j.erss.2020.101897

International Energy Agency. (2021). Net Zero by 2050: A roadmap for the global energy sector. https://www.iea.org/reports/net-zero-by-2050

International Labour Organization (ILO). (1989). *C169 - Indigenous and Tribal Peoples Convention, 1989 (No. 169)*. ILO.

Jaffar, A. (2020). Indigenous energy sovereignty: Models and pathways for equitable transitions. *Energy Research & Social Science, 69*, 101730. DOI: 10.1016/j.erss.2020.101730

Jasanoff, S., & Kim, S. H. (2015). *Dreamscapes of modernity: Sociotechnical imaginaries and the fabrication of power*. University of Chicago Press. DOI: 10.7208/chicago/9780226276663.001.0001

Jenkins, K., McCauley, D., Heffron, R., Stephan, H., & Rehner, R. (2016). Energy justice: A conceptual review. *Energy Research & Social Science, 11*, 174–182. DOI: 10.1016/j.erss.2015.10.004

Jorgensen, M. (2016). Sovereignty, wealth, culture, and control: The contemporary political economy of Indian gaming. *The American Behavioral Scientist, 60*(6), 713–733. DOI: 10.1177/0002764216632838

Joshi, A., & Pandey, A. (2025). Advancing Climate Finance for Sustainable Future: Integrating Human Capital, Climate Neutrality, and Emerging Technologies. *Climate Neutrality and Sustainable Ecosystems: Practical Approaches to Addressing Global Warming*, 231-246.

Kimmerer, R. W. (2013). *Braiding sweetgrass: Indigenous wisdom, scientific knowledge, and the teachings of plants*. Milkweed Editions.

Krupa, J. (2012). Blazing a new path forward: A case study on the renewable energy initiatives of the Pic River First Nation. *Environmental Development, 3*, 109–122. DOI: 10.1016/j.envdev.2012.05.003

LaDuke, W. (2020). *To be a water protector: The rise of the Wiindigoo slayers*. Fernwood Publishing.

Magni, G. (2017). Indigenous knowledge and implications for the sustainable development agenda. *European Journal of Education*, *52*(4), 437–447. DOI: 10.1111/ejed.12238

McCully, P. (2001). *Silenced rivers: The ecology and politics of large dams*. Zed Books.

McGregor, D. (2014). Traditional knowledge and water governance: The ethic of responsibility. *Alternative*, *10*(5), 493–507. DOI: 10.1177/117718011401000505

McGregor, D. (2018). Mino-mnaamodzawin: Achieving Indigenous environmental justice in Canada. *Environment and Society*, *9*(1), 7–24. DOI: 10.3167/ares.2018.090102

Mitchell, T. (2011). *Carbon democracy: Political power in the age of oil*. Verso.
Mitchell, T. (2011). *Carbon democracy: Political power in the age of oil*. Verso.
Mutu, M. (2019). *The Treaty of Waitangi and Māori rights in Aotearoa New Zealand*. Auckland University Press.

Newell, P., & Mulvaney, D. (2013). The political economy of the 'just transition.'. *The Geographical Journal*, *179*(2), 132–140. DOI: 10.1111/geoj.12008

Pandey, A. (2025). Blockchain in Industrial Supply Chains: Tracing the Transparency. In *Immersive Technology for the Gig Economy: Transformative Business Practices* (pp. 273-294). IGI Global Scientific Publishing.

Pandey, A., & Goyal, M. (2025). Leveraging Technology for Climate Resilience in Urban Areas. In *Revolutionizing Urban Development and Governance With Emerging Technologies* (pp. 137–162). IGI Global Scientific Publishing. DOI: 10.4018/979-8-3373-1375-7.ch005

Pandey, A., & Joseph, J. (2025). Integral Role of Blockchain and Artificial Intelligence in Sustainable Economic Development. In *Driving Socio-Economic Growth With AI and Blockchain* (pp. 271–290). IGI Global Scientific Publishing. DOI: 10.4018/979-8-3693-8664-4.ch012

Papillon, M., & Rodon, T. (2017). Proponent-Indigenous agreements and the implementation of the right to free, prior, and informed consent in Canada. *Environmental Impact Assessment Review*, *62*, 216–224. DOI: 10.1016/j.eiar.2016.06.009

Parlee, B. (2015). Avoiding the resource curse: Indigenous communities and Canada's oil sands. *World Development, 74*, 425–436. DOI: 10.1016/j.worlddev.2015.03.004

Parsons, M., Fisher, K., & Nalau, J. (2016). Alternative approaches to co-production: Building adaptive capacity for disaster risk management. *Climate Risk Management, 12*, 1–10. DOI: 10.1016/j.crm.2016.01.001

Powys Whyte, K., McGregor, D., & Talaga, A. (2019). Indigenous lessons about sustainability are not just for "all humanity.". *Nature Communications, 10*(1), 1–3. DOI: 10.1038/s41467-019-11971-9 PMID: 30602773

Quijano, A. (2000). Coloniality of power, Eurocentrism, and Latin America. *International Sociology, 1*(3), 533–580. DOI: 10.1177/0268580900015002005

Rodríguez-Garavito, C. (2019). Ethnicity.gov: Global governance, indigenous peoples, and the right to prior consultation in social minefields. *Indiana Journal of Global Legal Studies, 18*(1), 263–305. DOI: 10.2979/indjglolegstu.18.1.263

Schlosberg, D., & Collins, L. B. (2014). From environmental to climate justice: Climate change and the discourse of environmental justice. *Wiley Interdisciplinary Reviews: Climate Change, 5*(3), 359–374. DOI: 10.1002/wcc.275

Scott, D. N. (2020). Fossil capital and the law: Toward a just transition. *Osgoode Hall Law Journal, 57*(3), 845–878.

Scott, J. C. (2017). *Against the grain: A deep history of the earliest states*. Yale University Press. DOI: 10.2307/j.ctv1bvnfk9

Simpson, L. B. (2017). *As we have always done: Indigenous freedom through radical resistance*. University of Minnesota Press. DOI: 10.5749/j.ctt1pwt77c

Sovacool, B. K. (2021). Who are the victims of low-carbon transitions? Towards a political ecology of climate change mitigation. *Energy Research & Social Science, 73*, 101916. DOI: 10.1016/j.erss.2021.101916

Sovacool, B. K., Hook, A., Martiskainen, M., & Baker, L. (2020). The decarbonisation divide: Contextualizing landscapes of low-carbon exploitation and toxicity in Africa. *Global Environmental Change, 60*, 102028. DOI: 10.1016/j.gloenvcha.2019.102028

Tilt, B. (2015). *Dams and development in China: The moral economy of water and power*. Columbia University Press.

United Nations. (2007). *United Nations Declaration on the Rights of Indigenous Peoples*. United Nations General Assembly.

United Nations Environment Programme. (2022). Emissions gap report 2022. https://www.unep.org/resources/emissions-gap-report-2022

United Nations Framework Convention on Climate Change (UNFCCC). (2015). *The Paris Agreement*. UNFCCC.

Whyte, K. (2018). Indigenous science (fiction) for the Anthropocene: Ancestral dystopias and fantasies of climate change crises. *Environment and Planning. E, Nature and Space*, *1*(1–2), 224–242. DOI: 10.1177/2514848618777621

Whyte, K. P. (2017). Indigenous climate change studies: Indigenizing futures, decolonizing the Anthropocene. *English Language Notes*, *55*(1–2), 153–162. DOI: 10.1215/00138282-55.1-2.153

KEY TERMS AND DEFINITIONS

Authority: Indigenous peoples' innate right to self-determination in matters of energy and development, as well as their inherent authority to govern themselves, their lands, and their resources.

Equity Across Generations: The moral precept that in order to guarantee that future generations inherit a healthy environment and equitable opportunity, current generations must use resources responsibly and sustainably.

Equity: A balanced and equitable distribution of resources, opportunities, and advantages that guarantees marginalized groups—including Indigenous peoples—equal access and involvement in energy transitions.

Green Colonialism: Renewable energy projects that occupy or exploit Indigenous lands without honouring their rights, sovereignty, or cultural values are perpetuating colonial patterns of exploitation.

Justice for Climate Change: An ethical and political strategy that ensures that the people most impacted—often low-income and Indigenous communities—are at the center of solutions by connecting the effects of climate change to issues of inequality.

Justice for Energy: A framework for equitable energy decision-making, benefit and burden sharing, and rights acknowledgment, especially for vulnerable or excluded groups.

Native American Wisdom: Indigenous communities have evolved a comprehensive system of knowledge, customs, and cultural values over many generations, with a focus on sustainability, reciprocity, and stewardship.

Prior, Informed, and Free Consent (FPIC): Assuring true involvement in decision-making by granting Indigenous peoples the ability to agree or not to projects that impact their communities, territories, and resources.

Transition of Energy: The worldwide transition from fossil fuel-based energy systems to sustainable and renewable alternatives, motivated by the necessity for low-carbon development and the urgency of climate change.

Chapter 11 Spatio-Temporal Modeling and Forecasting in Environmental, Energy, and Demographic Systems: A Unified AI-Driven Framework

J. Vijaya
https://orcid.org/0000-0002-9948-8573
IIITNR, India

Jajam Nagaraju
Tirumala Engineering College, India

J. Manokaran
https://orcid.org/0000-0002-1251-3896
Sri Sakthi Institute of Engineering and Technology, Coimbatore, India

ABSTRACT

In today's data-driven world, modeling and forecasting environmental, energy, and demographic systems has become crucial for addressing some of the most pressing global challenges such as pollution, energy management, climate change, and disaster risk reduction. These systems are inherently dynamic, evolving across both space and time, and are influenced by a complex interplay of geographic, economic, and social factors. Traditional forecasting techniques often struggle to account for such spatio-temporal complexity, making room for the emergence of artificial intelligence (AI) and machine learning as powerful tools in this domain. By integrating spatial

DOI: 10.4018/979-8-3373-5302-9.ch011

and temporal dimensions, AI-powered systems can detect patterns, predict trends, and offer timely, location-specific insights that combine the strengths of different time series and LSTM algorithms. This advancement is particularly valuable for governments, researchers, and industries aiming to make proactive decisions for sustainable development, efficient resource allocation, and disaster preparedness.

INTRODUCTION

In the era of rapid digitalization and environmental awareness, modeling and forecasting systems that change across both time and space have become critical for addressing large-scale global challenges. Environmental conditions, energy demands, population growth, and natural disasters are deeply interconnected, influenced by geographic, economic, climatic, and behavioral factors. Traditional models that rely on fixed patterns or linear trends often fail to capture the complex and dynamic nature of these systems. Artificial intelligence (AI) and machine learning techniques are emerging as essential tools for spatio-temporal modeling, enabling more accurate predictions and more effective decision-making by integrating spatial data (geographical features, land use, regional connectivity) and temporal data (hourly, daily, seasonal trends). These systems support sustainable growth, safety, and policy-making across multiple domains.

Air Quality Forecasting

Rapid urbanization, industrialization, and vehicular growth have caused sharp rises in air pollution, especially in developing countries. AI models can analyze historical pollutant data (PM2.5, PM10, NO_2, CO) from sensors and satellite imagery, identify seasonal trends, spatial distribution patterns, and real-time spikes, and use meteorological variables like wind speed, humidity, and temperature to forecast air quality index (AQI) levels. Such forecasts enable timely health advisories, emission control policies, and cleaner transport planning, as demonstrated in cities like Beijing and Delhi (Abbasi et al., 2025) (Deveer & Minet, 2025) (Gupta et al., 2023).

Energy Systems

As nations integrate renewable energy sources like solar and wind into their grids, accurate forecasting is essential for stability and efficiency. Spatially distributed generation is affected by geography and weather, which change over time. AI models trained on historical generation, weather forecasts, and consumption patterns can predict energy output and demand at different spatial resolutions, optimizing dispatch, reducing fossil fuel reliance, and aiding decarbonization strategies (Alqasim, 2022) (Peng et al., 2022) (Beura et al., 2024).

Electricity Consumption

Demand fluctuates sharply with population growth and digital lifestyles. Utilities use spatio-temporal data to anticipate load curves, and AI models process historical load, user behavior, time-of-day, and climate data to forecast electricity needs. These forecasts help prevent outages, cut costs, and improve energy efficiency. In smart cities, predictive systems adjust lighting, HVAC, and storage based on demand forecasts, enabling effective demand-side management (Bilgili & Pinar, 2023; Li et al., 2022; Chi, 2022).

Earthquake Forecasting

While exact prediction remains challenging, spatio-temporal models can forecast likelihood and identify risk zones using seismic data, historical records, tectonic movement, and geological variables. Machine learning detects patterns in tremors, foreshocks, and stress distribution, classifies risk zones, and supports disaster preparedness. Early warning systems using AI can give seconds to minutes of notice, reducing casualties and damage (Arul & Kareem, 2021; Mallouhy et al., 2019; Berhich et al., 2022).

Birth Rate Forecasting

Accurate estimation of birth patterns is vital for healthcare planning, maternal services, immunization programs, and education policy. Despite abundant historical records, many models fail to leverage demographic and socioeconomic data fully or to capture non-linear, seasonal, and behavioral patterns. Incorporating these factors with advanced AI methods can improve precision and utility for population planning (Bheemanna et al., 2023; Kulkarni et al., 2021; Onambele et al., 2023; Suneer, 2024; Comolli & Vignoli, 2021).

Methods

Time series modeling is central to these forecasting efforts. Classical approaches such as ARIMA (AutoRegressive Integrated Moving Average) work well for stable, linear trends, while deep learning models like Long Short-Term Memory (LSTM) networks and Gated Recurrent Units (GRUs) handle complex, non-linear, and multivariate data. Combining these with convolutional neural networks (CNNs) enables spatial pattern analysis. Hybrid and ensemble models, attention mechanisms, and graph neural networks (GNNs) further enhance accuracy. Data preprocessing—normalization, imputation, noise filtering, feature selection—is essential, and evaluation metrics such as MAE, RMSE, precision, recall, and F1-score measure performance depending on the task. In conclusion Spatio-temporal modeling and forecasting are key to environmental sustainability, energy efficiency, and population resilience. As real-time, high-resolution spatial and temporal data become more available and algorithms more adaptive, AI-driven systems will continue to expand their role in enabling smarter, data-informed decisions across public and private sectors.

BACKGROUND

One of the foremost applications of AI-based spatio-temporal modeling lies in forecasting air quality across urban and rural regions. As cities expand and industrial activity increases, pollutants such as PM2.5, nitrogen dioxide, carbon monoxide, and ozone reach dangerously high levels, affecting public health and ecosystems. Accurate air quality forecasting can help inform citizens, assist urban planners, and guide environmental regulations. AI models analyze historical pollutant data, satellite imagery, traffic density, weather variables, and emission sources to identify pollution patterns and predict future AQI levels. These predictions enable authorities to issue early warnings during hazardous episodes and implement targeted interventions, such as restricting vehicle movement or temporarily halting industrial emissions. In regions like South Asia and East Asia, where seasonal air pollution events recur frequently, spatio-temporal forecasting systems have already demonstrated their ability to reduce exposure risks and support environmental governance. In recent years, research on time-series forecasting of air quality parameters has been highly active (Jayaraman, 2025). Numerous modeling strategies ranging from sophisticated machine learning techniques to conventional statistical methods have been proposed. For instance, Emeç & Yurtsever (2025) evaluated several time-series models, such as ARIMA, for predicting Delhi's daily air quality index and found that ARIMA performed reasonably well for short-term forecasts. Similarly, Khobragade & Ahirwar (2022) showed that incorporating seasonality into ARIMA models improved forecasting accuracy for PM2.5 concentrations in Beijing. Galán-Madruga et al., (2025) combined artificial neural networks with ARIMA, demonstrating that their hybrid approach outperformed individual models. Xayasouk et al., (2020) employed Long Short-Term Memory (LSTM) networks to forecast PM2.5 and compared their performance to traditional time-series models, concluding that LSTMs provided superior results. Despite these advances, a comprehensive analysis that considers temporal dependencies, data stationarity, and model performance across diverse contexts is still needed. Further research has continued to highlight the value of deep learning methods. Méndez et al., (2023) applied LSTM neural networks—well suited for time-series data—to predict AQI using historical pollutant and meteorological variables. Their study demonstrated that LSTMs, by capturing complex temporal dependencies, significantly outperformed regression-based models in both accuracy and adaptability. Meanwhile, Liu et al., (2019) focused on real-time AQI classification using supervised machine learning algorithms, such as Random Forest and Support Vector Machines (SVM). Their dataset included concentrations of PM2.5, NO, and O, with Random Forest achieving 92.5% classification accuracy. This approach proved particularly effective for quick assessments of air quality categories (e.g., “Good,” “Moderate,” “Unhealthy”), which can trigger timely alerts and policy actions. The European Space Agency (ESA) research team also employed Sentinel-5P satellite imagery to remotely estimate AQI by monitoring pollutants such as NO, SO, and PM2.5. Their satellite-derived estimates showed strong agreement with ground-based observations, confirming the value of remote sensing for large-scale air quality monitoring. Long before the surge in AI-driven environmental research, Wang et al., (2022) conducted a landmark cohort study across multiple U.S. cities to examine the long-term health impacts of air pollution. The findings were sobering: for every 10 $\mu g/m^3$ increase in PM2.5 concentration, there was a measurable rise in cardiopulmonary mortality. This study underscored the real-world consequences of poor air quality—not merely as an environmental concern but as a critical public health issue. It added a human dimension to AQI research, reminding us that behind every data point are individuals whose health and well-being are directly affected.

Another critical domain where spatio-temporal forecasting is transforming operations is renewable energy generation and grid management. Renewable energy sources like solar and wind are inherently intermittent and dependent on weather conditions. Accurate prediction of their power output requires integrating weather forecasts, historical energy data, spatial positioning of infrastructure, and time-of-day factors. AI models, particularly deep learning architectures, analyze this multidimensional data to predict how much energy a solar or wind plant will produce at a given time and place. Such forecasting improves operational efficiency, balances energy supply and demand, reduces reliance on fossil fuels, and supports storage planning and energy trading (Nguyen, 2025). In Europe and North America, several power companies already employ AI-driven spatio-temporal models for renewable integration and real-time energy management. Energy consumption forecasting is equally critical for modern power grids. It allows utilities to meet demand optimally, maximize efficiency, and minimize waste. Forecasts also inform demand response programs, which adjust electricity usage in real time to avoid grid overloads and reduce costly emergency generation. Moreover, accurate forecasting enables the integration of variable renewable resources, such as solar and wind, ensuring grid stability. Time series analysis is especially powerful here, as it captures historical consumption patterns, trends, seasonality, and temporal relationships. This provides actionable insights into both short-term fluctuations and long-term usage patterns in residential and commercial environments influenced by weather and lifestyle factors. In this research, multiple forecasting models are evaluated for short-term electricity demand prediction, with the aim of identifying the most precise and reliable method for residential consumption. By comparing traditional statistical approaches with modern machine learning techniques, the study highlights the strengths and weaknesses of various methods and advances intelligent energy management systems. The literature demonstrates this diversity: Bhatt et al., (2024) introduced Facebook Prophet for scalable forecasting; Helli et al., (2022) emphasized ensemble methods; Jain et al., (2022) applied regression for institutional buildings; Karthick et al., (2024) proposed hybrid models for improved accuracy; Khan et al., (2021) explored feature selection and optimization; Luo & Li (2023) reviewed ARIMA, SVM, and neural networks; and Mohapatra (2022) combined data-driven and physical models. Together, these works underscore the evolution of electricity consumption forecasting techniques.

Electricity consumption forecasting is a domain where precise, real-time predictions are critical. Urban infrastructure, industrial operations, and residential usage patterns vary with location, weather, time, and social behavior. AI models trained on historical data, temperature, holidays, and time intervals can predict demand fluctuations at city, district, or building levels. Spatio-temporal models provide granular insights that enable dynamic load balancing, cost optimization, and outage prevention. Smart meters, IoT sensors, and real-time monitoring systems further enhance forecasting, helping utilities reduce waste and ensure reliable supply (Riady, 2023) (Sau, 2024). During extreme weather events such as heatwaves or cold snaps accurate demand forecasts are essential to prevent blackouts and guide strategic decisions, including activating reserves or adjusting supply schedules. Recent studies highlight the growing role of deep learning in this field. Elsaraiti (2021) explored building automation systems using LSTM, GRU, and Bi-LSTM, achieving competitive results with the PJM Interconnection LLC dataset. Gonzalez (2022) demonstrated the versatility of RNNs, CNNs, and XGBoost in short-term load and HVAC forecasting, confirming the adaptability of deep models across applications. Hadjout (2022) applied Prophet, LSTM, and Transformer architectures, showing mixed results and emphasizing the need for context-specific approaches. Similarly, Klyuev (2022) reported state-of-the-art performance using Auto-Regressive models, CNNs, and stacked LSTM/GRU frameworks. Despite progress, challenges remain, including the requirement for high-quality datasets, better interpretability, and ethical considerations. The literature suggests that integrating advanced AI models with robust datasets and evaluation metrics (e.g., MAE, RMSE) can unlock higher accuracy in electricity forecasting, supporting grid optimization and sustainable energy management. This sets the foundation for the present study's comparative evaluation of LSTM and GRU models on the PJM dataset.

A growing area of application for spatio-temporal AI frameworks is earthquake modeling and seismic risk forecasting. Earthquakes, though difficult to predict precisely, leave rich spatial and temporal signatures in ground movement, foreshocks, aftershocks, and tectonic stress patterns. By combining seismic sensor data, geological surveys, and historical earthquake records, AI models can identify potential hotspots and estimate the likelihood of future events. While they cannot predict the exact time and location of an earthquake, these models are invaluable for long-term risk assessment and early-warning systems. For instance, recurring microseismic activity in a fault zone can indicate rising probabilities of a larger event in the coming weeks or months. Such forecasts inform infrastructure reinforcement, emergency preparedness, and insurance planning (Nurtas et al, 2024) (Reddy et al., 2023) (Ridzwan & Yusoff, 2023).

In some regions, AI-enabled early warning systems already provide seconds to minutes of advance notice, helping to mitigate damage and save lives. The effectiveness of such systems depends on robust algorithms that can capture complex temporal and spatial relationships. Traditional models like ARIMA (AutoRegressive Integrated Moving Average) remain useful for stationary, univariate data due to their simplicity and interpretability. However, they fall short in handling nonlinear interactions or multivariate, non-stationary time series typical of seismic and environmental processes. Advanced deep learning models, particularly Long Short-Term Memory (LSTM) networks, address these limitations . LSTMs are designed to learn long-term dependencies in sequential data and perform well with noisy, multivariate signals. By selectively retaining or discarding information through memory gates, they adapt to changing conditions and capture long-range temporal effects, making them well-suited for complex applications such as seismic risk forecasting, air quality prediction, and dynamic energy demand modeling (Debnath et al., 2021) (Galkina & Grafeeva, 2019) (Jafari et al., 2024) (Kruthi et al., 2023).

Literature on birth rate prediction demonstrates the use of both classical statistical methods and modern machine learning approaches. Traditional time series models such as ARIMA and its seasonal extension SARIMA have been widely employed for forecasting in fields like public health, economics, and environmental modeling, particularly when data is stationary. Exponential Smoothing techniques have also been explored for short-term prediction due to their simplicity and ease of implementation. However, these conventional models often struggle with datasets characterized by high-order nonlinearities and complex temporal patterns. Recent studies have therefore turned to advanced models, including Long Short-Term

Memory (LSTM) networks and ensemble learning approaches such as XGBoost, which offer superior performance in capturing sequential dependencies and handling noisy, high-dimensional data. For instance, Cruz et al., (2025) applied ARIMA for global birth rate prediction, while Esmaeili & Abbasi-Shavazi (2024) worked on automating ARIMA for economic indicators. Huang et al., (2022) compared statistical and machine learning models, noting the strengths of the latter but also issues of interpretability. Kumar & Meena (2025) applied time series forecasting to predict births in Mexico for health service planning, and the United Nations emphasized the policy relevance of accurate fertility forecasting. Despite these contributions, much of the existing work is limited to narrow sets of models or lacks integration of contemporary neural architectures. Addressing this gap, the present study contrasts classical and advanced forecasting techniques within a unified evaluation framework.

For even more sophisticated analysis, hybrid models are used—combining ARIMA with LSTM, or pairing LSTM with CNNs)to add spatial awareness. Such hybrid systems first extract spatial features through convolutional layers and then process the sequential data through LSTM layers to make precise spatio-temporal forecasts. In electricity demand forecasting, this might mean learning both how energy flows through a power grid (spatial) and how usage changes throughout the day and year (temporal). Graph Neural Networks (GNNs) are another emerging solution, especially effective in capturing spatial relationships across complex networks like road systems, electricity grids, or seismic sensor arrays. GNNs learn how the structure and interconnections between spatial nodes affect behavior, enabling predictions that account for proximity, flow, and influence. These models are often enhanced by attention mechanisms, which dynamically weight the importance of different features in time and space, improving both accuracy and interpretability. Before training any of these models, data preprocessing steps such as normalization, feature selection, noise filtering, and imputation of missing values are essential to ensure consistency and quality. After modeling, results are evaluated using performance metrics like Root Mean Squared Error (RMSE), Mean Absolute Error (MAE), precision, recall, and F1-score, depending on whether the output is a continuous forecast or a binary classification of risk (e.g., anomaly vs. normal). Visualization tools such as heatmaps, animated timelines, and interactive dashboards help translate these forecasts into decision-making tools for stakeholders. With the increasing availability of open-source geospatial data, IoT sensors, and high-performance computing, spatio-temporal modeling and forecasting is becoming accessible to a wider range of users and institutions, supporting real-time, location-aware decision systems across environmental, energy, and demographic sectors.

MAIN FOCUS OF THE CHAPTER

Time series algorithms have diverse applications across critical domains. They are employed in air quality forecasting, electricity consumption, and renewable energy generation to ensure efficiency and reliability. In earthquake studies, they help assess seismic risks, while in demographic research, they are used for accurate birth rate prediction. These applications highlight the versatility of time series methods in addressing real-world challenges which is shown in Figure 1.

Figure 1. Proposed framework for time series methods in addressing real-world challenges

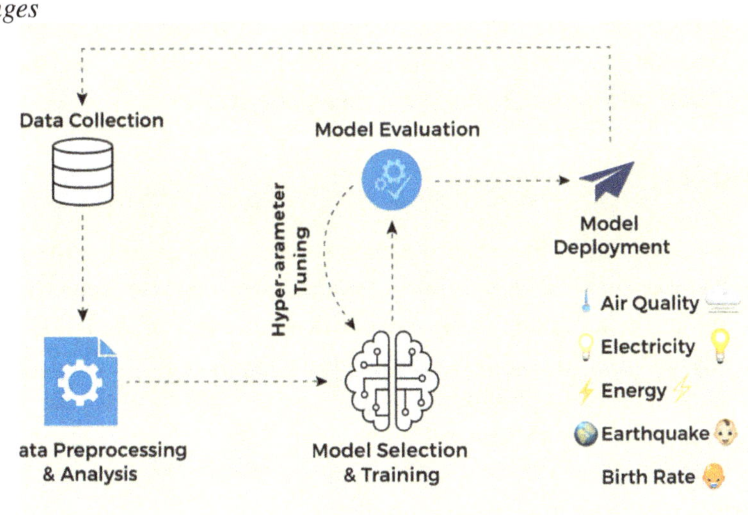

Forecasting on Air Quality Index

This section describes a method for time series predicting the daily concentration of PM2.5, an indicator for the Air Quality Index (AQI). First, data from air monitoring stations is collected and preprocessed: pollution trends are visualized, the data is resampled and averaged on a daily basis, and the PM2.5 concentration time series is analyzed using ARIMA model fitting to predict future values. The predicted PM2.5 concentrations are compared with the actual values, residual errors are analyzed, and forecast accuracy is assessed using RMSE (Root Mean Square Error). Based on ARIMA results, the model shows improved accuracy in predicting time-series air quality data compared to other forecasting methodologies (Sarkar et al., 2022) (Wang, J., 2022). This research aids in monitoring the dynamics of air quality and supports the development of policies related to the environment and public health. Air quality is a significant factor affecting human health, and monitoring its levels is crucial for environmental safety. The system evaluates and forecasts AQI values using pollutants such as PM2.5, PM10, NO, SO, CO, and O. AQI values are classified into standard categories—Good (0–50), Moderate (51–100), Unhealthy for Sensitive Groups (101–150), Unhealthy (151–200), Very Unhealthy (201–300), and Hazardous (301 and above)—to help users understand potential health risks. By analyzing patterns and predicting future trends, the model provides valuable insights for urban planning, public health management, and environmental monitoring. It also has the potential for real-time applications, offering timely alerts to help mitigate the effects of air pollution. With further development, this system can be extended into real-time monitoring platforms and integrated with mobile or web applications for broader accessibility, serving city planners, healthcare officials, and the general public by highlighting high-risk areas and time periods.

Data Collection

The dataset used in this investigation consists of historical air quality measurements collected from official air monitoring stations. The primary focus is on PM2.5 concentrations, as this pollutant is the most critical indicator of the Air Quality Index (AQI). Along with PM2.5, additional pollutants such as PM10, NO_2, SO_2, CO, and O_3 are also included to provide a comprehensive view of urban air pollution. **Attributes:** Each record in the dataset typically includes:

i. Timestamp (date and time of measurement)
ii. Location (city/station ID)
iii. PM2.5 concentration ($\mu g/m^3$)
iv. PM10 concentration ($\mu g/m^3$)
v. NO_2, SO_2, CO, and O_3 concentrations ($\mu g/m^3$ or ppm)
vi. Meteorological parameters (temperature, humidity, wind speed, etc., where available)

Data Preprocessing

To ensure the reliability and accuracy of the dataset, several preprocessing actions were performed:

i. **Handling Missing Values:** Gaps in the dataset due to sensor errors or transmission failures were treated using interpolation techniques or, in severe cases, by removing the affected records.

ii. **Resampling:** Hourly air quality measurements were resampled and averaged to daily values to capture long-term pollution trends.
iii. **Outlier Detection and Treatment:** Extreme values caused by sensor malfunctions or unusual events were identified using statistical methods (e.g., Z-score, IQR method) and corrected or removed.
iv. **Data Transformation:** The dataset was reformatted into a time series structure with continuous timestamps, ensuring compatibility with ARIMA and other forecasting models.

The cleaned and preprocessed data was then used for exploratory analysis and subsequent model development.

Exploratory Data Analysis (EDA)

Exploratory Data Analysis was conducted to uncover the underlying patterns and characteristics of PM2.5 time series data. The following analyses and visualizations were performed:

i. **Time Series Plot:** Visualization of daily PM2.5 concentrations over time to detect overall trends and anomalies.
ii. **Rolling Mean and Standard Deviation:** Used to examine long-term trends and seasonal fluctuations in the data.
iii. **Histograms and Box Plots:** Employed to analyze the distribution, variability, and presence of outliers in PM2.5 values.
iv. **Seasonal Decomposition:** Time series was decomposed into trend, seasonality, and residual components to better understand temporal dynamics.

These analyses provided key insights into the variability and seasonality of PM2.5 concentrations, guiding the selection and tuning of forecasting models.

Stationarity Testing

The results of the Augmented Dickey-Fuller (ADF) test are presented in Table 1. We reject the null hypothesis because the series is stationary, as indicated by the p-value of 0.011 (less than 0.05). This implies that the ARIMA model does not require the use of differencing. To capture short-term trends, we still take into account a differencing term based on the ACF and PACF analysis.

Table 1. Augmented Dickey-Fuller test results

Statistic	Value
ADF Statistic	-3.385
p-value	0.011
Critical Value (1%)	-3.439
Critical Value (5%)	-2.865
Critical Value (10%)	-2.569

ACF and PACF Analysis

Figures 2 show the ACF and PACF plots of the PM2.5 time series, respectively.

Figure 2. ACF and PACF plots of the PM2.5

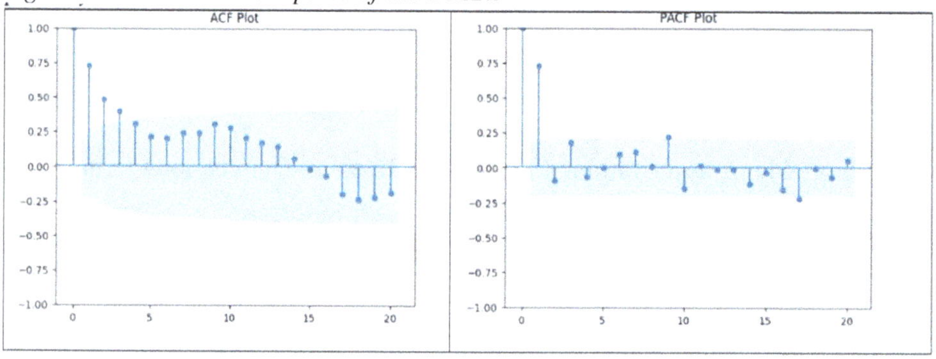

Model Evaluvation

In this study, we built and evaluated five different forecasting models to compare their predictive performance. The models are: Naïve Model, ARIMA (AutoRegressive Integrated Moving Average), SARIMA (Seasonal ARIMA), SARIMA – Trial 1, SARIMA – Trial 2. The performance of these models was assessed using the Root Mean Squared Error (RMSE) metric, where a lower value indicates better predictive accuracy which is shown in Figure 3.

Figure 3. Model comparison based on RMSE VALUESMODEL

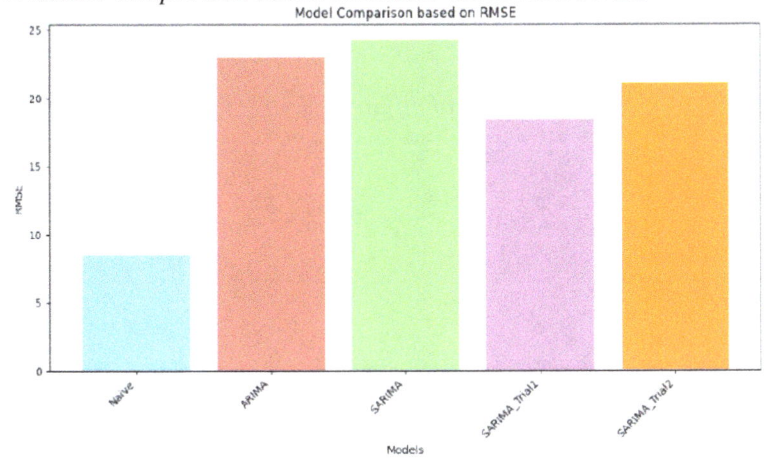

The performance of these models was assessed using the Root Mean Squared Error (RMSE) metric, where a lower value indicates better predictive accuracy.

i. **Naïve Model** achieved the lowest RMSE (~8–9), outperforming all other models despite its simplicity.
ii. **ARIMA** showed a higher RMSE (~23), indicating relatively poor performance.
iii. **SARIMA** recorded the highest RMSE (~24–25), suggesting it struggled to capture the data patterns effectively.
iv. **SARIMA Trial 1** performed better than ARIMA and SARIMA, with RMSE around 18–19.
v. **SARIMA Trial 2** had RMSE close to 21, slightly worse than Trial 1 but still an improvement over ARIMA and SARIMA.

Figure 4 presents a comparison of forecasting results from five models—Naïve, ARIMA, SARIMA, SARIMA Trial 1, and SARIMA Trial 2—against the actual test data for PM2.5 levels. The blue line represents the actual observed values, showing noticeable fluctuations and downward trends across the test period. In contrast, the forecasted lines from ARIMA, SARIMA, and their trial variants (dashed lines) remain relatively high and fail to capture the sharp declines seen in the actual data. Among them, SARIMA Trial 1 (purple dashed line) tracks the overall movement slightly better than others, though still with visible deviations. The Naïve forecast (black dashed line), while simplistic, maintains closer alignment with the actual test data in terms of level, supporting the earlier RMSE comparison that highlighted its superior accuracy. Overall, the figure illustrates that more complex models struggled to capture the actual trend, while the Naïve model provided the most reliable short-term predictions.

Figure 4. Forecasting comparisons of different models vs actual PM2.5 test data

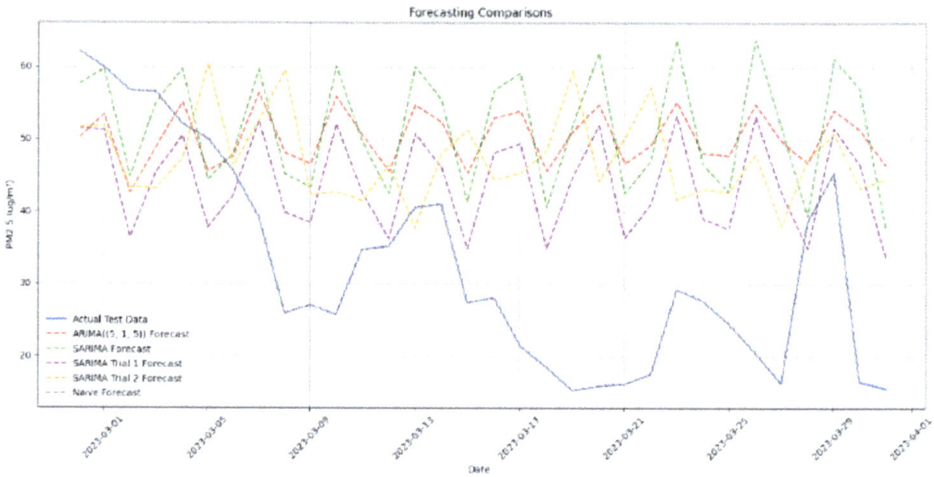

Forecasting Energy Consumption

Accelerated world energy consumption, combined with the increasing use of renewable power sources, highlights the importance of accurate energy consumption predictions. Time-series analysis is invaluable for identifying and forecasting energy-use trends, ensuring rational resource utilization, effective power regulation, and overall progress in energy management. Traditional statistical methods in time-series forecasting worked well for broad applications but struggle to capture complex, non-linear interdependencies in high-dimensional data. This limitation has driven the adoption of advanced deep learning techniques (Sun, 2024) (Vora, 2024) . This work presents key contributions to energy consumption forecasting using the PJM Interconnection LLC dataset, which records hourly power consumption. Predictive models are developed and tested based on Long Short-Term Memory (LSTM) and Gated Recurrent Unit (GRU) architectures. The study enhances understanding of deep learning methods in energy systems and provides insights for researchers, energy experts, and policymakers. A major contribution of this project lies in the preprocessing and analysis of energy consumption data. Missing values, outliers, and non-stationarity are carefully managed to ensure robust model training. Normalization and feature engineering further capture temporal patterns such as seasonality and trends, which are essential for accurate predictions.

Figure 5. Proposed model energy consumption

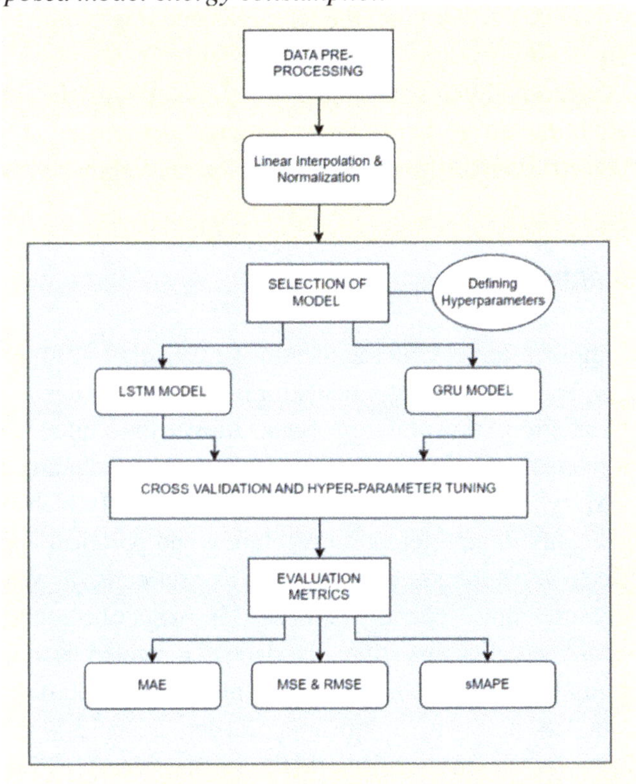

The preprocessing pipeline developed in this study serves as a reusable template for similar time-series datasets in energy and related fields. Beyond data preparation, the research implements LSTM and GRU models specifically for energy consumption forecasting. By tuning hyperparameters and applying cross-validation, both models achieve high predictive accuracy, measured through metrics such as Mean Absolute Error (MAE), Root Mean Squared Error (RMSE), and R-squared. LSTM demonstrates strong capability in capturing long-term dependencies, while GRU offers greater computational efficiency, together providing a balanced view of performance in energy forecasting. These models enable reliable hourly energy demand predictions, supporting applications like load balancing and demand response planning. A key contribution of the work is the comparative analysis of LSTM and GRU. Their performance is systematically evaluated, highlighting trade-offs between predictive accuracy and computational complexity. This comparison offers practical guidance for selecting models based on requirements such as real-time forecasting or resource-constrained environments. The study also emphasizes real-world implications for sustainable energy management. By anticipating demand fluctuations, the models enhance grid stability, support renewable energy integration, and reduce reliance on fossil fuels. Furthermore, the research encourages broader adoption of deep learning in future grid development and energy policy planning by openly sharing model design and assessment insights. Collectively, these contributions advance the state of the art in energy forecasting and demonstrate the potential of data-driven approaches to improve efficiency, resilience, and sustainability in global energy systems.

Dataset Description

This project employs the PJM Interconnection LLC Energy Consumption Dataset to develop and evaluate forecasting models for hourly energy demand. PJM Interconnection LLC, a regional transmission organization in the United States, manages electricity across states in the Eastern Interconnection grid, including Pennsylvania, New Jersey, and Maryland. The dataset, publicly available and widely used in energy forecasting studies, offers high temporal resolution and practical applicability for identifying intricate consumption patterns. The dataset consists of hourly energy consumption measurements, typically expressed in megawatt-hours (MWh), collected over several years. For this study, the dataset is treated as univariate, using only energy consumption values. This design simplifies the forecasting problem and emphasizes the models' ability to capture temporal dependencies without relying on external covariates. As with most real-world time-series data, the PJM dataset presents challenges such as sporadic missing values due to recording errors and outliers caused by extreme events (e.g., surges or blackouts). Preprocessing addresses these issues by applying linear interpolation for missing values and statistical thresholds (values beyond three standard deviations) to identify and correct outliers. To support model training, the data is normalized to the [0, 1] range and structured into supervised learning sequences (e.g., 24-hour windows) where past consumption predicts future demand. With tens of thousands of hourly records, the dataset is partitioned into 70% training, 20% validation, and 10% testing sets, ensuring sufficient data for both model development and performance evaluation. The richness of this dataset enables comprehensive exploration of energy consumption forecasting and supports practical applications in grid optimization and renewable energy integration.

Methodology

This study applies LSTM and GRU models to forecast energy consumption using the PJM Interconnection LLC dataset, which records hourly power usage across multiple regions. The dataset provides a solid basis for analyzing temporal trends and seasonal patterns in energy demand. The primary goal is to design forecasting models that are both accurate and informative, offering comparative insights into the performance of LSTM and GRU architectures for energy forecasting. By addressing challenges such as non-stationarity and variability in energy consumption data, the research contributes to developing more intelligent and resilient energy systems. Accurate forecasts are essential for optimizing grid operations, enabling demand response strategies, and supporting renewable energy integration. They also allow utility providers to anticipate peak loads, minimize operational inefficiencies, and reduce carbon emissions, thereby aligning with broader sustainability goals. The choice of LSTM and GRU models is motivated by their proven effectiveness in handling sequential data. LSTM demonstrates strong ability in capturing long-term dependencies, while GRU offers computational efficiency, making it particularly valuable for faster training and deployment is shown in Figure 6.

Figure 6. Working of LSTM and GRU

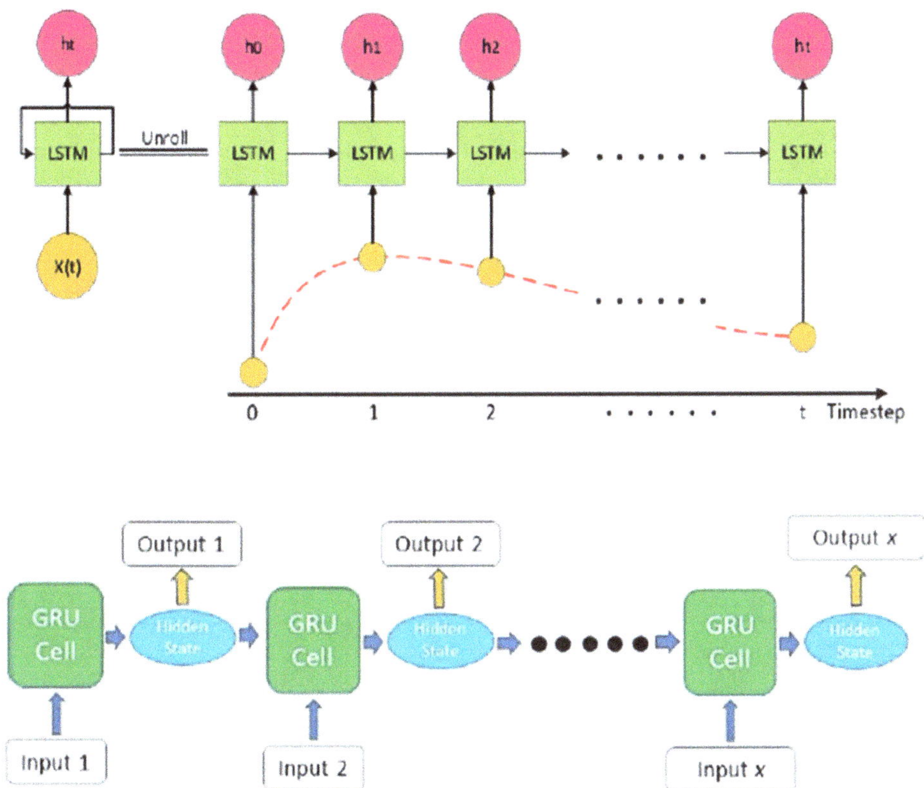

This work systematically evaluates their strengths and limitations, providing practical guidance on their applicability in real-world energy management scenarios.

Through comprehensive preprocessing, rigorous model training, and performance evaluation, the study delivers decision-relevant insights for researchers, energy practitioners, and policymakers.

Model Evaluvation

This research applies these models to the PJM Interconnection LLC Energy Consumption Dataset, which provides extensive hourly consumption data across the Eastern Interconnection grid in the United States. The dataset's richness makes it an ideal testbed for investigating advanced forecasting models. By leveraging LSTM and GRU, the project seeks to address real-world challenges faced by energy providers, such as predicting peak demand, minimizing energy losses, and planning for renewable energy integration. The evaluation results of the forecasting models are presented in the table, comparing LSTM and GRU across multiple performance metrics. The LSTM model consistently outperforms the GRU model, achieving lower error values in MAE (0.0452 vs. 0.0492), MSE (0.0054 vs. 0.0068), and RMSE (0.0735 vs. 0.0824). Similarly, LSTM records a lower symmetric Mean Absolute Percentage Error (sMAPE) of 29.2% compared to 31.1% for GRU. These findings suggest that LSTM is more effective in capturing long-term dependencies and providing accurate predictions, while GRU, though computationally efficient, shows slightly higher errors. Overall, the results highlight the superiority of LSTM for this energy forecasting task, particularly in terms of precision and reliability is shown in Table 2.

Table 2. LSTM and GRU model evaluvation

Evaluation Metric	**LSTM Model**	**GRU Model**
MAE	0.0452	0.0492
MSE	0.0054	0.0068
RMSE	0.0735	0.0824
sMAPE(in %)	29.2	31.1

Figure 7 compare actual energy consumption with predictions from LSTM and GRU models across three states—AEP, DEOK, and NI. In all cases, both models closely follow the actual consumption trends, effectively capturing periodic fluctuations and seasonal patterns. However, LSTM predictions align more smoothly with actual values, particularly during peak and trough points, whereas GRU shows slight deviations, especially under sharp demand variations. For instance, in the NI dataset, GRU exhibits noticeable lag during sudden consumption drops, while LSTM maintains better consistency. These results reinforce the earlier quantitative findings, confirming that while both models are capable of modeling temporal energy dynamics, LSTM provides superior accuracy and robustness in handling complex demand fluctuations.

Figure 7. LSTM and GRU model evaluvation

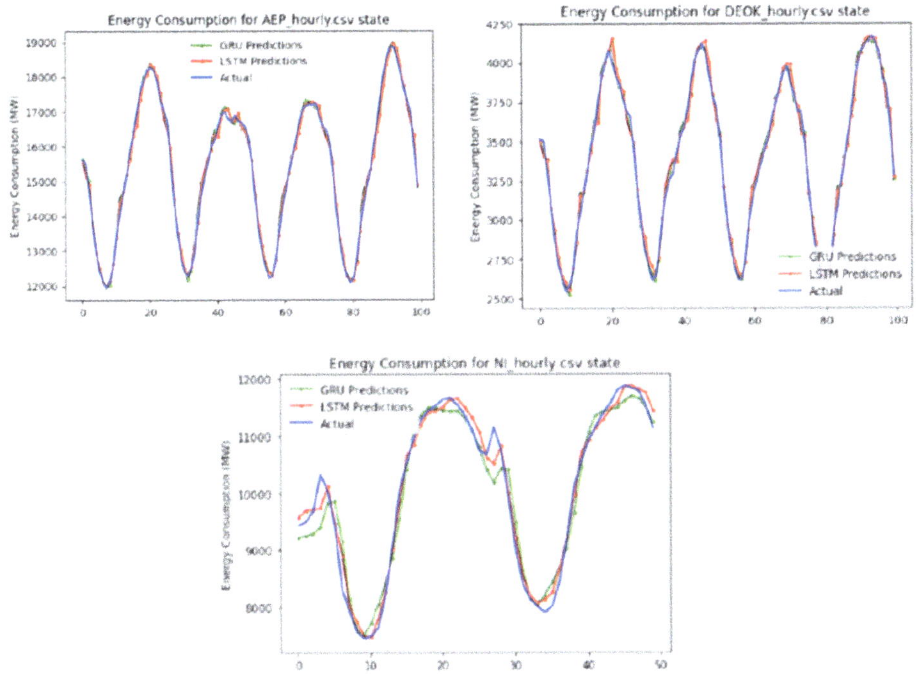

Forecasting Household Electicity Consumption

We evaluate and compare the performance of four forecasting models: ARIMA, SARIMA, Facebook Prophet, and XGBoost. The dataset, collected from smart meters installed in residential homes, is resampled into 3-hour intervals to enhance computational efficiency without significant information loss (Pinto et al., 2021) (Torres et al., 2022) (Luo et al., 2024).

Dataset Description

The data used in this research are obtained from the UCI Machine Learning Repository and are a comprehensive and detailed record of household electricity usage. The data was gathered from a single domestic household in Sceaux, France, and covers a contiguous time period between December 2006 and November 2010. The raw measurements were taken at an initial high temporal resolution of one-minute intervals, allowing for the detail of domestic power consumption with high granularity to be captured. Nevertheless, for the sake of limiting the analysis to recent and stable consumption behavior due to computational considerations, the data was limited to the year 2010. To further boost computational tractability and enhance readability, the data was resampled at 3-hour intervals via simple averaging. This resampling procedure enabled smoothing of minute-level noise while maintaining underlying trends and seasonality of the household energy consumption.

Each record includes:

i. **Date** and **Time**: Combined into a single timestamp.
ii. **Globalactivepower**: Total household power consump- tion in kilowatts (target variable).
iii. **Globalreactivepower**: Reactive power consumed.
iv. **Voltage**: Average voltage.
v. **Globalintensity**: Current intensity.
vi. **Submetering1, 2, 3**: Power consumption of kitchen, laundry room, and HVAC respectively.

Missing values in the data were represented by question marks ('?') and were handled by applying forward-fill imputa- tion, which fills missing values with the last non-missing value seen. This method maintains the temporal coherence of the data without creating artificial spikes or drops. All concerned columns were also cast to proper numeric data types to make them compatible with time series and machine learning algorithms. As a whole, this data set represents realistic, high- definition residential electricity usage and constitutes a solid base to assess numerous forecast models within the context of practical energy consumption.

Data Preprocessing and Feature Engineering

To prepare the dataset for modeling, several preprocessing techniques were applied to ensure data quality and optimize the data for time series forecasting. All numerical columns—including power consumption, voltage, and sub-metering values—were cast to appropriate numeric data types to facilitate accurate computation. Missing values, represented by question marks ('?'), were handled using forward-fill imputation, which replaces missing entries with the most recent valid observation, thereby maintaining temporal consistency and avoiding artificial spikes. The dataset was limited to the year 2010 to focus on stable and recent consumption patterns. To reduce computational complexity while preserving key trends and seasonality, the data was resampled to 3-hour intervals using mean aggregation. Additional preprocessing steps included normalization of numerical features to bring them within a comparable scale, which is particularly important for distance-based machine learning models like XGBoost. Outlier detection and smoothing were also performed to mitigate the effect of abnormal fluctuations in power readings. Time-based feature engineering was applied by extracting components such as hour of day, day of week, and weekend indicators from the timestamp. These features helped models capture cyclical consumption behaviors and improve forecast accuracy.

Methodology

ARIMA. ARIMA was utilized in this research to fit household electricity demand because of its ability to account for inherent patterns like trends and autocorrelations. ARIMA(2,1,2) model specification was adopted through examination of autocorrelation function (ACF), partial autocorrelation function (PACF), and model selection criteria such as Akaike Information Criterion (AIC). Here, d = 1 signifies that the data was differenced once to eliminate trends and gain stationarity—providing stable statistical properties over time. The p = 2 and q = 2 parameters signify that the model employs the previous two lagged values and the previous two forecast errors to make predictions. This combination enables ARIMA to efficiently model short-term temporal dependencies and adapt to changes in consumption.

SARIMA. In this research, we used a SARIMA(2,1,2)(1,1,1) setting to represent household electricity usage. The non- seasonal part (2,1,2) responds to short-term patterns and autocorrelation, and the seasonal part (1,1,1) addresses daily cycles. The 8-seasonal period represents a complete day, according to our 3-hour interval data (3 hours × 8 = 24 hours). This framework enables the model to capture both normal daily patterns and general consumption trends, enhancing forecast accuracy for datasets with high seasonality.

Facebook Prophet. Prophet is a versatile and robust time series forecasting model created by Facebook to work with real-world data with intricate patterns. One of Prophet's strengths is that it can model non-linear trends with change points, and it automatically detects changes in data behavior over time. Besides, it embeds both yearly and daily seasonality via Fourier series and includes user-specified holidays or other special events and is particularly suited for applications governed by recurring cycles or outside influences. Prophet is also extremely resistant to missing values and outliers, which are usually prevalent in real-world datasets, and does not involve extensive preprocessing. In this work, Prophet was trained on domestic electricity usage data resampled to 3-hour intervals. With its default hyperparameters, the model was able to pick up the typical daily usage patterns while being robust to gaps or outliers in the data, providing an easy-to-use yet efficient tool for energy demand forecasting.

XGBoost. In this research, XGBoost was used to forecast household electricity usage by relying on its capability to capture complex, non-linear relationships. Because XGBoost does not automatically consider temporal ordering, lag features were designed to bring in past con- sumption values as predictors so the model can essentially learn from prior time steps. Six lagged values (t-1 to t-6) were specifically created, each representing the power consumption during the previous six 3-hour intervals. All features were standardized with standard normalization before training to achieve consistency in input ranges and facilitate convergence. This preprocessing assisted the model in concentrating on significant patterns instead of raw differences in magnitude. XGBoost was subsequently trained on this structured, lagged data, allowing it to pick up on short-term dependencies and subtle patterns within the time series, making it a viable contender to conventional statistical forecasting models (Baur et al., 2024).

Model Evaluvation

The performance comparison of traditional and machine learning-based models highlights notable differences in forecasting accuracy is shown in Table 3. Among the baseline approaches, ARIMA achieves the lowest RMSE (0.4185) and MAE (0.3844), demonstrating relatively better performance than SARIMA, Prophet, and XGBoost. While SARIMA and Prophet produce comparable error levels, with RMSE values of 0.5982 and 0.5837 respectively, their MAE values (0.4307 and 0.4603) indicate slightly reduced accuracy. XGBoost, though widely recognized for handling structured data, records the highest error rates in this context, with an RMSE of 0.6143 and MAE of 0.4614. These results suggest that while traditional statistical methods like ARIMA can capture general consumption patterns reasonably well, they struggle with complex temporal dependencies compared to advanced deep learning models, reinforcing the need for more sophisticated architectures such as LSTM and GRU for improved forecasting.

Table 3. Model accuracy was assessed using RMSE and MAE over the final 30 days of the dataset

Evaluation Metric	LSTM Model	GRU Model
ARIMA	0.71	0.58
SARIMA	0.59	0.43
PROPHET	0.58	0.46
XG BOOST	0.61	0.461

Figure 8 compares actual energy consumption with forecasts from XGBoost, SARIMA, ARIMA, and Prophet models during November 17–21, 2010. All models capture the overall demand patterns, but their accuracy varies during periods of sharp fluctuation. XGBoost shows the closest alignment with actual values, though minor lags and overshooting occur. SARIMA performs well in modeling seasonality but tends to underestimate during sudden spikes. ARIMA produces smoother forecasts but struggles to capture high peaks, leading to underrepresentation of extreme values. Prophet, while effective in modeling trend and seasonality, demonstrates moderate deviations during volatile demand periods, showing less adaptability compared to XGBoost. Overall, the figure reinforces that while classical statistical and shallow machine learning models can capture broad consumption patterns, they fall short in precision and robustness compared to advanced deep learning methods such as LSTM and GRU.

Figure 8. Chosen forecastmodel vs actual (2010-11-17 to 2010-11-21)

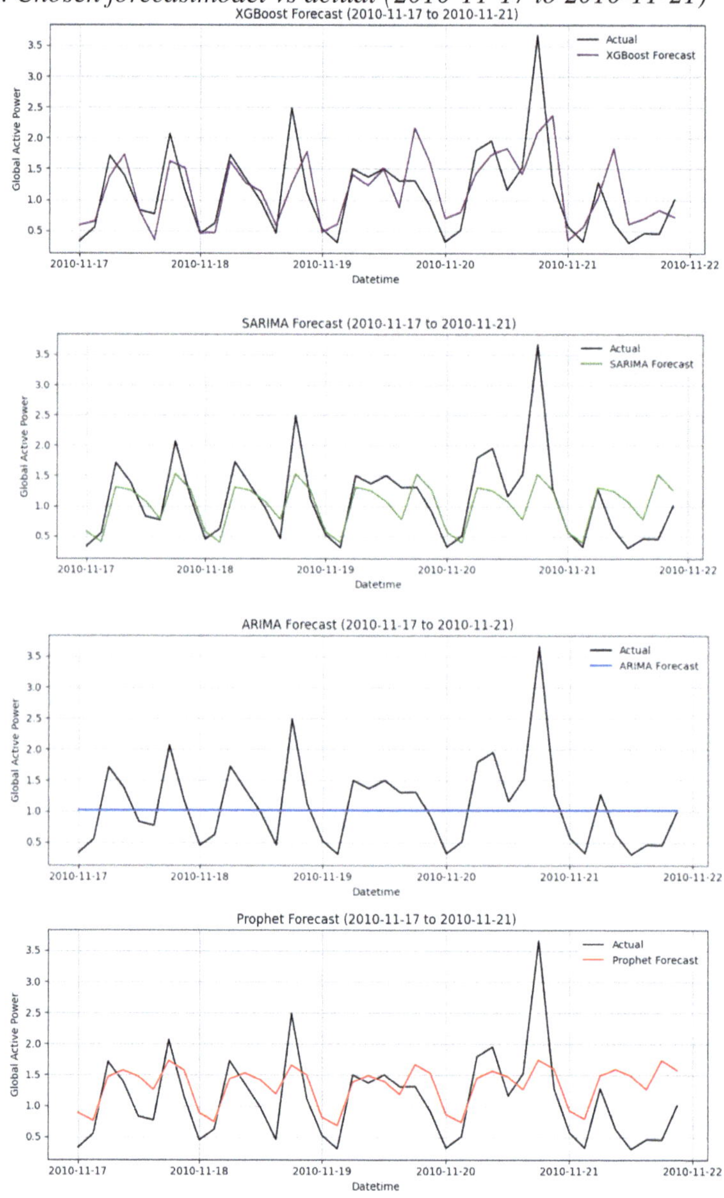

The comparative analysis conducted in this study demonstrates that advanced time series forecasting models, specifically Facebook Prophet and SARIMA, provide superior performance in predicting household electricity consumption compared to traditional ARIMA and machine learning-based XGBoost models. Key findings include:

Forecasting Earthquake Magnitude

Earthquakes are one of the most devastating natural disasters, usually leading to disastrous loss of lives and largescale destruction of infrastructure. The uncertainties and unforeseen nature of seismic activity present great challenges to early warning and disaster preparedness . The conventional systems of earthquake prediction have depended heavily on historic seismic activity data and geological analysis, which, although useful, are constrained in their capacity for timely and accurate forecasts. Over the past few years, data-driven methods have created new paths for seismic analysis. Of these, Machine Learning (ML) has attracted a lot of interest because it can identify patterns, process complex data, and make accurate predictions. ML methods have shown potential in several geophysical applications, such as earthquake classification, risk analysis, and magnitude estimation (Shrote et al., 2024) (Wang, C et al., 2022). This paper introduces QuakeSense, an earthquake magnitude forecasting machine learning model that estimates earthquake magnitudes based on past seismic information. Based on features like latitude, longitude, depth, and timestamp attributes, the system introduced here tries to offer a stable regression-based approach towards estimating the probable magnitude of seismic activity. Different models like Linear Regression and Random Forest Regressor are compared based on accuracy, error measures, and feature importance is shown in Figure 9. The envisaged work is designed in Google Colab utilizing a self-cleansed dataset and subsequently rolled out in web-based form employing Replit for real-time demonstrable applicability. The aim of QuakeSense is, finally, to aid early warnings and assist in disaster mitigation frameworks through the provisioning of a guaranteed magnitude prediction solution capable of issuing notifications to officials and the population at large in anticipation of ensuing seismic danger.

Figure 9. System architecture of Earth quake prediction

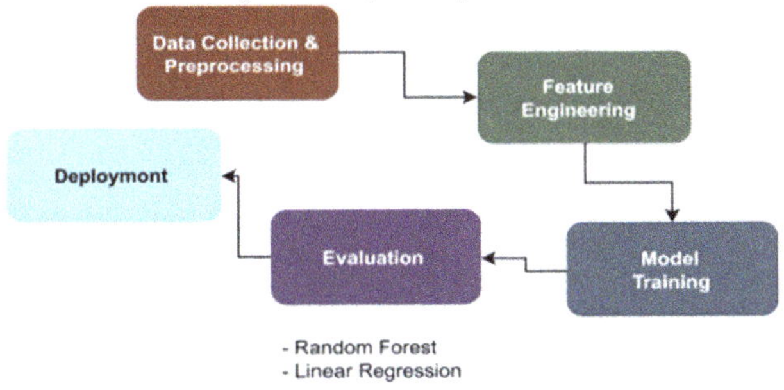

Data Collection and Preprocessing

The dataset for this project consists of past earthquake records, which have attributes like date, time, latitude, longitude, depth, and magnitude. As the raw data had non-standard headers and irregular spacing, a special parser was created to clean and format the dataset accordingly. Unnecessary fields were dropped, and features were extracted and normalized to make them consistent. The cleaned dataset was then divided into training and test sets in the ratio of 80:20. The dataset employed in this research was obtained from a publicly released earthquake record and encompassed a total of 18,000+ records prior to preprocessing. The data consists of seismic activity recorded along with corresponding spatial and temporal information. Upon cleaning and preprocessing, the final dataset employed in the training and testing of models consisted of 14,424 training samples and 3,606 testing samples. Each record in the dataset contains the following features: Preprocessing steps involved: - Elimination of unnecessary headers or rows with data Corruption - Conversion of data types - Missing value handling (though the cleaned dataset had none) - Splitting into training and test sets (80/20 split).

Methodology

Various machine learning models, such as Linear Regression and Random Forest Regressor, are trained with the preprocessed data. Random Forest was selected as the final model based on performance metrics

Model Evaluvation

The planned system was validated on a collection of historical records of earthquakes containing primary featuresnincluding latitude, longitude, depth, and time properties. Two of the regression types—Linear Regression and Random Forest Regressor—were applied and tested upon an 80-20 split train-test validation. The systems were compared via standard performance parameters of regression type: Mean Absolute Error (MAE), Mean Squared Error (MSE),and Coefficient of Determination (R2 Score) is shown in Table 4.

Table 4. LR and RF model evaluvation

Evaluation Metric	LR	RF
MAE	0.31	0.31
MSE	0.18	0.17
R2 SCORE	0.0072	0.0493

The Linear Regression model, as simple and easy to interpret, performed relatively weakly: These results demonstrate that the model had difficulty recognizing complex patterns within the data and producing predictions slightly better than mean magnitude baseline. The Random Forest model performed markedly better than Linear Regression on all the evaluation measures: Even though the R2 score is quite low, the decrease in MAE and MSE shows better prediction stability and accuracy. The model also gives insights into feature importance, with spatial features like latitude, longitude, and depth contributing most to earthquake magnitude prediction.

Forecasting Daily Birth Rate

The suggested framework emphasizes the prediction of daily birth rates through a combination of statistical and machine learning models. It starts with preprocessing data and stationarity tests through the ADF test, then trains five models— ARIMA, SARIMA, ETS, LSTM, andXGBoost—each identifying different temporal patterns. RMSE is employed for model evaluation, and the best model is chosen for future prediction (Tzitiridou-Chatzopoulou et al., 2024). This method provides a balanced analysis of birth patterns through multiple modeling approaches. The process of our birth rate prediction framework is serially structured to move from original data to interpretable and accurate predictions for the future. We start with loading and exploring the dataset, followed by several transformation and analytical processes to pre-process the data for forecasting. Several statistical and machine learning models are thereafter deployed and tested to determine the best way to proceed with final deployment is shown in Figure 10.

Figure 10. System architecture of daily birth rate prediction

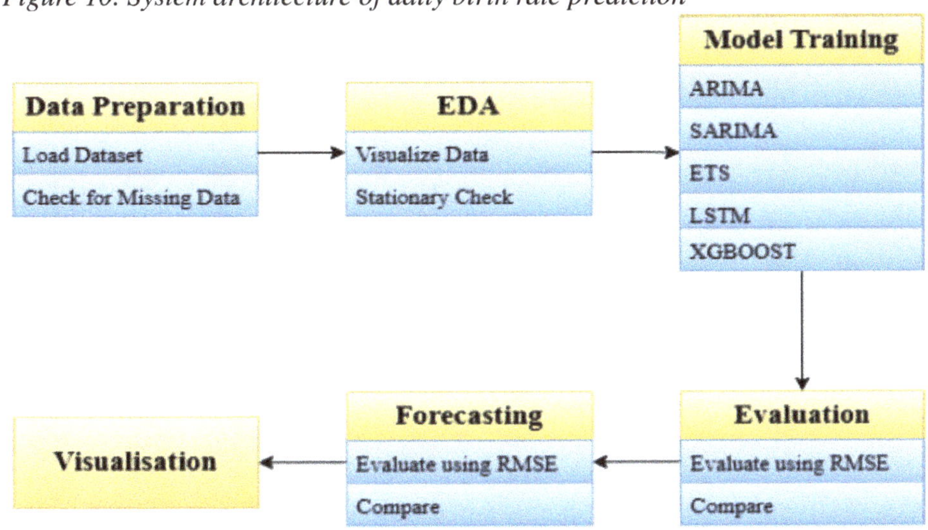

Data Collection and Preprocessing

The **Births2015** dataset, sourced from Kaggle (https://www.kaggle.com/datasets/manjusher/births2015), contains the number of births recorded each day throughout the year 2015. It consists of 365 entries, with two columns: *date* (in YYYY-MM-DD format) and *births* (daily birth count), forming a univariate time series with a strict daily frequency. For preprocessing, the date column was converted into datetime format and set as the index. Additionally, basic cleaning, stationarity checks, and visualizations were carried out to ensure data quality and readiness. This dataset was then utilized for training and evaluating multiple forecasting models to analyze daily birth trends.

Model Evaluvation

The plot in Figure 11 a shows the rolling mean and rolling standard deviation over the original series. It is clear that both the rolling mean (orange line) and standard deviation (greenline) both show identifiable trends and fluctuations over time. These trends suggest that the series is not stationary, since its statistical characteristics—mean and variance—are not stable This finding is in line with the ADF test outcome, which had a large p-value and did not reject the null hypothesis of nonstationarity. The plot in Figure 11b shows the rolling statistics for the differenced series of the time series. After the differencing process, the rolling mean and standard deviation seem to stabilize as time progresses, implying that the variance and mean are fairly stable. This visual observation is indicative that the series has reached stationarity. The result of the ADF test further verifies the same, where a highly negative value of the test statistic along with a p-value of almost zero strongly rejects the null hypothesis of non-stationarity.

Figure 11. Rolling statistics for non stationary vs stationary data (after differencing)

The ACF plot in Figure 12 a shows the relation between the lagged values of the time series and the original time series. For the series that has been differenced, the ACF demonstrates a decaying pattern, typical of an AR or ARMA process. The large peaks at early lags followed by a tapering effect indicate the presence of seasonality and, potentially, an MA component. This narrative assists in determining the right q (MA) and seasonal q (Q) parameters when setting up models such as SARIMA.

Figure 12. ACF Vs PACF plot

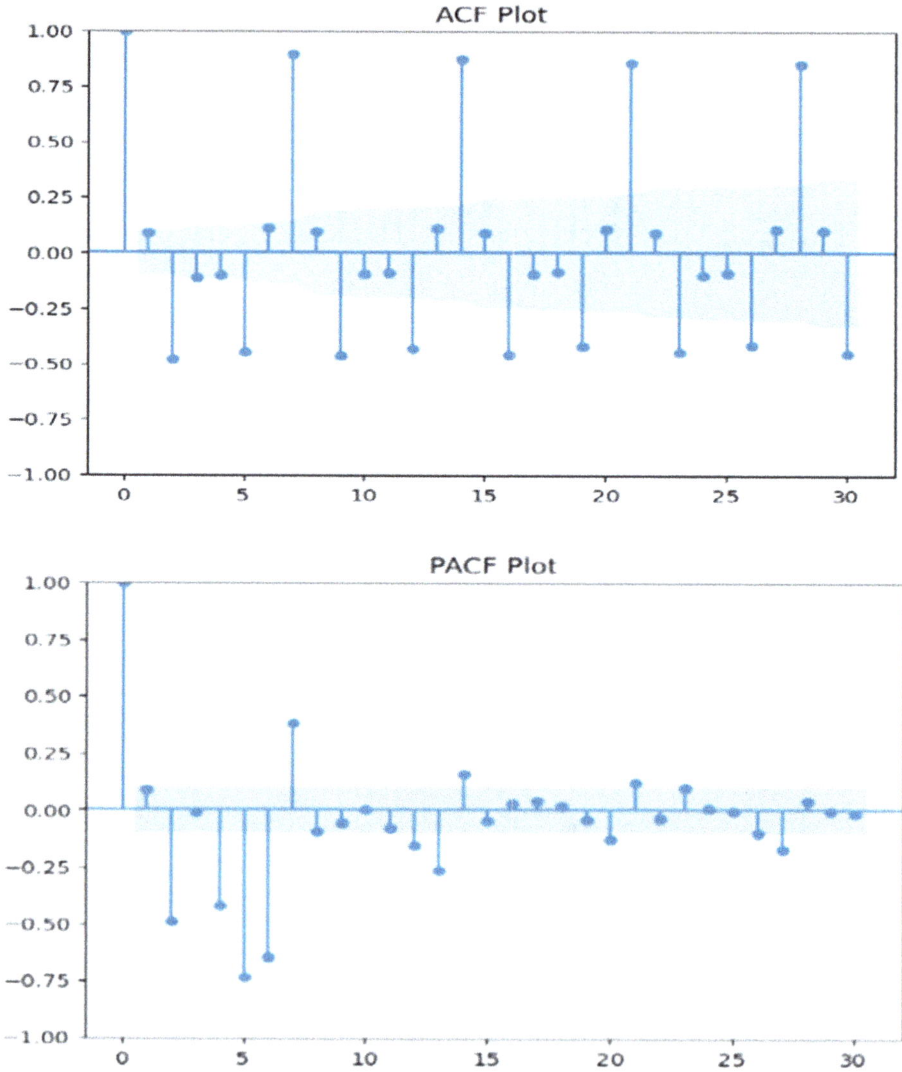

The PACF plot in Figure 12 b demonstrates the relationship between the series and its lags after stripping away the effect of intermediate lags. The presence of a distinct cutoff following the initial lags suggests the inclusion of an AR term (p) within the model. Particularly, the existence of a peak at lag 1 with a sharp fall subsequently implies an AR(1) process. It is a significant piece of information for the identification of p (AR) and seasonal P parameters while performing time series modeling.

Figure 13 shows birth prediction, which was conducted based on five models: ARIMA, SARIMA, ETS, LSTM, and XGBoost. SARIMA performed best with the minimum RMSE (1226) with good fit to both trend and seasonality. XGBoost came second best (1293), reflecting strong pattern learning. ETS worked moderately well (1408), while LSTM exhibited greater error (1819), perhaps from too little data or tuning. ARIMA reflected maximum RMSE (2071), indicating its lack of application with seasonal terms.

Figure 13. Daily birth forecasting with 5 models

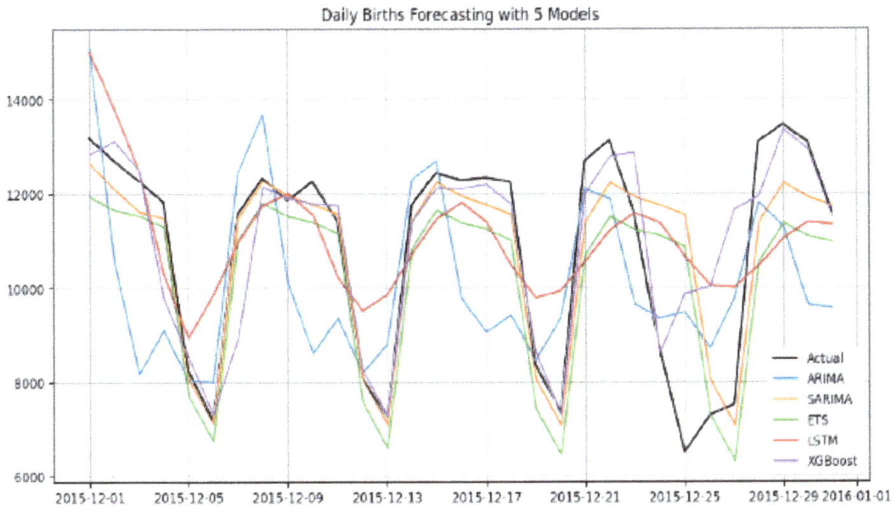

Table 5 presents the summarised comparative performance of all five forecasting models based on their RMSE values. SARIMA demonstrates superior performance with the lowest error rate of 1226.54, followed by XGBoost and ETS. The basic ARIMA model shows the highest error, confirming its limitations in handling seasonal time series data.

Table 4. Model performance comparison (RMSE values)

Model	RMSE
SARIMA7	1226.54
XGBoost	1293.00
ETS	1408.00

Model	RMSE
LSTM	1819.00
ARIMA(5,1,0)	2071.00

FUTURE RESEARCH DIRECTIONS

Air Quality Index

For future work, we suggest the following directions: • Incorporate external factors such as meteorological variables (temperature, humidity, wind speed) to improve forecasting accuracy • Explore ensemble methods that combine multiple forecasting approaches • Implement advanced machine learning techniques such as LSTM networks for comparison with statistical models • Extend the analysis to other air pollutants and geographical locations• Develop an integrated air quality forecasting system for real-time predictions and alerts These extensions would further enhance our understanding of air quality dynamics and improve the accuracy of forecasting models, (Yu, 2025; Zhang, 2022).

Household Electricity Consumption

Future research can explore the integration of external variables such as weather data, appliance-level usage, and occupancy patterns to further enhance the accuracy of electricity consumption forecasts. The incorporation of deep learning models, such as LSTM and Transformer-based architectures, could also be investigated to capture long-term dependencies and non-linear trends more effectively. Additionally, real-time forecasting using edge computing and IoT integration can be explored to support dynamic demand response systems in smart grids. Expanding the dataset to include diverse geographic regions and household profiles will also aid in developing more generalized and scalable forecasting models. Finally, the development of hybrid models combining statistical and machine learning approaches could offer a robust solution for electricity demand prediction in complex urban environments, (Xiang, 2020; Zhang, 2024).

Future enhancements could include:

- Incorporation of external variables (e.g., temperature, holidays).
- Use of deep learning models (LSTM, GRU).
- Expanding to multi-household or regional datasets.

Energy Consumption

In summary, this project emphasizes the revolutionary capacity of LSTM and GRU for energy consumption prediction, providing scalable solutions for developing smart grids and efficient energy use. Linking deep learning with real energy challenges opens the door to data-driven innovations that help build a resilient and sustainable energy future, (Grandon, 2024; Qureshi, 2024).

Earthquake Magnitude Prediction

Possible future developments would involve the integration of real-time seismic sensors, investigating deep learning architectures, and applying the model to multi region datasets for more generalized prediction. This research demonstrates the capability of data-driven methods in complementing conventional seismological techniques and aiding in more robust earthquake risk management systems (Shrote et al., 2024) (Wang, C et al., 2022) (Zhang, 2023).

Birth Rate Prediction

Birth Rate Prediction could involve the integration of real-time demographic and healthcare data streams (e.g., hospital records, census updates, or policy interventions), exploring advanced deep learning architectures such as LSTMs, Transformers, and hybrid models, and extending the analysis to multi-country or multi-regional datasets for more generalized and transferable insights. Incorporating external socio-economic and environmental factors—such as education levels, healthcare access, or seasonal effects—could also enhance predictive accuracy. This research highlights the potential of data-driven forecasting methods to complement traditional demographic techniques, offering valuable tools for policy planning, healthcare resource allocation, and long-term population studies, (Zhang, 2022).

CONCLUSION

In conclusion, this collection of studies highlights the versatility and effectiveness of both classical and deep learning approaches in solving diverse time series forecasting problems across domains such as air quality monitoring, energy consumption prediction, seismic risk management, and demographic analysis. The ARIMA(2,1,1) model for PM2.5 concentration forecasting demonstrated that even traditional statistical models, when rigorously tested and validated, can provide accurate and actionable insights for public health planning and air quality management.

Similarly, Prophet and SARIMA proved highly effective in forecasting household electricity usage, with Prophet excelling in robustness against irregularities and long-term forecasting, while SARIMA offered superior accuracy in capturing daily cyclical behavior. Extending into advanced deep learning, the use of LSTM and GRU models for energy demand forecasting on the PJM Interconnection dataset showcased the ability of recurrent architectures to capture complex temporal dependencies, with LSTM providing marginally higher accuracy and GRU offering efficiency and scalability for real-time systems. In earthquake magnitude prediction, the QuakeSense framework leveraged Random Forests to outperform linear models, underscoring the potential of ensemble learning in high-stakes applications such as disaster risk reduction. Finally, the forecasting of daily birth rates using ARIMA, SARIMA, ETS, LSTM, and XGBoost reaffirmed the importance of model selection tailored to data characteristics, with SARIMA standing out for its ability to handle both trend and seasonality effectively.

Taken together, these studies emphasize that no single model universally dominates across tasks; instead, model performance depends on the specific forecasting objectives, data properties, and application constraints. They also demonstrate how data-driven methods can complement traditional domain expertise, offering powerful decision-support tools in fields as varied as healthcare, environmental management, population studies, and disaster preparedness. Future work should focus on hybrid modeling strategies, integration of real-time data streams, and the development of scalable, interpretable forecasting frameworks to further enhance accuracy, applicability, and trust in predictive analytics.

REFERENCES

Abbasi, M. T., Alesheikh, A. A., Lotfata, A., & Azizi, Z. (2025). Hybrid graph convolutional networks for air quality prediction: A systematic review of foundations, challenges, and opportunities. *International Journal of Environmental Science and Technology*, *22*(13), 1–36. https://link.springer.com/article/10.1007/s13762-025-06599-8. DOI: 10.1007/s13762-025-06599-8

Alqasim, A. R. (2022). Using Regression Analysis for Predicting Energy Consumption in Dubai Police. https://repository.rit.edu/theses/11177/

Arul, M., & Kareem, A. (2021). Applications of shapelet transform to time series classification of earthquake, wind and wave data. *Engineering Structures*, *228*, 111564. https://www.sciencedirect.com/science/article/pii/S0141029620341651. DOI: 10.1016/j.engstruct.2020.111564

Baur, L., Ditschuneit, K., Schambach, M., Kaymakci, C., Wollmann, T., & Sauer, A. (2024). Explainability and interpretability in electric load forecasting using machine learning techniques–a review. *Energy and AI*, *16*, 100358. https://www.sciencedirect.com/science/article/pii/S2666546824000247

Berhich, A., Belouadha, F. Z., & Kabbaj, M. I. (2022). A location-dependent earthquake prediction using recurrent neural network algorithms. *Soil Dynamics and Earthquake Engineering*, *161*, 107389. https://www.sciencedirect.com/science/article/pii/S026772612200238X. DOI: 10.1016/j.soildyn.2022.107389

Beura, S., Mohanty, S., Mohapatra, S. K., Panda, N., & Acharya, A. A. (2024). Effective deep learning techniques for prediction of energy consumption in smart buildings. In *2024 International Conference on Advancements in Smart, Secure and Intelligent Computing (ASSIC)* (pp. 1-6). IEEE. https://ieeexplore.ieee.org/abstract/document/10507898

Bhatt, C., Barthwal, A., Shoyab, R., Khamboj, P., & Singh, T. (2024). Optimizing Short-Term Load Forecasting for Energy Management Systems: Integrative Time Series Analysis and Advanced Neural Network Methodologies. In *2024 10th International Conference on Advanced Computing and Communication Systems (ICACCS)* (Vol. 1, pp. 401-406). IEEE. https://ieeexplore.ieee.org/abstract/document/10716879/

Bheemanna, N., Megeri, M., & Budihal, H. H. (2023). Forecast of demographic variables using the ARIMA Model in India. *Int J Stat Appl Math*, *8*(5S), 1009–1018. https://www.mathsjournal.com/special-issue/2023/vol8/issue5S/S-8-5-117. DOI: 10.22271/maths.2023.v8.i5Sn.1322

Bilgili, M., & Pinar, E. (2023). Gross electricity consumption forecasting using LSTM and SARIMA approaches: A case study of Türkiye. *Energy*, *284*, 128575. https://www.sciencedirect.com/science/article/pii/S0360544223019692. DOI: 10.1016/j.energy.2023.128575

Chi, D. (2022). Research on electricity consumption forecasting model based on wavelet transform and multi-layer LSTM model. *Energy Reports*, *8*, 220–228. https://www.sciencedirect.com/science/article/pii/S235248472200169X. DOI: 10.1016/j.egyr.2022.01.169

Comolli, C. L., & Vignoli, D. (2021). Spreading uncertainty, shrinking birth rates: A natural experiment for Italy. *European Sociological Review*, *37*(4), 555–570. DOI: 10.1093/esr/jcab001

Cruz, A. V. D., Bayan, S. O., Mabborang, R. C., & Lamprea, C. T. (2025). Forecasting Number of Births in the Philippines Using Seasonal Autoregressive Integrated Moving Average (SARIMA) Model. *European Journal of Information Technologies and Computer Science*, *5*(3), 1–10. https://www.ej-compute.org/index.php/compute/article/view/129. DOI: 10.24018/compute.2025.5.3.129

Debnath, P., Chittora, P., Chakrabarti, T., Chakrabarti, P., Leonowicz, Z., Jasinski, M., Gono, R., & Jasińska, E. (2021). Analysis of earthquake forecasting in India using supervised machine learning classifiers. *Sustainability (Basel)*, *13*(2), 971. https://www.mdpi.com/2071-1050/13/2/971. DOI: 10.3390/su13020971

Deveer, L., & Minet, L. (2025). Real-time air quality prediction using traffic videos and machine learning. *Transportation Research Part D, Transport and Environment*, *142*, 104688. https://www.sciencedirect.com/science/article/pii/S1361920925000987. DOI: 10.1016/j.trd.2025.104688

Elsaraiti, M., Ali, G., Musbah, H., Merabet, A., & Little, T. (2021, April). Time series analysis of electricity consumption forecasting using ARIMA model. In *2021 IEEE Green technologies conference (GreenTech)* (pp. 259-262). IEEE. https://ieeexplore.ieee.org/abstract/document/9458543

Emeç, M. U. R. A. T., & Yurtsever, M. (2025). A novel ensemble machine learning method for accurate air quality prediction. *International Journal of Environmental Science and Technology*, *22*(1), 459–476. https://link.springer.com/article/10.1007/s13762-024-05671-z. DOI: 10.1007/s13762-024-05671-z

Esmaeili, N., & Abbasi-Shavazi, M. J. (2024). Forecasting number of births and sex ratio at birth in Iran using deep neural network and ARIMA: Implications for policy evaluations. *Journal of Population Research (Canberra, A.C.T.)*, *41*(4), 26. DOI: 10.1007/s12546-024-09348-9

Galán-Madruga, D., de Lourdes Berríos Cintrón, M., Broomandi, P., Oleniacz, R., & Cáceres, J. O. (2025). Long-term performance of air quality networks: Implications in health and environmental management. *International Journal of Environmental Science and Technology*, •••, 1–26. https://link.springer.com/article/10.1007/s13762-025-06526-x. DOI: 10.1007/s13762-025-06526-x

Galkina, A., & Grafeeva, N. (2019). Machine learning methods for earthquake prediction: A survey. In *Proceedings of the fourth conference on software engineering and information management (SEIM-2019), Saint Petersburg, Russia* (Vol. 13, p. 25). https://www.researchgate.net/profile/Natalia-Grafeeva/publication/

Gonzalez, D., Patricio, M. A., Berlanga, A., & Molina, J. M. (2022). Variational autoencoders for anomaly detection in the behaviour of the elderly using electricity consumption data. *Expert Systems: International Journal of Knowledge Engineering and Neural Networks*, *39*(4), e12744. https://onlinelibrary.wiley.com/doi/full/10.1111/exsy.12744. DOI: 10.1111/exsy.12744

Grandón, T. G., Schwenzer, J., Steens, T., & Breuing, J. (2024). Electricity demand forecasting with hybrid classical statistical and machine learning algorithms: Case study of Ukraine. *Applied Energy*, *355*, 122249. https://www.sciencedirect.com/science/article/pii/S0306261923016136. DOI: 10.1016/j.apenergy.2023.122249

Gupta, N. S., Mohta, Y., Heda, K., Armaan, R., Valarmathi, B., & Arulkumaran, G. (2023). Prediction of air quality index using machine learning techniques: A comparative analysis. *Journal of Environmental and Public Health*, *2023*(1), 4916267. https://onlinelibrary.wiley.com/doi/full/10.1155/2023/4916267. DOI: 10.1155/2023/4916267

Hadjout, D., Torres, J. F., Troncoso, A., Sebaa, A., & Martínez-Álvarez, F. (2022). Electricity consumption forecasting based on ensemble deep learning with application to the Algerian market. *Energy*, *243*, 123060. https://www.sciencedirect.com/science/article/pii/S0360544221033090. DOI: 10.1016/j.energy.2021.123060

Helli, S. S., Tanberk, S., & Demir, O. (2022). *Forecasting energy consumption using deep learning in smart cities. In 2022 International Conference on Artificial Intelligence of Things (ICAIoT)*. IEEE., https://ieeexplore.ieee.org/abstract/document/10121846

Huang, B., Zheng, S., Ma, B., Yang, Y., Zhang, S., & Jin, L. (2022). Using deep learning to predict the outcome of live birth from more than 10,000 embryo data. *BMC Pregnancy and Childbirth*, *22*(1), 36. DOI: 10.1186/s12884-021-04373-5 PMID: 35034623

Jafari, A., Fox, G., Rundle, J. B., Donnellan, A., & Ludwig, L. G. (2024). Time series foundation models and deep learning architectures for earthquake temporal and spatial nowcasting. *GeoHazards*, *5*(4), 1247–1274. https://www.mdpi.com/2624-795X/5/4/59. DOI: 10.3390/geohazards5040059

Jain, M., AlSkaif, T., & Dev, S. (2022). Are deep learning models more effective against traditional models for load demand forecasting? In *2022 International Conference on Smart Energy Systems and Technologies (SEST)* (pp. 1-6). IEEE. https://ieeexplore.ieee.org/abstract/document/9898424

Jayaraman, S. T, N., S, A., & G, S. (2025). Enhancing urban air quality prediction using time-based-spatial forecasting framework. *Scientific Reports, 15*(1), 4139. https://www.nature.com/articles/s41598-024-83248-z

Karthick, K., Dharmaprakash, R., & Sathya, S. (2024). Predictive modeling of energy consumption in the steel industry using CatBoost regression: A data-driven approach for sustainable energy management. *International Journal of Robotics and Control Systems*, *4*(1), 33–49. https://pubs2.ascee.org/index.php/IJRCS/article/view/1234. DOI: 10.31763/ijrcs.v4i1.1234

Khan, A. N., Iqbal, N., Rizwan, A., Ahmad, R., & Kim, D. H. (2021). An ensemble energy consumption forecasting model based on spatial-temporal clustering analysis in residential buildings. *Energies*, *14*(11), 3020. https://www.mdpi.com/1996-1073/14/11/3020. DOI: 10.3390/en14113020

Khobragade, P. P., & Ahirwar, A. V. (2022). Source apportionment of SPM by positive matrix factorization and PM2. 5 analysis in an urban industrial area. *World Journal of Engineering*, *19*(4), 446–458. https://www.emerald.com/wje/article/19/4/446/458644/Source-apportionment-of-SPM-by-positive-matrix. DOI: 10.1108/WJE-11-2020-0550

Klyuev, R. V., Morgoev, I. D., Morgoeva, A. D., Gavrina, O. A., Martyushev, N. V., Efremenkov, E. A., & Mengxu, Q. (2022). Methods of forecasting electric energy consumption: A literature review. *Energies*, *15*(23), 8919. https://www.mdpi.com/1996-1073/15/23/8919. DOI: 10.3390/en15238919

Kruthi, N., Kumar, A. A., Aromal, A. J., & Ravindra, C. (2023). QuakeML: ML-Based Earthquake Prediction and Warning System. In *2023 IEEE Fifth International Conference on Advances in Electronics, Computers and Communications (ICAECC)* (pp. 1-5). IEEE. https://ieeexplore.ieee.org/abstract/document/10560276

Kulkarni, M., Mulay, C., Marathe, S., & Itankar, P. (2021). Earthquake prediction using machine learning. *International Journal of Advance Research, Ideas and Innovations in Technology, 7*(3).

Kumar, A., & Meena, B. (2025). A comparative analysis of the Holt and ARIMA models for predicting the future total fertility rate in India. *Life Cycle Reliability and Safety Engineering*, *14*(1), 117–126. DOI: 10.1007/s41872-024-00287-1

Li, J., Luo, Y., & Wei, S. (2022). Long-term electricity consumption forecasting method based on system dynamics under the carbon-neutral target. *Energy*, *244*, 122572. https://www.sciencedirect.com/science/article/pii/S0360544221028218. DOI: 10.1016/j.energy.2021.122572

Liu, H., Li, Q., Yu, D., & Gu, Y. (2019). Air quality index and air pollutant concentration prediction based on machine learning algorithms. *Applied Sciences (Basel, Switzerland)*, *9*(19), 4069. https://www.mdpi.com/2076-3417/9/19/4069. DOI: 10.3390/app9194069

Luo, H., & Li, X. (2023). A hybrid model of CNN-BiLSTM and XGBoost for HVAC systems energy consumption prediction. In *2023 5th International Conference on Industrial Artificial Intelligence (IAI)* (pp. 1-6). IEEE. https://ieeexplore.ieee.org/abstract/document/10327594

Luo, S., Wang, B., Gao, Q., Wang, Y., & Pang, X. (2024). Stacking integration algorithm based on CNN-BiLSTM-Attention with XGBoost for short-term electricity load forecasting. *Energy Reports*, *12*, 2676–2689. https://www.sciencedirect.com/science/article/pii/S0275531923002581. DOI: 10.1016/j.egyr.2024.08.078

Mallouhy, R., Abou Jaoude, C., Guyeux, C., & Makhoul, A. (2019). Major earthquake event prediction using various machine learning algorithms. In *2019 international conference on information and communication technologies for disaster management (ICT-DM)* (pp. 1-7). IEEE. https://ieeexplore.ieee.org/abstract/document/9032983

Méndez, M., Merayo, M. G., & Núñez, M. (2023). Machine learning algorithms to forecast air quality: A survey. *Artificial Intelligence Review*, *56*(9), 10031–10066. https://link.springer.com/article/10.1007/s10462-023-10424-4. DOI: 10.1007/s10462-023-10424-4 PMID: 36820441

Mohapatra, S. K., Mishra, S., & Tripathy, H. K. (2022). Energy consumption prediction in electrical appliances of commercial buildings using LSTM-GRU model. In *2022 International Conference on Advancements in Smart, Secure and Intelligent Computing (ASSIC)* (pp. 1-5). IEEE. https://ieeexplore.ieee.org/abstract/document/10088334

Nguyen, T. P. (2025). AIoT-based indoor air quality prediction for building using enhanced metaheuristic algorithm and hybrid deep learning. *Journal of Building Engineering*, *105*, 112448. https://www.sciencedirect.com/science/article/pii/S2352710225006850. DOI: 10.1016/j.jobe.2025.112448

Nurtas, M., Zhantaev, Z., & Altaibek, A. (2024). Earthquake time-series forecast in Kazakhstan territory: Forecasting accuracy with SARIMAX. *Procedia Computer Science*, *231*, 353–358. https://www.sciencedirect.com/science/article/pii/S1877050923022287. DOI: 10.1016/j.procs.2023.12.216

Onambele, L., Guillen-Aguinaga, S., Guillen-Aguinaga, L., Ortega-Leon, W., Montejo, R., Alas-Brun, R., Aguinaga-Ontoso, E., Aguinaga-Ontoso, I., & Guillen-Grima, F. (2023). Trends, projections, and regional disparities of maternal mortality in Africa (1990–2030): An ARIMA Forecasting approach. *Epidemiologia*, *4*(3), 322–351. https://www.mdpi.com/2673-3986/4/3/32. DOI: 10.3390/epidemiologia4030032 PMID: 37754279

Peng, L., Wang, L., Xia, D., & Gao, Q. (2022). Effective energy consumption forecasting using empirical wavelet transform and long short-term memory. *energy, 238*, 121756. https://www.sciencedirect.com/science/article/pii/S0360544221020041

Pinto, T., Praça, I., Vale, Z., & Silva, J. (2021). Ensemble learning for electricity consumption forecasting in office buildings. *Neurocomputing*, *423*, 747–755. https://www.sciencedirect.com/science/article/pii/S0925231220307372. DOI: 10.1016/j.neucom.2020.02.124

Qureshi, M., Arbab, M. A., & Rehman, S. U. (2024). Deep learning-based forecasting of electricity consumption. *Scientific Reports*, *14*(1), 6489. https://www.nature.com/articles/s41598-024-56602-4. DOI: 10.1038/s41598-024-56602-4 PMID: 38499617

Reddy, R., Monisha, V., Ramya, P., Deeksha, H. C., & Lavanya, A. (2023, June). Earthquake magnitude prediction using deep learning techniques. In *2023 International Conference on Sustainable Computing and Smart Systems (ICSCSS)* (pp. 159-164). IEEE. https://ieeexplore.ieee.org/abstract/document/10169746/

Riady, S. R., & Apriani, R. (2023, February). Multivariate time series with Prophet Facebook and LSTM algorithm to predict the energy consumption. In *2023 International Conference on Computer Science, Information Technology and Engineering (ICCoSITE)* (pp. 805-810). IEEE. https://ieeexplore.ieee.org/abstract/document/10127735

Ridzwan, N. S. M., & Yusoff, S. H. M. (2023). Machine learning for earthquake prediction: A review (2017–2021). *Earth Science Informatics*, *16*(2), 1133–1149. https://link.springer.com/article/10.1007/s12145-023-00991-z. DOI: 10.1007/s12145-023-00991-z

Sarkar, N., Gupta, R., Keserwani, P. K., & Govil, M. C. (2022). Air quality index prediction using an effective hybrid deep learning model. *Environmental Pollution, 315*, 120404. https://www.sciencedirect.com/science/article/pii/S0269749122016189. DOI: 10.1016/j.envpol.2022.120404 PMID: 36240962

Sau, A., & Singh, A. (2024). Forecasting Energy Consumption Patterns: Weather-Driven Insights from Time Series Analysis. In *2024 IEEE Students Conference on Engineering and Systems (SCES)* (pp. 1-6). IEEE. https://ieeexplore.ieee.org/abstract/document/10652321

Shrote, P., Dasarwar, P., Dongre, S., & Khobragade, P. (2024). Earthquake Prediction Through Machine Learning Approach. In *2024 4th International Conference on Technological Advancements in Computational Sciences (ICTACS)* (pp. 542-546). IEEE. https://ieeexplore.ieee.org/abstract/document/10840652

Sun, B., Zhang, Q., Mao, H., & Li, Z. (2024). Validation of a statistical-dynamic framework for predicting energy consumption: A study on vehicle energy conservation equation. *Energy Conversion and Management, 307*, 118330. https://www.sciencedirect.com/science/article/pii/S0196890424002711. DOI: 10.1016/j.enconman.2024.118330

SUNEER. S. (2024). *PREDICTIVE ANALYSIS OF BIRTH RATES IN INDIA* (Doctoral dissertation, St. Teresa's College (autonomous) Ernakulam). http://117.239.78.102:8080/jspui/bitstream/123456789/4484/1/SIFNA.pdf

Torres, J. F., Martínez-Álvarez, F., & Troncoso, A. (2022). A deep LSTM network for the Spanish electricity consumption forecasting. *Neural Computing & Applications, 34*(13), 10533–10545. https://link.springer.com/article/10.1007/s00521-021-06773-2. DOI: 10.1007/s00521-021-06773-2 PMID: 35153386

Tzitiridou-Chatzopoulou, M., Zournatzidou, G., & Kourakos, M. (2024). Predicting future birth rates with the use of an adaptive machine learning algorithm: A forecasting experiment for scotland. *International Journal of Environmental Research and Public Health, 21*(7), 841. https://www.mdpi.com/1660-4601/21/7/841. DOI: 10.3390/ijerph21070841 PMID: 39063418

Vora, D., Zade, N., Patwa, P., Kushwaha, A., Agrawal, G., & Sharma, A. (2024). Temporal Energy Consumption Exploration and Time Series Forecasting of Electricity Dynamics. In *2024 IEEE Pune Section International Conference (PuneCon)* (pp. 1-6). IEEE. https://ieeexplore.ieee.org/abstract/document/10895847

Wang, C., Li, C., Yong, S., Wang, X. A., & Yang, C. (2022). Time Series and Non-Time Series Models of Earthquake Prediction Based on AETA Data: 16-Week Real Case Study. *Applied Sciences (Basel, Switzerland)*, *12*(17), 8536. https://www.mdpi.com/2076-3417/12/17/8536. DOI: 10.3390/app12178536

Wang, J., Li, X., Jin, L., Li, J., Sun, Q., & Wang, H. (2022). An air quality index prediction model based on CNN-ILSTM. *Scientific Reports*, *12*(1), 8373. https://www.nature.com/articles/s41598-022-12355-6. DOI: 10.1038/s41598-022-12355-6 PMID: 35589914

Xayasouk, T., Lee, H., & Lee, G. (2020). Air pollution prediction using long short-term memory (LSTM) and deep autoencoder (DAE) models. *Sustainability (Basel)*, *12*(6), 2570. https://www.mdpi.com/2071-1050/12/6/2570. DOI: 10.3390/su12062570

Xiang, L., Xie, T., & Xie, W. (2020). Prediction model of household appliance energy consumption based on machine learning. In *Journal of Physics: Conference Series* (Vol. 1453, No. 1, p. 012064). IOP Publishing. https://iopscience.iop.org/article/10.1088/1742-6596/1453/1/012064/meta

Yu, C., Wang, F., Wang, Y., Shao, Z., Sun, T., Yao, D., & Xu, Y. (2025). Mgsfformer: A multi-granularity spatiotemporal fusion transformer for air quality prediction. *Information Fusion*, *113*, 102607. https://www.sciencedirect.com/science/article/pii/S1566253524003853. DOI: 10.1016/j.inffus.2024.102607

Zhang, H., Sun, Z., Wang, Z., Zhang, Q., Pan, R., & Cao, S. (2022, August). Analysis of Influencing Factors of Birth Rate and Prediction of Policy Scenario. In *China National Conference on Big Data and Social Computing* (pp. 309-331). Singapore: Springer Nature Singapore. DOI: 10.1007/978-981-19-7532-5_20

Zhang, Q., Han, Y., Li, V. O., & Lam, J. C. (2022). Deep-AIR: A hybrid CNN-LSTM framework for fine-grained air pollution estimation and forecast in metropolitan cities. *IEEE Access : Practical Innovations, Open Solutions*, *10*, 55818–55841. https://ieeexplore.ieee.org/abstract/document/9780279. DOI: 10.1109/ACCESS.2022.3174853

Zhang, X., Zhang, Z., Liu, Y., Xu, Z., & Qu, X. (2024). A review of machine learning approaches for electric vehicle energy consumption modelling in urban transportation. *Renewable Energy*, *234*, 121243. https://www.sciencedirect.com/science/article/pii/S0960148124013119. DOI: 10.1016/j.renene.2024.121243

Zhang, Z., & Wang, Y. (2023). A spatiotemporal model for global earthquake prediction based on convolutional LSTM. *IEEE Transactions on Geoscience and Remote Sensing*, *61*, 1–12. https://ieeexplore.ieee.org/abstract/document/10210326. DOI: 10.1109/TGRS.2023.3302316

ADDITIONAL READING

Lévy, L. N. (2024). *Advanced clustering and AI-driven decision support systems for smart energy management* (Doctoral dissertation, Université Paris-Saclay). https://theses.hal.science/tel-04634793/

Moloi, K., Addo, K., & Mnkandla, E. (2025). Spatio-temporal; Artificial Intelligence; Multi-Hazard Risk Mapping; Renewable Energy Sitting; Multi-objective Optimization. https://www.preprints.org/frontend/manuscript/e3a84e5403285650d89a43d2cff07b34/download_pub

Rahmani, M. (2024). Next-Generation Air Pollution Forecasting: Integrating AI, Spatiotemporal Dynamics, and Privacy-Ensuring Approaches for Urban Areas (Doctoral dissertation, Université de Lille). https://theses.hal.science/tel-05068744

KEY TERMS AND DEFINITIONS

Time Series Analysis: A method of analyzing data points collected or recorded at specific time intervals. It's used to detect trends, seasonality, and forecast future values—crucial in anomaly detection and predictive maintenance.

Spatio-Temporal Analysis: The study of phenomena that vary across both space and time, enabling insights into patterns, trends, and dependencies in multidimensional datasets related to environment, energy, and population.

AI-Driven Predictive Modeling: The application of artificial intelligence techniques such as machine learning, deep learning, and neural networks to analyze past data and predict future outcomes within spatial and temporal contexts.

Chapter 12

The Triple Dividend: Economic Prosperity, Social Equity, and Climate Resilience in the Age of Energy Transition

Akanksha Singh Fouzdar
https://orcid.org/0000-0002-2274-679X
GLA University, India

Ankit Saxena
https://orcid.org/0000-0002-1916-2662
GLA University, India

ABSTRACT

This chapter explores the concept of the Triple Dividend—economic prosperity, social equity, and climate resilience in the context of the global energy transition. It examines how inclusive and sustainable energy policies can generate synergistic benefits across economic growth, social justice, and environmental sustainability. Through global case studies, policy analysis, and theoretical frameworks, the chapter provides actionable recommendations for integrating the triple dividend into energy and climate governance.

INTRODUCTION

Humanity faces a number of crises in the twenty-first century, including growing economic inequality, pervasive environmental degradation, and enduring social inequalities, all of which call for immediate and coordinated action. Addressing these major issues separately has not worked because they are intricately linked. Consequently, the global energy transition from fossil fuel dependence to renewable

DOI: 10.4018/979-8-3373-5302-9.ch012

and sustainable energy systems has become a potent lever to unlock what academics and policy organizations refer to as the Triple Dividend: climate resilience, social equity, and economic prosperity (UNEP, 2023; Queku & Seidu, 2025). Policymakers, scholars, and development professionals are encouraged by this paradigm to acknowledge the mutual advantages of inclusive, low-carbon transitions, (Raworth, 2017).

A break from the conventional siloed approach to development is represented by the triple dividend concept. In the past, climate mitigation measures were frequently viewed as expensive compromises that might obstruct economic expansion. Evidence from numerous international case studies and new research, however, point to the opposite. Strategically planned investments in clean energy can improve energy security, create a large number of job opportunities, draw in innovation-driven capital, and lower long-term public health costs by enhancing air quality, (ILO, 2022; OECD, 2021). These financial benefits are crucial results that support the sound economic reasoning behind sustainable policies; they are not just byproducts of climate action, (Sovacool et al., 2020).

In addition to economics, social equity has received a lot of attention lately. By increasing access to clean energy, lowering energy poverty, and making sure that underserved communities are not left behind, the energy transition presents an opportunity to address systemic injustices. The International Labour Organization and international civil society support the idea of a "just transition," which emphasizes equity as a crucial component of sustainability that cannot be compromised. This includes empowering women and indigenous groups in renewable energy programs, protecting workers' rights during industrial restructuring, and democratizing energy systems through local ownership models, (Cui et al., 2025; Dammak et al., 2025).

Climate resilience, or the ability of ecosystems, economies, and communities to withstand shocks and adjust to a changing climate, is the third element of the triple dividend. Investing in renewable infrastructure is important for decarbonization and for creating more resilient systems that can endure disruptions, as extreme weather events are becoming more frequent. Nature-based solutions, decentralized renewable energy sources, and resilient energy grids increase adaptive capacity while lowering emissions, (IPCC, 2022).Particularly in vulnerable areas where livelihoods are already being threatened by climate risks and inequality is already being exacerbated, these environmental benefits also have a positive impact on the economy and society.

In many nations, the triple dividend is still more of an ideal than a practical solution, despite its potential. Policy fragmentation, conflicting short-term political interests, a lack of funding in developing nations, and poor institutional coordination are some of the implementation challenges (Agency, 2021). For instance, although national climate plans frequently establish lofty objectives, they might overlook equitable distribution or neglect to incorporate strategies for economic growth. In a similar vein, a lot of development initiatives overlook resilience planning, leaving them vulnerable to potential threats.

However, there is progress. There are promising models that try to balance the three dividends. Bangladesh's climate-resilient solar microgrids, the European Union's Green Deal, and South Africa's Just Energy Transition Investment Plan (JET-IP) serve as examples of how cross-sector cooperation and integrated planning can produce observable co-benefits. These instances demonstrate that the triple dividend is not just a theoretical concept but rather an attainable result, given inclusive, data-driven, and context-specific policies.

The purpose of this chapter is to examine the triple dividend framework as a useful tool for analysis and as a tactical approach. It explores the theoretical underpinnings, examines global empirical data, and assesses the financial and policy tools that have aided or impeded advancement(Healy & Barry, 2017). By doing this, the chapter adds to the expanding body of knowledge about how integrated energy transition strategies can lead to revolutionary changes in the social, economic, and environmental spheres. The triple dividend framework provides a potent road map for sustainable and cooperative development as the window for climate action closes (Fankhauser et al., 2016).

CONCEPTUAL FRAMEWORK

In the context of the global energy transition, the idea of the "Triple Dividend" stems from the understanding that social justice, economic growth, and climate resilience are not mutually exclusive. Rather, when properly crafted, decarbonization-related policies and investments can yield mutually beneficial outcomes in all three areas. The just transition paradigm, green growth models, and sustainable development theory are some of the theoretical and policy frameworks that are used to further explain the triple dividend in this section, (Hallegatte et al., 2019).

Sustainable Development Theory: The Foundational Logic

The Brundtland Commission (1987) defined sustainable development as development that satisfies current needs without jeopardizing the capacity of future generations to satisfy their own. This concept is at the core of the triple dividend. The interdependence of social inclusion, environmental sustainability, and economic efficiency is acknowledged by this framework by default, (Newell & Mulvaney, 2013).

Sustainable development is both a strategic necessity and a normative objective in the context of energy transitions. It makes the assumption that low-carbon routes must also result in economic opportunity and human well-being. According to Liu et al. (2025), climate action and sustainable development are inextricably linked, especially in developing nations where energy access and poverty reduction are top priorities.

The Global Center on Adaptation (2023) warns that only 7% of tracked climate finance is currently devoted to adaptation, despite growing vulnerabilities. Their recommendations include mainstreaming adaptation into national budgets, creating dedicated sovereign resilience bonds, and developing metrics for adaptive capacity across regions. Blended finance instruments and concessional support are deemed essential to de-risk private investment in adaptation infrastructure.

Just Transition: Embedding Social Equity

The Just Transition model, which was created by labor movements and later approved by the International Labour Organization, is the second pillar of the conceptual framework (ILO, 2022). This model highlights the need for equitable distribution of the costs and rewards of climate action. In the transition away from fossil fuels, it promotes worker protections, reskilling initiatives, inclusive planning procedures, and participatory governance.

The energy transition poses short-term risks like job displacement, regional economic decline, and energy insecurity for vulnerable populations, even though it is necessary for long-term environmental survival, (Hallegatte et al., 2019). By incorporating social protections into climate policies, a Just Transition framework helps reduce these risks. Cui et al. (2025) assert that incorporating social equity mechanisms can increase public support for decarbonization and avert backlash.

Green Growth Models: Aligning Markets and Ecology

The triple dividend framework's third pillar is provided by green growth models. By reorienting the focus from extractive industrial practices to innovation-driven, resource-efficient, and climate-compatible growth, these models seek to redefine economic growth metrics (OECD, 2021). According to the theory of "green growth," countries can separate GDP growth from environmental deterioration by implementing targeted investments, regulations, and innovative ideas(Stern, 2007).

According to the World Bank (2022), investments in green infrastructure yield significant returns, especially when planned with co-benefits like increased employment, decreased risk of disasters, and better health outcomes. However, these returns might avoid underprivileged communities if they are not purposefully designed. As a result, green growth works best when paired with strategies that emphasize equity and resilience.

Integration through the Triple Dividend Lens

The triple dividend approach is an integrative strategy that requires policy coherence and systems thinking, making it more than the sum of its parts. It requires governments and interested parties to take into account:

- Economic results: industrial competitiveness, employment, and GDP growth
- Social outcomes: empowerment, access, equity, and inclusion
- Environmental results: resilience, biodiversity, and emissions reduction

Queku and Seidu (2025) assert that African countries investigating carbon pricing schemes have the chance to use tax income to fund social protection and green job initiatives, thus reaping all three benefits. In a similar vein, Dammak et al. (2025) support the multi-capital integration approach in supply chains, placing equal emphasis on financial metrics as well as environmental, social, and human capital.

Policy Coherence and Governance Challenges

Although the triple dividend presents a compelling vision, strong institutional capacity and policy integration are necessary for its actual implementation. Contradictory results may arise from fragmented policy approaches in which social, economic, and climate policies are developed separately. A carbon tax that does not provide low-income households with compensation, for instance, might lower emissions but increase inequality (UNEP, 2023).

Additionally, the framework emphasizes the necessity of metrics and instruments capable of measuring multifaceted results. Economic efficiency is frequently given precedence over social or environmental benefits in current evaluation systems. Therefore, creating indicators that are sensitive to the triple dividend is essential for directing implementation and guaranteeing accountability(Stern, 2007).

ECONOMIC PROSPERITY THROUGH ENERGY TRANSITION

It is common to frame the global transition to renewable energy as an environmental necessity. Nonetheless, its economic potential is becoming more widely acknowledged as a key factor influencing regional and national policy. This section examines how, with careful planning, the energy transition can foster economic prosperity by fostering innovation, competitiveness, job creation, and energy security. The triple dividend framework's first dividend is represented by these results.

Job Creation and Labor Market Transformation

The creation of new jobs in industries like solar energy, wind power, electric vehicle manufacturing, energy efficiency retrofits, and grid modernization is a major economic advantage of the shift to clean energy. Over 13.7 million people were employed worldwide in the renewable energy sector in 2022, and under a net-zero pathway, this number is expected to rise to over 38 million by 2030, according to the International Renewable Energy Agency (IRENA, 2023).

These jobs, which provide temporary job boosts, are typically more labor-intensive during the construction phase. However, skills alignment, training and education investments, and strong labor protections are necessary for long-term job stability, especially in the Global South where informal employment is common (ILO, 2022). Significantly, estimates show that the solar PV and wind industries are already outperforming fossil fuels in terms of job multipliers, creating two to five times as many jobs for every dollar invested (Bhattacharya et al., 2022).

Energy Transition as a Competitiveness Strategy

Energy transitions are changing global economic competitiveness in addition to creating jobs. First-mover advantages in international trade and investment flows are increasingly being gained by nations that are at the forefront of clean technology. For instance, China now controls more than 70% of the global supply chain for solar panels and batteries, making it the industry leader (Liu et al., 2025). In a similar vein, the EU's Green Deal Industrial Plan seeks to strengthen technological sovereignty and lessen reliance on foreign fossil fuels by making calculated investments in green industries(Newell & Mulvaney, 2013).

Additionally, domestic clean energy development improves macroeconomic stability by lowering exposure to fluctuations in the world's energy prices. For instance, nations like India and Kenya that import a lot of fossil fuels have combined solar and wind to lessen trade deficits and the burden of subsidies (Queku & Seidu, 2025).

Decentralized Energy and Local Economic Revitalization

Reviving rural and underdeveloped areas with decentralized renewable energy systems is another way that energy transition promotes prosperity. In addition to offering electricity, mini-grids and community solar projects promote the expansion of digital services, small enterprises, and agricultural processing (World Bank, 2022).

Solar home systems and solar irrigation pumps have helped rural communities in Bangladesh and Nepal earn more money while lowering their reliance on diesel.

These systems support inclusive growth by producing both environmental and localized economic benefits (UNEP, 2023).

Innovation, Investment, and Industrial Transformation

Another source of prosperity is green innovation. The necessity of encouraging research and development (R&D) in sustainable technologies, supporting industrial retrofitting, and funding clean tech startups is becoming more widely acknowledged by governments. The OECD (2021) reports that nations that make investments in green R&D see increases in value-added manufacturing, export growth, and long-term productivity gains.

The financial markets are reacting accordingly. By 2023, the cumulative issuance of green bonds worldwide had reached $2.5 trillion, with more than 40% of new bonds going toward clean energy infrastructure (Bhattacharya et al., 2022). This suggests a growing interest in sustainable finance as well as an understanding of clean energy as a sector with strategic growth potential (Hallegatte et al., 2019).

Challenges and Equity in Economic Gains

Even though energy transitions promise prosperity, they can also cause economic disruption, particularly in areas that rely heavily on fossil fuels. Strategies for economic diversification and proactive governance are necessary in light of job losses in the coal, oil, and gas industries, especially in rural areas. Without these, the shift might incite opposition and worsen regional disparities.

Moreover, the distribution of economic gains is not uniform. While many low-income countries face structural obstacles like high capital costs, limited technology transfer, and weak policy coherence, countries with strong institutional capacity and access to green finance are better positioned to benefit (Cui et al., 2025). Therefore, regional coordination and deliberate assistance from international organizations are necessary to achieve economic dividends.

ADVANCING SOCIAL EQUITY

The shift to clean energy is essentially a social justice issue as well as an economic and environmental undertaking. The triple dividend framework's notion of social equity highlights the need for equitable distribution of the costs and benefits of climate action among populations, geographical areas, and generations. Even though the world is moving more quickly toward renewable energy, how inclusive and participatory the transition is will determine whether inequality is improved or worsened. With an emphasis on energy access, distributional justice, gender inclusion, and community empowerment, this section examines how energy policies can be formulated to advance social equity, (Hallegatte et al., 2019).

Energy Poverty and Access to Clean Energy

Lack of access to modern, reasonably priced, and dependable energy services is known as energy poverty, and it continues to be a major obstacle to equity, especially in the Global South. Nearly 675 million people globally did not have access to electricity as of 2022, with more than 80% of those affected living in Sub-Saharan Africa (IEA, 2023). Energy transitions will exacerbate current inequalities rather than lessen them if this gap is not addressed.

A workable and scalable way to address this inequality is to switch to decentralized renewable energy systems, like solar home systems and mini-grids. For example, India's Saubhagya Scheme and Kenya's Last Mile Connectivity Project have linked millions of rural homes to off-grid solar systems. These initiatives have supported small business operations, increased women's safety through lighting, and allowed kids to study at night. But affordability is still a major problem. Low-income households might not be able to afford clean energy systems in the absence of subsidies or concessional financing. Therefore, deliberate public investment, affordability frameworks, and inclusive financing models are necessary for a socially equitable transition, (Raworth, 2017).

Gender Justice in Energy Transition

Gender equity is significantly impacted by the energy transition. As the primary consumers of household energy, women are disproportionately affected by energy poverty, including time spent gathering fuelwood, the negative health effects of indoor air pollution, and limited mobility, especially in rural and marginalized communities. In addition, they continue to be underrepresented in positions of decision-making in the energy industry and are not eligible for financial, training, or employment benefits, (Raworth, 2017).

Just 32% of workers in the global renewable energy sector are women, and even fewer hold technical or leadership roles, according to IRENA (2022). Systemic obstacles such as unequal access to education, discrimination in the job market, and gender-blind policy design are reflected in this gender gap. Incorporating gender equity into energy transition policies requires more than just quotas; it also calls for equitable land rights, financial inclusion for female entrepreneurs, gender-sensitive planning, and investment in STEM education for girls (Cui et al., 2025).

Results are always better when women are actively involved in energy governance. Research indicates that energy projects involving cooperatives led by women have greater success rates, better repayment records, and longer-lasting effects. Gender-responsive strategies can increase the transition's overall benefits and are also in line with SDG 5.

Just Transition for Workers and Regions

Workers and areas reliant on extractive industries run the risk of being economically marginalized as fossil fuel industries decline. This can lead to severe social backlash, labor unrest, and increased regional inequality if left unchecked. To avoid mass displacement, a socially equitable transition must incorporate economic diversification, retraining, compensation, and job placement services.

By integrating community involvement and labor protections into climate policy, the Just Transition framework provides a road map for overcoming these obstacles (ILO, 2022). Phased coal plant closures with negotiated worker support mechanisms have already been implemented by nations like Canada and Germany. With a commitment of $8.5 billion from international partners, South Africa's Just Energy Transition Partnership (JETP) promotes rural infrastructure, SME development, and the retraining of coal workers in impacted provinces in the Global South. (Queku & Seidu, 2025).

These models emphasize that in energy transitions, justice must be proactive rather than reactive. Social cohesion and climate progress will be hampered if people are not invested in alongside technologies.

Indigenous Rights and Energy Democracy

Indigenous communities have historically been shut out of the advantages and decision-making processes associated with energy development in many parts of the world. The extractive models of fossil fuels have occasionally been mimicked by renewable energy projects, resulting in land grabs, evictions, or the destruction of cultural heritage.

Nonetheless, new forms of energy democracy are beginning to take shape, in which renewable energy resources are owned and managed by local communities, including indigenous groups. Indigenous-owned hydro and solar projects in Canada and Alaska are improving resource sovereignty, lowering reliance on diesel, and generating jobs. These models place a strong emphasis on co-governance, open benefit-sharing, and Free, Prior, and Informed Consent (FPIC). In order to advance energy justice, it is imperative that legal protections, customary land rights, and inclusive energy financing mechanisms be upheld (Dammak et al., 2025).

Distributional Justice and Urban-Rural Equity

The gap between urban and rural areas is another significant equity issue. Infrastructure and mobility systems for clean energy tend to benefit urban areas first, while rural communities fall behind. Planning that purposefully directs investment into under-resourced areas is necessary for distributional justice, which is the equitable distribution of advantages and burdens.

European city-led projects, like the Green Social Housing initiatives in Barcelona and Paris, demonstrate how smart meters, rooftop solar, and insulation retrofits can help low-income citizens. Climate-resilient solar microgrids in Bangladesh and Nepal, meanwhile, are sustaining irrigation, reducing the risk of flooding, and supplying electricity to vital facilities like schools and clinics (UNEP, 2023). These illustrations show that interventions with an equity focus can have multifaceted benefits in different regions.

CLIMATE RESILIENCE AND ENVIRONMENTAL GAINS

Although social justice and economic prosperity are important aspects of the energy transition, climate resilience is the third and possibly most pressing pillar. The ability of ecological and human systems to foresee, absorb, recover from, and adjust to shocks and stresses associated with climate change is known as climate resilience. In addition to lowering emissions, this section examines how clean energy systems, nature-based remedies, and sustainable land-use plans improve biodiversity, adaptive capacity, and long-term environmental sustainability.

Mitigation: Reducing Greenhouse Gas Emissions

The energy transition's ability to significantly lower greenhouse gas (GHG) emissions is its most obvious environmental benefit. As of 2022, the energy sector accounted for more than 73% of global greenhouse gas emissions, making it the largest contributor to global emissions (IPCC, 2023). The quickest route to decarbonization is to gradually replace coal, oil, and gas with wind, solar, hydro, and geothermal energy.

Measurable emissions reductions have already started to appear in nations that have quickly implemented renewable energy technologies. Denmark, for example, has reduced its carbon footprint by more than 40% since 1990 by obtaining more than 50% of its electricity from wind energy (OECD, 2021). In a similar vein, Costa Rica has maintained its forest cover and biodiversity while achieving over 98%

renewable electricity for a number of years running, mostly from hydropower and geothermal sources.

Clean energy systems also lessen co-pollutants that are associated with respiratory illnesses, like sulfur dioxide, nitrogen oxides, and particulate matter. This promotes resilience and human well-being by improving environmental quality and producing co-benefits for public health (UNEP, 2023).

Adaptation: Resilient Infrastructure and Systems

Resilience concentrates on the effects of climate change, whereas mitigation tackles its underlying causes. If properly designed, clean energy infrastructure can greatly increase resilience. For instance, centralized fossil fuel grids are more susceptible to severe weather disruptions than decentralized renewable energy systems, like solar rooftops and microgrids. They can keep running on their own in the event of a grid outage, guaranteeing the continuation of vital services like communications, healthcare, and water supply (World Bank, 2022).

Climate-resilient solar microgrids have assisted coastal communities in Bangladesh in keeping their electricity running during floods and cyclones. These systems greatly lower mortality and economic disruption by powering hospitals, emergency shelters, and early warning systems (Dammak et al., 2025). In a similar vein, schools and evacuation centers powered by solar energy have been incorporated into post-disaster reconstruction initiatives in the Philippines to promote quicker recovery.

Food security is also supported by resilient energy infrastructure. In East Africa, farmers can minimize post-harvest losses while adjusting to unpredictable temperature fluctuations and rainfall thanks to solar-powered irrigation and cold storage systems. Vulnerable populations and ecosystems benefit over the long run from these investments in adaptive capacity.

Nature-Based Solutions and Ecosystem Services

The natural environment must be viewed as a partner in resilience as well as something to be protected during the energy transition. Reforestation, wetland restoration, and agroecology are examples of nature-based solutions (NbS) that absorb carbon and improve ecosystem services. These tactics are essential for reaching net-zero goals and enhancing climate resilience.

Mangrove restoration, for example, has been shown to be more economical than engineered seawalls in nations like Indonesia and Kenya when it comes to reducing storm surges and storing significant amounts of carbon. These ecosystems sustain local communities' livelihoods, stabilize coastlines, and promote biodiversity (UNEP,

2023). Land-use efficiency and co-benefits can be maximized by combining NbS with renewable energy sources like agrovoltaics or floating solar on reservoirs.

However, ill-conceived renewable energy projects can result in ecological damage, such as the loss of biodiversity from massive solar farms or the deforestation for hydropower. To guarantee that climate actions do not unintentionally harm the environment, environmental planning and cumulative impact assessments are essential (Liu et al., 2025).

Climate Finance and Environmental Justice

Particularly in the Global South, investments in climate resilience continue to be woefully insufficient. Less than 7% of global climate finance is currently allocated to adaptation, according to the Climate Policy Initiative (2023). Strong public-private partnerships, debt-for-nature exchanges, and grant-based funding models from multilateral organizations like the Adaptation Fund and Green Climate Fund are necessary to close this gap.

Resilience also depends critically on environmental justice. Although they have made the least contribution to climate change, marginalized communities—many of which are located in climate-vulnerable areas—suffer the most from its consequences. Building systemic resilience requires making sure these communities have access to climate-resilient housing, early warning systems, and participatory planning tools (ILO, 2022).

Monitoring, Metrics, and Governance

Reliable climate risk metrics and strong governance frameworks are essential for constructing climate resilience. To measure resilience gains, evaluate vulnerability, and track the environmental effects of clean energy deployment, governments must create data systems. Progress is being measured using instruments like the Environmental Performance Index (EPI) and the Climate Resilience Index (CRI). Environmental resilience is also improved by participatory governance. Communities are better equipped to respond to climate threats and safeguard local ecosystems when they have more autonomy and information access. Consequently, climate adaptation can become more context-sensitive and long-lasting by coordinating local knowledge systems with national planning procedures (Queku & Seidu, 2025).

SYNERGIES AND TRADE-OFFS

The triple dividend framework is a useful tool for creating integrated energy transition plans, but putting it into practice frequently requires difficult trade-offs and complex synergies. Theoretically, initiatives that support the use of green energy should concurrently enhance equity, strengthen resilience, and improve economic results. However, when one pillar is given priority over others, conflicts may arise in real life. In order to make sure that decisions made in the short term do not jeopardize long-term sustainable development goals, this section critically analyzes how synergies can be increased and trade-offs managed.

Synergies: The Power of Integration

The triple dividend's capacity to promote cross-sector co-benefits holds its greatest promise. For example, governments frequently increase energy access, lower emissions, and generate a sizable number of local jobs when they invest in extensive renewable energy infrastructure. These concurrent social, environmental, and economic benefits show how climate action can support a variety of policy objectives.

One of the biggest solar plants in the world, the Noor Solar Complex in Morocco, serves as a prime example. In addition to providing electricity to more than a million households, the project has created jobs, decreased dependency on imported fossil fuels, and benefited local companies by granting them contracts for proc urement (World Bank, 2022).

Similarly, Kenya's geothermal energy investment has reduced electricity costs, increased rural electrification, and improved grid stability all while lowering carbon emissions and increasing resilience to energy disruptions brought on by climate change (UNEP, 2023).

When cross-sectoral governance and systems thinking are used to design policies, these positive feedback loops are most apparent. For example, combining investments in renewable energy with training and education initiatives produces skilled workers, enhancing social mobility and long-term economic competitiveness (ILO, 2022). Similarly, green urban planning that combines public transportation and energy-efficient housing lowers emissions, improves air quality, and benefits public health all at once (OECD, 2021).

Trade-offs and Unintended Consequences

Notwithstanding these possible benefits, if energy transition policies are not properly handled, they could have unforeseen repercussions. Aggressive decarbonization goals, for instance, may lead to job losses in communities that rely heavily on fossil fuels, escalating regional inequality and political opposition. The social equity dividend might be foregone in favor of environmental benefits if just transition plans are not adequately funded (Cui et al., 2025).

Furthermore, some renewable energy projects, especially in ecologically sensitive or indigenous territories, may result in biodiversity loss or land conflicts. Local protests and livelihood losses have resulted from the displacement of agricultural communities in Brazil and India due to solar and wind projects (Dammak et al., 2025). These incidents highlight the fact that, absent inclusive planning and environmental protections, environmental benefits are not always socially equitable.

Climate financing has historically prioritized mitigation (e.g., solar farms, carbon markets) over adaptation (e.g., flood defenses, resilient housing) due to perceived return on investment; however, this undermines resilience goals in countries most vulnerable to climate change, such as small island developing states (Climate Policy Initiative, 2023). This is another important trade-off.

Managing Competing Priorities: Institutional and Policy Tools

To navigate these trade-offs, policy coherence and institutional alignment are critical. Governments must avoid siloed decision-making by fostering collaboration between environmental agencies, finance ministries, labor unions, local governments, and civil society. National development plans and climate policies should be co-designed to reflect shared priorities across sectors.

Several tools can help facilitate this integration:

- ***Cost-Benefit Analyses with Co-Benefit Valuation:*** These allow decision-makers to evaluate policies not just by cost, but by the full range of social and environmental benefits.
- ***Environmental and Social Impact Assessments (ESIA):*** These ensure that energy infrastructure projects consider potential trade-offs in advance.
- ***Integrated National Financing Frameworks (INFFs):*** This help align national budgets with the Sustainable Development Goals (SDGs), including climate and equity objectives.

The European Union's Green Deal is one example of an integrated approach that balances climate ambition with economic support mechanisms, including a Just Transition Fund that provides €17.5 billion to regions most affected by decarbonization (OECD, 2021).

Political Economy and Power Asymmetries

The political economy of energy transitions also shapes synergies. Changes that threaten their economic foundation may encounter resistance from actors with vested interests, such as utility monopolies, fossil fuel companies, or even governments that depend on extractive revenues. In the meantime, the workers and communities most impacted by the shift frequently lack the political clout to change its course.

Priorities may be distorted by this power imbalance, favoring urban or elite groups while excluding vulnerable groups. As Queku & Seidu (2025) note, in many African contexts, carbon pricing mechanisms have been introduced without redistributive components, resulting in increased energy costs for the poor. Therefore, managing trade-offs is not just a technical task—it is a matter of governance, participation, and accountability.

Toward Triple Dividend Metrics and Evaluation

Lastly, the absence of integrated metrics to evaluate triple dividend performance is one of the reasons trade-offs continue to exist. Social and environmental benefits are often overlooked by traditional cost-effectiveness models used in climate policy. Governments and investors can make better decisions by creating triple-bottom-line reporting and multi-criteria decision-making tools.

New frameworks for measuring impact, like the World Bank's Resilience Rating System or the UNDP's Climate Public Expenditure and Institutional Review (CPEIR), provide ways to assess how well policies deliver on a variety of dividends. However, these tools must be integrated into public sector accountability systems and modified for local contexts (IPCC, 2023).

POLICY INSTRUMENTS AND GLOBAL BEST PRACTICES

It takes more than just conceptual clarity to achieve the Triple Dividend of social justice, economic prosperity, and climate resilience. To operationalize integrated outcomes, it calls for carefully crafted financial frameworks, regulatory actions, and policy tools. This section reviews global best practices that have produced quantifiable outcomes in a variety of areas and looks at important policy tools that support a fair and sustainable energy transition.

Carbon Pricing with Equity Mechanisms

One of the most popular policy tools for internalizing the environmental cost of emissions is carbon pricing. It generates a price signal that deters high-carbon activity and encourages low-carbon alternatives when it is implemented through carbon taxes or emissions trading systems (ETS).

However, carbon pricing needs to be combined with redistribution mechanisms that improve equity in order to support a triple dividend. For instance, low-income households receive direct dividends from British Columbia's carbon tax in the form of tax refunds. Comparably, South Africa is investigating progressive carbon pricing in conjunction with social protection initiatives for impacted communities and workers (Queku & Seidu, 2025).

Recycling revenue can help fund clean infrastructure investments, rural electrification, and job retraining, transforming potentially regressive policies into ones that benefit the poor. As demonstrated by the French Gilets Jaunes protests, carbon pricing runs the risk of escalating energy poverty and social unrest in the absence of such reforms (OECD, 2021).

Green Subsidies and Public Procurement

Public procurement and targeted subsidies can hasten the adoption of clean technologies and boost regional economies. Markets for renewable energy have grown in nations like China, India, and Germany thanks in large part to feed-in tariffs, tax breaks for solar installations, and green innovation grants.

Financial incentives are provided to domestic manufacturers through the Indian government's Production Linked Incentive (PLI) Scheme for solar photovoltaic manufacturing, which also supports employment, economic competitiveness, and climate goals. The U.S. Inflation Reduction Act (IRA) has implemented extensive clean energy tax credits, ensuring that climate investments yield both social and economic benefits by linking benefits to local content and labor conditions (Bhattacharya et al., 2022).

Green building regulations and low-emission transportation contracts are examples of public procurement policies that increase demand for sustainable goods and services. These tools directly support social equity when they are in line with inclusion criteria (e.g., minority contractors, women-owned businesses) (ILO, 2022).

Climate Finance for Equity and Resilience

Having access to climate finance is essential for achieving integrated results, especially in low-income and climate-vulnerable countries. Adaptation and equity-focused interventions are still woefully underfunded, despite an increase in global climate finance flows (Climate Policy Initiative, 2023).

Grant-based mechanisms for decentralized energy access, clean cooking solutions, and community adaptation have been introduced by organizations such as the Green Climate Fund (GCF) and Climate Investment Funds (CIF). By avoiding conventional middlemen and guaranteeing local ownership, the GCF's Enhanced Direct Access model enables national entities typically NGOs or regional banks to manage funds directly.

Using blended finance to reduce the risk of private investment in socially beneficial projects is another excellent practice. Over 3 million rural households in Kenya have benefited from market-based last-mile access solutions made possible by concessional loans in the off-grid solar sector (World Bank, 2022).

Legal and Regulatory Instruments for Justice

To ensure that energy transitions are not only green but just, countries must embed justice considerations in their legal and regulatory frameworks. These may include:

- Just Transition legislation (e.g., Spain’s Just Transition Strategy) mandating social dialogue, labor guarantees, and retraining.
- Land-use planning laws protecting indigenous territories and natural habitats from green colonialism.
- Environmental and Social Impact Assessments (ESIA) with mandatory community consultations. In Canada, the Impact Assessment Act requires large energy projects to demonstrate how they will promote Indigenous participation and climate mitigation, creating a legal basis for inclusive project governance (Dammak et al., 2025).

Integrated Planning and Institutional Capacity

It is a multi-level governance challenge to achieve the triple dividend. The effectiveness of policy tools depends on the organizations that create and carry them out. It is crucial to connect national goals to regional and local initiatives through integrated energy and climate planning. Among the best practices are:

- **South Africa's Just Energy Transition Investment Plan (JET-IP):** A donor-supported, nationally driven framework that incorporates social safeguards, stakeholder mapping, and financial modeling.
- **Costa Rica's Decarbonization Plan:** Provides a long-term outlook with cross-sectoral coordination between the ministries of education, forest-

ry, energy, and transportation, bolstered by data dashboards and public participation.

As crucial as the tools themselves is developing institutional capacity in planning, monitoring, and stakeholder engagement. Knowledge transfer and localized innovation can be accelerated through South-South collaboration, policy learning networks, and technical assistance (IPCC, 2023).

Measuring Impact and Scaling What Works

Policy tools must be assessed for their multifaceted impact in order to replicate and scale success. Governments and donors can measure not only the reduction of emissions or the creation of jobs, but also the improvements in equity and resilience through the use of tools such as the Triple Dividend Impact Assessment (TDIA) and Multi-Criteria Decision Analysis (MCDA). To standardize metrics and ease scaling, international platforms like the Partnership for Action on Green Economy (PAGE) and NDC Partnership provide templates, peer learning, and technical tools. These strategies gain political traction and longevity when they are incorporated into national development frameworks.

REFERENCES

Agency, I. E. (2021). *Net Zero by 2050: A Roadmap for the Global Energy Sector*. https://www.iea.org/reports/net-zero-by-2050

Agency, I. E. (2021). *Net Zero by 2050: A Roadmap for the Global Energy Sector*. https://www.iea.org/reports/net-zero-by-2050

Bhattacharya, A., Kharas, H., & Plant, M. (2022). *Financing a big investment push in emerging markets for sustainable recovery*. Brookings Institution. https://www.brookings.edu/articles/financing-a-big-investment-push-in-emerging-markets-for-sustainable-recovery/

Climate Policy Initiative. (2023). *Global Landscape of Climate Finance 2023*. https://www.climatepolicyinitiative.org

Cui, D., Zhang, T., Yu, C. W., & Li, J. (2025). Equal distribution of new energy in urban low-carbon renewal: Opportunities and challenges. *Indoor and Built Environment*, *34*(7), 1279–1283. DOI: 10.1177/1420326X241309772

Dammak, K., Zouari, A., & Sidhom, L. (2025). *Sustainability and Multi-Capital Integration in Agri-Food Supply Chains: A Systematic Review*. ResearchSquare. https://www.researchsquare.com/article/rs-6115997/latest

Fankhauser, S., Bowen, A., Calel, R., Dechezleprêtre, A., Grover, D., Rydge, J., & Sato, M. (2016). Who will win the green race? In search of environmental competitiveness and innovation. *Global Environmental Change*, *44*, 28–38.

Fankhauser, S., Bowen, A., Calel, R., Dechezleprêtre, A., Grover, D., Rydge, J., & Sato, M. (2016). Who will win the green race? In search of environmental competitiveness and innovation. *Global Environmental Change*, *44*, 28–38.

Hallegatte, S., Rentschler, J., & Rozenberg, J. (2019). *Lifelines: The Resilient Infrastructure Opportunity*. World Bank Publications. DOI: 10.1596/978-1-4648-1430-3

Hallegatte, S., Rentschler, J., & Rozenberg, J. (2019). *Lifelines: The Resilient Infrastructure Opportunity*. World Bank Publications. DOI: 10.1596/978-1-4648-1430-3

Healy, N., & Barry, J. (2017). Politicizing energy justice and energy system transitions: Fossil fuel divestment and a "just transition.". *Energy Policy*, *108*, 451–459. DOI: 10.1016/j.enpol.2017.06.014

Healy, N., & Barry, J. (2017). Politicizing energy justice and energy system transitions: Fossil fuel divestment and a "just transition.". *Energy Policy*, *108*, 451–459. DOI: 10.1016/j.enpol.2017.06.014

ILO. (2022). *Just Transition Guidelines for a Greener Future.* https://www.ilo.org/global/topics/green-jobs/WCMS_432859/lang--en/index.htm

IPCC. (2022). *Climate Change 2022: Impacts, Adaptation and Vulnerability.* https://www.ipcc.ch/report/ar6/wg2/

IPCC. (2022). *Climate Change 2022: Impacts, Adaptation and Vulnerability.* https://www.ipcc.ch/report/ar6/wg2/

IPCC. (2023). *Synthesis Report: Climate Change 2023.* Intergovernmental Panel on Climate Change. https://www.ipcc.ch/report/ar6/syr/

IRENA. (2023). *Renewable Energy and Jobs – Annual Review 2023.* International Renewable Energy Agency. https://www.irena.org/publications

Liu, J., Fang, Y., Liu, T., ... (2025). *"Policy synergy" or "strive alone"? "Dual-pilot" policy to enable renewable energy technology innovations.* Clean Technologies and Environmental Policy. https://link.springer.com/article/10.1007/s10098-025-03236-3

Newell, P., & Mulvaney, D. (2013). The political economy of the 'just transition.'. *The Geographical Journal, 179*(2), 132–140. DOI: 10.1111/geoj.12008

Newell, P., & Mulvaney, D. (2013). The political economy of the 'just transition.'. *The Geographical Journal, 179*(2), 132–140. DOI: 10.1111/geoj.12008

OECD. (2021). *The Economic Benefits of Climate Action.* OECD Publishing. https://www.oecd.org/environment/the-economic-benefits-of-climate-action-9789264307691-en.htm

Queku, Y. N., & Seidu, B. A. (2025). *Targeting in Africa: Readiness and Challenges.* In *Clean Energy and Development.* https://books.google.com/books?id=LSVEEQAAQBAJ

Raworth, K. (2017). *Doughnut Economics: Seven Ways to Think Like a 21st-Century Economist.* Chelsea Green Publishing.

Raworth, K. (2017). *Doughnut Economics: Seven Ways to Think Like a 21st-Century Economist.* Chelsea Green Publishing.

Sovacool, B. K., Burke, M., Baker, L., Kotikalapudi, C. K., & Wlokas, H. L. (2020). New frontiers and conceptual frameworks for energy justice. *Energy Policy, 105,* 677–691. DOI: 10.1016/j.enpol.2017.03.005

Sovacool, B. K., Burke, M., Baker, L., Kotikalapudi, C. K., & Wlokas, H. L. (2020). New frontiers and conceptual frameworks for energy justice. *Energy Policy*, *105*, 677–691. DOI: 10.1016/j.enpol.2017.03.005

Stern, N. (2007). *The Economics of Climate Change: The Stern Review*. Cambridge University Press. DOI: 10.1017/CBO9780511817434

UNEP. (2023). *Making Peace with Nature*. United Nations Environment Programme. https://www.unep.org/resources/making-peace-nature

World Bank. (2022). *Climate-Resilient and Inclusive Cities*. https://documents.worldbank.org/en/publication/documents-reports/documentdetail/503201643479926123

Chapter 13

Application of Marketing Analytics on Green Marketing Mix

Gagan Gulati
Manipal University Jaipur, India

Mahesh Chandra Joshi
Manipal University Jaipur, India

Amit Kumar
https://orcid.org/0000-0002-1686-3279
Manipal University Jaipur, India

ABSTRACT

Technology has transformed managerial decision-making, particularly through analytical tools that support real-time adjustments in marketing strategies based on data. Marketing analytics is vital for informing organizational actions. Concurrently, sustainable development and green marketing are increasingly significant for organizational longevity and environmental protection. Understanding green product attributes, pricing, availability, and promotion is crucial. Utilizing technology in the green marketing mix aids organizations in grasping customer preferences and market trends. The chapter discusses green marketing's importance, detailing elements of the green marketing mix, the role of marketing analytics in decision-making, and the related green Ps, as well as the challenges and opportunities within this domain.

DOI: 10.4018/979-8-3373-5302-9.ch013

INTRODUCTION

Today, all the life forms on the planet are facing severe issue of environmental pollution (Mahmoud, 2018). The entire world is facing various environmental problems like climate changes, global fall of natural resources, depletion of ozone layer deforestation etc (Attih, 2025). Due to irrational use resources, the environmental problems have risen that creates the threat for the living things (Türk, b., & erciş, a., 2017). Environmental sustainability has gain importance in today's world due to environmental degradation caused by different environmental problems like global warming, environmental pollution, greenhouse effect and different type pollution (Air, Water) etc, that has also enhanced consumer sensitivity towards buying products that are green (Kumar et al., 2020; Purwanti et al., 2019). Still one of the major concerns for the humans or the world are problems related to environment, where ecological unbalances, air pollution, greenhouse effect are few of them, (Sharma, 2011).

The world is becoming more conscious towards the environment due to environmental problems. Buyers are ready for bringing the change in their habits, which is also impacting the working of the companies, so the companies must respond in the sustainable manner. (Papadas et al., 2019). Environmental damage due to excessive pollution, challenges due to globalization, has raised the public concern towards the traditional products and this concern has made the marketeers to think for environmentally friendly products, adopting new business models and altering their strategic approach towards business activities. It also emphasizes on the marketing the products and services in the environmentally friendly manner. This form of marketing is known as green marketing. It can also be used for gaining competitive advantage (Purwanti et al., 2019) (Andronie, M et al., 2019).

The needs as well as the wants of the humas are not limited, but the resources are scarce and very limited. It becomes important for the companies to use the available resources in the most effective manner, reducing the wastages as much as possible, (Papadas et al., 2019). To protect and preserve the environment various actions have been taken like legal measures, taxes etc., (Türk & Erciş, 2017).

With the change of consumer lifestyles, supply chains, more regulatory and environment pressures as well as with the coming of information and communication technologies the marketing has transformed (Ofori, D., 2021). By changing their buying and consumption habits consumers perform an important part in protection of environment, even companies must also take initiatives in educating customers about the environmental conditions (Mark Bekweri EDEH, 2020). The green consumer behaviour has emerged as new field of research for the marketers and the researchers. Consumers are thinking towards green choices while making the decisions (Kaur, S. et al., (2025).

"Green" has become an important buzz word today. Environmental issues are crucial and plays a significant role in the modern businesses, that's why green marketing is very significant for the contemporary market (Vijai & Anitha, 2020). Various environmental problems like global warming, pollution etc are raising concerns about how goods and services are manufactured and distributed, companies are becoming more sensitive towards these problems and looking forward for better practices (Mishra & Sharma, 2014). Varied set of environmental, economic, social, marketing and corporate factors impact the green competitiveness of the companies (Chygryn et al., 2024).

GREEN MARKETING

Green marketing includes various range of activities focused at protecting the environment, consumer rights, and fulfilling consumer wants and needs. It helps in fighting against methods which are misleading as well as protect the customers or the consumers from the harmful practices adopted by the manufacturers, (Attih, 2025).

Green marketing or eco-friendly marketing is becoming more crucial over the world these days due to consumers concern about the environment. It helps the organization in multiple ways like targeting at to new markets, gaining advantage, long -term survival and growth of an organization, improves the quality of environment etc. (Sawant, A. V., 2020). Globalization, excessive pollution are some of the challenges faced by the organization, which are the reasons for thinking about new business model and different ways of performing business activities (Andronie et al., 2019).

The term green marketing is the made up with the two words, that is green and the marketing, where green focuses on the environment friendliness or having the minimum unfavourable influence on the environment on the other side marketing means satisfying the requirements of the buyers in the profitable manner. It signifies that the product is produced in the environment friendly way, or it has been marketed in the eco-friendly manner having quality, good performance, better pricing and convenience for the customers. It focusses on providing multiple environmental benefits, (Sarbjit & Sanjay, 2017).

Green marketing can also be termed as ecological or the environmental marketing. Which states that it is a holistic marketing concept, where each activity from production to consumption to the disposal is done in such a manner that it has insignificant or minimum or no adverse impact on the environment, (Mishra & Sharma, 2014). In simple words, green marketing means eco-friendly activities (Vijai & Anitha, 2020).

Green marketing incorporates a wide range of activities, that includes modification of product, changes to the manufacturing process, changes in packaging style, as well as modification in the way companies do advertising (Polonsky, 2008).

Green marketing is a comprehensive approach to marketing that emphasizes environmentally responsible production and consumption. It has gained importance due to increasing awareness of issues like global warming, non-biodegradable waste, and harmful pollutants (Singh, 2012).

Green marketing involves a wide variety of practices such as altering products, updating production methods, redesigning packaging, and adjusting advertising strategies to promote environmental sustainability. Green marketing forms an essential part of the holistic marketing approach, especially for industries that rely heavily on natural resources—such as fishing, food processing, tourism, and adventure sports. These sectors are particularly vulnerable to environmental changes, which can significantly impact their operations. Recognizing this, many global companies across various industries are adopting green marketing strategies. Central to this movement are the three core principles of environmental sustainability: Reduce, Reuse, and Recycle. (Mishra & Sharma, 2014).

Green marketing is defined as the response of marketing to the ecological effects. It impacts the way of the design, production, packaging, labelling, use, and disposal of products or services. It is also perceived by the various business leaders that green marketing is profitable. The purpose of green marketing is to take care of environmental problems in their marketing activities. The reason behind this is that marketers can provide valuable and useful information about the green properties of the product to the consumer, which will help them in their purchase decision (Solaiman, Osman, & Halim, 2015). Green marketing is a combination of diverse activities like modification in product, alteration in production process, different packaging style or advertising method, (Mahmoud, 2018).

The process of management that looks towards identifying, anticipating, and fulfilling the requirements of the buyers and society in a profitable and sustainable manner is known as green marketing, (Welford, 2000).

Green marketing or ecological marketing is a tactical approach directed at guarding the environment for coming generations. It is a complex concept that requires careful planning and evaluation by organizations to determine its feasibility. Although green marketing is still in its early stages, but still, it holds significant long-term benefits. While its immediate implementation may present challenges, its adoption can enhance a company's sustainability and reputation over time. Moreover, it is a key strategy to combat pollution and environmental problems like climate change. It demands strong regulations, increased consumer responsibility, and true dedication from businesses. Many buyers are ready for paying extra for environment-friendly products, but excessive and unstable consumption patterns remain a concern. While one company alone can't drive large-scale change, businesses that focus on sustainability can lead by example and encourage their partners and customers to do the same. For green marketing to become widespread, a collective shift in attitudes and behaviours is essential. (Hooda & Jaggarwal, 2021).

Through sustainable marketing, there is control over the entire value chain to address customers and endorse the business model that helps optimize resources and eliminate waste, (Andronie et al., 2019). Green marketing as a holistic strategy that focuses on developing, communicating and providing services/products with the minor impact on the environment, (Papadas, 2021)

Promotion of eco-friendly products like Phosphate-free products, toxin-free food or refillable containers is consider as green marketing, (Giwa-Amu, 2022; Khan et al., 2019) sees green marketing as adopting practices which are fruitful for the environment as well as for the human, means the marketing practices should be favorable for both human mankind and environment.

The commitment of organization towards green products and customer support are where the future of green marketing lies, (Solaiman, Osman, & Halim, 2015). Moreover, producing eco-friendly products through clean production process and marketing of them will result in sustainable utilization by consumers, (Dangelico, & Vocalelli, 2017).

The key concept of green marketing i.e. the sustainability and holism that seems to appear easy but when put into the practice it is incredibly difficult. The main reason for that is conventional management wisdom's that put emphasis on the speciality and reductionism. Furthermore, the eco-friendly marketing tries to move marketing theory and practice away from the economic hyperspace in which it has formed and bring it back to earth and reality (Peattie, K., 2016).

The various activities included in green marketing are pricing, manufacturing, promotion of products and services, these help in fulfilling the environmental needs of the consumers, enhancing the intention of the consumers to purchase (Ahmed et al., 2025).

RESEARCH METHODOLOGY

For this study, the secondary data is used to identify the prominent practices related to marketing analytics on the green marketing mix. Around eighty-three research papers were analyzed and studied to recognize the meaning, definition, factors, green marketing meaning and definition, green marketing mix and its elements, marketing analytics, digital marketing and challenges as well as opportunities.

OBJECTIVES OF THE CHAPTER

a) Understand the concept of green marketing mix
b) Explore the various elements of green marketing mix
c) Understand the implementation of marketing analytics on the elements of green marketing mix.
d) Analyze the challenges and opportunities in the green marketing mix

CONCEPT OF GREEN MARKETING MIX

Green marketing can be one of the strategies adopted by the marketeers, this strategy is known for their environmental benefits focusing on fulfilling the needs of the customers through manufacturing of goods or products that have minor adverse impact on the environment, (Firdiansyah, 2021). It requires very conscious, targeted and an open effort by the companies towards social and environmental problems, considering the different stakeholders of the organization, (Bednárik, 2000). Further to influence the customers more and more organizations are implementing eco-friendly strategies, which is also helping in improving the negative effects of pollution and environmental footprints, (Sultan et al., 2023). Further green marketing has gain utmost importance in the contemporary business marketplace, (Mishra & Sharma, 2014). Green marketing mix is a concept based on the primary marketing mix. As a crucial marketing concept, it comprises of four elements, (Pitaloka et al., 2023).

One of the most important parts of business strategy is the green marketing, it helps in the sustainable development. The main purpose of the green marketing is to include eco-issues in their marketing endeavours and to promote responsible consumption. Traditional marketing mix does not accurately fit into sustainable development requirements, which creates the need for the green marketing mix adoption. Green marketing mix is known as marketing of environmentally friendly products, that considers the protection of environment from the start to end i.e. manufacturing to promotion, (Risteska, 2023).

The change of buyer focus on green marketing mix is not a distinctive idea Where most companies are looking towards adopting strategies that are green in nature to reduce environmental footprints as well as for gaining customer confidence, (Sultan et al., 2023).

The demand of the green products is mainly dependent upon how the companies communicate and educate consumers about the ecological elements in the product. The better the company can communicate more is the scope of the green products, (Bednárik, 2000).

The notion of the green marketing mix is rooted in the use of environmentally responsible strategies of marketing that is parallel to traditional marketing approaches. However, what sets it apart is its firm dedication to ecological sustainability. The green marketing mix seeks to balance business goals with the urgent need to conserve natural resources and reduce ecological harm, (Silaban et al. 2021).

Green marketing mix is comprised of different tools and strategies use by organizations to promote their eco- friendly services or products. It utilizes the traditional 4Ps of marketing but with a focus on environmental stewardship and sustainability, (Nguyen-Viet, 2023). Through this companies tries to fulfil the requirements of the customers, (Astuti et al., 2021).

Marketing activities that reflect the green values in its implementation is called green marketing mix. It involves the implementation of the four greens, green place, green price, green promotion and the green product, (Purwanti et al., 2019).

Company can attain competitive position in the market through effective marketing mix based on the needs and desires of the consumers, (Mahmoud, 2019). The combination of environmentally friendly strategies adopted by firms in fulfilling the needs of consumers, consisting of the four components is known as green marketing mix, (Mahmoud et al., 2024).

They have highlighted that customer commitment is impacted by the green marketing mix. The important factors are brand loyalty, perceived quality, quality brand association. So, to foster customer commitment green 7Ps must be prioritized, (Nguyen et al., 2025).

ELEMENTS OF GREEN MARKETING MIX

To make the economy greener, there is requirement of the new set of products, technologies, process etc, that are greener in nature. Companies must bring tremendous change in each of the element or component of the marketing mix for example for manufacturing product they must go for the green technology reducing the wastages in the production, saving electricity etc. On the other hand, they must look the sustainability from economic, ethical, social and environmental point of view, (Bednárik, 2000).

GREEN PRODUCT

In green marketing mix product plays a key role, it is the core of it. It plays a crucial role in the marketing strategy of the business or the organization, (Mahmoud,

2019). In simple terms the goods that are friendly for the environment and secure for the consumption or usage are known as green products, (Tsai et al., 2020).

The products manufactured by the companies should be a solution to environmental problems like global warming, air pollution, waste content etc. The products must have least toxic waste and should not endanger the lives of any creatures for example experiments on animal, (Pitaloka et al., 2023).

A product, whether intangible in nature or tangible in nature is green when it provides added benefits or value to the customers while reducing the negative impact on the environment, (Nwankwo & Kanyangale, 2023). As per Ebhote and Izedonmi (2021) any product/goods or service can be said green when they are safe for the consumption and do not harm any humans, basically they are environment friendly products possessing qualities that are not harmful or do not possess any risks for the people, society or ecological unit at large. The core idea of the product that is green in nature focus on to minimize, prevent or eliminate the adverse environmental of the elements of the earth i.e. water, air or soil, (Dauda et al., 2024).

Some of the traits of the green products includes: they are grown originally, they have natural ingredients, they can be recycled or some of their parts can be recycled, they have chemical that is not harmful for the nature or using those chemicals only that are approved, having eco-friendly packaging that means the containers can be refill, it can be reuse for the other purposes etc. Further promotion of the products which are eco-friendly are very important and essential for saving the natural resources and sustainable development, (Mishra & Sharma, 2014).

While designing the green products sustainability must be considered. Through the entire new product development process sustainability must be kept in mind, so from idea generation stage to commercialization sustainability must be at the core. Recycling, water saving, energy saving etc must be considered while developing green products, (Goyal & Pahwa, 2018).

The green product is par with the conventional product having one major difference and that is, it less harmful or not at all harmful for the environment, its composition is done in such a way that there is no negative effect to the environment during its entire lifecycle, (Mahmoud, 2019). Let's consider the example of Coca-Cola. The firm has spent a massive amount of funds in the different type of eco-friendly activities like packaging, recycling etc., (Polonsky, 2008).

Green products are those developed using environmentally sustainable technologies and practices, designed to minimize harm to the natural world. Promoting such products and the technologies behind them is significant for conserving natural resources and supporting long-term sustainable development. Green products can be identified by several characteristics, including being organically produced, non-toxic, recyclable, reusable, or biodegradable, not being involved in animal testing and being safe for the environment and utilizing eco-conscious packaging, such as refillable or reusable containers, (Sarbjit & Sanjay, 2017).

Labelling an important part of the product, also requires modification, here companies need to move from simple labels to the eco-labels, that clearly states that it is an environmentally friendly product or the safe product. In context to labelling companies may choose whether they want to label the product by themselves or want to go for the third party. Further companies also must decide whether they want to opt single labelling for the country or there will be modifications as per different parts of the country. The labelling needs to check according to the country. Another vital aspect of the product is the packaging, it serves many functions like promotion and protection of the product, ease to carry it etc. Here also companies must adopt greener packaging or methods which have minimum wastes (Solaiman, Osman, & Halim, 2015).

The most crucial element of the green marketing strategy is the product; it is center of the green marketing mix, (Fan & Zeng, 2011). Green product also called as the eco-friendly product are those products which do not damage the environment in any of its process i.e. production, usage or the disposal, (Risteska, 2023).

GREEN PRICE

In green marketing mix, one of the most flexible elements is green pricing, it can be easily modified or align with respect to the dynamic market conditions. As per the changes in the business environment the companies can modify their pricing strategies according with not much concern, (Alapa, Ugbe, & Alakali, 2024).

The pricing of the product is influenced by many factors like costs associated with the production, costs associated with the promotion, market competition etc. The companies which implement green marketing to their products they are relatively expensive than those who do not apply green concept, (Pitaloka et al., 2023).

The additional costs consumers pay for the eco-friendly products are known as green prices, (NguyenViet, 2023). On the other hand, there are researchers like Silaban et al. (2021) due to reusability and recyclability of the green products they might be priced lower in comparison to the traditional or the conventional products.

Green products take into consideration the consumer welfare, productivity as well as environmental benefits, leading to higher prices. Some buyers are ready to pay extra money for the green products due to their additional product value. Let's consider an example of Wall Mart, where customers are ready to shed extra penny because they provide the recyclable shopping bag, (Sarbjit & Sanjay, 2017).

One of the key considerations in setting prices for eco-friendly products is understanding whether the use of sustainable practices and green technologies leads to increased production costs, and to what extent these can be passed on to consumers. It is also important to assess how willing customers are to accept the price difference associated with environmentally responsible goods. The ongoing development of new sustainable materials, resources, and technologies often results in higher operational or material expenses for companies. However, these added costs can be offset through savings in areas such as reduced energy and material use, minimized packaging, lower waste management expenses, and even the creation of additional revenue streams through by-product utilization. In sustainability-focused pricing, the emphasis can shift from initial costs to long-term value—for example, while energy-efficient bulbs may be more expensive upfront, they offer reduced energy expenses over time, (Bednárik, 2000).

Companies have several pricing strategies available for environmentally friendly products or services, including:

- **Premium (Elite) Pricing:** This involves setting a high price for innovative green products that use the latest technology and offer significant environmental advantages. This strategy highlights the product's superior quality and ecological benefits, targeting consumers who prioritize sustainability, (Bednárik, 2000).
- **Penetration Pricing:** Often used for mass-market organic products, this approach sets a lower initial price to encourage consumer adoption and familiarity. It helps position the product as a viable alternative to conventional goods and can boost market share. Companies often pair this with strategies aimed at reducing production costs to sustain profitability at lower prices, (Bednárik, 2000).
- In situations where the production of green products involves high costs that cannot be transferred to consumers, businesses may cover the shortfall through profits from other products. This enables them to keep prices reasonable without compromising on their environmental goals. This method involves increasing the price of a green product to reflect the added ecological value it provides to society. The price premium serves to support broader environmental outcomes and helps convey the product's positive impact, (Bednárik, 2000).

The consumer gets attracted towards the green products because of many reasons like high quality, better value, safer products for children etc, but consumer are only willing to buy these products when they are affordable, (Risteska, 2023). The price of the product depends upon the environmental and economic costs. It may include various aspects like discounts for recycling packaging, (Sen, 2024).

Green Place

Applying environmentally friendly principles for various range of activities like material procurement, distribution, etc., (Sohail, 2017). Selecting the most suitable place for their market offerings makes it a valuable element of the marketing mix, (Ahmed et al., 2023).

It talks about various ways of reducing the carbon footprints, transportation emissions etc. For example, rather than giving emphasis or the importance to the imported product the license to manufacture it should be provided by the importing country, it will help in reducing cost like shipping, reduce carbon emissions etc., (Goya & Pahwa, 2018).

Green distribution involves streamlining logistics and supply chain operations to reduce emissions from transportation, with the primary goal of decreasing the carbon footprint, (Solaiman, Osman, & Halim, 2015).

Green distribution is also be called as green placement. Distribution or the placement means the various actions taken by the organizations for making the services or the products accessible to customers. This P is responsible for making the products easily available to customers as per their requirements. It deals with selecting the appropriate channels for delivering the products to the ultimate consumers. Here companies must keep in mind that the channel must not be harmful for the environment, or the practices adopted for distribution by the companies must be eco-friendly, (Alabo & Anyasor, 2021).

The choice of location should align with the brand image a company aims to convey. It should also help distinguish the business from its opponents. This can be accomplished through in-store promotions, pleasant visual merchandising, and the use of recycled materials to draw attention to both environmental and other value-driven benefits, (Sawant, 2020).

Though very few customers will take an extra effort to buy green products but being first to provide green product provide an extra advantage to the marketeers. Green shipping helps in protection of environment as shipping effects air, water etc leading to water pollution, oil pollution, air pollution etc., (Risteska, 2023).

The main aim of the green logistics is to reduce the eco damage associated with every single activity of the supply chain, it includes green distribution channel, green information system, green packaging etc., (Yeşil & Projeleri, 2019).

GREEN PROMOTION

The source of product information is the promotion. Companies must follow environmentally friendly principles during their promotional activities for example use of digital platforms for the promotion rather than using paper form of promotion, (Pitaloka et al., 2023). Green promotion means organizing the various promotional tools such as website, presentations, marketing materials, presentations, signatures etc by keeping the focus on planet, people and profits, (Sarbjit & Sanjay, 2017). It is a way through which firm creates awareness and share the information about their services or products to the target markets through various tools like personal selling, direct marketing, interactive marketing or the internet marketing, (Attih, 2025).

The companies should exclusively promote their sustainable products and services, as it is an important element of any business activity. Further sustainability values followed by the firm should be clearly visible to the people or to the society or world at large depending upon type of the organization. For example, if an organization sell products globally than such organization must bring awareness about its sustainable practices globally. In addition to it organizations who do not follow such practices and even provide unfavourable working environment for their employees must not claim that they follow sustainable practices. It may impact the company's reputation and its credibility, (Goyal & Pahwa, 2018).

Researchers have also discussed about different type of green promotions, the one that addresses the biophysical environment relationship with the service/product, the one that promote the green lifestyle by showcasing the green characteristics of the product or service and the one that promotes environmental brand image, highlighting how much environment friendly a brand is, (Goyal & Pahwa, 2018)

Marketing communication or the promotion is considered most controversial in comparison to the other elements of green marketing mix. Sometimes companies just advertise or claim that their product is green but in actual it is reverse, this is called as greenwashing. This highlights the major reason for doubting the green products by the consumers. They become sceptical that whether the product is genuinely green or not. So, it is very important for the companies to follow the certain guidelines or rules firstly make sure that the product characteristics that is advertised must have significant environmental impact. Secondly, the characteristics must be clearly state how it relates to the green product. Thirdly it must provide clear information to the consumer so that they can make their own judgments as well compare them with other. Fourthly the advertising tool must be chosen carefully, so that consumers can understand the message provided by organization clearly, (Bednárik, 2000).

Moving forward there are certain points that need to be care taken off by the companies with respect to the green promotion like providing detail information about green products where the customers are not too much well versed with this concept as well as where customers feel difficult to understand. Avoid unsolicited emails, as it reduces the trust of the customers. Companies promoting product through sponsorship or through the sales promotion must always use the trusted channels for that, moreover public relations is a more trustworthy tool. Further using innovative strategies for promoting green products or services motivates customers to purchase these products, (Bednárik, 2000).

Green advertising refers to promotional content that directly or indirectly highlights the connection between a product or service and the natural environment. It may encourage an eco-friendly lifestyle, with or without focusing on a specific offering, and often aims to portray the company as environmentally responsible, (Samarasinghe, 2012). Green promotion focuses on creating messages devised to resonate with the values and ambitions of environmentally aware consumers, (Mahmoud, 2018). Public relations, direct marketing, advertising, on-site promotions etc are the various methods used by the organizations for the promotion of the products and the services. Green marketers who are smart will give importance to the sustainable communication tools, (Sawant, 2020).

Green promotion utilizes marketing techniques that highlight the environmental advantages and sustainable features of a company's products, services, or overall operations. It focuses on conveying to customers and stakeholders how the organization's offerings support environmental protection and sustainability objectives. This approach not only appeals to eco-conscious consumers but also fosters trust and long-term loyalty by showcasing the company's genuine dedication to sustainability, (Alapa, Ugbe, & Alakali, 2024).

Organizations may include number of methods to communicate about the green features of the products as well as service they offer. It will help the organization to form the green brand image, that will help the consumers in their green purchase decision, (Sen, 2024).

MARKETING ANALYTICS

In the area of marketing, the research which is primary in nature focused on the qualitative or the quantitate methods to provide the insights about the customers. With the rise and growth in technology, especially data mining, marketing analytics holds an important role and becomes essential tool and must be looked an important part of the marketing research. Furthermore, marketers with the help of analytics can easily analyze the customer data and understand the customers at the different touch points throughout their association with the business, (Hauser, 2007).

Marketing analytics is becoming very prevalent among different organizations in the digital corporate world of the 21st century, (Matharu, 2023). Marketing Analytics is an evolving discipline within marketing science, expanding into a specialized sphere of competence comparable to other sub-fields of marketing, aligned under the wider conceptual umbrella of data science and formed by the developments of the 4th Industrial Revolution, (Dar, Rana, & Nasir, 2023). With the growth of digitization and the surge of data, the unification of Marketing Analytics and Artificial Intelligence (AI) has become a transformative factor, reforming strategic decision-making in marketing, (Bella, 2024). AI and human intelligence collaboration brings and important benefit to the domain of marketing, (Petrescu & Krishen, 2023). Marketing analytics is one of the major potential drivers of the competitive advantage, (Cao, Tian, & Blankson, 2022). Market -sensing capability of the firm is reflected in how they use marketing analytics (Cao, Duan, & El Banna, 2019; Benoit, Lessmann, & Verbeke, 2020).

Marketing analytics is defined as the systematic process of collecting, measuring, analyzing, and interpreting data to guide marketing strategies and improve organizational outcomes. In the present era, marked by the rising importance of digitalization and data-driven methods, marketing analytics has become an important tool for companies of different scales and across varied sectors. Marketing analytics has become a very crucial tool for the businesses or marketers to gain a comprehensive understanding of customer behaviour, consumer preferences, and latest trends. Due to the rapid advancements in the technology, it is crucial for marketers to remain updated on latest technologies like as artificial intelligence (AI), virtual reality (VR), machine learning (ML), and augmented reality (AR), to effectively utilize marketing analytics and enhance business growth. To engage customers meaningfully and to properly design targeted marketing strategies the marketers must give importance to social media, personalization and predictive analytics, (Nan, Yuping, & Kongjue, 2024).

IMPLEMENTATION OF MARKETING ANALYTICS ON THE ELEMENTS OF GREEN MARKETING MIX

With the rise of digitalization, there is a requirement of rising need for tracking consumer behaviour data for example online clickstreams etc or offline sensors for movement tracking of shoppers, (Kakatkar & Spann, 2019). A holistic approach should be considered for marketing analytics. The various domain areas which can be benefitted from marketing analytics are B2B, consumer behaviour and innovation. They have also highlighted the most widely applied research methods i.e. regression, modelling, semantic analysis, social network analysis and data mining, (Iacobucci et al. 2019). In relation with marketing, business analytics combines customer related data with the market, quantitative analysis, technology and various models related to computers to equip managers with relevant information for better decision making.

One of the most important area that is covered is relationship management specially with the customers, (Chahal, Jyoti, & Wirtz, 2018).

Marketing information and data plays a crucial role for the successful implementation of green marketing principles in an organization, effective methods in this area are strongly associated with the extensive usage of latest tools and technologies for instance big data. Where big data means the collection of various kinds of information that might be unstructured, structured or semi-structured, heterogeneous and complex, (Andronie et al., 2019).

Digital green accounting practical and scientific issue was discussed, from the prospective of digital green accounting, with the relation to marketing mix, associated with e-business organizations in favor of sustainable development. The outcome stated that digitalisation transforms the marketing mix, (Abdullaeva. et al., 2025)

Through the rising dependence on digital platforms, it has become highly important for the organizations to utilize marketing analytics, which also highlights the importance of adopting it by the businesses for staying competitive and for attaining the success, (Erwin et al., 2021; Meisaroh et al., 2022). In the digital world, marketing analytics is an important part of companies digital marketing strategy. Social media, email, mobile devices are some of the digital channels use in digital marketing for promoting services or products, (Hidayati et al., 2024).

Reaching wider audience, social media marketing, search engine optimization (SEO), engaging with customers, pay-per-click (PPC) advertising, content marketing, email marketing, engaging customers in real time, etc are significant contribution of digital marketing, (Piranda et al., 2022); Khanom, 2023; Elizabet, 2022).

Measuring campaign effectiveness, market trends, optimizing strategies, making data driven decisions, providing valuable insights, customer likings etc are some of its benefits. Furthermore, various metrics as customer engagement, website traffic, conversion rates can be used for deeper understanding of consumer behaviour. With the help of these metrics' companies can identify their specific target audience, segments, drive engagement etc and assign resources accordingly leading to better outcome for the business, (Hidayati et al., 2024; Erwin et al., 2021; Meisaroh et al., 2022).

According to Raichand (2024), some of the strategies that can be used for sustainable digital marketing are:

a) Optimizing digital advertisement expenditure by targeting the right audience. Through higher conversion opportunities, it helps in lowering the carbon footprint and expanding the ROI.
b) Companies can brand themselves as a sustainable brand, they can share their journey of transformation, discussing their challenges and achievements. Transparency will help in gaining trust and confidence of the audience.

c) Exploiting social media platforms for building brand sustainability message. Companies may use videos, stories, posts etc for raising environmental issues awareness and the methods of sustainability.
d) Organizations can also take advantage of content marketing. They can make content related to eco-friendly habits, build community. It helps in deepening the connections between the company and its audience.
e) The website designing and hosting should also be eco-friendly. Choosing services which are based on renewable energy, efficient coding etc.

CHALLENGES AND OPPORTUNITIES IN THE GREEN MARKETING MIX

The green marketing practices has created both challenges and opportunities for the organizations or the businesses. Some companies are entering into having a separate product line for the green products and some still are ignoring, (Mishra & Sharma, 2014).

Challenges

a) Firstly, companies are boasting that they are going green, but the reality is little far from that. There is no proper standardization to verify these claims. Further there is a requirement for having the control board for licencing, labelling etc. So, there is a necessity of standardization (M. K., Pandey, N., & Sajid, R., 2015).
b) Looking towards the Indian scenario, urban buyers are more aware about the green products, and out of them few are ready to pay higher prices for the green products, on the other side it's a new concept for most people. It creates the need for educating about environment issues to the masses. Further, the process may require more time and effort by the companies, (Pandey & Sajid, 2015).
c) The organizations must face lots of issues in implementing and adopting green strategies. The common challenges associated with it are the expenditures companies must incur in procuring green technology, green power, Research and development programmes etc, making it difficult for the companies to adopt greener practices. As the companies must bear more cost, it also impacts the profits of the companies, (Sawant, 2020). Cost is also incurred with various digital marketing practices, constant learning and adaptation, as well as on the eco-friendly technologies, upgrading of technology also requires lot of investment, (Raichand, 2024). As green marketing comprises of green energy, green technology, green services, green products or energy marketing etc, so

the companies adopting green marketing must spend ample amount of money on research and development resulting in cost overruns, (Hooda & Jaggarwal, 2021).

d) In a rush to become green, sometimes organization make false claims, these claims may damage the brand image, (Raichand, 2024).
e) The companies also must avoid the problem or the issue of green myopia, offering green product is not suffice, there should be customer satisfaction also, (Pandey, & Sajid, 2015).
f) The major challenge for humanity is to explore sustainable ways for producing, consuming and living, (Peattie, 2016).
g) There is no proper authentication of claims made by the marketers. Consumers need to be educated. The shareholders have to keep patience and trust on green movement. Motivating the customers to buy the green products, even at premium price, (Shetty, 2019).
h) AR (Augmented Reality) and VR (Virtual Reality) in marketing analytics, both can create an interactive and immersive experience for their customers and make them interested in the brand, (Nan, Yuping, & Kongjue 2024; Rane, Choudhary, & Rane, 2023).
i) AR and VR also help in visualizations of the products, helping customers to interact with the product in the virtual environment, it also helps in better customer experience, (Nan, Yuping, & Kongjue, 2024)

Opportunities

a) Organizations who deal with green products have competitive advantage in comparison to other firms who do not deal with eco-friendly products. Companies have to build innovative products using green technology, material etc, which sets them apart from their competition, (Polonsky, 2008; Raichand, 2024).
b) It helps in the long-term growth of the organizations as well as it increases the profitability of the organizations through subsidies by Government for promoting greener products, (Sarbjit & Sanjay, 2017).
c) It helps the companies in fulfilling their corporate social responsibility, (Sarbjit & Sanjay, 2017).
d) Firms may develop an effective process for production because of changes in raw material or using better energy saving methods etc which may help companies to enhance their efficiency, increase their profits etc., (Lal, 2015).
e) As per Raichand (2024) companies by truly committing towards sustainability may strengthen their relationships with their customers. It helps in building customer loyalty and reaching wider audience.

f) It helps in building as well improving the brand image significantly, further it also acts as a differentiator in this competitive industry, (Raichand, 2024).

MANAGERIAL IMPLICATIONS

This chapter is very beneficial for the academicians, researchers, students and the industries. It highlights the importance of adopting the greener practices in the organizations, it also talks about the various ways in which companies can adopt the greener practices, further it highlights about various digital marketing metrics that can be used to analyze the impact of green practices on the sales, profits etc. Finally at the end the chapter throws lights on the various challenges of adopting greener practices, as well as the opportunities for adopting the greener practices. Companies looking the various challenges can modify their strategies accordingly.

Conclusion

Green marketing includes those activities that are not harmful to the environment. It focuses on designing the services and facilities that help in fulfilling the needs as well as the desire of the consumer without harming or affecting the environment. There is a significant impact of green marketing on the image of the brand, that impact consumer decision-making behaviour It is also said that there is a mismatch between the consumer perceptions and the market perception, due to competitive pressures. Though consumers are motivated to buy environmentally friendly products but lack of proper knowledge of environmental issues among consumers may impact the consumer decision making and companies also need to promote the product in different manner. (Mark Bekweri EDEH., 2020).

Although consumers are increasingly aware of the advantages of eco-friendly products, green marketing remains a relatively new idea. There is a pressing need to educate the public about environmental challenges and the benefits of sustainable consumption. Reaching the broader population with green initiatives will require significant time, effort, and consistent awareness campaigns, (Sarbjit & Sanjay, 2017).

As green marketing focuses on environmental responsibility, it must also consider economic viability. Marketers need to understand the broader implications of adopting green strategies. It's a misconception to assume that consumers are indifferent to environmental concerns or unwilling to pay more for sustainable products. In reality, businesses must innovate to enhance product performance, build stronger customer loyalty, and justify premium pricing. As green marketing is still developing, further research is essential to unlock its full potential. Both individual consumers and industrial clients play a vital role by encouraging companies to integrate environmental values into their operations, ultimately helping to minimize their ecological footprint, (Lal, 2015).

Implementing sustainable strategies by the firm is not just a fad, it's a positive change, it is the way the companies interact with their customers, it's the way companies take care of the planet. It's not just talking about sustainable marketing strategies it's all about implementing those strategies by the organizations. This just not only give opportunity to the brand to stand out, but it also helps in building the strong connections with their customers and make them loyal towards the brand. Green marketing is important and will become more important in the future. Brand who has adopted greener practices may perform far better in future in comparison to those who are still thinking to embrace greener practices in their business activities. There will be more opportunities for the brand who are more open to greener practices creating threats for their competitors. Adoption of green marketing practices help the companies to achieve the competitive advantage. Businesses and investors must approach green marketing as a long-term strategic investment. Since this is still an emerging field, it will naturally face a gradual adoption curve. Patience and sustained commitment are essential for realizing its long-term benefits. So, adopting sustainable marketing strategies are not just good for the companies or humans it is good for the planet.

REFERENCES

Abdullaeva, F., Kutusheva, N., Zholdubaeva, S., & Tubalets, A. A. (2025). Digital Green Accounting in the Marketing Mix of E-Business for Its Sustainable Development. In *Green Accounting and Digitalization of E-Business for Sustainable Development* (pp. 91–112). IGI Global Scientific Publishing. DOI: 10.4018/979-8-3693-8362-9.ch006

Ahmed, I. M., Abd Eljalil, S. G., Mohamed, A. M., & Abouzied, F. Y. (2025). Applying Tourist Green Marketing Mix Strategy: Challenges and Opportunities (Applied in Luxor City). *International Journal of Tourism and Hospitality Management, 8*(1), 108–124.

Ahmed, R. R., Streimikiene, D., Qadir, H., & Streimikis, J. (2023). Effect of green marketing mix, green customer value, and attitude on green purchase intention: Evidence from the USA. *Environmental Science and Pollution Research International, 30*(5), 11473–11495. DOI: 10.1007/s11356-022-22944-7

Alabo, R. L., & Anyasor, O. M. (2021). Green marketing and sustainability of breweries in Nigeria. *European Journal of Business and Innovation Research, 9*(1), 1–30.

Alapa, F. I., Ugbe, A. U., & Alakali, T. F. (2024). Green marketing mix (GMX) and sustainability of small and medium scale manufacturing enterprises in North Central Nigeria. International Journal of Entrepreneurship. *Technology and Innovation, 1*(2).

Andronie, M., Gârdan, D. A., Dumitru, I., Gârdan, I. P., Andronie, I. E., & U ă, C. (2019). Integrating the principles of green marketing by using big data. Good practices. *Amfiteatru Economic, 21*(50), 258–269. DOI: 10.24818/EA/2019/50/258

Astuti, R., Deoranto, P., Wicaksono, M. L. A., & Nazzal, A. (2021, April). Green marketing mix: An example of its influences on purchasing decision. [). IOP Publishing.]. *IOP Conference Series. Earth and Environmental Science, 733*(1), 012064. DOI: 10.1088/1755-1315/733/1/012064

Attih, O. B. (2025). GREEN MARKETING MIX AND CONSUMER PURCHASE DECISION OFPRODUCTS IN UYO METROPOLIS, AKWA IBOM STATE, NIGERIA. *INTERNATIONAL JOURNAL OF BUSINESS EDUCATION AND MANAGEMENT STUDIES, 2*(1), 21–36.

Bednárik, L. S. J. (2000). Green marketing as opportunity for innovation. of Innovations, 2(3), 494.

Bella, K. M. J. (2024). A study on marketing analytics and artificial intelligence. *SELP Journal of Social Science-A Blind Review & Refereed Quarterly Journal, XV*(55), 53–57.

Benoit, D. F., Lessmann, S., & Verbeke, W. (2020). On realising the utopian potential of big data analytics for maximising return on marketing investments. *Journal of Marketing Management, 36*(3-4), 233–247. DOI: 10.1080/0267257X.2020.1739446

Cao, G., Duan, Y., & El Banna, A. (2019). A dynamic capability view of marketing analytics: Evidence from UK firms. *Industrial Marketing Management, 76*, 72–83. DOI: 10.1016/j.indmarman.2018.08.002

Cao, G., Tian, N., & Blankson, C. (2022). Big data, marketing analytics, and firm marketing capabilities. *Journal of Computer Information Systems, 62*(3), 442–451. DOI: 10.1080/08874417.2020.1842270

Chahal, H., Jyoti, J., & Wirtz, J. (2018). Business analytics: Concept and applications. In *Understanding the Role of Business Analytics: Some Applications* (pp. 1–8). Springer Singapore.

Chygryn, O., Lyulyov, O., Pimonenko, T., & Illiashenko, A. (2024, January). Companies' green competitiveness: Justifying the role of marketing determinants. In AIP Conference Proceedings (Vol. 3033, No. 1). AIP Publishing.

Dangelico, R. M., & Vocalelli, D. (2017). "Green Marketing": An analysis of definitions, strategy steps, and tools through a systematic review of the literature. *Journal of Cleaner Production, 165*, 1263–1279. DOI: 10.1016/j.jclepro.2017.07.184

Dar, I., Rana, A. U. R., & Nasir, I. (2023). Marketing analytics theoretical development and trends: The journey of defining marketing analytics. *Research Journal for Societal Issues, 5*(1), 54–70. DOI: 10.56976/rjsi.v5i1.57

Dauda, M., Mustapha, B., & Sani, S. A. (2024). Green marketing strategy and its effect on the performance of small and medium business enterprises (SMEs). International Journal of Accounting. *Finance and Administrative Research, 1*(2), 229–239.

Ebhote, O., & Izedonmi, F. O. I. (2021). Green marketing practices in Nigeria: Challenges and prospect. *Nigerian Journal of Business Administration, 15*(1 & 2), 112–125.

Elizabet, M. (2022). DIGITAL MARKETING TRAINING: Strategies to Increase Sales in Indonesian Market Place. *Jurnal PKM Manajemen Bisnis, 2*(2), 63–72. DOI: 10.37481/pkmb.v2i2.484

Erwin, E., Suade, Y. K. M., & Poernomo, W. (2021). Analyzing Digital Marketing, Green Marketing, Networking and Product Innovation on Sustainability Business Performance, Silk Cluster in Polewali-Mandar, West Sulawesi. International Journal of Economics [IJEBAR]. *Business and Accounting Research*, *5*(3), 2850–2871.

Fan, H., & Zeng, L. (2011). Implementation of green marketing strategy in China: A study of the green food industry.

Firdiansyah, R., Mohamed, M., Yusoff Yusliza, M., Saputra, J., & Muhammad, Z. (2021, March). A Review of green marketing strategy literature: Mini review approach. In Proceedings of the 11th Annual International Conference on Industrial Engineering and Operations Management Singapore: IEOM Society International (Vol. 21030, pp. 5092-5108).

Giwa-Amu, O. (2022). Effect of green marketing on customer satisfaction. Sapientia Global Journal of Arts. *Humanities and Development Studies*, *5*(2), 77–86.

Goyal, M., & Pahwa, M. S. (2018). Green Marketing Mix A Model towards Sustainability. *International Journal on Computer Science and Engineering*, *6*(09), 23–27.

Hauser, W. J. (2007). Marketing analytics: the evolution of marketing research in the twenty-first century. *Direct marketing: an international journal, 1*(1), 38-54.

Hidayati, A., Susanti, E., Jamalong, A., Ginting, D., Suwanto, W., & Arifin, A. (2024). Marketing analytics in the era of digital-based marketing strategy. *Jurnal Ilmiah Ilmu Terapan Universitas Jambi*, *8*(1), 61–75. DOI: 10.22437/jiituj.v8i1.31908

. Hooda, G., & Jaggarwal, S. (2021). Green marketing in India: Challenges and opportunities. Anvesak–UGC Care listed Journal, ISSN, 0378-4566.

Iacobucci, D., Petrescu, M., Krishen, A., & Bendixen, M. (2019). The state of marketing analytics in research and practice. *Journal of Marketing Analytics*, *7*(3), 152–181. DOI: 10.1057/s41270-019-00059-2

Kakatkar, C., & Spann, M. (2019). Marketing analytics using anonymized and fragmented tracking data. *International Journal of Research in Marketing*, *36*(1), 117–136. DOI: 10.1016/j.ijresmar.2018.10.001

Kaur, S., Kumar, A., Sharma, P., Gulati, G., & Kaur, A. (2025). Green Consumer Behaviour: A Bibliometric Analysis and Insight on the Factors Influencing Green Product Adoption. In Green Economics and Strategies for Business Sustainability (pp. 289-314). IGI Global Scientific Publishing. DOI: 10.4018/979-8-3693-8949-2.ch012

Khan, S. Z., Yang, Q., & Waheed, A. (2019). Investment in intangible resources and capabilities spurs sustainable competitive advantage and firm performance. *Corporate Social Responsibility and Environmental Management*, *26*(2), 285–295. DOI: 10.1002/csr.1678

Khanom, M. T. (2023). Using social media marketing in the digital era: A necessity or a choice. International Journal of Research in Business and Social Science (2147-4478), 12(3), 88–98.

Kumar Phookan, N., Sarmah, B., & Sarika Devi, T. (2020). Exploring determinants of green marketing practices in food service industry to attain environmental sustainability: A conceptual analysis. In Proceedings of the 5th International Conference on Computers & Management Skills (ICCM 2019)| North Eastern Regional Institute of Science & Technology (NERIST), Nirjuli, Arunachal Pradesh, India. DOI: 10.2139/ssrn.3517420

Lal, B. S. (2015). Green marketing: Opportunities and issues. *International Journal (Toronto, Ont.)*.

Mahmoud, M. A., Seidu, A. S., Tweneboah-Koduah, E. Y., & Ahmed, A. S. (2024). Green marketing mix and repurchase intention: The role of green knowledge. *African Journal of Economic and Management Studies*, *15*(3), 501–518. DOI: 10.1108/AJEMS-04-2023-0137

Mahmoud, T. O. (2018). Impact of green marketing mix on purchase intention. International Journal of Advanced and applied sciences, 5(2), 127-135.

Mahmoud, T. O. (2019). Green marketing: A marketing mix concept. International Journal of Electrical. *Electronics and Computers*, *4*(1), 20–26.

Mark Bekweri, E. D. E. H. (2020). Green Marketing As A Marketing Tool And Consumer Purchase Behavior. Journal of Global Social Sciences, Volume 1, Number 3, 37-54/ISSN: Print 2735-9344 [Nile University of Nigeria]. *Online (Bergheim)*, •••, 2735–9352.

Matharu, M. (2023). Marketing Analytics Concept, Applications, Opportunities, and Challenges Ahead. *Data Analytics and Business Intelligence*, 170-186.

Meisaroh, F., Sudarmiatin, S., & Hermawan, A. (2022). the Importance of Digital Marketing Strategy Emos Marketplace During Covid-19 Pandemic Period. *International Conference of Business and Social Sciences*, 430–439.

Mishra, P., & Sharma, P. (2014). Green marketing: Challenges and opportunities for business. BVIMR Management Edge, 7(1).

Nan, Z., Yuping, C., & Kongjue, Z. (2024). The Future of Marketing Analytics: Trends and Emerging Technologies. [IJABMR]. *International Journal of Advances in Business and Management Research, 1*(3), 23–32. DOI: 10.62674/ijabmr.2024.v1i03.003

Nguyen, K. M., Dinh, L. T. K., Ngo, T. T., Phan, T. T. T., Nguyen, P. D. H., Tran, H. P. M., & Nguyen, N. T. (2025). The Impact of 7Ps Green Marketing Mix on Customer Commitment and Willingness to Pay a Premium Price: Evidence from Vietnam. *Journal of Promotion Management, 31*(4), 1–42. DOI: 10.1080/10496491.2025.2484712

Nguyen-Viet, B. (2023). The impact of green marketing mix elements on green customer based brand equity in an emerging market. *Asia-Pacific Journal of Business Administration, 15*(1), 96–116. DOI: 10.1108/APJBA-08-2021-0398

Nwankwo, C., & Kanyangale, M. (2023). Effect of sustainable marketing strategies on consumption of beer brands in Anambra State. Acta Universitatis Danubius. Œconomica, 19(2), 188-205.

Ofori, D. (2021). Opportunities and challenges of green marketing. Green Marketing in Emerging Markets: Strategic and Operational Perspectives, 251-276.

Papadas, K. (2021). Green Marketing Strategy. In *Industry, Innovation and Infrastructure* (pp. 450–461). Springer International Publishing. DOI: 10.1007/978-3-319-95873-6_121

Papadas, K. K., Avlonitis, G. J., Carrigan, M., & Piha, L. (2019). The interplay of strategic and internal green marketing orientation on competitive advantage. *Journal of Business Research, 104*, 632–643. DOI: 10.1016/j.jbusres.2018.07.009

Peattie, K. (2016). Green marketing. In *The marketing book* (pp. 573–597). Routledge.

Petrescu, M., & Krishen, A. S. (2023). Hybrid intelligence: Human–AI collaboration in marketing analytics. *Journal of Marketing Analytics, 11*(3), 263–274. DOI: 10.1057/s41270-023-00245-3

Piranda, D. R., Sinaga, D. Z., Putri, E. E., & Asahan, U. (2022). Online Marketing Strategy in Facebook. Journal of Humanities. *Social Sciences and Business, 1*(2), 79–86.

Pitaloka, E., Suyoto, Y. T., Febriyanti, A., & Sukarno, D. A. R. (2023). The Impact of Green Marketing Mix on Brand Image of Unilever Indonesia. *KnE Social Sciences*, 65-76.

Polonsky, M. J. (2008). An introduction to green marketing. *Global Environment: Problems and Policies, 2*(1), 1–10.

Purwanti, I., Abadi, M. D., Suyanto, U. Y., & Ahmad, S. T. I. E. K. (2019). Green marketing: Strategy for gaining sustainable competitive advantage in industry 4.0. *Manajemen Bisnis*, *9*(2), 176–193. DOI: 10.22219/jmb.v9i2.10039

Raichand, M. (2024, February 23). Understanding green marketing: How to include sustainability in your marketing strategy. LinkedIn. https://www.linkedin.com/pulse/understanding-green-marketing-how-include-your-mehek-raichand-auhbf/

Rane, N., Choudhary, S., & Rane, J. (2023). Metaverse marketing strategies: enhancing customer experience and analysing consumer behaviour through leading-edge Metaverse technologies, platforms, and models. *Platforms, and Models (November 3, 2023).*

Risteska, L. (2023). Green marketing mix and promotion of responsible consumption. *Knowledge-International Journal*, *57*(1), 121–126.

. Samarasinghe, R. (2012). Is social psychological model sufficient: empirical research gaps for understanding green consumer attitudinal behavior. International journal of advanced research in management and social sciences, 1(4), 28-54.

Sarbjit, K., & Sanjay, K. B. (2017). Opportunities and challenges of Green Marketing in India.

Sawant, A. V. (2020). challenge And Opportunities Before Green Marketing in India. Journal of marketing Economics, 7(4).

Sen, T. (2024)The influence of green marketing mix on green purchase decision: the mediating role of green brand image.

Sharma, Y. (2011). changing consumer behaviour with respect to green. international journal of multidisciplinary research, 152- 162.

Shetty, N. (2019). Green Marketing Strategy: Opportunities And Challenges For Marketers. Advance and Innovative Research, 421.

Silaban, S. A., Inulingga, S., & Fadli, M. (2021). The effect of green marketing on purchase decisions and brand image as intervening variables: Case Study of Pt. Sari Coffee Indonesia- Starbucks Focal Point Medan. *International Journal of Research and Review*, *8*(1), 159–169.

Singh, S. (2012). Green marketing: Challenges and strategy in the changing scenario. *International Journal of Advanced Research in Management and Social Sciences*, *1*(6), 164–172.

Sohail, M. S. (2017). Green marketing strategies: How do they influence consumer-based brand equity? *Journal for Global Business Advancement*, *10*(3), 229–243. DOI: 10.1504/JGBA.2017.084607

Solaiman, M., Osman, A., & Halim, M. S. B. A. (2015). Green marketing: A marketing mix point of view. *International Journal of Business and Technopreneurship*, *5*(1), 87–98.

Sultan, M. F., Tunio, M. N., Shaukat, G., & Asim, M. (2023). Green Marketing Mix From the Perspective of Service Sector: Leveraging Marketing of Services With Green-SIVA Marketing Mix Elements. In Entrepreneurship and Green Finance Practices (pp. 267-283). Emerald Publishing Limited.

Tsai, P. H., Lin, G. Y., Zheng, Y. L., Chen, Y. C., Chen, P. Z., & Su, Z. C. (2020). Exploring the effect of Starbucks' green marketing on consumers' purchase decisions from consumers' perspective. *Journal of Retailing and Consumer Services*, *56*, 102162. DOI: 10.1016/j.jretconser.2020.102162

Türk, B., & Erciş, A. (2017). Elements Of The Sustainable Marketing Mix. *Studies on Sustainability Research*, 105-125.

Vijai, C., & Anitha, P. (2020). The importance of green marketing. *International Journal of Future Generation Communication and Networking*, *13*(3), 4137–4142.

Welford, R. (2000). *Hijacking Environmentalism*. Earthscan.

Yeşil, Y. P. K. İ. K., & Projeleri, L. (2019). Global Green Logistics Projects In Green Marketing Mix. *The Sciences*, *5*(21), 1269–1288.

Compilation of References

Abbasi, M. T., Alesheikh, A. A., Lotfata, A., & Azizi, Z. (2025). Hybrid graph convolutional networks for air quality prediction: A systematic review of foundations, challenges, and opportunities. *International Journal of Environmental Science and Technology*, *22*(13), 1–36. https://link.springer.com/article/10.1007/s13762-025-06599-8. DOI: 10.1007/s13762-025-06599-8

Abdullaeva, F., Kutusheva, N., Zholdubaeva, S., & Tubalets, A. A. (2025). Digital Green Accounting in the Marketing Mix of E-Business for Its Sustainable Development. In *Green Accounting and Digitalization of E-Business for Sustainable Development* (pp. 91–112). IGI Global Scientific Publishing. DOI: 10.4018/979-8-3693-8362-9.ch006

Aboltins, R., & Blumberga, D. (2019). Key factors for successful implementation of energy efficiency policy instruments: A theoretical study and the case of Latvia. *Environmental and Climate Technologies*, *23*(2), 187–206. DOI: 10.2478/rtuect-2019-0063

Abou Taleb, M., Gibson, B., & Hovey, M. (2015). Fifty years of Sustainability Accounting: Does accounting for income in business sustainability really exist?. *International Journal of Accounting and Financial Reporting*.

Abram, S., Atkins, E., Dietzel, A., Jenkins, K., Kiamba, L., Kirshner, J., Kreienkamp, J., Parkhill, K., Pegram, T., & Santos Ayllón, L. M. (2022). Just Transition: A whole-systems approach to decarbonisation. *Climate Policy*, *22*(8), 1033–1049. DOI: 10.1080/14693062.2022.2108365

Adams, C. A., & Frost, G. R. (2008)., Integrating sustainability reporting into management practices. [). No longer published by Elsevier.]. *Accounting Forum*, *32*(4), 288–302. DOI: 10.1016/j.accfor.2008.05.002

ADMIE. (2024). Tender for the Corinth-Cos electricity interconnection begins, Independent Electricity Transmission System Operator, https://www.admie.gr/kentro-typoy/deltia-typoy/xekina-o-diagonismos-gia-tin-ilektriki-diasyndesi-korinthoy-ko

ADMIE. (2025a), Ten – year Network Development Plan, Independent Electricity Transmission System Operator, https://www.admie.gr/en/grid/development/ten-year-development-plan

ADMIE. (2025b), Electrical Interconnection between Crete and Attica in operation mode, https://www.admie.gr/kentro-typoy/deltia-typoy/se-leitoyrgia-i-ilektriki-diasyndesi-kritis-attikis

Adun, H., Ishaku, H. P., Jazayeri, M., Dagbasi, M., Olusola, B., Okoye, T., & Dike, G. C. (2022). Decarbonization of EU energy sector: Techno-feasibility analysis of 100% renewables by 2050 in Cyprus. *Clean Technologies and Environmental Policy*, *24*(9), 2801–2824. DOI: 10.1007/s10098-022-02356-4

Afanasiev, A., Kandinskaia, O., Eminidou, S., & Sousanis, A. (2025). Hydrogen Valley in Cyprus: Insights and Strategies for Citizen Engagement. *Sustainability (Basel)*, *17*(2), 804. DOI: 10.3390/su17020804

Agathokleous, R. A., & Kalogirou, S. A. (2021). PV roofs as the first step towards 100% RES electricity production for Mediterranean islands: The case of Cyprus. *Smart Energy*, *4*, 100053. DOI: 10.1016/j.segy.2021.100053

Agency, I. E. (2021). *Net Zero by 2050: A Roadmap for the Global Energy Sector*. https://www.iea.org/reports/net-zero-by-2050

Aggelopoulos, E., & Lampropoulos, I. (2023). The effect of organic growth and acquisitions on performance of retailing networks in an alternating external operating environment. *Benchmarking*, *31*(1), 1–28. DOI: 10.1108/BIJ-02-2022-0087

Aggelopoulos, E., Lampropoulos, I., & Paraschi, E. P. (2023). An empirical analysis of evaluating the operational and economic efficiency of retail stores under an adverse operating environment. *MDE. Managerial and Decision Economics*, *44*(5), 2713–2741. DOI: 10.1002/mde.3844

Ahmad, M. (2024). Community-Driven Finance Unveiling the Dynamics of DeFi Governance. *Advances in Finance, Accounting, and Economics*, •••, 33–48. DOI: 10.4018/979-8-3693-3346-4.ch003

Ahmed, I. M., Abd Eljalil, S. G., Mohamed, A. M., & Abouzied, F. Y. (2025). Applying Tourist Green Marketing Mix Strategy: Challenges and Opportunities (Applied in Luxor City). *International Journal of Tourism and Hospitality Management*, *8*(1), 108–124.

Ahmed, R. R., Streimikiene, D., Qadir, H., & Streimikis, J. (2023). Effect of green marketing mix, green customer value, and attitude on green purchase intention: Evidence from the USA. *Environmental Science and Pollution Research International*, *30*(5), 11473–11495. DOI: 10.1007/s11356-022-22944-7

Alabo, R. L., & Anyasor, O. M. (2021). Green marketing and sustainability of breweries in Nigeria. *European Journal of Business and Innovation Research*, *9*(1), 1–30.

Alamgir, M., Campbell, M. J., Sloan, S., Goosem, M., Clements, G. R., Mahmoud, M. I., & Laurance, W. F. (2017). Economic, socio-political and environmental risks of road development in the tropics. *Current Biology*, *27*(20), R1130–R1140. DOI: 10.1016/j.cub.2017.08.067 PMID: 29065299

Alapa, F. I., Ugbe, A. U., & Alakali, T. F. (2024). Green marketing mix (GMX) and sustainability of small and medium scale manufacturing enterprises in North Central Nigeria. International Journal of Entrepreneurship. *Technology and Innovation*, *1*(2).

Alexander, S., & Floyd, J. (2020). The political economy of deep decarbonization: Tradable Energy Quotas for energy descent futures. *Energies*, *13*(17), 4304. DOI: 10.3390/en13174304

Alqasim, A. R. (2022). Using Regression Analysis for Predicting Energy Consumption in Dubai Police. https://repository.rit.edu/theses/11177/

Alsaigh, R., Mehmood, R., & Katib, I. (2022). AI Explainability and Governance in Smart Energy Systems: A Review. *ArXiv (Cornell University)*. https://doi.org// arxiv.2211.00069DOI: 10.48550

Alvis, A. D., Luce, C. H., & Istanbulluoglu, E. (2022). How does traffic affect erosion of unpaved forest roads? *Environmental Reviews*, *31*(1), 182–194. DOI: 10.1139/er-2022-0032

Ambu-Saidi, B., Fung, C. Y., Turner, K., & Lim, A. S. S. (2024). A critical review on training evaluation models: A search for future agenda. *Journal of Cognitive Sciences and Human Development.*, *10*(1). Advance online publication. DOI: 10.33736/jcshd.6336.2024

Andersson, M. B. Claudio; Morgan, Julian (2020). Climate change and the macro economy. In (Vol. 243). ISBN 978-92-899-4249-2: European Central Bank (ECB) Occasional Paper.

Andrei, M. (2023). *The role of industrial energy management in the transition toward sustainable energy systems: Exploring practices, knowledge dynamics and policy evaluation*. Linkopings Universitet.

Andronie, M., Gârdan, D. A., Dumitru, I., Gârdan, I. P., Andronie, I. E., & U ă, C. (2019). Integrating the principles of green marketing by using big data. Good practices. *Amfiteatru Economic, 21*(50), 258–269. DOI: 10.24818/EA/2019/50/258

Apostolopoulos, N., Makris, I., Apostolopoulos, S., & Dimitrakopoulos, P. (2024). Resilience of rural micro-businesses in an adverse entrepreneurial environment: adapting to the energy crisis. *Journal of Enterprising Communities: People and Places in the Global Economy*, (ahead-of-print).

Apostolopoulos, N., Makris, I., Apostolopoulos, S., & Liargovas, P. (2025). The Role of Rural Entrepreneurship in the Redevelopment of Rural Areas Previously Involved in Resource Extraction. In *Rural Entrepreneurship: Harvesting Ideas and Sowing New Seeds* (pp. 145-156). Emerald Publishing Limited. https://doi.org/DOI: 10.1108/S2040-724620250000020011

Apostolopoulos, N., Kakouris, A., Liargovas, P., Borisov, P., Radev, T., Apostolopoulos, S., Daskou, S., & Anastasopoulou, E. E. (2023). Just transition policies, power plant workers and green entrepreneurs in Greece, Cyprus and Bulgaria: Can education and retraining meet the challenge? *Sustainability (Basel), 15*(23), 16307. DOI: 10.3390/su152316307

Apostolopoulos, N., Liargovas, P., & Apostolopoulos, S. (2025). European Transnational Cooperation Programmes Involving the Academic Community and Their Contributions to Sustainable Entrepreneurship. In *Real-World Tools and Scenarios for Entrepreneurship Exploration* (pp. 1–24). IGI Global.

Apostolopoulos, N., Makris, I., Apostolopoulos, S., & Dimitrakopoulos, P. (2023). Resilience of rural micro-businesses in an adverse entrepreneurial environment: Adapting to the energy crisis. *Journal of Enterprising Communities: People and Places in the Global Economy, 18*(5), 1023–1044. DOI: 10.1108/JEC-08-2023-0144

Aravind, A. R., Santhi, G. B., Patil, S. T., P, S., Sharma, G., Dhamone, J., & Ragu Nathan, S. (2024). Blockchain Technology in Energy Markets: Enabling Peer-to-Peer Energy Trading. *E3S Web of Conferences, 591*, 06002. DOI: 10.1051/e3sconf/202459106002

Arcelay, I., Goti, A., Oyarbide-Zubillaga, A., Akyazi, T., Alberdi, E., & Garcia-Bringas, P. (2021). Definition of the Future Skills Needs of Job Profiles in the Renewable Energy Sector. *Energies, 14*(9), 2609. https://www.mdpi.com/1996-1073/14/9/2609. DOI: 10.3390/en14092609

Arjmand, A., Kiadaliri, H., Kazemnezhad, F., & Eshagh Nimvari, M. (2023). Effects of forest roads on vegetation biodiversity and soil characteristics in Hyrcanian forests. *Nordic Journal of Botany, 2023*(12), e04039. DOI: 10.1111/njb.04039

Armstrong, H. W., Giordano, B., Kizos, T., Macleod, C., Olsen, L. S., & Spilanis, I. (2012). The European Regional Development Fund and island regions: An evaluation of the 2000-06 and 2007-13 programs. *Island Studies Journal*, *7*(2), 177–198. DOI: 10.24043/isj.267

Arul, M., & Kareem, A. (2021). Applications of shapelet transform to time series classification of earthquake, wind and wave data. *Engineering Structures*, *228*, 111564. https://www.sciencedirect.com/science/article/pii/S0141029620341651. DOI: 10.1016/j.engstruct.2020.111564

Astuti, R., Deoranto, P., Wicaksono, M. L. A., & Nazzal, A. (2021, April). Green marketing mix: An example of its influences on purchasing decision. []. IOP Publishing.]. *IOP Conference Series. Earth and Environmental Science*, *733*(1), 012064. DOI: 10.1088/1755-1315/733/1/012064

Athens Chamber of Commerce and Industry, (2024). Island Fragmentation Fund: €5 billion investment in interconnections, renewable energy, water supply and reduction of electricity bills, Athens Chamber of Commerce and Industry (in Greek)

Atkinson, R. D. (2022). *Why America Should Compete to Win in Advanced Industries*.

Attih, O. B. (2025). GREEN MARKETING MIX AND CONSUMER PURCHASE DECISION OFPRODUCTS IN UYO METROPOLIS, AKWA IBOM STATE, NIGERIA. *INTERNATIONAL JOURNAL OF BUSINESS EDUCATION AND MANAGEMENT STUDIES*, *2*(1), 21–36.

Auffret, A. G., & Lindgren, E. (2020). Roadside diversity in relation to age and surrounding source habitat: Evidence for long time lags in valuable green infrastructure. *Ecological Solutions and Evidence*, *1*(1), e12005. DOI: 10.1002/2688-8319.12005

Aung, K. T., Razak, R. A., & Nazry, N. N. M. (2021). Establishing validity and reliability of semi-structured interview questionnaire in developing risk communication module: A pilot study. *Edunesia: Jurnal Ilmiah Pendidikan*, *2*(3), 600–606. DOI: 10.51276/edu.v2i3.177

Aurelle, D., Thomas, S., Albert, C., Bally, M., Bondeau, A., Boudouresque, C. F., Cahill, A. E., Carlotti, F., Chenuil, A., Cramer, W., Davi, H., De Jode, A., Ereskovsky, A., Farnet, A. M., Fernandez, C., Gauquelin, T., Mirleau, P., Monnet, A. C., Prévosto, B., & Fady, B. (2022). Biodiversity, climate change, and adaptation in the Mediterranean. *Ecosphere*, *13*(4), e3915. DOI: 10.1002/ecs2.3915

Ayyoob, A., & Sajeev, A. (2024b). Diversity, Equity, and Inclusion: Employees Perspectives on Sustainable Business Practices. *Routledge EBooks*, 348–367. DOI: 10.4324/9781003487180-27

Ayyoob, A., & Sajeev, A. (2024c). Navigating Sustainability: Assessing the Imperative of ESG Considerations in Achieving SDGs. *ESG Frameworks for Sustainable Business Practices*, 53–84. DOI: 10.4018/979-8-3693-3880-3.ch003

Ayyoob, A., & Sajeev, A. (2024d). Role of Social Entrepreneurship in Sustainable Development. *Advances in Business Strategy and Competitive Advantage Book Series*, 193–224. DOI: 10.4018/979-8-3693-4338-8.ch008

Ayyoob, A., & Sajeev, A. (2025). Resilient and Sustainable Organizations. IGI Global EBooks, 35–58. DOI: 10.4018/979-8-3373-0608-7.ch003

Ayyoob, A., & Sajeev, A. (2024a). Corporate Sustainability: The Role of Consumer Skepticism, Transparency, and Marketing in Greenwashing Practices. *Corporate Governance Insight*, *6*(2), 51–68. DOI: 10.58426/cgi.v6.i2.2024.51-68

Azhgaliyeva, D., Kapoor, A., & Liu, Y. (2020). Green bonds for financing renewable energy and energy efficiency in South-East Asia: A review of policies. *Journal of Sustainable Finance & Investment*, *10*(2), 113–140. DOI: 10.1080/20430795.2019.1704160

Babayomi, O. O., Dahoro, D. A., & Zhang, Z. (2022). Affordable clean energy transition in developing countries: Pathways and technologies. *iScience*, *25*(5).

Baker, S., & De Silva, L. (2021). Co-governance of renewable energy projects with Indigenous communities. *Energy Research & Social Science*, *79*, 102160. DOI: 10.1016/j.erss.2021.102160

Balcilar, M., Özkan, O., Usman, O., Saint Akadırı, S., & Zambrano-Monserrate, M. A. (2025). A global shift: How modern technologies are powering the energy transition in the face of climate change. *Journal of Environmental Management*, *384*, 125610. DOI: 10.1016/j.jenvman.2025.125610 PMID: 40315652

Bali Swain, R., Karimu, A., & Gråd, E. (2022). Sustainable development, renewable energy transformation and employment impact in the EU. *International Journal of Sustainable Development and World Ecology*, *29*(8), 695–708. DOI: 10.1080/13504509.2022.2078902

Ballis, A., Paravantis, J. A., & Moschovou, T. (2018). Assessing the tourism potential of the Greek islands of South Aegean. In *2018 9th International Conference on Information, Intelligence, Systems and Applications (IISA)* (pp. 1-7). IEEE. DOI: 10.1109/IISA.2018.8633653

Bañales, S. (2020). The enabling impact of digital technologies on distributed energy resources integration. *Journal of Renewable and Sustainable Energy*, *12*(4), 045301. DOI: 10.1063/5.0009282

Banerjee, S. B. (2019). Decolonizing sustainability. *Organization Studies*, *40*(3), 437–456. DOI: 10.1177/0170840618818992

Bargh, M. (2019). Indigenous peoples and renewable energy: Rights and recognition. *Indigenous Policy Journal*, *30*(2), 1–16.

Barone, G., Buonomano, A., Forzano, C., Giuzio, G. F., & Palombo, A. (2021). Increasing renewable energy penetration and energy independence of island communities: A novel dynamic simulation approach for energy, economic, and environmental analysis, and optimization. *Journal of Cleaner Production*, *311*, 127558. DOI: 10.1016/j.jclepro.2021.127558

Barta, C. (2008). *ENSCO: The First 20 Years*. Elsevier.

Barthelmie, R. J., & Pryor, S. C. (2021). Climate change mitigation potential of wind energy. *Climate (Basel)*, *9*(9), 136. DOI: 10.3390/cli9090136

Basavanna, Chavan, G., YM, D., & Verma, N. A. (2024). Smart Grid Using IOT. *International Journal of Advanced Research in Science Communication and Technology*, 580–587. DOI: 10.48175/IJARSCT-22769

Bashir, M. F., Pata, U. K., & Shahzad, L. (2025). Linking climate change, energy transition and renewable energy investments to combat energy security risks: Evidence from top energy consuming economies. *Energy*, *314*, 134175. DOI: 10.1016/j.energy.2024.134175

Bashmakov, I., Myshak, A., Bashmakov, V. A., Bashmakov, V. I., Borisov, K., Dzedzichek, M., Lunin, A., Lebedev, O., & Shishkina, T. (2023). Russian energy balance, energy efficiency, and energy-related GHG emission accounting system. *Energy Efficiency*, *16*(6), 67. DOI: 10.1007/s12053-023-10132-6

Batas-Bjelić, I., Rajaković, N., Ćosić, B., & Duić, N. (2015). A realistic EU vision of a lignite-based energy system in transition: Case study of Serbia. *Thermal Science*, *19*(2), 371–382. DOI: 10.2298/TSCI140613118B

Baumgartner, R. J. (2014). Managing corporate sustainability and CSR: A conceptual framework combining values, strategies and instruments contributing to sustainable development. *Corporate Social Responsibility and Environmental Management*, *21*(5), 258–271. DOI: 10.1002/csr.1336

Baur, L., Ditschuneit, K., Schambach, M., Kaymakci, C., Wollmann, T., & Sauer, A. (2024). Explainability and interpretability in electric load forecasting using machine learning techniques–a review. *Energy and AI*, *16*, 100358. https://www.sciencedirect.com/science/article/pii/S2666546824000247

Bazilian, M. D., Carley, S., Konisky, D., Zerriffi, H., Pai, S., & Handler, B. (2021). Expanding the scope of just transitions: Towards localized solutions and community-level dynamics. *Energy Research & Social Science*, *80*, 102245. DOI: 10.1016/j.erss.2021.102245

Bebbington, A., Bebbington, D. H., Sauls, L. A., Rogan, J., Agrawal, S., Gamboa, C., & Verdum, R. (2018). Resource extraction and infrastructure threaten forest cover and community rights. *Proceedings of the National Academy of Sciences of the United States of America*, *115*(52), 13164–13173. DOI: 10.1073/pnas.1812505115 PMID: 30509982

Beccali, M., Bonomolo, M., Ciulla, G., Galatioto, A., & Brano, V. L. (2015). Improvement of energy efficiency and quality of street lighting in South Italy as an action of Sustainable Energy Action Plans. The case study of Comiso (RG). *Energy*, *92*, 394–408. DOI: 10.1016/j.energy.2015.05.003

Becker, S., Naumann, M., & Moss, T. (2018). Between coproduction and commons: understanding initiatives to reclaim urban energy provision in Berlin and Hamburg. In *Coproducing Water, Energy and Waste Services* (pp. 63–85). Routledge. DOI: 10.4324/9781351238526-4

Bednárik, L. S. J. (2000). Green marketing as opportunity for innovation. of Innovations, 2(3), 494.

Belgian, Federal, & Government. (2018). Visienota: Belgisch Interfederaal Energiepact. In.

Bella, K. M. J. (2024). A study on marketing analytics and artificial intelligence. *SELP Journal of Social Science-A Blind Review & Refereed Quarterly Journal*, *XV*(55), 53–57.

Benavides, D., Arévalo, P., Ortega, A. C., Sánchez-Sutil, F., & Villa-Ávila, E. (2023). *Energy Management Model for a Remote Microgrid Based on Demand-Side Energy Control*. https://doi.org/DOI: 10.20944/preprints202311.1125.v1

Benoit, D. F., Lessmann, S., & Verbeke, W. (2020). On realising the utopian potential of big data analytics for maximising return on marketing investments. *Journal of Marketing Management*, *36*(3-4), 233–247. DOI: 10.1080/0267257X.2020.1739446

Berauer, B. J., Wilfahrt, P. A., Arfin-Khan, M. A., Eibes, P., Von Heßberg, A., Ingrisch, J., Schloter, M., Schuchardt, M. A., & Jentsch, A. (2019). Low resistance of montane and alpine grasslands to abrupt changes in temperature and precipitation regimes. *Arctic, Antarctic, and Alpine Research*, *51*(1), 215–231. DOI: 10.1080/15230430.2019.1618116

Berg, W., Nowak, S., Kelly, M., Vaidyanathan, S., Shoemaker, M., Chittum, A., DiMascio, M., & Kallakuri, C. (2016). The 2016 state energy efficiency scorecard.

Berglund, H. (2007). Researching entrepreneurship as lived experience. In *Handbook of qualitative research methods in entrepreneurship*. Edward Elgar Publishing., DOI: 10.4337/9781847204387.00011

Bergmeier, E., Konstantinou, M., Tsiripidis, I., & Sýkora, K. V. (2009). Plant communities on metalliferous soils in northern Greece. *Phytocoenologia*, *39*(4), 411–438. DOI: 10.1127/0340-269X/2009/0039-0411

Berhich, A., Belouadha, F. Z., & Kabbaj, M. I. (2022). A location-dependent earthquake prediction using recurrent neural network algorithms. *Soil Dynamics and Earthquake Engineering*, *161*, 107389. https://www.sciencedirect.com/science/article/pii/S026772612200238X. DOI: 10.1016/j.soildyn.2022.107389

Beura, S., Mohanty, S., Mohapatra, S. K., Panda, N., & Acharya, A. A. (2024). Effective deep learning techniques for prediction of energy consumption in smart buildings. In *2024 International Conference on Advancements in Smart, Secure and Intelligent Computing (ASSIC)* (pp. 1-6). IEEE. https://ieeexplore.ieee.org/abstract/document/10507898

Bhatt, C., Barthwal, A., Shoyab, R., Khamboj, P., & Singh, T. (2024). Optimizing Short-Term Load Forecasting for Energy Management Systems: Integrative Time Series Analysis and Advanced Neural Network Methodologies. In *2024 10th International Conference on Advanced Computing and Communication Systems (ICACCS)* (Vol. 1, pp. 401-406). IEEE. https://ieeexplore.ieee.org/abstract/document/10716879/

Bhattacharya, A., Kharas, H., & Plant, M. (2022). *Financing a big investment push in emerging markets for sustainable recovery*. Brookings Institution. https://www.brookings.edu/articles/financing-a-big-investment-push-in-emerging-markets-for-sustainable-recovery/

Bhattacharya, S. C., & Jana, C. (2009). Renewable energy in India: Historical developments and prospects. *Energy*, *34*(8), 981–991.

Bheemanna, N., Megeri, M., & Budihal, H. H. (2023). Forecast of demographic variables using the ARIMA Model in India. *Int J Stat Appl Math*, *8*(5S), 1009–1018. https://www.mathsjournal.com/special-issue/2023/vol8/issue5S/S-8-5-117. DOI: 10.22271/maths.2023.v8.i5Sn.1322

Bhide, A., & Monroy, C. R. (2011).. . *Renewable & Sustainable Energy Reviews*, *15*(2), 1057–1066.

Bilgili, M., & Pinar, E. (2023). Gross electricity consumption forecasting using LSTM and SARIMA approaches: A case study of Türkiye. *Energy*, *284*, 128575. https://www.sciencedirect.com/science/article/pii/S0360544223019692. DOI: 10.1016/j.energy.2023.128575

Blechinger, P., Seguin, R., Cader, C., Bertheau, P., & Breyer, C. (2014). Assessment of the global potential for renewable energy storage systems on small islands. *Energy Procedia*, *46*, 325–331. DOI: 10.1016/j.egypro.2015.01.071

Boccard, N. (2009). Capacity factor of wind power realized values vs. estimates. *Energy Policy*, *37*(7), 2679–2688. DOI: 10.1016/j.enpol.2009.02.046

Bocquillon, P., & Evrard, A. (2022). Energy Governance in France. In Knodt, M., & Kemmerzell, J. (Eds.), *Handbook of Energy Governance in Europe* (pp. 647–666). Springer International Publishing., DOI: 10.1007/978-3-030-43250-8_10

Boukas, N., & Chourides, P. (2016). Niche tourism in Cyprus: Conceptualising the importance of social entrepreneurship for the sustainable development of islands. *International Journal of Leisure and Tourism Marketing*, *5*(1), 26–43. DOI: 10.1504/IJLTM.2016.077128

Bratman, E. (2014). Contradictions of green development: Human rights and environmental norms in light of Belo Monte Dam activism. *Journal of Latin American Studies*, *46*(2), 261–289. DOI: 10.1017/S0022216X14000042

Bray, R., Mejía Montero, A., & Ford, R. (2022). Skills deployment for a "just" net zero energy transition. *Environmental Innovation and Societal Transitions*, *42*, 395–410. DOI: 10.1016/j.eist.2022.02.002

Brezar, A., & Skopeliti, C. (2025). *Portugal and parts of France hit by massive power outage*.

Briera, T., & Lefèvre, J. (2024). Reducing the cost of capital through international climate finance to accelerate the renewable energy transition in developing countries. *Energy Policy*, *188*, 114104.

Briggs, C., Atherton, A., Gill, J., Langdon, R., Rutovitz, J., & Nagrath, K. (2022). Building a "Fair and Fast" energy transition? Renewable energy employment, skill shortages and social licence in regional areas. *Renewable and Sustainable Energy Transition, 2*, 100039. https://www.sciencedirect.com/science/article/pii/S2667095X2200023X

Brisbois, M. C. (2022). *Governing Decentralized Electricity: Taking a Participatory Turn*. 325–346. DOI: 10.1002/9783527831425.ch15

Brodny, J., Tutak, M., & Grebski, W. (2024). Empirical Assessment of the Efficiency of Poland's Energy Transition Process in the Context of Implementing the European Union's Energy Policy. *Energies, 17*(11), 2689. DOI: 10.3390/en17112689

Bruelheide, H., Jandt, U., Fernández, N., Marmol-Guijarro, A., Smets, B., Buchhorn, M., Giagnacovo, L., Milli, G., Jimenez-Alfaro, B., & Álvarez-Martínez, J. M. (2024). D5. 2 Past-to-present EBV modelled datasets and status indicator for selected terrestrial habitats in the Habitats Directive. *ARPHA* Preprints, *5*, e128158. DOI: 10.3897/arphapreprints.e128158

Brugge, D., & Goble, R. (2002). The history of uranium mining and the Navajo people. *American Journal of Public Health, 92*(9), 1410–1419. DOI: 10.2105/AJPH.92.9.1410 PMID: 12197966

Bugeja, A., Azzopardi, B., & Loizou, E. (2023). A Comparative Analysis of Electric Mobility Operations in the Island States: A Case Study of Malta and Cyprus. *Journal of Energy: Energija, 72*(2.), 15-19. https://doi.org/DOI: 10.37798/2023722456

Burrow, S. (2017). *Just Transition. A Report for the OECD*. Just Transition Center.

Cachia, M., & Millward, L. (2011). The telephone medium and semi-structured interviews: A complementary fit. *Qualitative Research in Organizations and Management, 6*(3), 265–277. DOI: 10.1108/17465641111188420

Calcaterra, M., Aleluia Reis, L., Fragkos, P., Briera, T., de Boer, H. S., Egli, F., & Tavoni, M. (2024). Reducing the cost of capital to finance the energy transition in developing countries. *Nature Energy, 9*(10), 1241–1251.

Caldera, U., Gulagi, A., Jayasinghe, N., & Breyer, C. (2023). Looking island wide to overcome Sri Lanka's energy crisis while gaining independence from fossil fuel imports. *Renewable Energy, 218*, 119261. DOI: 10.1016/j.renene.2023.119261

Cameron, A., Claeys, G., Midões, C., & Tagliapietra, S. (2020). *How good is the European Commission's Just Transition Fund proposal?* (No. 2020/04). Bruegel Policy Contribution.

Cantarero, M. M. V. (2020). Of renewable energy, energy democracy, and sustainable development: A roadmap to accelerate the energy transition in developing countries. *Energy Research & Social Science, 70*, 101716.

Cao, G., Duan, Y., & El Banna, A. (2019). A dynamic capability view of marketing analytics: Evidence from UK firms. *Industrial Marketing Management, 76*, 72–83. DOI: 10.1016/j.indmarman.2018.08.002

Cao, G., Tian, N., & Blankson, C. (2022). Big data, marketing analytics, and firm marketing capabilities. *Journal of Computer Information Systems, 62*(3), 442–451. DOI: 10.1080/08874417.2020.1842270

Cappa, E., Lamperti, F., & Pallante, G. (2024). *Creating jobs out of the green: The employment effects of the energy transition* (No. 2024/21). LEM Working Paper Series.

Carley, S., Evans, T. P., & Konisky, D. M. (2018). Adaptation, culture, and the energy transition in American coal country. *Energy Research & Social Science, 37*, 133–139. DOI: 10.1016/j.erss.2017.10.007

Central Electricity Authority. (2023). *Report on optimal generation capacity mix for 2030*. Government of India.

Central Electricity Authority. (2024). *Annual Report 2024*. New Delhi: Central Electricity Authority. https://cea.adgstaging.in/annual-report/?lang=en

CERA. (2022),Electricity, https://www.cera.org.cy/engb/ilektrismos/details/ilektrismos-adeiodotisi

Chahal, H., Jyoti, J., & Wirtz, J. (2018). Business analytics: Concept and applications. In *Understanding the Role of Business Analytics: Some Applications* (pp. 1–8). Springer Singapore.

Chailleux, S., & Hourcade, R. (2021). Introduction. Local energy policies: Renewal under constraint. *Nature Sciences Sociétés, 29*(1), 3–12. DOI: 10.1051/nss/2021018

Cha, J. M. (2017). A just transition: Why transitioning workers into a new clean energy economy should be at the center of climate change policies. *Fordham Environmental Law Review, 29*(2), 196–220.

Cha, J. M., & Pastor, M. (2022). Just transition: Framing, organizing, and power-building for decarbonization. *Energy Research & Social Science, 90*, 102588. DOI: 10.1016/j.erss.2022.102588

Chakraborty, S., Sadhu, P. K., & Goswami, U. (2016).. . *Problemy Ekorozwoju, 11*(2), 75–80.

Chaskos, D. C., Bartzokas, A., & Pnevmatikos, J. D. (2012). Wind energy in NW Greece. In Helmis, C. G., & Nastos, P. T. (Eds.), *Advances in Meteorology, Climatology and Atmospheric Physics* (pp. 43–50). Springer., DOI: 10.1007/978-3-642-29172-2_7

Chawla, Y., Shimpo, F., & Sokołowski, M. M. (2022). *Artificial intelligence and information management in the energy transition of India: Lessons from the global IT heart*. Digital Policy. *Regulation & Governance, 24*(1), 17–29.

Chen, C., Chen, W.-A., Greig, J., & Robinson, C. (2024). When concentrated disadvantage happens: Exploring the nexus of energy and internet insecurities among vulnerable households. *iScience*, *27*(6), 109895. DOI: 10.1016/j.isci.2024.109895 PMID: 38827408

Cheung, G., Davies, P. J., & Bassen, A. (2019). In the transition of energy systems: What lessons can be learnt from the German achievement? *Energy Policy*, *132*, 633–646. DOI: 10.1016/j.enpol.2019.05.056

Chi, D. (2022). Research on electricity consumption forecasting model based on wavelet transform and multi-layer LSTM model. *Energy Reports*, *8*, 220–228. https://www.sciencedirect.com/science/article/pii/S235248472200169X. DOI: 10.1016/j.egyr.2022.01.169

Chilvers, J., Bellamy, R., Pallett, H., & Hargreaves, T. (2021). A systemic approach to mapping participation with low-carbon energy transitions. *Nature Energy*, *6*(3), 250–259. DOI: 10.1038/s41560-020-00762-w

Chouvardas, D., Karatassiou, M., Tsioras, P., Tsividis, I., & Palaiochorinos, S. (2022). Spatiotemporal changes (1945–2020) in a grazed landscape of northern Greece, in relation to socioeconomic changes. *Land (Basel)*, *11*(11), 1987. DOI: 10.3390/land11111987

Chua, K. J., Yang, W. M., Er, S. S., & Ho, C. A. (2014). Sustainable energy systems for a remote island community. *Applied Energy*, *113*, 1752–1763. DOI: 10.1016/j.apenergy.2013.09.030

Chygryn, O., Lyulyov, O., Pimonenko, T., & Illiashenko, A. (2024, January). Companies' green competitiveness: Justifying the role of marketing determinants. In AIP Conference Proceedings (Vol. 3033, No. 1). AIP Publishing.

Clarke, L., Curtis, M., Eisenberg, A., Grubert, E., Haggerty, J. H., James, A., Jensen, N., Kaufman, N., Krause, E., Raimi, D., Tingley, D., & Weber, J. (2024). *A research agenda for economic resilience in fossil fuel–dependent communities*. Environmental Research Energy., DOI: 10.1088/2753-3751/ad6d70

Clifton, A., Barber, S., Stökl, A., Frank, H., & Karlsson, T. (2022). Research challenges and needs for the deployment of wind energy in atmospherically complex locations. *Wind Energy Science Discussions*, *7*, 2231–2254. DOI: 10.5194/wes-7-2231-2022

Climate Policy Initiative. (2023). *Global Landscape of Climate Finance 2023*. https://www.climatepolicyinitiative.org

Colchester, M., & Ferrari, M. F. (2007). *Making FPIC—free, prior and informed consent—work: Challenges and prospects for indigenous peoples*. Forest Peoples Programme.

Community, E. (2014). Energy Efficiency Directive in the Energy Community. In *Justification Note*.

Comolli, C. L., & Vignoli, D. (2021). Spreading uncertainty, shrinking birth rates: A natural experiment for Italy. *European Sociological Review*, *37*(4), 555–570. DOI: 10.1093/esr/jcab001

Copernicus Land Monitoring Service. (2021). OBSERVER: Copernicus Land Monitoring Service. *The Copernicus Land Monitoring Service: Measuring the effectiveness of protecting. Natura, 2000*, •••. https://www.copernicus.eu/en/news/news/observer-copernicus-land-monitoring-service-measuring-effectiveness-protecting-natura

Corbach, M. (2012). *Die deutsche Stromwirtschaft und der Emissionshandel* (Vol. 5). ibidem-Verlag/ibidem Press.

Corntassel, J. (2012). Re-envisioning resurgence: Indigenous pathways to decolonization and sustainable self-determination. *Decolonization*, *1*(1), 86–101.

Cortez, M. A. A., & Kelly, A. (2025). The Impact of ESG on Corporate Financial Performance: The Case of Global Energy Companies. *Asia-Pacific Social Science Review*, *25*(1), 2. DOI: 10.59588/2350-8329.1555

Creswell, J. W., & Creswell, J. D. (2017). *Research design: Qualitative, quantitative, and mixed methods approaches*. Sage publications.

Cruz, A. V. D., Bayan, S. O., Mabborang, R. C., & Lamprea, C. T. (2025). Forecasting Number of Births in the Philippines Using Seasonal Autoregressive Integrated Moving Average (SARIMA) Model. *European Journal of Information Technologies and Computer Science*, *5*(3), 1–10. https://www.ej-compute.org/index.php/compute/article/view/129. DOI: 10.24018/compute.2025.5.3.129

Cui, D., Zhang, T., Yu, C. W., & Li, J. (2025). Equal distribution of new energy in urban low-carbon renewal: Opportunities and challenges. *Indoor and Built Environment*, *34*(7), 1279–1283. DOI: 10.1177/1420326X241309772

Curtis, E. M., O'Kane, L., & Park, R. J. (2024). Workers and the green-energy transition: Evidence from 300 million job transitions. *Environmental and Energy Policy and the Economy*, *5*(1), 127–161. DOI: 10.1086/727880

Curto, D., Franzitta, V., Trapanese, M., & Cirrincione, M. (2020). A preliminary energy assessment to improve the energy sustainability in the small islands of the Mediterranean Sea. *Journal of Sustainable Development of Energy. Water and Environment Systems*, *8*(4), 735–753.

Cyprus Electricity Authority. (2022). Electricity generation from Renewable Sources,(Greek),https://www.eac.com.cy/EL/EAC/Sustainability/Pages/Electricity Production.aspx

Cyprus Energy Agency, (2019). Public consultation on the National Energy and Climate Plan 2021-2030

Cyprus Human Resources Development Authority, (2018). Green skills needs identification in the Cyprus economy 2017-2027, Directorate of Research and Planning, (in Greek)

da Cunha, Í. G. F., Policarpo, R. V. S., de Oliveira, P. C. S., Abdala, E. C., & do Nascimento Rebelatto, D. A. (2025). A systematic review of ESG indicators and corporate performance: Proposal for a conceptual framework. *Future Business Journal*, *11*(1), 106. DOI: 10.1186/s43093-025-00539-1

Dalle Fratte, M., & Cerabolini, B. E. (2023). Extending the interpretation of Natura 2000 habitat types beyond their definition can bias their conservation status assessment: An example with species-rich Nardus grasslands (6230*). *Ecological Indicators*, *156*, 111113. DOI: 10.1016/j.ecolind.2023.111113

Dammak, K., Zouari, A., & Sidhom, L. (2025). *Sustainability and Multi-Capital Integration in Agri-Food Supply Chains: A Systematic Review*. ResearchSquare. https://www.researchsquare.com/article/rs-6115997/latest

Damtoft, N. F., van Liempd, D., & Lueg, R. (2025). Sustainability performance measurement–a framework for context-specific applications. *Journal of Global Responsibility*, *16*(1), 162–201. DOI: 10.1108/JGR-05-2023-0082

Dangelico, R. M., & Vocalelli, D. (2017). "Green Marketing": An analysis of definitions, strategy steps, and tools through a systematic review of the literature. *Journal of Cleaner Production*, *165*, 1263–1279. DOI: 10.1016/j.jclepro.2017.07.184

Daoutis, C., & Lempesi, A. (2023). The effect of the forest road network on grassland ecosystems with the contribution of the SWOT-AHP method. In: *E3S Web of Conferences, 4th International Conference on Environmental Design (ICED2023)*, 20-22 October 2023, Athens, Greece, 436, 09007, EDP Sciences. https://doi.org/ DOI: 10.1051/e3sconf/202343609007

Dar, I., Rana, A. U. R., & Nasir, I. (2023). Marketing analytics theoretical development and trends: The journey of defining marketing analytics. *Research Journal for Societal Issues*, *5*(1), 54–70. DOI: 10.56976/rjsi.v5i1.57

Datta, A., Coates, S., Rossiter, A., & Krishnamoorti, R. (2024). Reskilling and Upskilling for Decarbonization: Analyzing Micro-Credential Programs for Energy Workforce Development. *The Journal of Continuing Higher Education*, •••, 1–20. DOI: 10.1080/07377363.2024.2377777

Dauda, M., Mustapha, B., & Sani, S. A. (2024). Green marketing strategy and its effect on the performance of small and medium business enterprises (SMEs). International Journal of Accounting. *Finance and Administrative Research*, *1*(2), 229–239.

David, M., & Schulte-Römer, N. (2021). Phasing out and in: System transition through disassociation in the German energy transition–The case of light and coal. *Energy Research & Social Science*, *80*, 102204. DOI: 10.1016/j.erss.2021.102204

Deal, G. (2020). The Just Transition Mechanism: Making Sure No One Is Left Behind.

Debnath, P., Chittora, P., Chakrabarti, T., Chakrabarti, P., Leonowicz, Z., Jasinski, M., Gono, R., & Jasińska, E. (2021). Analysis of earthquake forecasting in India using supervised machine learning classifiers. *Sustainability (Basel)*, *13*(2), 971. https://www.mdpi.com/2071-1050/13/2/971. DOI: 10.3390/su13020971

Delagrammatikas, G., & Roukanas, S. (2023). Offshore wind farm in the Southeast Aegean Sea and energy security. *Energies*, *16*(13), 5208. DOI: 10.3390/en16135208

Delipetrou, P., Sarika, M., Michopoulos, P., & Karetsos, G. (2015). *Mountain grassland composition, structure, and ecology in Mt. Oiti and Mt. Kallidromo. Deliverable A.5.1. for the project LIFE11 NAT/GR/2014 - ForOpenForests*. Hellenic Agricultural Organization Dimiter - Institute of Mediterranean Forest Ecosystems and Forest Products Technology, National and Kapodistrian University of Athens - HSPN.

Dengler, J., Bergmeier, E., Willner, W., & Chytrý, M. (2013). Towards a consistent classification of European grasslands. *Applied Vegetation Science*, *16*(3), 518–520. Advance online publication. DOI: 10.1111/avsc.12041

Denzin, N. K., & Lincoln, Y. S. (Eds.). (2011). *The Sage handbook of qualitative research*. Sage Publication.

DEPEEP, (2024). Energy mix 2023, file:///C:/Users/ A /Desktop/ ENERGY - MIX -2023.pdf

Deshaies, M. (2020). Metamorphosis of Mining Landscapes in the Lower Lusatian Lignite Basin (Germany): New uses and new image of a mining region. *Les Cahiers de la recherche architecturale urbaine et paysagère*, (7). https://doi.org/ DOI: 10.4000/craup.4018

Deshmukh, R., Phadke, A., & Callaway, D. S. (2021)... *Proceedings of the National Academy of Sciences of the United States of America*, *118*(13), e2008128118.

Deveer, L., & Minet, L. (2025). Real-time air quality prediction using traffic videos and machine learning. *Transportation Research Part D, Transport and Environment*, *142*, 104688. https://www.sciencedirect.com/science/article/pii/S1361920925000987. DOI: 10.1016/j.trd.2025.104688

Di Vaio, A., Chhabra, M., Zaffar, A., & Balsalobre-Lorente, D. (2025). Accounting and Accountability in the Transition to Zero-Carbon Energy for Climate Change: A Systematic Literature Review. *Business Strategy and the Environment*, *34*(5), 5925–5946. DOI: 10.1002/bse.4282

Di Vaio, A., Van Engelenhoven, E., Chhabra, M., & Garofalo, A. (2024). Decarbonization of waste management practices and GHG accounting for energy transition: Evidence from European electricity corporations' reporting. *Environment, Development and Sustainability*, *27*(7), 1–25. DOI: 10.1007/s10668-024-04629-y

Diffendorfer, J. E., Vanderhoof, M. K., & Ancona, Z. H. (2022). Wind turbine wakes can impact down-wind vegetation greenness. *Environmental Research Letters*, *17*(10), 104025. DOI: 10.1088/1748-9326/ac8da9

Dimitrakopoulos, P., Makris, I., & Apostolopoulos, S. (2024). The Impact of Major Energy Indexes on the Economic Growth of EU Countries, According to the Contribution of Industry to GDP. In *International Conference on Applied Economics* (pp. 59-74). Cham: Springer Nature Switzerland.

Dimitrakopoulos, P., Makris, I., & Apostolopoulos, S. (2024). The Impact of Major Energy Indexes on the Economic Growth of EU Countries, According to the Contribution of Industry to GDP. In *International Conference on Applied Economics* (pp. 59-74). Cham: Springer Nature Switzerland. https://doi.org/DOI: 10.1007/978-3-031-76658-9_4

Dimitrakopoulos, P., Makris, I., & Apostolopoulos, S. (2024, July). The Impact of Major Energy Indexes on the Economic Growth of EU Countries, According to the Contribution of Industry to GDP. In *International Conference on Applied Economics* (pp. 59-74). Cham: Springer Nature Switzerland.

Dimopoulos, P., Bergmeier, E., & Fischer, P. (2005). Monitoring and conservation status assessment of habitat types in Greece: Fundamentals and exemplary cases. *Annali di Botanica, 5*, 7–20.

Dimopoulos, P., Bergmeier, E., & Fischer, P. (2006). Natura 2000 habitat types of Greece evaluated in the light of distribution, threat and responsibility. *Biology and Environment, 106B*(3), 175–187. DOI: 10.1353/bae.2006.0015

Din, J., & Su, H. (2024). Blockchain-Enabled Smart Grids for Optimized Electrical Billing and Peer-to-Peer Energy Trading. *Energies, 17*(22), 5744. DOI: 10.3390/en17225744

Dong, K., Dong, X., & Jiang, Q. (2020). How renewable energy consumption lower global CO2 emissions? Evidence from countries with different income levels. *World Economy, 43*(6), 1665–1698. DOI: 10.1111/twec.12898

Doulos, I. G., Katsoulakos, N. M., & Kaliampakos, D. C. (2019). The Electricity Market Structure in Greece and the Paradox of Renewable Energy Sources. *Journal of Power and Energy Engineering, 7*(4), 1–26. DOI: 10.4236/jpee.2019.74001

Dunlap, A. (2021). *Renewable energy: A critical introduction*. Routledge.

Dupont, C., Moore, B., Boasson, E. L., Gravey, V., Jordan, A., Kivimaa, P., Kulovesi, K., Kuzemko, C., Oberthür, S., Panchuk, D., Rosamond, J., Torney, D., Tosun, J., & von Homeyer, I. (2024). Three decades of EU climate policy: Racing toward climate neutrality? *Wiley Interdisciplinary Reviews: Climate Change, 15*(1), e863. https://doi.org/https://doi.org/10.1002/wcc.863. DOI: 10.1002/wcc.863

Dupont, E., Germain, M., & Jeanmart, H. (2021). Feasibility and Economic Impacts of the Energy Transition. *Sustainability (Basel), 13*(10), 5345. DOI: 10.3390/su13105345

Dupont, E., Koppelaar, R., & Jeanmart, H. (2020). Global available solar energy under physical and energy return on investment constraints. *Applied Energy, 257*, 113968. DOI: 10.1016/j.apenergy.2019.113968

Ebhote, O., & Izedonmi, F. O. I. (2021). Green marketing practices in Nigeria: Challenges and prospect. *Nigerian Journal of Business Administration, 15*(1 & 2), 112–125.

Edem, K. (2024). Economic impact of digital twins on renewable energy investments. *Engineering Science & Technology Journal, 5*(7), 2232–2247. DOI: 10.51594/estj.v5i7.1318

Edem, K. (2024). Peer-to-peer energy trading: Innovations, regulatory challenges, and the future of decentralized energy systems. *World Journal of Advanced Research and Reviews, 24*(2), 172–186. DOI: 10.30574/wjarr.2024.24.2.3324

EEA. (2017). *Climate change mitigation: Trends and projections in Europe*. European Environment Agency.

Egli, F., Polzin, F., Sanders, M., Schmidt, T., Serebriakova, A., & Steffen, B. (2022). Financing the energy transition: Four insights and avenues for future research. *Environmental Research Letters, 17*(5), 051003. DOI: 10.1088/1748-9326/ac6ada

Eklund, M., Khalilpour, K., Voinov, A., & Hossain, M. J. (2023). Understanding the community in community microgrids: A conceptual framework for better decision-making. *Energy Research & Social Science, 104*, 103260. DOI: 10.1016/j.erss.2023.103260

Elias, R. J., & Victor, D. G. (2005). Energy transitions in developing countries: a review of concepts and literature. Program on Energy and Sustainable Development, Working Paper. Stanford: Stanford University.

Elizabet, M. (2022). DIGITAL MARKETING TRAINING: Strategies to Increase Sales in Indonesian Market Place. *Jurnal PKM Manajemen Bisnis, 2*(2), 63–72. DOI: 10.37481/pkmb.v2i2.484

Elsaraiti, M., Ali, G., Musbah, H., Merabet, A., & Little, T. (2021, April). Time series analysis of electricity consumption forecasting using ARIMA model. In *2021 IEEE Green technologies conference (GreenTech)* (pp. 259-262). IEEE. https://ieeexplore.ieee.org/abstract/document/9458543

Ember Energy. (2025). Greece In Figures-data, https://github.com/GreeceInFigures/greeceinfigures-data/tree/main/energy

Ember.org. (2024). "EU Electricity trends", https://ember-energy.org/latest-insights/european-electricity-review-2024/eu-electricity-trends/#electricity-demand

Emeç, M. U. R. A. T., & Yurtsever, M. (2025). A novel ensemble machine learning method for accurate air quality prediction. *International Journal of Environmental Science and Technology, 22*(1), 459–476. https://link.springer.com/article/10.1007/s13762-024-05671-z. DOI: 10.1007/s13762-024-05671-z

Energiewende, A. (2018). *Energiewende 2030: The Big Picture; Megatrends, Targets, Strategies and a 10-Point Agenda for the Second Phase of Germany's Energy Transition. Impulse*. Agora Energiewende.

Energy Institute. (2025). *Statistical Review of World Energy 2025*. London: Energy Institute. https://www.energyinst.org/statistical-review/home

Energy, M. o. (2017). 2017–2020 National energy efficiency action plan. In. Spain.

EPRS European Parliamentary Research Service. Briefing, (2023), https://www.europarl.europa.eu/RegData/etudes/BRIE/2023/754623/EPRS_BRI(2023754623_EN.pdf European Commission (2020), Technical support for implementing the European Green DealLuxembourg: Publications Office of the European Union, 2020 https://reform-support.ec.europa.eu/what-we-do/green-transition_en

Erwin, E., Suade, Y. K. M., & Poernomo, W. (2021). Analyzing Digital Marketing, Green Marketing, Networking and Product Innovation on Sustainability Business Performance, Silk Cluster in Polewali-Mandar, West Sulawesi. International Journal of Economics [IJEBAR]. *Business and Accounting Research*, *5*(3), 2850–2871.

Esmaeili, N., & Abbasi-Shavazi, M. J. (2024). Forecasting number of births and sex ratio at birth in Iran using deep neural network and ARIMA: Implications for policy evaluations. *Journal of Population Research (Canberra, A.C.T.)*, *41*(4), 26. DOI: 10.1007/s12546-024-09348-9

Europa.eu. (2017). Political declaration on clean energy for EU islands, https://energy.ec.europa.eu/system/files/2017-05/170505_political_declaration_on_clean_energy_for_eu_islands-_final_version_16_05_20171_0.pdf

European Commission, (2022b). "Cohesion Open Data Platform".

European Commission. (2019). "The European Green deal", https://commission.europa.eu/strategy-and-policy/priorities-2019-2024/european-green-deal_en, accessed 9/7/2025

European Commission. (2020). A strong social Europe for just transitions. COM (2020) 14 Final, https://ec.europa.eu/commission/presscorner/detail/en/qanda_20_20

European Commission. (2020a) Regulation of the European Parliament and of the Council establishing a European Union Just Transition Fund. Available at https://eur-lex.europa.eu/legal-content/EN/TXT/?uri=CELEX%3A52020PC0022 (accessed 14 January 2020).

European Commission. (2020b) The European Green Deal Investment Plan and Just Transition Mechanism explained. Available at https://ec.europa.eu/commission/presscorner/detail/el/qanda_20_24 (accessed 14 January 2020)

European Commission. (2022). *Press release: European Green Deal – Just Transition Mechanism and funding sources.* European Commission. https://ec.europa.eu/commission/presscorner/detail/en/ip_22_3711

European Commission. (2022a). EU Cohesion Policy: more than €1 billion for Cyprus for economic and social development and a fair green transition in 2021-2027, Brussels, Press release, https://ec.europa.eu/commission/presscorner/api/files/document/print/en/ip_22_4326/IP_22_4326_EN.pdf

European Commission. (2022c). Commission participates in launch of EuroAsia Electricity Interconnector, Commission participates in launch of EuroAsia Electricity Interconnector - European Commission European Parliament, (2025). Cyprus's climate action strategy, https://www.europarl.europa.eu/RegData/etudes/BRIE/2025/769495/EPRS_BRI(2025)769495_EN.pdf

European Commission. (2024). Executive Vice-President Šefčovič signs agreement between Greece, European Investment Bank and the Commission on a Decarbonisation Fund for Greek islands, https://climate.ec.europa.eu/news-your-voice/news/executive-vice-president-sefcovic-signs-agreement-between-greece-european-investment-bank-and-2024-11-21_en

European Commission. National energy and climate plans - NECP. (2025). In *Energy Union.* Retrieved July 29, 2025, from https://energy.ec.europa.eu/strategy/energy-union_en#national-energy-and-climate-plans

European Environment Agency. (2020). State of Nature in the EU – Results from reporting under the nature directives 2013–2018. EEA Report No DOI: 10/2020. 10.2800/088178

European Parliament. (2022). Report on EU islands and cohesion policy: current situation and future challenges, Report - A9-0144/2022, https://www.europarl.europa.eu/doceo/document/A-9-2022-0144_EN.html

Eurostat (2024). Electricity production, consumption and market overview https://ec.europa.eu/eurostat/statistics-explained/index.php?title=Electricity_production,_consumption_and_market_overview

Evangelinos, K., Nikolaou, I., & Leal Filho, W. (2015). The effects of climate change policy on the business community: A corporate environmental accounting perspective. *Corporate Social Responsibility and Environmental Management, 22*(5), 257–270. DOI: 10.1002/csr.1342

Evans, D. (2012). Building the European union's Natura 2000 network. *Nature Conservation, 1,* 11–26. DOI: 10.3897/natureconservation.1.1808

Fafaliou, I., & Polemis, M. L. (2010). Liberalisation of the European natural gas market: Myth or reality? Evidence from Greece. *Research Topics in Agricultural and Applied Economics*, *1*, 168–183.

Fan, H., & Zeng, L. (2011). Implementation of green marketing strategy in China: A study of the green food industry.

Fankhauser, S., Bowen, A., Calel, R., Dechezleprêtre, A., Grover, D., Rydge, J., & Sato, M. (2016). Who will win the green race? In search of environmental competitiveness and innovation. *Global Environmental Change*, *44*, 28–38.

Farooquee, A. A., & Shrimali, G. (2016).. . *Energy Policy*, *95*, 518–528.

Fawcett, T., & Hampton, S. (2020). Why & how energy efficiency policy should address SMEs. *Energy Policy*, *140*, 111337. https://doi.org/https://doi.org/10.1016/j.enpol.2020.111337. DOI: 10.1016/j.enpol.2020.111337

Fawcett, T., Rosenow, J., & Bertoldi, P. (2019). Energy efficiency obligation schemes: Their future in the EU. *Energy Efficiency*, *12*(1), 57–71. DOI: 10.1007/s12053-018-9657-1

Fernandez, R. (2018). Crisis on Spain's Environmental Policy. *The impact of the economic crisis on European environmental policy*, 152.

Fetzel, T., Petridis, P., Noll, D., Singh, S. J., & Fischer-Kowalski, M. (2018). Reaching a socio-ecological tipping point: Overgrazing on the Greek island of Samothraki and the role of European agricultural policies. *Land Use Policy*, *76*, 21–28. DOI: 10.1016/j.landusepol.2018.04.042

Filipović, S., Lior, N., & Radovanović, M. (2022). The green deal–just transition and sustainable development goals Nexus. *Renewable & Sustainable Energy Reviews*, *168*, 112759. DOI: 10.1016/j.rser.2022.112759

Filis, E., Kyrkas, D., Mantzos, N., Dimitrakopoulos, P. G., Fotiadis, G., & Konstantinou, M. (2024). Grassland flora of ultramafic areas in Northern Pindus (Greece). *Ecological Research*, *39*(4), 531–542. DOI: 10.1111/1440-1703.12469

Firdiansyah, R., Mohamed, M., Yusoff Yusliza, M., Saputra, J., & Muhammad, Z. (2021, March). A Review of green marketing strategy literature: Mini review approach. In Proceedings of the 11th Annual International Conference on Industrial Engineering and Operations Management Singapore: IEOM Society International (Vol. 21030, pp. 5092-5108).

Fometescu, A., Hategan, C. D., & Pitorac, R. I. (2024). How Responsible Are Energy and Utilities Companies in Terms of Sustainability and Economic Development? *Energies, 17*(23), 6209. DOI: 10.3390/en17236209

Formuszewicz, R., & Gawlikowska-Fyk, A. (2014). New EU energy and climate framework: challenges for Poland and Germany.

Fossey, E., Harvey, C., McDermott, F., & Davidson, L. (2002). Understanding and evaluating qualitative research. *The Australian and New Zealand Journal of Psychiatry, 36*(6), 717–732. DOI: 10.1046/j.1440-1614.2002.01100.x PMID: 12406114

Fotis, G., Maris, T. I., & Mladenov, V. (2025). Risks, Obstacles and Challenges of the Electrical Energy Transition in Europe: Greece as a Case Study. *Sustainability (Basel), 17*(12), 5325. DOI: 10.3390/su17125325

Fotopoulos, N., Goulas, C., & Karra, V. (2015). Trade Union Education and Life Long Learning in Modern Greece. *World Journal of Educational Research (Los Angeles, Calif.), 2*(1), 15. Advance online publication. DOI: 10.22158/wjer.v2n1p15

Fotopoulou, M., Pediaditis, P., Skopetou, N., Rakopoulos, D., Christopoulos, S., & Kartalidis, A. (2024). A review of the energy storage systems of non-interconnected European islands. *Sustainability (Basel), 16*(4), 1572. DOI: 10.3390/su16041572

Franzitta, V., & Curto, D. (2017). Sustainability of the renewable energy extraction close to the Mediterranean Islands. *Energies, 10*(3), 283. DOI: 10.3390/en10030283

Gabderakhmanova, T., & Marinelli, M. (2022). Multi-energy system demonstration pilots on geographical Islands: An overview across Europe. *Energies, 15*(11), 3908. DOI: 10.3390/en15113908

Galán-Madruga, D., de Lourdes Berríos Cintrón, M., Broomandi, P., Oleniacz, R., & Cáceres, J. O. (2025). Long-term performance of air quality networks: Implications in health and environmental management. *International Journal of Environmental Science and Technology*, •••, 1–26. https://link.springer.com/article/10.1007/s13762-025-06526-x. DOI: 10.1007/s13762-025-06526-x

Galkina, A., & Grafeeva, N. (2019). Machine learning methods for earthquake prediction: A survey. In *Proceedings of the fourth conference on software engineering and information management (SEIM-2019), Saint Petersburg, Russia* (Vol. 13, p. 25). https://www.researchgate.net/profile/Natalia-Grafeeva/publication/

Gambhir, A., Green, F., & Pearson, P. J. (2018). Towards a just and equitable low-carbon energy transition. *Grantham Institute Briefing Paper, 26*. Advance online publication. DOI: 10.1080/14693062.2019.1657379

Garcia-Casals, X., Ferroukhi, R., & Parajuli, B. (2019). Measuring the socio-economic footprint of the energy transition. *Energy Transitions*, *3*(1-2), 105–118. DOI: 10.1007/s41825-019-00018-6

Garcia-Torea, N., Larrinaga, C., & Luque-Vílchez, M. (2023). Bridging the understanding of sustainability accounting and organizational change. *Organization & Environment*, *36*(1), 17–38. DOI: 10.1177/10860266221083339

Gartner, W. B., & Birley, S. (2002). Introduction to the special issue on qualitative methods in entrepreneurship research. *Journal of Business Venturing*, *17*(5), 387–395. DOI: 10.1016/S0883-9026(01)00077-5

Gawlik, L. (2018). The Polish power industry in energy transformation process. *Mineral Economics*, *31*(1-2), 229–237. DOI: 10.1007/s13563-017-0128-5

Gebhardt, M., Thun, T. W., Seefloth, M., & Zülch, H. (2023). Managing sustainability - Does the integration of environmental, social and governance key performance indicators in the internal management systems contribute to companies' environmental, social and governance performance? *Business Strategy and the Environment*, *32*(4), 2175–2192. DOI: 10.1002/bse.3242

Gema, S. B. (2023). Financing the energy transition: The new paradigm for renewable energy investors. In *The Palgrave Handbook of Zero Carbon Energy Systems and Energy Transitions* (pp. 1–44). Springer International Publishing., DOI: 10.1007/978-3-030-74380-2_4-1

Genave, A. (2019). Energy vulnerability in the Southwest Indian Ocean islands. *Journal of the Indian Ocean Region*, *15*(1), 40–57. DOI: 10.1080/19480881.2019.1560760

Genc, T. S., & Kosempel, S. (2023). Energy transition and the economy: A review article. *Energies*, *16*(7), 2965. DOI: 10.3390/en16072965

Georgiou, A., Bednarova, L., & Kudelas, D. (2024). Climate Change Risk and Financial Performance in Cyprus: Management Perceptions. *Acta Montanistica Slovaca*, *29*(4), 920–932. Advance online publication. DOI: 10.46544/AMS.v29i4.11

Georgiou, P. N., Mavrotas, G., & Diakoulaki, D. (2011). The effect of islands' interconnection to the mainland system on the development of renewable energy sources in the Greek power sector. *Renewable & Sustainable Energy Reviews*, *15*(6), 2607–2620. DOI: 10.1016/j.rser.2011.03.007

Giampietro, M., Mayumi, K., & Sorman, A. (2011). *The metabolic pattern of societies: Where economists fall short*. Routledge. DOI: 10.4324/9780203635926

Giannakopoulos, C., Psiloglou, B., Lemesios, G., Xevgenos, D., Papadaskalopoulou, C., Karali, A., Varotsos, K. V., Zachariou-Dodou, M., Moustakas, K., Ioannou, K., Petrakis, M., & Loizidou, M. (2016). Climate change impacts, vulnerability and adaptive capacity of the electrical energy sector in Cyprus. *Regional Environmental Change*, *16*(7), 1891–1904. DOI: 10.1007/s10113-015-0885-z

Gielen, D., Boshell, F., Saygin, D., Bazilian, M. D., Wagner, N., & Gorini, R. (2021). The role of renewable energy in the global energy transformation. *Energy Strategy Reviews*, *24*, 38–50.

Gillespie, R. G. (2009). Encyclopedia of islands. *University of California, 1074.*

Gill-Wiehl, A., Miles, S., Wu, J., & Kammen, D. M. (2022). Beyond customer acquisition: A comprehensive review of community participation in mini grid projects. *Renewable & Sustainable Energy Reviews*, *153*, 111778. DOI: 10.1016/j.rser.2021.111778

Giwa-Amu, O. (2022). Effect of green marketing on customer satisfaction. Sapientia Global Journal of Arts. *Humanities and Development Studies*, *5*(2), 77–86.

Gkonis, N., Arsenopoulos, A., Stamatiou, A., & Doukas, H. (2020). Multi-perspective design of energy efficiency policies under the framework of national energy and climate action plans. *Energy Policy*, *140*, 111401. https://doi.org/https://doi.org/10.1016/j.enpol.2020.111401. DOI: 10.1016/j.enpol.2020.111401

Goh, K. C., Kurniawan, T. A., Zainordin, N., Arifah, I. D. C., Abas, M. A., Masrom, M. A. N., Mohamed, S., Omar, R., Khoo, S. L., Gui, H. C., Toh, T. C., & Onn, C. W. (2024). Expediting decarbonization in energy, waste, and water sector through digitalization in sustainable smart cities (SSC): Case-studies in Malaysia and China based on Industry 5.0 paradigm. *Sustainable Cities and Society*, *117*, 105969. DOI: 10.1016/j.scs.2024.105969

Gonzalez, D., Patricio, M. A., Berlanga, A., & Molina, J. M. (2022). Variational autoencoders for anomaly detection in the behaviour of the elderly using electricity consumption data. *Expert Systems: International Journal of Knowledge Engineering and Neural Networks*, *39*(4), e12744. https://onlinelibrary.wiley.com/doi/full/10.1111/exsy.12744. DOI: 10.1111/exsy.12744

González, F. J. P. (2018). El céntimo sanitario: Potencial recaudatorio frente a recaudación real. *Anales de ASEPUMA*, (26), 15.

Goodrick, D., & Rogers, P. J. (2015). Qualitative data analysis. *Handbook of practical program evaluation*, 561-595. https://doi.org/DOI: 10.1002/9781119171386.ch22

Gopal, J. N., Muthukaruppasamy, S., Rajagopal, V., Thomas, & Parimalarsundar, E. (2024). Artificial Intelligence Governance and Comprehensibility in Renewable Energy Systems. *Chapman and Hall/CRC EBooks*, 73–88. DOI: 10.1201/9781003457176-5

Gordon, J. (2022). *Energy cultures: Technology, justice, and everyday life*. Cambridge University Press.

Gorski, A.-T., Gligorea, I., Gorski, H., & Oancea, R. (2023). Navigating the Disruptive Challenges and Opportunities of Digital Transformation in the Labour Market: Upskilling and Reskilling for the Fourth Industrial Revolution. *International Conference Knowledge Based Organization, 29*(3), 23–29. DOI: 10.2478/kbo-2023-0071

Goswami, A., Bandyopadhyay, K. R., & Kumar, A. (2017). Exploring the nature of rural energy transition in India: Insights from case studies of eight villages in Bihar. *International Journal of Energy Sector Management, 11*(3), 463–479.

Gougou, F., & Persico, S. (2020). Deciding together: The Citizens' convention on climate and the democratic challenge. In. Books & Ideas.net.

Government Committee SDAM. (2020). Just Transition Development Plan of Lignite Areas, (Greek), https://www.sdam.gr/sites/default/files/consultation/Master_Plan_Public_Consultation_ENG.pdf

Government Gazette, (2016). Law 4414 "New support scheme for power plants using renewable energy sources and high-efficiency cogeneration of electricity and heat - Provisions for the legal and operational separation of the supply and distribution sectors in the natural gas market and other provisions," Government Gazette A' 149/9-8-2016, (in Greek).

Government Gazette, (2018). Law 4546 "Transposition into Greek law of Directive 2014/89/EU "establishing a framework for maritime spatial planning" and other provisions", Government Gazette A 101/12.06.2018, (in Greek).

Government Gazette. (2025). GReco Islands Charter: Determining the conditions for co-financing GReco Islands projects under Article 21 of Law No. 4936/2022 from the NSRF Programmes 2021-2027 and establishment of the "GRecoISLAND" label, Sheet No. 2119/5.5.2025, (in Greek), https://www.espa.gr/elibrary/YA6422715042025-FEK2119B05052025_Charta-GReco-islands.pdf

Govindan, K., & Shankar, M. (2016).. . *International Journal of Energy Sector Management, 10*(2), 266–282.

Goyal, M., & Pahwa, M. S. (2018). Green Marketing Mix A Model towards Sustainability. *International Journal on Computer Science and Engineering, 6*(09), 23–27.

Grandón, T. G., Schwenzer, J., Steens, T., & Breuing, J. (2024). Electricity demand forecasting with hybrid classical statistical and machine learning algorithms: Case study of Ukraine. *Applied Energy*, *355*, 122249. https://www.sciencedirect.com/science/article/pii/S0306261923016136. DOI: 10.1016/j.apenergy.2023.122249

Greek Ministry of Energy. (2021), Territorial plan for a just and developmental transition of the North and South Aegean Islands & Crete (SDAM), sdam.gr, https://sdam.gr/sites/default/files/2021-05/ΕΔΑΦΙΚΟ%20ΣΧΕΔΙΟ%20ΝΗΣΩΝ%20_ΔΙ-ΑΒΟΥΛΕΥΣΗ%2027042021.pdf

Groppi, D., Garcia, D. A., Basso, G. L., Cumo, F., & De Santoli, L. (2018). Analysing economic and environmental sustainability related to the use of battery and hydrogen energy storages for increasing the energy independence of small islands. *Energy Conversion and Management*, *177*, 64–76. DOI: 10.1016/j.enconman.2018.09.063

Guarino, R., Vranakis, M., Rodríguez-Rojo, M. P., Giuga, L., & Pasta, S. (2020). Grasslands and Shrublands of the Mediterranean Region. In Goldstein, M., & Della Sala, D. (Eds.), *Encyclopedia of the World's Biomes* (Vol. 2, pp. 638–655). Oxford Academic Press, Elsevier., DOI: 10.1016/B978-0-12-409548-9.12119-0

Gudynas, E. (2011). Buen Vivir: Today's tomorrow. *Development*, *54*(4), 441–447. DOI: 10.1057/dev.2011.86

Gu, H., Wang, J., Lin, Q., & Gong, Q. (2014). Automatic contour-based road network design for optimized wind farm micrositing. *IEEE Transactions on Sustainable Energy*, *6*(1), 281–289. DOI: 10.1109/TSTE.2014.2369432

Gulagi, A., Bogdanov, D., & Breyer, C. (2017). The demand for storage technologies in energy transition pathways towards 100% renewable energy for India. *Energy Procedia*, *135*, 37–50.

Gulagi, A., Bogdanov, D., & Breyer, C. (2018). The role of storage technologies in energy transition pathways towards achieving a fully sustainable energy system for India. *Journal of Energy Storage*, *17*, 525–539.

Gulluscio, C., Puntillo, P., Luciani, V., & Huisingh, D. (2020). Climate change accounting and reporting: A systematic literature review. *Sustainability (Basel)*, *12*(13), 5455. DOI: 10.3390/su12135455

Gunasundari, K. (2016). A study on quality of worklife among mine workers with special reference to Neyveli Lignite Corporation. *International Journal of Advanced Research and Development*, *1*(7), 13–15.

Gupta, N. S., Mohta, Y., Heda, K., Armaan, R., Valarmathi, B., & Arulkumaran, G. (2023). Prediction of air quality index using machine learning techniques: A comparative analysis. *Journal of Environmental and Public Health, 2023*(1), 4916267. https://onlinelibrary.wiley.com/doi/full/10.1155/2023/4916267. DOI: 10.1155/2023/4916267

Gürtler, K., Postpischil, R., & Quitzow, R. (2019). The dismantling of renewable energy policies: The cases of Spain and the Czech Republic. *Energy Policy, 133*, 110881. DOI: 10.1016/j.enpol.2019.110881

Gururaja, H. S., Hebbar, A., Poojary, A. S., Bharadwaj, A., & Rakshitha, B. R. (2024). Decentralized Energy Trading for Grids Using Blockchain for Sustainable Smart Cities. *Deleted Journal, 2*(02), 134–141. DOI: 10.47392/IRJAEH.2024.0025

Gusc, J., Bosma, P., Jarka, S., & Biernat-Jarka, A. (2022). The big data, artificial intelligence, and blockchain in true cost accounting for energy transition in Europe. *Energies, 15*(3), 1089. DOI: 10.3390/en15031089

Hadjigeorgiou, I. (2011). Past, present and future of pastoralism in Greece. *Pastoralism: research, policy and practice, 1*, 1-22. https://doi.org/DOI: 10.1186/2041-7136-1-24

Hadjipetrou, S., & Kyriakidis, P. (2024). High-resolution wind speed estimates for the Eastern Mediterranean Basin: A statistical comparison against coastal meteorological observations. *Wind, 4*(4), 311–341. DOI: 10.3390/wind4040016

Hadjout, D., Torres, J. F., Troncoso, A., Sebaa, A., & Martínez-Álvarez, F. (2022). Electricity consumption forecasting based on ensemble deep learning with application to the Algerian market. *Energy, 243*, 123060. https://www.sciencedirect.com/science/article/pii/S0360544221033090. DOI: 10.1016/j.energy.2021.123060

Haldar, S. (2022). A landscape level analysis of entrepreneurship and sustainable energy transitions: Evidences from Gujarat, India. *Sustainable Development (Bradford), 30*(4), 489–502. DOI: 10.1002/sd.2244

Haldar, S., Peddibhotla, A., & Bazaz, A. (2023). Analysing intersections of justice with energy transitions in India-A systematic literature review. *Energy Research & Social Science, 98*, 103010.

Hallegatte, S., Rentschler, J., & Rozenberg, J. (2019). *Lifelines: The Resilient Infrastructure Opportunity*. World Bank Publications. DOI: 10.1596/978-1-4648-1430-3

Hällgren, M., & Maaninen-Olsson, E. (2005). Deviations, ambiguity and uncertainty in a project-intensive organization. *Project Management Journal, 36*(3), 17–26. DOI: 10.1177/875697280503600303

Hällgren, M., & Wilson, T. L. (2007). Mini-muddling: Learning from project plan deviations. *Journal of Workplace Learning*, *19*(2), 92–107. DOI: 10.1108/13665620710728475

Han, J., Zhang, L., & Li, Y. (2022). Spatiotemporal analysis of rural energy transition and upgrading in developing countries: The case of China. *Applied Energy*, *307*, 118225.

Hanke, F., Guyet, R., & Feenstra, M. (2022). Energy communities' social role in a just energy transition. *Elsevier EBooks*, 195–208. DOI: 10.1016/B978-0-323-91135-1.00027-4

Hanna, R., Heptonstall, P., & Gross, R. (2024). Job creation in a low carbon transition to renewables and energy efficiency: A review of international evidence. *Sustainability Science*, *19*(1), 125–150. DOI: 10.1007/s11625-023-01440-y

Hanson, G. H. (2023). *Local labor market impacts of the energy transition: Prospects and policies* (No. w30871). National Bureau of Economic Research. https://doi.org/DOI: 10.3386/w30871

Harichandan, S., Kar, S. K., Bansal, R., Mishra, S. K., Balathanigaimani, M. S., & Dash, M. (2022). Energy transition research: A bibliometric mapping of current findings and direction for future research. *Cleaner Production Letters*, *3*, 100026. DOI: 10.1016/j.clpl.2022.100026

Harris, M. (2021). Māori approaches to energy: Geothermal development and Indigenous sovereignty. *Energy Policy*, *156*, 112403. DOI: 10.1016/j.enpol.2021.112403

Hassink, J., Hulsink, W., & Grin, J. (2016). Entrepreneurship in agriculture and healthcare: Different entry strategies of care farmers. *Journal of Rural Studies*, *43*, 27–39. DOI: 10.1016/j.jrurstud.2015.11.013

Hatziargyriou, N., Margaris, I., Stavropoulou, I., Papathanassiou, S., & Dimeas, A. (2017). Noninterconnected Island systems: The Greek case. *IEEE Electrification Magazine*, *5*(2), 17–27. DOI: 10.1109/MELE.2017.2685739

Hauser, W. J. (2007). Marketing analytics: the evolution of marketing research in the twenty-first century. *Direct marketing: an international journal, 1*(1), 38-54.

Hayawi, K., Shahriar, S., & Al-Ali, A. R. (2024). *Ethical Considerations in AI Applications in Smart Grid.* DOI: 10.20944/preprints202412.1883.v1

Healy, N., & Barry, J. (2017). Politicizing energy justice and energy system transitions: Fossil fuel divestment and a "just transition.". *Energy Policy*, *108*, 451–459. DOI: 10.1016/j.enpol.2017.06.014

HEDNO. (2023). The Electricity Market of Non-Interconnected Islands in Greece, Islands Network Operation Department, Argiro Magkanioti Head of Electricity Generation Development and Management Section, https://clean-energy-islands.ec.europa.eu/system/files/2023-01/HEDNO%20-%20Operator%20of%20NIIs.pdf

Heffron, R. J., & McCauley, D. (2017). The concept of energy justice across the disciplines. *Energy Policy*, *105*, 658–667. DOI: 10.1016/j.enpol.2017.03.018

Heffron, R. J., & McCauley, D. (2018). What is the 'just transition'? *Geoforum*, *88*, 74–77. DOI: 10.1016/j.geoforum.2017.11.016

Hellenic Association for Energy Economics. (2023). Assessing the energy transition of the Greek island, *IN FOCUS,* Issue 4, at https://www.haee.gr/media/5374/haee_2023_infocus_issue_4_standard_final.pdf]

Hellenic Data Service. (2018). *Natura 2000 Network and Protected Areas [Data set]*. https://data.hellenicdataservice.gr/dataset/289263ce-04be-4e88-941c-81ea79da7169

Hellenic Republic. (2020). *National Development Program 2021-2025.* Hellenic Republic - Ministry of Development and Investment. (in Greek)

Hellenic Statistical Authority. (2025). Greece in numbers (in Greek) https://www.statistics.gr/documents/20181/18573742/GreeceInFigures_2025Q1_GR.pdf/871e88cc-0188-a27b-0a6a-3bc8e03f56b3

Hellenic Wind Energy Scientific Association (ELETAEN). (2024). *HWEA Status & Challenges for the Supply Chain for Offshore Wind in Greece* [PDF report]. Retrieved from https://eletaen.gr/wp-content/uploads/2024/09/2024-09-02-OW-Supply-Chain-Greece-ELETAEN-consolidated-report.pdf

Hellenic Wind Energy Scientific Association (ELETAEN). (2025). *HWEA Wind Energy Statistics* [PDF report]. Retrieved from https://eletaen.gr/wp-content/uploads/2025/07/2025-07-09-eletaen-statistic-s1-f-2025.pdf

Helli, S. S., Tanberk, S., & Demir, O. (2022). *Forecasting energy consumption using deep learning in smart cities. In 2022 International Conference on Artificial Intelligence of Things (ICAIoT).* IEEE., https://ieeexplore.ieee.org/abstract/document/10121846

Henderson, C. (2013). *Aboriginal power: Clean energy and the future of Canada's First Peoples*. Rainforest Editions.

Henry, M. S., Bazilian, M. D., & Markuson, C. (2020). Just transitions: Histories and futures in a post-COVID world. *Energy Research & Social Science*, *68*, 101668. DOI: 10.1016/j.erss.2020.101668 PMID: 32839696

Herring, H. (2006). Energy efficiency—A critical view. *Energy*, *31*(1), 10–20. DOI: 10.1016/j.energy.2004.04.055

Hess, M. C., Gómez-Ruiz, P. A., Morellato, L. P. C., & Buisson, E. (2022). Phenological patterns of herbaceous Mediterranean plant communities in spring. *Plant Ecology and Evolution*, *155*(2), 207–220. DOI: 10.5091/plecevo.86335

Hidayati, A., Susanti, E., Jamalong, A., Ginting, D., Suwanto, W., & Arifin, A. (2024). Marketing analytics in the era of digital-based marketing strategy. *Jurnal Ilmiah Ilmu Terapan Universitas Jambi*, *8*(1), 61–75. DOI: 10.22437/jiituj.v8i1.31908

Hoicka, C. E., Savic, K., & Campney, A. (2021). Reconciliation through renewable energy? A survey of Indigenous communities, involvement, and peoples in Canada. *Energy Research & Social Science*, *74*, 101897. DOI: 10.1016/j.erss.2020.101897

Hollway, W., & Jefferson, T. (2012). Doing qualitative research differently: A psychosocial approach. https://doi.org/DOI: 10.4135/9781526402233

Hopwood, A. G., & Unerman, J. (Eds.). (2010). *Accounting for sustainability: Practical insights*. Earthscan. DOI: 10.4324/9781849776332

Hristov, I., & Chirico, A. (2019). The role of sustainability key performance indicators (KPIs) in implementing sustainable strategies. *Sustainability (Basel)*, *11*(20), 5742. DOI: 10.3390/su11205742

Hrnčić, B., Pfeifer, A., Jurić, F., Duić, N., Ivanović, V., & Vušanović, I. (2021). Different investment dynamics in energy transition towards a 100% renewable energy system. *Energy*, *237*, 121526. DOI: 10.1016/j.energy.2021.121526

Huang, B., Zheng, S., Ma, B., Yang, Y., Zhang, S., & Jin, L. (2022). Using deep learning to predict the outcome of live birth from more than 10,000 embryo data. *BMC Pregnancy and Childbirth*, *22*(1), 36. DOI: 10.1186/s12884-021-04373-5 PMID: 35034623

Hubbard, G. (2009). Measuring organizational performance: Beyond the triple bottom line. *Business Strategy and the Environment*, *18*(3), 177–191. DOI: 10.1002/bse.564

Iacobucci, D., Petrescu, M., Krishen, A., & Bendixen, M. (2019). The state of marketing analytics in research and practice. *Journal of Marketing Analytics*, *7*(3), 152–181. DOI: 10.1057/s41270-019-00059-2

Icaza, D., Vallejo-Ramirez, D., Guerrero Granda, C., & Marín, E. (2024). Challenges, roadmaps and smart energy transition towards 100% renewable energy markets in American Islands: A review. *Energies*, *17*(5), 1059. DOI: 10.3390/en17051059

ICTPLUS. (2024). Hellenic Privatisation Fund: how the Island Resettlement Fund will work (in Greek) https://www.ictplus.gr/taiped-pos-tha-leitourgei-to-tameio-apanthrakopoiisis-nision/

IEA. (2015). Policy pathway-Accelerating energy efficiency in small and medium-sized enterprises. In: International Energy Agency Paris, France.

IEA. (2016). *Energy policies of IEA countries: Belgium 2016 review*. OECD/IEA.

IEA. (2021). Energy system of The Netherlands. In.

IEA. I. (2023). CO2 Emissions in 2022. *IEA, Paris*.

Ikevuje, A. H., Kwakye, J. M., Ekechukwu, D. E., Ogundipe, O. B., & Esiri, A. E.Augusta Heavens IkevujeJephta Mensah KwakyeDarlington Eze Ekechuk-wuOlorunshogo Benjamin OgundipeAndrew Emuobosa Esiri. (2023). Energy justice: Ensuring equitable access to clean energy in underprivileged communities. *Magna Scientia Advanced Research and Reviews*, *8*(2), 211–218. DOI: 10.30574/msarr.2023.8.2.0097

Iliopoulou, C., Kepaptsoglou, K., & Schinas, O. (2018). Energy supply security for the Aegean islands: A routing model with risk and environmental considerations. *Energy Policy*, *113*, 608–620. DOI: 10.1016/j.enpol.2017.11.032

ILO. (2022). *Just Transition Guidelines for a Greener Future*. https://www.ilo.org/global/topics/green-jobs/WCMS_432859/lang--en/index.htm

Inderwildi, O., Zhang, C., Wang, X., & Kraft, M. (2020). The impact of intelligent cyber-physical systems on the decarbonization of energy. *Energy & Environmental Science*, *13*(3), 744–771. DOI: 10.1039/C9EE01919G

Indrajayanthan, V., & Mohanty, N. K. (2022). Assessment of clean energy transition potential in major power-producing states of India using multi-criteria decision analysis. *Sustainability*, *14*(3), 1166.

In, S. Y., Monk, A. H. B., & Knox-Hayes, J. (2020). Financing Energy Innovation: The Need for New Intermediaries in Clean Energy. *Sustainability (Basel)*, *12*(24), 10440. DOI: 10.3390/su122410440

International Energy Agency (IEA). (2021). *World Energy Outlook 2021*. IEA.

International Energy Agency (IEA). (2022). *World Energy Outlook 2022*. IEA. International Energy Agency (IEA). (2023). *Tracking Clean Energy Progress 2023*. IEA.

International Energy Agency (IEA). (2023a). *Electricity Market Report 2023*. IEA.

International Energy Agency (IEA). (2023b). *Global Energy Review 2023.* IEA.

International Energy Agency. (2011). *Energy Policies of IEA Countries: Greece 2011 Review*. OECD/IEA. Retrieved from https://www.iea.org/reports/energy-policies-of-iea-countries-greece-2011-review

International Energy Agency. (2021). Net Zero by 2050: A roadmap for the global energy sector. https://www.iea.org/reports/net-zero-by-2050

International Energy Agency. (2024). India and the IEA. https://www.iea.org/countries/india

International Energy Agency. (2025). Global Energy Review 2025, https://iea.blob.core.windows.net/assets/5b169aa1-bc88-4c96-b828-aaa50406ba80/GlobalEnergyReview2025.pdf

International Labour Organization (ILO). (1989). *C169 - Indigenous and Tribal Peoples Convention, 1989 (No. 169).* ILO.

International Renewable Energy Agency (IRENA). (2023). *World Energy Transitions Outlook 2023.* IRENA.

International Renewable Energy Agency. (2023). *Renewable energy prospects for India.* IRENA.

International Renewable Energy Agency. (2025). *Renewable energy statistics 2025.* Abu Dhabi: IRENA. https://www.irena.org/

International Solar Alliance. (2024). *Annual report. Gurugram.* ISA Secretariat.

Ioannidis, A., & Chalvatzis, K. J. (2017). Energy supply sustainability for Island Nations: A study on 8 Global Islands. *Energy Procedia*, *142*, 3028–3034. DOI: 10.1016/j.egypro.2017.12.440

Ioannidis, A., Chalvatzis, K. J., Li, X., Notton, G., & Stephanides, P. (2019). The case for islands' energy vulnerability: Electricity supply diversity in 44 global islands. *Renewable Energy*, *143*, 440–452. DOI: 10.1016/j.renene.2019.04.155

Ioannidis, F., Kosmidou, K., & Papanastasiou, D. (2023). Public awareness of renewable energy sources and Circular Economy in Greece. *Renewable Energy*, *206*, 1086–1096. DOI: 10.1016/j.renene.2023.02.084

IPCC. (2022). *Climate Change 2022: Impacts, Adaptation and Vulnerability.* https://www.ipcc.ch/report/ar6/wg2/

IPCC. (2023). *Synthesis Report: Climate Change 2023.* Intergovernmental Panel on Climate Change. https://www.ipcc.ch/report/ar6/syr/

IPTO, (2023). IPTO is the new project promoter of the Greece-Cyprus-Israel Interconnection, IPTO is the new project promoter of the Greece-Cyprus-Israel Interconnection I IPTO

IPTO. (2024). Public consultation: Preliminary draft of the Ten Year Development Plan of HETS, for the period 2025-2034, https://www.admie.gr/en/anakoinoseis/announcement/public-consultation-preliminary-draft-ten-year-development-plan-hets

Iram, R., Zhang, J., Erdogan, S., Abbas, Q., & Mohsin, M. (2020). Economics of energy and environmental efficiency: Evidence from OECD countries. *Environmental Science and Pollution Research International*, *27*(4), 3858–3870. DOI: 10.1007/s11356-019-07020-x PMID: 31823262

IRENA. (2023). *Renewable Energy and Jobs – Annual Review 2023*. International Renewable Energy Agency. https://www.irena.org/publications

Irfan, M. (2021)... *Renewable Energy*, *179*, 65–74.

Irfan, M., Zhao, Z. Y., Ikram, M., Gilal, N. G., Li, H., & Rehman, A. (2020). Assessment of India's energy dynamics: Prospects of solar energy. *Journal of Renewable and Sustainable Energy*, *12*(5).

Jadhav, D. S., & Sarangi, S. (2025). Corporate sustainability and performance: A bibliometric review for two decades. *Benchmarking*, *32*(4), 1459–1483. DOI: 10.1108/BIJ-07-2023-0467

Jafari, A., Fox, G., Rundle, J. B., Donnellan, A., & Ludwig, L. G. (2024). Time series foundation models and deep learning architectures for earthquake temporal and spatial nowcasting. *GeoHazards*, *5*(4), 1247–1274. https://www.mdpi.com/2624-795X/5/4/59. DOI: 10.3390/geohazards5040059

Jaffar, A. (2020). Indigenous energy sovereignty: Models and pathways for equitable transitions. *Energy Research & Social Science*, *69*, 101730. DOI: 10.1016/j.erss.2020.101730

Jain, M., AlSkaif, T., & Dev, S. (2022). Are deep learning models more effective against traditional models for load demand forecasting? In *2022 International Conference on Smart Energy Systems and Technologies (SEST)* (pp. 1-6). IEEE. https://ieeexplore.ieee.org/abstract/document/9898424

Jain, V., & Mitra, A. (2024). Ethical Implications of Artificial Intelligence in Promoting Sustainable Clean Energy Solutions. *Practice, Progress, and Proficiency in Sustainability*, •••, 96–118. DOI: 10.4018/979-8-3693-6567-0.ch006

Jairaj, B., & Kumar, P. (2019). An assessment of India's energy transition: Paris and beyond. *Jindal Global Law Review, 10*(1), 35–47.

Jana, S. K., Ghosh, M., & Karmakar, A. K. (2019). State-Level Status of Renewable Energy Development in India. In *Handbook of Research on Economic and Political Implications of Green Trading and Energy Use* (pp. 237–251). IGI.

Jana, S. K., & Singh, K. (2022). Progress and determinants of renewable energy development in India. In *Eco-Friendly and Agile Energy Strategies and Policy Development* (pp. 190–203). IGI Global.

Janeková, J., Fabianová, J., Onofrejová, D., Puskas, E., & Busa, M. (2017). Implementation of deviation analysis method in the utilisation phase of the investment project: A case study. *Polish. Journal of Management Studies, 15*(1), 99–109. Advance online publication. DOI: 10.17512/pjms.2017.15.1.10

Janghorban, R., Roudsari, R. L., & Taghipour, A. (2014). Skype interviewing: The new generation of online synchronous interview in qualitative research. *International Journal of Qualitative Studies on Health and Well-being, 9*(1), 24152. DOI: 10.3402/qhw.v9.24152 PMID: 24746247

Janišová, M., Bystrica, B., Vrahnakis, M., Dengler, J., & Rūsiņa, S. (2010). In *The European Dry Grassland Group. Bulletin, 6*, •••. https://edgg.org/publ/bulletin/Bulletin_EDGG_06.pdf

Jankalová, M., & Jankal, R. (2024). Review of Sustainability Accounting Terms. *Administrative Sciences, 14*(7), 137. DOI: 10.3390/admsci14070137

Jankowska, K. (2016). Poland's clash over energy and climate policy: green economy or grey status quo? In *The European Union in International Climate Change Politics* (pp. 169–182). Routledge.

Jaramillo-Nieves, L., & Del Río, P. (2010). Contribution of renewable energy sources to the sustainable development of islands: An overview of the literature and a research agenda. *Sustainability (Basel), 2*(3), 783–811. DOI: 10.3390/su2030783

Jasanoff, S., & Kim, S. H. (2015). *Dreamscapes of modernity: Sociotechnical imaginaries and the fabrication of power*. University of Chicago Press. DOI: 10.7208/chicago/9780226276663.001.0001

Javadian, G., Dobratz, C., Gupta, A., Gupta, V. K., & Martin, J. A. (2020). Qualitative research in entrepreneurship studies: A state-of-science. *The Journal of Entrepreneurship, 29*(2), 223–258. DOI: 10.1177/0971355720930564

Jayaraman, S. T, N., S, A., & G, S. (2025). Enhancing urban air quality prediction using time-based-spatial forecasting framework. *Scientific Reports, 15*(1), 4139. https://www.nature.com/articles/s41598-024-83248-z

Jebali, E., Essid, H., & Khraief, N. (2017). The analysis of energy efficiency of the Mediterranean countries: A two-stage double bootstrap DEA approach. *Energy, 134*, 991–1000. https://doi.org/https://doi.org/10.1016/j.energy.2017.06.063. DOI: 10.1016/j.energy.2017.06.063

Jędrych, E., Klimek, D., & Rzepka, A. (2021). Principles of Sustainable Management of Energy Companies: The Case of Poland. *Energies, 14*(8), 2042. DOI: 10.3390/en14082042

Jenkins, K., McCauley, D., Heffron, R., Stephan, H., & Rehner, R. (2016). Energy justice: A conceptual review. *Energy Research & Social Science, 11*, 174–182. DOI: 10.1016/j.erss.2015.10.004

Jiao, L., & Tan, Y. (2024). Social Equity and Renewable Energy Access in Developing Countries. *MEDAAD, 2024*, 27–33. DOI: 10.70470/MEDAAD/2024/005

Jin, W., Zhang, H., Liu, S., & Zhang, H. (2019). Technological innovation, environmental regulation, and green total factor efficiency of industrial water resources. *Journal of Cleaner Production, 211*, 61–69. https://doi.org/https://doi.org/10.1016/j.jclepro.2018.11.172. DOI: 10.1016/j.jclepro.2018.11.172

Johansson, M. T., & Thollander, P. (2018). A review of barriers to and driving forces for improved energy efficiency in Swedish industry–Recommendations for successful in-house energy management. *Renewable & Sustainable Energy Reviews, 82*, 618–628. DOI: 10.1016/j.rser.2017.09.052

John, V. (2024). Energy Islands in Greece: Are they Feasible in Aegean Sea? *International Journal of Advanced Multidisciplinary Research and Studies, 4*(2), 148–155. DOI: 10.62225/2583049X.2024.4.2.2454

Joi a, D., & Dobrotă, C. E. (2022). Aspects of European funding for the energy transition-Just transition. *Ovidius Univ. Ann. Ser. Econ. Sci, 22*(1), 289–297. DOI: 10.61801/OUAESS.2022.1.39

Jorgensen, M. (2016). Sovereignty, wealth, culture, and control: The contemporary political economy of Indian gaming. *The American Behavioral Scientist, 60*(6), 713–733. DOI: 10.1177/0002764216632838

Joshi, A., & Pandey, A. (2025). Advancing Climate Finance for Sustainable Future: Integrating Human Capital, Climate Neutrality, and Emerging Technologies. *Climate Neutrality and Sustainable Ecosystems: Practical Approaches to Addressing Global Warming*, 231-246.

Kakatkar, C., & Spann, M. (2019). Marketing analytics using anonymized and fragmented tracking data. *International Journal of Research in Marketing*, *36*(1), 117–136. DOI: 10.1016/j.ijresmar.2018.10.001

Kaldellis, J. K., Boulogiorgou, D., Kondili, E. M., & Triantafyllou, A. G. (2023). Green Transition and Electricity Sector Decarbonization: The Case of West Macedonia. *Energies*, *16*(16), 5970. DOI: 10.3390/en16165970

Kaldellis, J. K., Gkikaki, A., Kaldelli, E., & Kapsali, M. (2012). Investigating the energy autonomy of very small non-interconnected islands: A case study: Agathonisi, Greece. *Energy for Sustainable Development*, *16*(4), 476–485. DOI: 10.1016/j.esd.2012.08.002

Kallimanis, A. S., Tsiafouli, M. A., Pantis, J. D., Mazaris, A. D., Matsinos, Y., & Sgardelis, S. P. (2008). Arable land and habitat diversity in Natura 2000 sites in Greece. *Journal of Biological Research (Thessaloniki)*, *9*, 55–66.

Kalogirou, S. A. (2003). The energy subsidisation policies of Cyprus and their effect on renewable energy systems economics. *Renewable Energy*, *28*(11), 1711–1728. DOI: 10.1016/S0960-1481(03)00062-4

Kantabutra, S. (2024). Toward a sustainability performance management framework. *Heliyon*, *10*(13), e33729. DOI: 10.1016/j.heliyon.2024.e33729 PMID: 39071584

Kapoor, K., Pandey, K. K., Jain, A. K., & Nandan, A. (2014). Evolution of solar energy in India: A review. *Renewable & Sustainable Energy Reviews*, *40*, 475–487.

Karamountzou, S., & Vagiona, D. G. (2023). Suitability and sustainability assessment of existing onshore wind farms in Greece. *Sustainability (Basel)*, *15*(3), 2095. DOI: 10.3390/su15032095

Karampela, S., Kizos, T., & Spilanis, I. (2014). Accessibility of islands: Towards a new geography based on transportation modes and choices. *Island Studies Journal*, *9*(2), 293–306. DOI: 10.24043/isj.307

Karasmanaki, E., Ioannou, K., Katsaounis, K., & Tsantopoulos, G. (2020). The attitude of the local community towards investments in lignite before transitioning to the post-lignite era: The case of Western Macedonia, Greece. *Resources Policy*, *68*, 101781. DOI: 10.1016/j.resourpol.2020.101781

Kar, S. K., Sharma, A., & Roy, B. (2016). Solar energy market developments in India. *Renewable & Sustainable Energy Reviews, 62*, 121–133.

Karthick, K., Dharmaprakash, R., & Sathya, S. (2024). Predictive modeling of energy consumption in the steel industry using CatBoost regression: A data-driven approach for sustainable energy management. *International Journal of Robotics and Control Systems, 4*(1), 33–49. https://pubs2.ascee.org/index.php/IJRCS/article/view/1234. DOI: 10.31763/ijrcs.v4i1.1234

Karystianos, M. K., Pitas, C. N., Efstathiou, S. P., Tsili, M. A., Mantzaris, J. C., Leonidaki, E. A., Voumvoulakis, E. M., & Sakellaridis, N. G. (2021). Planning of Aegean Archipelago Interconnections to the Continental Power System of Greece, Energies, 14, 3818, Katsaprakakis, D. A., Proka, A., Zafirakis, D., Damasiotis, M., Kotsampopoulos, P., Hatziargyriou, N., ... & Xevgenos, D. (2022). Greek Islands' energy transition: From lighthouse projects to the emergence of energy communities. *Energies, 15*(16), 5996.

Katemliadis, I., & Markatos, G. (2021). Stakeholders' involvement in sustainability planning and implementation: The case of Cyprus. *Worldwide Hospitality and Tourism Themes, 13*(6), 709–718. DOI: 10.1108/WHATT-07-2021-0095

Kati, V., Kassara, C., Vrontisi, Z., & Moustakas, A. (2021). The biodiversity-wind energy-land use nexus in a global biodiversity hotspot. *The Science of the Total Environment, 768*, 144471. DOI: 10.1016/j.scitotenv.2020.144471 PMID: 33454485

Kati, V., Spiliopoulou, K., Stefanidis, A., & Kassara, C. (2025). Sacrificing Wilderness for Renewables? Land Artificialization from Inadequate Spatial Planning of Wind Energy in Evvoia, Greece. *Land (Basel), 14*(6), 1296. DOI: 10.3390/land14061296

Katsaprakakis, D. A., Proka, A., Zafirakis, D., Damasiotis, M., Kotsampopoulos, P., Hatziargyriou, N., Dakanali, E., Arnaoutakis, G., & Xevgenos, D. (2022). Greek Islands' energy transition: From lighthouse projects to the emergence of energy communities. *Energies, 15*(16), 5996. DOI: 10.3390/en15165996

Katsoufis, P. (2017). *Interconnection process of photovoltaic systems into the Electricity Authority of Cyprus's power grid using the method of self-production* (Master's thesis, Universitat Politècnica de Catalunya).

Katsoulakos, N. M. (2019). An overview of the Greek Islands' autonomous electrical systems: Proposals for a sustainable energy future. *Smart Grid and Renewable Energy, 10*(04), 55–82. DOI: 10.4236/sgre.2019.104005

Kaur, S., Kumar, A., Sharma, P., Gulati, G., & Kaur, A. (2025). Green Consumer Behaviour: A Bibliometric Analysis and Insight on the Factors Influencing Green Product Adoption. In Green Economics and Strategies for Business Sustainability (pp. 289-314). IGI Global Scientific Publishing. DOI: 10.4018/979-8-3693-8949-2.ch012

Kavadias, K. A., Zafirakis, D., Kondili, E., & Kaldellis, J. K. (2007). The contribution of renewables on reducing the electricity generation cost in autonomous island networks. In *2007 International Conference on Clean Electrical Power* (pp. 777-783). DOI: 10.1109/ICCEP.2007.384302

Kavitha, C. R., Varalatchoumy, M., Mithuna, H. R., Bharathi, K., Geethalakshmi, N. M., & Boopathi, S. (2023). Energy Monitoring and Control in the Smart Grid. *Advances in Bioinformatics and Biomedical Engineering Book Series*, 290–316. DOI: 10.4018/978-1-6684-6577-6.ch014

KEDE. (2024) New 2 billion fund for the carbonization of the Greek islands, Central Union of Greek Municipalities (in Greek) https://kede.gr/neo-tameio-2-dis-gia-tin-apanthrakopoiisi-ton-ellinikon-nision/

Keller, G. R., & Berry, J. (2007). Reduced Impact Logging Road Issues in Tropical Forests. *Transportation Research Record*, 1989-1(1), 98-106. DOI: 10.3141/1989-11

Kentas, G. (2023). Cyprus: Global Energy Crisis Is an Opportunity to Tackle Climate Change. In *Climate Change and the Future of Europe: Views from the Capitals* (pp. 19–22). Springer International Publishing., DOI: 10.1007/978-3-031-23328-9_5

Kerr, N., Gouldson, A., & Barrett, J. (2017). The rationale for energy efficiency policy: Assessing the recognition of the multiple benefits of energy efficiency retrofit policy. *Energy Policy*, *106*, 212–221. DOI: 10.1016/j.enpol.2017.03.053

Khan, A. N., Iqbal, N., Rizwan, A., Ahmad, R., & Kim, D. H. (2021). An ensemble energy consumption forecasting model based on spatial-temporal clustering analysis in residential buildings. *Energies*, *14*(11), 3020. https://www.mdpi.com/1996-1073/14/11/3020. DOI: 10.3390/en14113020

Khanom, M. T. (2023). Using social media marketing in the digital era: A necessity or a choice. International Journal of Research in Business and Social Science (2147-4478), 12(3), 88–98.

Khan, S. Z., Yang, Q., & Waheed, A. (2019). Investment in intangible resources and capabilities spurs sustainable competitive advantage and firm performance. *Corporate Social Responsibility and Environmental Management*, *26*(2), 285–295. DOI: 10.1002/csr.1678

Khobragade, P. P., & Ahirwar, A. V. (2022). Source apportionment of SPM by positive matrix factorization and PM2. 5 analysis in an urban industrial area. *World Journal of Engineering*, *19*(4), 446–458. https://www.emerald.com/wje/article/19/4/446/458644/Source-apportionment-of-SPM-by-positive-matrix. DOI: 10.1108/WJE-11-2020-0550

Kilintzis, P., Samara, E., Topaloglou, L., Avlogiaris, G., & Kafetzopoulos, D. (2023). The role of Green Public Procurements in energy transition: The case of Western Macedonia. *Journal of Innovation and Entrepreneurship*, *12*(1), 87. DOI: 10.1186/s13731-023-00354-4

Kim, J. E. (2019). Sustainable energy transition in developing countries: The role of energy aid donors. *Climate Policy*, *19*(1), 1–16.

Kim, J., Jaumotte, F., Panton, A. J., & Schwerhoff, G. (2025). Energy security and the green transition. *Energy Policy*, *198*, 114409. DOI: 10.1016/j.enpol.2024.114409

Kimmerer, R. W. (2013). *Braiding sweetgrass: Indigenous wisdom, scientific knowledge, and the teachings of plants*. Milkweed Editions.

Kindvall, O., Franzén, M., Askling, J., Forsman, A., & Johansson, V. (2022). Subsidized Common Agricultural Policy grazing jeopardizes the protection of biodiversity and Natura 2000 targeted species. *Animal Conservation*, *25*(5), 597–607. DOI: 10.1111/acv.12773

Kitivo, E. M., Kanga, A., & Mbae, J. (2021). Evaluation of training programmes: A review of selected models and approaches. *International Journal of Educational Theory and Practice*, *4*(1-4), 1–16.

Kleinschroth, F., Healey, J. R., Sist, P., Mortier, F., & Gourlet-Fleury, S. (2016). How persistent are the impacts of logging roads on Central African forest vegetation? *Journal of Applied Ecology*, *53*(4), 1127–1137. DOI: 10.1111/1365-2664.12661

Klyuev, R. V., Morgoev, I. D., Morgoeva, A. D., Gavrina, O. A., Martyushev, N. V., Efremenkov, E. A., & Mengxu, Q. (2022). Methods of forecasting electric energy consumption: A literature review. *Energies*, *15*(23), 8919. https://www.mdpi.com/1996-1073/15/23/8919. DOI: 10.3390/en15238919

Kocmanova, A., Nemecek, P., & Docekalova, M. (2012). Environmental, social and governance (ESG) key performance indicators for sustainable reporting.

Kocmanová, A., & Dočekalová, M. (2011). Corporate sustainability: Environmental, social, economic and corporate performance. *Acta Universitatis Agriculturae et Silviculturae Mendelianae Brunensis*, *59*(7), 203–208. DOI: 10.11118/actaun201159070203

Kokkoris, I., Dimitrellos, G., Kougioumoutzis, K., Laliotis, I., Georgiadis, T., & Tiniakou, A. (2014). The native flora of Mountain Panachaikon (Peloponnese, Greece): New records and diversity. *Journal of Biological Research (Thessaloniki), 21*(1), 9. DOI: 10.1186/2241-5793-21-9 PMID: 25984492

Koktsidis, P. I., Constantinou, C. M., & Joseph, J. (2021). Legal and Geopolitical Aspects of Energy Development in the Eastern Mediterranean and the EU's Strategic Involvement. In *The New Eastern Mediterranean Transformed: Emerging Issues and New Actors* (pp. 119–137). Springer International Publishing., DOI: 10.1007/978-3-030-70554-1_6

Kolokasidou, A., & Van Son, L. (2022). The EastMed Pipeline-How it is expected to impact the geostrategic position of Cyprus.

Kolovos, N., Georgakopoulos, A., Filippidis, A., & Kavouridis, C. (2002). Environmental effects of lignite and intermediate steriles coexcavation in the southern lignite field mine of Ptolemais, northern Greece. *Energy Sources, 24*(6), 561–573. DOI: 10.1080/00908310290086563

Komarova, A. V., Filimonova, I. V., & Kartashevich, A. A. (2022). Energy consumption of the countries in the context of economic development and energy transition. *Energy Reports, 8*, 683–690.

Komninou, M., Afouxenidis, A., Souliotis, N., & Karoulas, G. (2023). Windfarms, social acceptability, and the environmental divide in Greece. *Qeios, RZBUH5*. Advance online publication. DOI: 10.32388/RZBUH5

Kona, A., Monforti-Ferrario, F., Bertoldi, P., Baldi, M. G., Kakoulaki, G., Vetters, N., Thiel, C., Melica, G., Lo Vullo, E., Sgobbi, A., Ahlgren, C., & Posnic, B. (2021). Global Covenant of Mayors, a dataset of greenhouse gas emissions for 6200 cities in Europe and the Southern Mediterranean countries. *Earth System Science Data, 13*(7), 3551–3564. DOI: 10.5194/essd-13-3551-2021

Konstantinidis, E. I., Kompolias, D. G., & Botsaris, P. N. (2014). Viability analysis of an offshore wind farm in North Aegean Sea, Greece. *Journal of Renewable and Sustainable Energy, 6*(2), 023116. Advance online publication. DOI: 10.1063/1.4871484

Konstantopoulos, K., Moustakas, A., & Vogiatzakis, I. N. (2020). A spatially explicit impact assessment of road characteristics, road-induced fragmentation and noise on birds species in Cyprus. *Biodiversity (Nepean), 21*(1), 61–71. DOI: 10.1080/14888386.2020.1736154

Korakis, G., & Gerasimidis, A. (2010). Vegetation and habitat types. In Catsadorakis, G., & Källander, H. (Eds.), *The Dadia–Lefkimi–Souli Forest National Park, Greece: Biodiversity, Management and Conservation* (pp. 85–93). WWF Greece.

Koroneos, C., Fokaidis, P., & Moussiopoulos, N. (2005). Cyprus energy system and the use of renewable energy sources. *Energy, 30*(10), 1889–1901. DOI: 10.1016/j.energy.2004.11.011

Kosarkoska, D., & Joldeska, I. (2024). The impact of climate change and climate risk on the Accounting and Accounting professions.

Kotzebue, J. R., & Weissenbacher, M. (2020). The EU's Clean Energy strategy for islands: A policy perspective on Malta's spatial governance in energy transition. *Energy Policy, 139*, 111361. DOI: 10.1016/j.enpol.2020.111361

Kougias, I., Szabó, S., Nikitas, A., & Theodossiou, N. (2019). Sustainable energy modelling of non-interconnected Mediterranean islands. *Renewable Energy, 133*, 930–940. DOI: 10.1016/j.renene.2018.10.090

Kougioumoutzis, K., Kaloveloni, A., & Petanidou, T. (2022). Assessing climate change impacts on Island bees: The Aegean Archipelago. *Biology (Basel), 11*(4), 552. DOI: 10.3390/biology11040552 PMID: 35453751

Koukoura, Z. (2007). Natural grasslands—a case study in Greece. In *Nutrient Cycling in Terrestrial Ecosystems* (pp. 257–269). Springer Berlin Heidelberg. DOI: 10.1007/978-3-540-68027-7_9

Kourkouridis, D., Salepaki, A., Kasselouris, G., & Baltas, E. (2024). Regional and Local Development in Energy Transition Areas: The Case of Lignite Phase-Out in the Peloponnese. *Sustainability (Basel), 16*(14), 6153. DOI: 10.3390/su16146153

Kral, A. H., Malekinejad, M., Vaudrey, J., Martinez, A. N., Lorvick, J., McFarland, W., & Raymond, H. F. (2010). Comparing respondent-driven sampling and targeted sampling methods of recruiting injection drug users in San Francisco. *Journal of Urban Health, 87*(5), 839–850. DOI: 10.1007/s11524-010-9486-9 PMID: 20582573

Krechovská, M., & Procházková, P. T. (2014). Sustainability and its integration into corporate governance focusing on corporate performance management and reporting. *Procedia Engineering, 69*, 1144–1151. DOI: 10.1016/j.proeng.2014.03.103

Krippendorff, K. (2018). *Content analysis: An introduction to its methodology*. Sage publications., DOI: 10.4135/9781071878781

Krupa, J. (2012). Blazing a new path forward: A case study on the renewable energy initiatives of the Pic River First Nation. *Environmental Development, 3*, 109–122. DOI: 10.1016/j.envdev.2012.05.003

Kruthi, N., Kumar, A. A., Aromal, A. J., & Ravindra, C. (2023). QuakeML: ML-Based Earthquake Prediction and Warning System. In *2023 IEEE Fifth International Conference on Advances in Electronics, Computers and Communications (ICAECC)* (pp. 1-5). IEEE. https://ieeexplore.ieee.org/abstract/document/10560276

Kuhn, T. (2017). The revision of the German Fertiliser Ordinance in 2017.

Kuklina, V., Bilichenko, I., Bogdanov, V., Kobylkin, D., Petrov, A. N., & Shiklomanov, N. (2021). Informal road networks and sustainability of Siberian boreal forest landscapes: Case study of the Vershina Khandy taiga. *Environmental Research Letters, 16*(11), 115001. DOI: 10.1088/1748-9326/ac22bd

Kulkarni, M., Mulay, C., Marathe, S., & Itankar, P. (2021). Earthquake prediction using machine learning. *International Journal of Advance Research, Ideas and Innovations in Technology, 7*(3).

Kulkarni, S. H., & Anil, T. R. (2020). In *International Solutions to Sustainable Energy, Policies and Applications* (pp. 359-386). River Publishers.

Kumar Phookan, N., Sarmah, B., & Sarika Devi, T. (2020). Exploring determinants of green marketing practices in food service industry to attain environmental sustainability: A conceptual analysis. In Proceedings of the 5th International Conference on Computers & Management Skills (ICCM 2019)| North Eastern Regional Institute of Science & Technology (NERIST), Nirjuli, Arunachal Pradesh, India. DOI: 10.2139/ssrn.3517420

Kumar, A., & Meena, B. (2025). A comparative analysis of the Holt and ARIMA models for predicting the future total fertility rate in India. *Life Cycle Reliability and Safety Engineering, 14*(1), 117–126. DOI: 10.1007/s41872-024-00287-1

Kumar, S., Smith, S. R., Fowler, G., Velis, C., Kumar, S. J., Arya, S., & Cheeseman, C. (2017)... *Royal Society Open Science, 4*(3), 160764.

Kunsch, P. L., & Friesewinkel, J. (2014). Nuclear energy policy in Belgium after Fukushima. *Energy Policy, 66*, 462–474. https://doi.org/https://doi.org/10.1016/j.enpol.2013.11.035. DOI: 10.1016/j.enpol.2013.11.035

Kurniawan, A., & Utami, Z. R. (2025). Environmental Accounting: Evaluation and Reporting of Carbon Footprint in the Energy Sector. *FIDUSIA: JURNAL KEUANGAN DAN PERBANKAN, 8*(1).

Kuzemko, C., Blondeel, M., Dupont, C., & Brisbois, M. C. (2022). Russia's war on Ukraine, European energy policy responses & implications for sustainable transformations. *Energy Research & Social Science*, *93*, 102842. DOI: 10.1016/j.erss.2022.102842

Kyriazi, A., & Miró, J. (2023). Towards a socially fair green transition in the EU? An analysis of the Just Transition Fund using the Multiple Streams Framework. *Comparative European Politics*, *21*(1), 112–132. DOI: 10.1057/s41295-022-00304-6

Kythreotou, N., & Mesmeris, T. (2023). 8th National Communication and 5th Biennial report under the UNFCCC of Cyprus. Department of Environment, Ministry of Agriculture, Rural 237 Development and Environment. Nicosia. Available online at https://unfccc.int/sites/default/files/resource/8th-NatCom-V1.4_submited301222.pdf

Labanca, N., & Bertoldi, P. (2018). Beyond energy efficiency and individual behaviours: Policy insights from social practice theories. *Energy Policy*, *115*, 494–502. https://doi.org/DOI. DOI: 10.1016/j.enpol.2018.01.027

LaDuke, W. (2020). *To be a water protector: The rise of the Wiindigoo slayers*. Fernwood Publishing.

Lafortune, G., Fuller, G., Kloke-Lesch, A., Koundouri, P., & Riccaboni, A. (2024). Europe Sustainable Development Report 2023-2024: European Elections, Europe's Future and the Sustainable Development Goals.

Lal, B. S. (2015). Green marketing: Opportunities and issues. *International Journal (Toronto, Ont.)*.

Lal, R., & Kumar, S. (2022). Energy security assessment of small Pacific Island Countries–Sustaining the call for renewable energy proliferation. *Energy Strategy Reviews*, *41*, 100866. DOI: 10.1016/j.esr.2022.100866

Lal, S. R. S., Herbert, G. M. J., Arjunan, P., & Suryan, A. (2025). Advancements in renewable energy transition in India: A review. *Energy Sources. Part A, Recovery, Utilization, and Environmental Effects*, *47*(2), 2024921.

Lamberton, G. (2005). Sustainability accounting - a brief history and conceptual framework. [). No longer published by Elsevier.]. *Accounting Forum*, *29*(1), 7–26. DOI: 10.1016/j.accfor.2004.11.001

Lamnatou, C., Cristofari, C., & Chemisana, D. (2024). Renewable energy sources as a catalyst for energy transition: Technological innovations and an example of the energy transition in France. *Renewable Energy*, *221*, 119600. DOI: 10.1016/j.renene.2023.119600

Larsen, T. (2019). The Hegemony of Green Capitalism: A critique of the imaginary that the environmental crisis can be solved by capitalism.

Larson, C. D., & Rew, L. J. (2022). Restoration intensity shapes floristic recovery after forest road decommissioning. *Journal of Environmental Management, 319*, 115729. DOI: 10.1016/j.jenvman.2022.115729 PMID: 35853306

Lassalle, J. M., Martínez, D. F., & Fernández, L. V. (2022). Optimisation of hybrid renewable energy systems on islands: A review. *Island Studies Journal, 17*(1), 221–242. DOI: 10.24043/isj.167

Latinopoulos, D., & Kechagia, K. (2015). A GIS-based multi-criteria evaluation for wind farm site selection. A regional scale application in Greece. *Renewable Energy, 78*, 550–560. DOI: 10.1016/j.renene.2015.01.041

Lavidas, G. (2019). Energy and socio-economic benefits from the development of wave energy in Greece. *Renewable Energy, 132*, 1290e1300. DOI: 10.1016/j.renene.2018.09.007

Lawrence, A., Friedrich, F., & Beierkuhnlein, C. (2021). Landscape fragmentation of the Natura 2000 network and its surrounding areas. *PLoS One, 16*(10), e0258615. DOI: 10.1371/journal.pone.0258615 PMID: 34673802

Leiren, M. D., Szulecki, K., Rayner, T., & Banet, C. (2019). Energy security concerns versus market harmony: The Europeanisation of capacity mechanisms. *Politics and Governance, 7*(1), 92–104. DOI: 10.17645/pag.v7i1.1791

Lempesi, A., & Daoutis, C. (2024). The effect of trail paths in grassland ecosystems using a SWOT-AHP Hybrid method. In: *Proceedings of 24th International Multidisciplinary Scientific GeoConference SGEM 2024*, 29th June 2024 to 8th July 2024, Albena, Bulgaria, STEF92 Technology, ()DOI: 10.5593/sgem2024/3.1/s14.49

Lempesi, A., Eleftheriadou, A., Delivasi, Z., Psyllidou, A., Korakis, G., & Kyriazopoulos, A. P. (2017). Effects of grazing intensity on the regeneration of woody species in an oak woodland. *Notulae Botanicae Horti Agrobotanici Cluj-Napoca, 45*(2), 597–601. DOI: 10.15835/nbha45210908

Leppänen, T., & Liefferink, D. (2022). Agenda-setting, policy formulation, and the EU institutional context: The case of the Just Transition Fund. *European Policy Analysis, 8*(1), 51–67. DOI: 10.1002/epa2.1136

Li, B. (2023). The role of financial markets in the energy transition: An analysis of investment trends and opportunities in renewable energy and clean technology. *Environmental Science and Pollution Research International, 30*(43), 97948–97964. DOI: 10.1007/s11356-023-29014-6 PMID: 37599346

Li, J., Luo, Y., & Wei, S. (2022). Long-term electricity consumption forecasting method based on system dynamics under the carbon-neutral target. *Energy*, *244*, 122572. https://www.sciencedirect.com/science/article/pii/S0360544221028218. DOI: 10.1016/j.energy.2021.122572

Li, L., Ma, W., Duan, X., Wang, S., Wang, Q., Gu, H., & Wang, J. (2024). Effects of Wind Farm Construction on Soil Nutrients and Vegetation: A Case Study of Linxiang Wind Farm in Hunan Province. *Sustainability (Basel)*, *16*(15), 6350. DOI: 10.3390/su16156350

Liu, J., Fang, Y., Liu, T., ... (2025). *"Policy synergy" or "strive alone"? "Dual-pilot" policy to enable renewable energy technology innovations*. Clean Technologies and Environmental Policy. https://link.springer.com/article/10.1007/s10098-025-03236-3

Liu, H., Li, Q., Yu, D., & Gu, Y. (2019). Air quality index and air pollutant concentration prediction based on machine learning algorithms. *Applied Sciences (Basel, Switzerland)*, *9*(19), 4069. https://www.mdpi.com/2076-3417/9/19/4069. DOI: 10.3390/app9194069

Liu, W., Shen, Y., & Razzaq, A. (2023). How renewable energy investment, environmental regulations, and financial development derive renewable energy transition: Evidence from G7 countries. *Renewable Energy*, *206*, 1188–1197. DOI: 10.1016/j.renene.2023.02.017

Lo Iacono, V., Symonds, P., & Brown, D. H. (2016). Skype as a tool for qualitative research interviews. *Sociological Research Online*, *21*(2), 103–117. DOI: 10.5153/sro.3952

Loewen, B. (2022). Coal, green growth and crises: Exploring three European Union policy responses to regional energy transitions. *Energy Research & Social Science*, *93*, 102849. DOI: 10.1016/j.erss.2022.102849

Logan, K. G., Nelson, J. D., McLellan, B. C., & Hastings, A. (2020). Electric and hydrogen rail: Potential contribution to net zero in the UK. *Transportation Research Part D, Transport and Environment*, *87*, 102523. DOI: 10.1016/j.trd.2020.102523

Lovins, A. (2017). Energy efficiency. *Energy Economics*, *1*, 234–258.

Lugo, A. E., & Gucinski, H. (2000). Function, effects, and management of forest roads. *Forest Ecology and Management*, *133*(3), 249–262. DOI: 10.1016/S0378-1127(99)00237-6

Luo, H., & Li, X. (2023). A hybrid model of CNN-BiLSTM and XGBoost for HVAC systems energy consumption prediction. In *2023 5th International Conference on Industrial Artificial Intelligence (IAI)* (pp. 1-6). IEEE. https://ieeexplore.ieee.org/abstract/document/10327594

Luo, S., Wang, B., Gao, Q., Wang, Y., & Pang, X. (2024). Stacking integration algorithm based on CNN-BiLSTM-Attention with XGBoost for short-term electricity load forecasting. *Energy Reports*, *12*, 2676–2689. https://www.sciencedirect.com/science/article/pii/S0275531923002581. DOI: 10.1016/j.egyr.2024.08.078

Luthra, S., Kumar, S., Garg, D., & Haleem, A. (2015). Barriers to renewable/sustainable energy technologies adoption: Indian perspective. *Renewable & Sustainable Energy Reviews*, *41*, 762–776.

Lypiridi, D. (2021). Just Transition Mechanism and Lignite Phase-Out in Greece: Challenges and Proposals for Policy Adoption. https://doi.org/DOI: 10.12681/hapscpbs.29493

Magni, G. (2017). Indigenous knowledge and implications for the sustainable development agenda. *European Journal of Education*, *52*(4), 437–447. DOI: 10.1111/ejed.12238

Magoni, M. (2018). European energy policies and regional burden sharing: What are the effects on the territory of Lombardy? *City. Territory and Architecture*, *5*(1), 1–7. DOI: 10.1186/s40410-018-0086-4

Mahmoud, T. O. (2018). Impact of green marketing mix on purchase intention. International Journal of Advanced and applied sciences, 5(2), 127-135.

Mahmoud, M. A., Seidu, A. S., Tweneboah-Koduah, E. Y., & Ahmed, A. S. (2024). Green marketing mix and repurchase intention: The role of green knowledge. *African Journal of Economic and Management Studies*, *15*(3), 501–518. DOI: 10.1108/AJEMS-04-2023-0137

Mahmoud, T. O. (2019). Green marketing: A marketing mix concept. International Journal of Electrical. *Electronics and Computers*, *4*(1), 20–26.

Majeed, Y., Khan, M. U., Waseem, M., Zahid, U., Mahmood, F., Majeed, F., Sultan, M., & Raza, A. (2023). Renewable energy as an alternative source for energy management in agriculture. *Energy Reports*, *10*, 344–359. DOI: 10.1016/j.egyr.2023.06.032

Majid, M. (2020).. . *Energy, Sustainability and Society*, *10*(1), 1–36.

Makridou, G., Doumpos, M., & Lemonakis, C. (2024). Relationship between ESG and corporate financial performance in the energy sector: Empirical evidence from European companies. *International Journal of Energy Sector Management, 18*(4), 873–895. DOI: 10.1108/IJESM-01-2023-0012

Makris, I., Apostolopoulos, S., & Anastasopoulou, E. E. (2024). An Entrepreneurial Perspective on the Transition of Lignite Rural Areas to a New Regime within a Suffocating Timeframe. *Administrative Sciences, 14*(4), 64. DOI: 10.3390/admsci14040064

Makris, I., Apostolopoulos, S., & Dimitrakopoulos, P. (2024). Analysing supporting mechanisms, cultural shaping and incentives for circular economy: The case of local governments. *Interdisciplinary Environmental Review, 23*(3), 218–235. DOI: 10.1504/IER.2024.139254

Makris, I., Apostolopoulos, S., Giannopoulos, V., Dimitrakopoulos, P., & Charalampakis, P. (2025). The Impact of Formal and Informal Institutional Elements on Land Mobility Within Rural Greece. *Sustainability (Basel), 17*(10), 4412. DOI: 10.3390/su17104412

Makris, I., Babalos, V., & Dimitrakopoulos, P. (2022). A study of the energy efficiency formula for the development of economic progress policies in Greece. *Energy Sources. Part B, Economics, Planning, and Policy, 17*(1), 2007178. DOI: 10.1080/15567249.2021.2007178

Mallinis, G., Chrysafis, I., Lempesi, A., Kyriazopoulos, A. P., Korakis, G., Kokkoris, I. P., & Dimopoulos, P. (2020). Spatiotemporal changes of forest line and canopy cover between 1945-2019 at selected Greek mountains. In: *Eighth International Conference on Remote Sensing and Geoinformation of the Environment (RSCy2020)*, 26 August 2020, Paphos, Cyprus, 11524, 236-248, 2020 Society of Photo-Optical Instrumentation Engineers (SPIE). https://doi.org/DOI: 10.1117/12.2570735

Mallouhy, R., Abou Jaoude, C., Guyeux, C., & Makhoul, A. (2019). Major earthquake event prediction using various machine learning algorithms. In *2019 international conference on information and communication technologies for disaster management (ICT-DM)* (pp. 1-7). IEEE. https://ieeexplore.ieee.org/abstract/document/9032983

Maltz, E., Bi, H. H., & Bateman, M. (2018). Benchmarking sustainability performance: The next step in building sustainable business models. *Journal of Public Affairs, 18*(3), e1606. DOI: 10.1002/pa.1606

Manalu, C. I. B., Manurung, M., Muda, I., & Kesuma, S. A. (2023). Implementation of stakeholder theory in sustainability accounting: A literature review. *Brazilian Journal of Development, 9*(12), 31853–31871. DOI: 10.34117/bjdv9n12-087

Maneesriwongul, W., & Dixon, J. K. (2004). Instrument translation process: A methods review. *Journal of Advanced Nursing*, *48*(2), 175–186. DOI: 10.1111/j.1365-2648.2004.03185.x PMID: 15369498

Manta, O., Gouliamos, K., Kong, J., Li, Z., Ha, N. M., Mohanty, R. P., Yang, H., Pu, R., & Yue, X. G. (2020). The architecture of financial networks and models of financial instruments according to the "Just transition mechanism" at the European level. *Journal of Risk and Financial Management*, *13*(10), 235. DOI: 10.3390/jrfm13100235

Marczinkowski, H. M., Østergaard, P. A., & Mauger, R. (2022). Energy transitions on European islands: Exploring technical scenarios, markets and policy proposals in Denmark, Portugal and the United Kingdom. *Energy Research & Social Science*, *93*, 102824. DOI: 10.1016/j.erss.2022.102824

Marinakis, V., Flamos, A., Stamtsis, G., Georgizas, I., Maniatis, Y., & Doukas, H. (2020). The Efforts towards and Challenges of Greece's Post-Lignite Era: The Case of Megalopolis. *Sustainability (Basel)*, *12*(24), 10575. DOI: 10.3390/su122410575

Mark Bekweri, E. D. E. H. (2020). Green Marketing As A Marketing Tool And Consumer Purchase Behavior. Journal of Global Social Sciences, Volume 1, Number 3, 37-54/ISSN: Print 2735-9344 [Nile University of Nigeria]. *Online (Bergheim)*, •••, 2735–9352.

Markandya, A., Labandeira, X., & Ramos, A. (2014). Policy instruments to foster energy efficiency. In *Green Energy and Efficiency: An Economic Perspective* (pp. 93–110). Springer.

Marketos, T. N. (2021). Energy Geopolitics Crossovers in Central & Eastern Mediterranean at the Prize of the Energy Union Policy. *Aspects of the Energy Union: Application and Effects of European Energy Policies in SE Europe and Eastern Mediterranean*, 47-71. https://doi.org/DOI: 10.1007/978-3-030-55981-6_3

Marsh, D. M., Page, R. B., Hanlon, T. J., Corritone, R., Little, E. C., Seifert, D. E., & Cabe, P. R. (2008). Effects of roads on patterns of genetic differentiation in red-backed salamanders, Plethodon cinereus. *Conservation Genetics*, *9*(3), 603–613. DOI: 10.1007/s10592-007-9377-0

Martinez, S., & Vazquez, A. B. (2024). Mandatory ESG reporting and corporate performance. *Available at SSRN* 4358256.

Matharu, M. (2023). Marketing Analytics Concept, Applications, Opportunities, and Challenges Ahead. *Data Analytics and Business Intelligence*, 170-186.

Mauger, R. (2021). Cast Away? How EU Energy Law Provides for a Just Transition for EU Islands. *OGEL, 2021*(1), 1–16.

Mauger, R., Diestelmeier, L., & Nieuwenhout, C. (2024). Harnessing EU legal concepts for the energy transition on islands. *Journal of World Energy Law and Business, 17*(3), 167–183. DOI: 10.1093/jwelb/jwae002

Mavris, C. I. (2014). Social, cultural and environmental impacts of tourism in Cyprus: the need for sustainable development. *World Review of Entrepreneurship, Management and Sustainable Development 5, 10*(2-3), 197-215. https://doi.org/ DOI: 10.1504/WREMSD.2014.060388

McCauley, D., Pettigrew, K. A., Todd, I., & Milchram, C. (2023). Leaders and laggards in the pursuit of an EU just transition. *Ecological Economics, 205*, 107699. DOI: 10.1016/j.ecolecon.2022.107699

McCully, P. (2001). *Silenced rivers: The ecology and politics of large dams*. Zed Books.

McGregor, D. (2014). Traditional knowledge and water governance: The ethic of responsibility. *Alternative, 10*(5), 493–507. DOI: 10.1177/117718011401000505

McGregor, D. (2018). Mino-mnaamodzawin: Achieving Indigenous environmental justice in Canada. *Environment and Society, 9*(1), 7–24. DOI: 10.3167/ares.2018.090102

McKinsey & Company. (2024). "Global Energy perspective 2024" https://www.mckinsey.com/industries/energy-and-materials/our-insights/global-energy-perspective

Meckling, J., Galeazzi, C., Shears, E., Xu, T., & Anadon, L. D. (2022). Energy innovation funding and institutions in major economies. *Nature Energy, 7*(9), 876–885. DOI: 10.1038/s41560-022-01117-3

Meeus, L., Hafner, M., Azevedo, I., Marcantonini, C., & Glachant, J.-M. (2011). *Transition Towards a Low Carbon Energy System by 2050: What Role for the EU?* Ministry for Energy, Utilities, & Climate. (2019). *Cyber- og informationssikkerhedsstrategi for energisektorerne.*

Meisaroh, F., Sudarmiatin, S., & Hermawan, A. (2022). the Importance of Digital Marketing Strategy Emos Marketplace During Covid-19 Pandemic Period. *International Conference of Business and Social Sciences*, 430–439.

Melica, G., Treville, A., Franco De Los Rios, C., Baldi, M., Monforti-Ferrario, F., Palermo, V., & Bertoldi, P. (2022). Covenant of Mayors: 2021 assessment. *Climate change mitigation and adaptation at local level (European Commission. Joint Research Centre, JRC130957*, 2022.

Mencel, J., Mocek-Płóciniak, A., & Kryszak, A. (2022). Soil microbial community and enzymatic activity of grasslands under different use practices: A review. *Agronomy (Basel)*, *12*(5), 1136. DOI: 10.3390/agronomy12051136

Méndez, M., Merayo, M. G., & Núñez, M. (2023). Machine learning algorithms to forecast air quality: A survey. *Artificial Intelligence Review*, *56*(9), 10031–10066. https://link.springer.com/article/10.1007/s10462-023-10424-4. DOI: 10.1007/s10462-023-10424-4 PMID: 36820441

Meschede, H., Bertheau, P., Khalili, S., & Breyer, C. (2022). A review of 100% renewable energy scenarios on islands. *Wiley Interdisciplinary Reviews. Energy and Environment*, *11*(6), e450. DOI: 10.1002/wene.450

Meschede, H., Child, M., & Breyer, C. (2018). Assessment of sustainable energy system configuration for a small Canary island in 2030. *Energy Conversion and Management*, *165*, 363–372. DOI: 10.1016/j.enconman.2018.03.061

Meyer, B. C., Kirsten, F., Sattler, D., & Heinrich, J. (2020). The Land Use-Land Degradation Nexus in Mediterranean Landscapes:–Drivers of Changes And Key Processes at Selected Natura 2000 Sites of Crete, Greece. *Journal of Environmental Geography*, *13*(1-2), 13–24. DOI: 10.2478/jengeo-2020-0002

Mezirow, J. (2018). Transformative learning theory. In *Contemporary theories of learning* (pp. 114–128). Routledge., DOI: 10.4324/9781315147277-8

Milonc, D., Curto, D., Franzitta, V., Guercio, A., Cirrincione, M., & Mohammadi, A. (2022). An economic approach to size of a renewable energy mix in small islands. *Energies*, *15*(6), 2005. DOI: 10.3390/en15062005

Minachilis, K., Kantsa, A., Devalez, J., Vujic, A., Pauly, A., & Petanidou, T. (2023). High species turnover and unique plant–pollinator interactions make a hyperdiverse mountain. *Journal of Animal Ecology*, *92*(5), 1001–1015. DOI: 10.1111/1365-2656.13898 PMID: 36754546

Ministry of Agriculture. Rural Development and Environment, (2024). CYPRUS, National Greenhouse Gas Inventory 1990-2022, National Inventory Document under the United Nations Convention on Climate Change and the Paris Agreement, Ministry of Agriculture, Rural Development and Environment DepartmentofEnvironment, https://unfccc.int/sites/default/files/resource/CyprusNIR2024_vUNFCCC_v1.2.pdf

Ministry of Development and Investment (2020). National Development Program 2021–2025, (in Greek).

Ministry of Development and Investment. (2021) Just Transition Development Plan 2021-2027. NSDP 2014-2020. https://2014-2020.espa.gr/el/Documents/Kalliris_17-6.pdf

Ministry of Energy, Commerce and Industry, (2023). Revision of the National Energy and Climate Plan 2021-2030, (in Greek)

Ministry of Environment and Energy, (2021). Territorial Just Transition Development Plan for the North and South Aegean Islands and Crete (in Greek)

Ministry of Environment and Energy. (2018). *National Strategy for the Circular Economy*. Hellenic Republic. (in Greek)

Ministry of Environment and Energy. (2020). *Just Transition Development Plan of Lignite Areas, Just Development Transition Plan*; Brussels: European Commission. Availableonline: https://www.sdam.gr/sites/default/files/consultation/Master_Plan_Public_Consultation_ENG.pdf, (In Greek)

Ministry of Environment and Energy. (2021). Territorial Just Transition Territorial Plan for Megalopolis. Available online: https://www.sdam.gr/node/431, (In Greek)

Ministry of Environment and Energy. (2024). Establishment of an Island Resettlement Fund, (Greek), https://ypen.gov.gr/systinetai-tameio-apanthrakopoiisis-ton-nision/

Ministry of Finance. (2020). *Explanatory Report on the 2021 Budget, Hellenic Republic*. Ministry of Finance. (in Greek)

Ministry of Finance. (2022a). Just Transition Plan 2021-2027, Directorate General for Development, Ministry of Finance, (2022b). Partnership Agreement 2021-2027, Directorate General for Development, https://thalia.com.cy/wp-content/uploads/2022/07/%CE%A3%CE%95%CE%A3-2021-2027.pdf

Ministry of National Economy and Finance. (2024). Presentation of the GReco Islands initiative and Integrated Spatial Investments at the 15th Panhellenic Conference of Small Islands, (in Greek), https://www.espa.gr/el/Pages/NewsFS.aspx?item=1810

Ministry of New and Renewable Energy. (2015). *National offshore wind energy policy*. Government of India.

Ministry of New and Renewable Energy. (2023a). Jawaharlal Nehru National Solar Mission. https://mnre.gov.in

Ministry of New and Renewable Energy. (2023b). Scheme for development of solar parks and ultra-mega solar power projects. https://mnre.gov.in

Ministry of New and Renewable Energy. (2023c). PLI scheme for high efficiency solar PV modules. https://mnre.gov.in

Ministry of New and Renewable Energy. (2023d). PM-KUSUM scheme. https://mnre.gov.in

Ministry of New and Renewable Energy. (2023e). Wind power programme. https://mnre.gov.in

Ministry of New and Renewable Energy. (2023f). National project on biogas development. https://mnre.gov.in

Ministry of New and Renewable Energy. (2023g). National programme on improved chulha. https://mnre.gov.in

Ministry of New and Renewable Energy. (2023h). Bio-energy programme 2021–26. https://mnre.gov.in

Ministry of New and Renewable Energy. (2023i). National policy on biofuels. https://mnre.gov.in

Ministry of New and Renewable Energy. (2023j). National green hydrogen mission. https://mnre.gov.in

Ministry of New and Renewable Energy. (2023k). Renewable energy R&D and HRD schemes. https://mnre.gov.in

Mirasgedis, S., Sarafidis, Y., Georgopoulou, E., Lalas, D. P., & Papastavros, C. (2004). Mitigation policies for energy related greenhouse gas emissions in Cyprus: The potential role of natural gas imports. *Energy Policy*, *32*(8), 1001–1011. DOI: 10.1016/S0301-4215(03)00048-X

MISE. M. (2017). Strategia Energetica Nazionale. In: SEN.

Mishra, P., & Sharma, P. (2014). Green marketing: Challenges and opportunities for business. BVIMR Management Edge, 7(1).

Mitchell, G. (2021). Supercharged: The EuroAsia interconnector and Israel's pursuit of energy interdependence. *Mitvim Institute*, 1-8.

Mitchell, T. (2011). *Carbon democracy: Political power in the age of oil*. Verso.

Mizuno, O., Kikuno, T., Inagaki, K., Takagi, Y., & Sakamoto, K. (2000). Statistical analysis of deviation of actual cost from estimated cost using actual project data. *Information and Software Technology*, *42*(7), 465–473. DOI: 10.1016/S0950-5849(00)00092-6

Moesker, K., & Pesch, U. (2022). The just transition fund–Did the European Union learn from Europe's past transition experiences? *Energy Research & Social Science*, *91*, 102750. DOI: 10.1016/j.erss.2022.102750

Mohanty, S., Patra, P. K., Sahoo, S. S., & Mohanty, A. (2017). Forecasting of solar energy with application for a growing economy like India: Survey and implication. *Renewable & Sustainable Energy Reviews*, *78*, 539–553.

Mohapatra, S. K., Mishra, S., & Tripathy, H. K. (2022). Energy consumption prediction in electrical appliances of commercial buildings using LSTM-GRU model. In *2022 International Conference on Advancements in Smart, Secure and Intelligent Computing (ASSIC)* (pp. 1-5). IEEE. https://ieeexplore.ieee.org/abstract/document/10088334

Morabito, A., Musarella, C. M., & Spampinato, G. (2024). Diversity and ecological assessment of grasslands habitat types: A case study in the Calabria Region (Southern Italy). *Land (Basel)*, *13*(6), 719. DOI: 10.3390/land13060719

Morse, J. M., & Field, P. A. (2013). *Nursing research: The application of qualitative approaches*. Springer.

Mortensen, D. A., Rauschert, E. S., Nord, A. N., & Jones, B. P. (2009). Forest roads facilitate the spread of invasive plants. *Invasive Plant Science and Management*, *2*(3), 191–199. DOI: 10.1614/IPSM-08-125.1

Moustakas, A., Georgiakakis, P., Kret, E., & Kapsalis, E. (2023). Wind turbine power and land cover effects on cumulative bat deaths. *The Science of the Total Environment*, *892*, 164536. DOI: 10.1016/j.scitotenv.2023.164536 PMID: 37268139

Mo, W., Wang, Y., Zhang, Y., & Zhuang, D. (2017). Impacts of road network expansion on landscape ecological risk in a megacity, China: A case study of Beijing. *The Science of the Total Environment*, *574*, 1000–1011. DOI: 10.1016/j.scitotenv.2016.09.048 PMID: 27668852

Mutu, M. (2019). *The Treaty of Waitangi and Māori rights in Aotearoa New Zealand*. Auckland University Press.

Nadeem, T. B., Siddiqui, M., Khalid, M., & Asif, M. (2023). Distributed energy systems: A review of classification, technologies, applications, and policies. *Energy Strategy Reviews, 48*, 101096. https://doi.org/https://doi.org/10.1016/j.esr.2023.101096

Nahm, J. M., Miller, S. M., & Urpelainen, J. (2022). G20's US $14-trillion economic stimulus reneges on emissions pledges. *Nature*, *603*(7899), 28–31. DOI: 10.1038/d41586-022-00540-6 PMID: 35236968

Nanaki, E. A., Koroneos, C. J., & Xydis, G. A. (2016). Environmental impact assessment of electricity production from lignite. *Environmental Progress & Sustainable Energy*, *35*(6), 1868–1875. DOI: 10.1002/ep.12427

Nan, Z., Yuping, C., & Kongjue, Z. (2024). The Future of Marketing Analytics: Trends and Emerging Technologies. [IJABMR]. *International Journal of Advances in Business and Management Research*, *1*(3), 23–32. DOI: 10.62674/ijabmr.2024.v1i03.003

Nasirov, S., Girard, A., Peña, C., Salazar, F., & Simon, F. (2021). Expansion of renewable energy in Chile: Analysis of the effects on employment. *Energy*, *226*, 120410. DOI: 10.1016/j.energy.2021.120410

Nel, E., Marais, L., & Mqotyana, Z. (2023). The regional implications of just transition in the world's most coal-dependent economy: The case of Mpumalanga, South Africa. *Frontiers in Sustainable Cities*, *4*, 1059312. Advance online publication. DOI: 10.3389/frsc.2022.1059312

Neofytou, H., Sarafidis, Y., Gkonis, N., Mirasgedis, S., & Askounis, D. (2020). Energy Efficiency contribution to sustainable development: A multi-criteria approach in Greece. *Energy Sources. Part B, Economics, Planning, and Policy*, *15*(10-12), 572–604. DOI: 10.1080/15567249.2020.1849449

Neuendorf, K. A. (2017). *The content analysis guidebook*. sage. https://doi.org/ DOI: 10.4135/9781071873045

Newell, P., & Mulvaney, D. (2013). The political economy of the 'just transition.'. *The Geographical Journal*, *179*(2), 132–140. DOI: 10.1111/geoj.12008

Nguyen, K. M., Dinh, L. T. K., Ngo, T. T., Phan, T. T. T., Nguyen, P. D. H., Tran, H. P. M., & Nguyen, N. T. (2025). The Impact of 7Ps Green Marketing Mix on Customer Commitment and Willingness to Pay a Premium Price: Evidence from Vietnam. *Journal of Promotion Management*, *31*(4), 1–42. DOI: 10.1080/10496491.2025.2484712

Nguyen, M. P., & Ponomarenko, T. (2025). State Incentives for Solar Energy in the Context of Energy Transition in Developed and Developing Countries. *Energies*, *18*(5), 1227.

Nguyen, T. P. (2025). AIoT-based indoor air quality prediction for building using enhanced metaheuristic algorithm and hybrid deep learning. *Journal of Building Engineering*, *105*, 112448. https://www.sciencedirect.com/science/article/pii/S2352710225006850. DOI: 10.1016/j.jobe.2025.112448

Nguyen-Viet, B. (2023). The impact of green marketing mix elements on green customer based brand equity in an emerging market. *Asia-Pacific Journal of Business Administration*, *15*(1), 96–116. DOI: 10.1108/APJBA-08-2021-0398

Nikolaidis, P., & Poullikkas, A. (2017). A comparative overview of hydrogen production processes. *Renewable & Sustainable Energy Reviews*, *67*, 597–611. DOI: 10.1016/j.rser.2016.09.044

Nikolaidis, P., & Poullikkas, A. (2022). A thorough emission-cost analysis of the gradual replacement of carbon-rich fuels with carbon-free energy carriers in modern power plants: The case of Cyprus. *Sustainability (Basel)*, *14*(17), 10800. DOI: 10.3390/su141710800

Niraula, D., Wang, H., Wang, Z., & Wang, Y. (2024)... *Community-Based Microgrid Planning and Operation for Fostering Energy Justice*, *1–5*, 1–5. Advance online publication. DOI: 10.1109/NAPS61145.2024.10741738

Nolden, C. (2019). Transaction cost analysis of digital innovation governance in the UK energy market. *The Journal of Energy Markets*. DOI: 10.21314/JEM.2019.194

North Aegean Region, (2022). North Aegean Regional Programme 2021-2027, (in Greek)

North Aegean Region. (2021). *Strategic Operational Program for the North Aegean Region*. North Aegean Region. (in Greek)

Notton, G. (2015). Importance of islands in renewable energy production and storage: The situation of the French islands. *Renewable & Sustainable Energy Reviews*, *47*, 260–269. DOI: 10.1016/j.rser.2015.03.053

Nurhidayah, L., Alam, S., Utomo, N. A., & Suntoro, A. (2024). Indonesia's Just Energy Transition: The Societal Implications of Policy and Legislation on Renewable Energy. *Climate Law*, *14*(1), 36–66. DOI: 10.1163/18786561-bja10047

Nurtas, M., Zhantaev, Z., & Altaibek, A. (2024). Earthquake time-series forecast in Kazakhstan territory: Forecasting accuracy with SARIMAX. *Procedia Computer Science, 231*, 353–358. https://www.sciencedirect.com/science/article/pii/S1877050923022287. DOI: 10.1016/j.procs.2023.12.216

Nwankwo, C., & Kanyangale, M. (2023). Effect of sustainable marketing strategies on consumption of beer brands in Anambra State. Acta Universitatis Danubius. Œconomica, 19(2), 188-205.

Oberthür, S., Khandekar, G., & Wyns, T. (2021). Global governance for the decarbonization of energy-intensive industries: Great potential underexploited. *Earth System Governance*, *8*, 100072. DOI: 10.1016/j.esg.2020.100072

OECD, I. (2016). Energy and air pollution: world energy outlook special report 2016.

OECD. (2021). *The Economic Benefits of Climate Action*. OECD Publishing. https://www.oecd.org/environment/the-economic-benefits-of-climate-action-9789264307691-en.htm

Official Journal of the European Union, (2022). COMMISSION DELEGATED REGULATION (EU) 2022/564 of 19 November 2021 amending Regulation (EU) No 347/2013 of the European Parliament and of the Council as regards the Union list of projects of common interest, L 109/14

Official Journal of the European Union. (2021) Regulation (EU) 2021/1056 of the European Parliament and of the Council of 24 June 2021 establishing the Just Transition Fund. 30 June. Brussels: Publications Office of the European Union Official Journal of the European Union, (2016). Consolidated Versions of the Treaty on European Union and the treaty on the Functioning of the European Union (2016/C 202/01), https://eur-lex.europa.eu/resource.html?uri=cellar:9e8d52e1-2c70-11e6-b497-01aa75ed71a1.0006.01/DOC_1&format=PDF

Ofori, D. (2021). Opportunities and challenges of green marketing. Green Marketing in Emerging Markets: Strategic and Operational Perspectives, 251-276.

Oikonomou, D., Vrahnakis, M., Yiakoulaki, M., Xanthopoulos, G., & Kazoglou, Y. (2023). Grazing as a management tool in Mediterranean Pastures: A meta-analysis based on a literature review. *Land (Basel)*, *12*(7), 1290. DOI: 10.3390/land12071290

Olson-Hazboun, S. K. (2018). "Why are we being punished and they are being rewarded?" views on renewable energy in fossil fuels-based communities of the US west. *The Extractive Industries and Society*, *5*(3), 366–374. DOI: 10.1016/j.exis.2018.05.001

Onambele, L., Guillen-Aguinaga, S., Guillen-Aguinaga, L., Ortega-Leon, W., Montejo, R., Alas-Brun, R., Aguinaga-Ontoso, E., Aguinaga-Ontoso, I., & Guillen-Grima, F. (2023). Trends, projections, and regional disparities of maternal mortality in Africa (1990–2030): An ARIMA Forecasting approach. *Epidemiologia*, *4*(3), 322–351. https://www.mdpi.com/2673-3986/4/3/32. DOI: 10.3390/epidemiologia4030032 PMID: 37754279

Onukwulu, C., Odochi, M., & Eyo-Udo, L. (2023). Decentralized energy supply chain networks using blockchain and IoT. *International Journal of Scholarly Research in Multidisciplinary Studies, 2*(2), 066–085. DOI: 10.56781/ijsrms.2023.2.2.0055

Ordonez, J. A., Jakob, M., Steckel, J. C., & Ward, H. (2023). India's just energy transition: Political economy challenges across states and regions. *Energy Policy*, *179*, 113621.

Pablo-Romero, M. D. P., & Sánchez-Braza, A. (2015). Productive energy use and economic growth: Energy, physical and human capital relationships. *Energy Economics*, *49*, 420–429. DOI: 10.1016/j.eneco.2015.03.010

Pachauri, S., & Jiang, L. (2008). The household energy transition in India and China. *Energy Policy*, *36*(11), 4022–4035.

Pacheco, A., Monteiro, J., Santos, J., Sequeira, C., & Nunes, J. (2022). Energy transition process and community engagement on geographic islands: The case of Culatra Island (Ria Formosa, Portugal). *Renewable Energy*, *184*, 700–711. DOI: 10.1016/j.renene.2021.11.115

Palaiologou, P., Kalabokidis, K., Haralambopoulos, D., Feidas, H., & Polatidis, H. (2011). Wind characteristics and mapping for power production in the Island of Lesvos, Greece. *Computers & Geosciences*, *37*(7), 962–972. DOI: 10.1016/j.cageo.2010.05.025

Pal, M., Dass, R., & Nehra, V. (2023)... *International Journal of Sustainable Energy*, *42*(1), 152–181.

Panagiotopoulos, P. P., & Roukanas, S. A. (2024). Navigating the Greek Energy Crisis through a Multidimensional Approach: A Review Article. *Energies*, *17*(16), 3915. DOI: 10.3390/en17163915

Panayiotou, G., & Kalogirou, S. A. (2009). The introduction of Hydrogen Economy in Cyprus. In *Conference on the promotion of Distributed Renewable Energy Sources in the Mediterranean region*. https://doi.org/DOI: 10.1016/j.renene.2016.05.043

Pandey, A. (2025). Blockchain in Industrial Supply Chains: Tracing the Transparency. In *Immersive Technology for the Gig Economy: Transformative Business Practices* (pp. 273-294). IGI Global Scientific Publishing.

Pandey, A., & Goyal, M. (2025). Leveraging Technology for Climate Resilience in Urban Areas. In *Revolutionizing Urban Development and Governance With Emerging Technologies* (pp. 137–162). IGI Global Scientific Publishing. DOI: 10.4018/979-8-3373-1375-7.ch005

Pandey, A., & Joseph, J. (2025). Integral Role of Blockchain and Artificial Intelligence in Sustainable Economic Development. In *Driving Socio-Economic Growth With AI and Blockchain* (pp. 271–290). IGI Global Scientific Publishing. DOI: 10.4018/979-8-3693-8664-4.ch012

Pandey, P., & Sharma, A. (2021). Knowledge politics, vulnerability and recognition-based justice: Public participation in renewable energy transitions in India. *Energy Research & Social Science*, *71*, 101824.

Pandey, S., Singh, V. S., Gangwar, N. P., Vijayvergia, M. M., Prakash, C., & Pandey, D. N. (2012). Determinants of success for promoting solar energy in Rajasthan, India. *Renewable & Sustainable Energy Reviews*, *16*(6), 3593–3598.

Papachristou, D., (2025). Presentation at the event organized by the Waste, Energy and Water Regulatory Authority in Thessaloniki in the framework of the 'Helexpo Dialogues' conference, the timetable for the electrical interconnection of the islands in March 2025 (in Greek).

Papadas, K. (2021). Green Marketing Strategy. In *Industry, Innovation and Infrastructure* (pp. 450–461). Springer International Publishing. DOI: 10.1007/978-3-319-95873-6_121

Papadas, K. K., Avlonitis, G. J., Carrigan, M., & Piha, L. (2019). The interplay of strategic and internal green marketing orientation on competitive advantage. *Journal of Business Research*, *104*, 632–643. DOI: 10.1016/j.jbusres.2018.07.009

Papademetriou, C., Ragazou, K., Belias, D., Anastasiadou, S., Garefalakis, A., & Prodromou, A. (2025). The Importance of Green Human Resources: Green Ambidexterity and Environmental Performance. In K. Ragazou, A. Garefalakis, C. Papademetriou, & A. Samara (Eds.), *Sustainability Through Green HRM and Performance Integration* (pp. 83-100). IGI Global Scientific Publishing. https://doi.org/DOI: 10.4018/979-8-3693-5981-5.ch004

Papademetriou, C., Anastasiadou, S., Belias, D., & Ragazou, K. (2025). Integrating sustainability into human resource management: Building a greener workforce for the future. *Sustainability (Basel)*, *17*(3), 1113. DOI: 10.3390/su17031113

Papadis, E., & Tsatsaronis, G. (2020). Challenges in the decarbonization of the energy sector. *Energy*, *205*, 118025. DOI: 10.1016/j.energy.2020.118025

Papapostolou, C. M., Tzanes, G. T., & Kondili, E. M. (2019). Optimisation of the integrated water–energy systems in small group of islands communities. [). Elsevier.]. *Computer-Aided Chemical Engineering*, *46*, 1735–1740. DOI: 10.1016/B978-0-12-818634-3.50290-3

Papillon, M., & Rodon, T. (2017). Proponent-Indigenous agreements and the implementation of the right to free, prior, and informed consent in Canada. *Environmental Impact Assessment Review*, *62*, 216–224. DOI: 10.1016/j.eiar.2016.06.009

Parlee, B. (2015). Avoiding the resource curse: Indigenous communities and Canada's oil sands. *World Development*, *74*, 425–436. DOI: 10.1016/j.worlddev.2015.03.004

Parsons, M., Fisher, K., & Nalau, J. (2016). Alternative approaches to co-production: Building adaptive capacity for disaster risk management. *Climate Risk Management*, *12*, 1–10. DOI: 10.1016/j.crm.2016.01.001

Pathak, V., Jena, B., & Kalra, S. (2013). Qualitative research. *Perspectives in Clinical Research*, *4*(3), 192. DOI: 10.4103/2229-3485.115389 PMID: 24010063

Patton, M. Q. (2014). *Qualitative research & evaluation methods: Integrating theory and practice*. Sage publications.

Paul, S., Panja, A., & Jha, P. (2025). *Biofuels and Sustainability*, 237-252.

Pavlis, E. (2025). Large-Scale Wind Energy Farm Development and its Impacts on the Landscape: A Review of the Greek Case. *Journal of Landscape Ecology*, *18*(3), 24–61. DOI: 10.2478/jlecol-2025-0019

Pavloudakis, F., Karlopoulos, E., Roumpos, C., & Koukouzas, N. (2021). Energy transition and the future of lignite mining in the region of Western Macedonia, Greece. In *Zlatibor: 10th International conference coal 2021* (pp. 13-16).

Pavloudakis, F., Karlopoulos, E., & Roumpos, C. (2023). Just transition governance to avoid socio-economic impacts of lignite phase-out: The case of Western Macedonia, Greece. *The Extractive Industries and Society*, *14*, 101248. DOI: 10.1016/j.exis.2023.101248

Pavloudakis, F., Roumpos, C., Karlopoulos, E., & Koukouzas, N. (2020). Sustainable rehabilitation of surface coal mining areas: The case of Greek lignite mines. *Energies*, *13*(15), 3995. DOI: 10.3390/en13153995

Peattie, K. (2016). Green marketing. In *The marketing book* (pp. 573–597). Routledge.

Pegkas, P. (2024). Energy consumption and human capital: Does human capital stimulate renewable energy? The case of Greece. *Journal of the Knowledge Economy*, *15*(4), 1–28. DOI: 10.1007/s13132-024-01770-x

Pelz, S., Chindarkar, N., & Urpelainen, J. (2020). Energy Access for Marginalized Communities: Evidence from Rural North India, 2015-2018. SSRN *Electronic Journal*.

Peñalvo-López, E., Andrada-Monrós, C., León-Martínez, V., & Valencia-Salazar, I. (2024). Assessing energy transition in Mediterranean islands. A review. *E-Prime-Advances in Electrical Engineering. Electronics and Energy*, *9*, 100719.

Peng, L., Wang, L., Xia, D., & Gao, Q. (2022). Effective energy consumption forecasting using empirical wavelet transform and long short-term memory. *energy, 238*, 121756. https://www.sciencedirect.com/science/article/pii/S0360544221020041

Pepe, J. M. (2024). Europe and the Emerging Geopolitics of Electricity Grids.

Peterson, J. A., Reisinger, H. S., Schwartz, R. P., Mitchell, S. G., Kelly, S. M., Brown, B. S., & Agar, M. H. (2008). Targeted sampling in drug abuse research: A review and case study. *Field Methods*, *20*(2), 155–170. DOI: 10.1177/1525822X08314988

Petrescu, M., & Krishen, A. S. (2023). Hybrid intelligence: Human–AI collaboration in marketing analytics. *Journal of Marketing Analytics*, *11*(3), 263–274. DOI: 10.1057/s41270-023-00245-3

Phillips, B. B., Bullock, J. M., Osborne, J. L., & Gaston, K. J. (2020). Ecosystem service provision by road verges. *Journal of Applied Ecology*, *57*(3), 488–501. DOI: 10.1111/1365-2664.13556

Pilavachi, P. A., Kalampalikas, N. G., Kakouris, M. K., Kakaras, E., & Giannakopoulos, D. (2009). The energy policy of the Republic of Cyprus. *Energy*, *34*(5), 547–554. DOI: 10.1016/j.energy.2008.08.007

Pinczynski, M., Kasperowicz, R., Azzopardi, B., Bilan, Y., & Štreimikienė, D. (2024). Malta's low carbon transition towards sustainability. *Sustainable Development (Bradford)*, *32*(5), 5120–5128. DOI: 10.1002/sd.2967

Pinto, T., Praça, I., Vale, Z., & Silva, J. (2021). Ensemble learning for electricity consumption forecasting in office buildings. *Neurocomputing*, *423*, 747–755. https://www.sciencedirect.com/science/article/pii/S0925231220307372. DOI: 10.1016/j.neucom.2020.02.124

Piranda, D. R., Sinaga, D. Z., Putri, E. E., & Asahan, U. (2022). Online Marketing Strategy in Facebook. Journal of Humanities. *Social Sciences and Business*, *1*(2), 79–86.

Pitaloka, E., Suyoto, Y. T., Febriyanti, A., & Sukarno, D. A. R. (2023). The Impact of Green Marketing Mix on Brand Image of Unilever Indonesia. *KnE Social Sciences*, 65-76.

Pollin, R., & Callaci, B. (2019). The economics of just transition: A framework for supporting fossil fuel–dependent workers and communities in the United States. *Labor Studies Journal*, *44*(2), 93–138. DOI: 10.1177/0160449X18787051

Polonsky, M. J. (2008). An introduction to green marketing. *Global Environment: Problems and Policies*, *2*(1), 1–10.

Popescu, C., Panait, M., Palazzo, M., & Siano, A. (2022). Energy transition in European Union—Challenges and opportunities. *Energy Transition: Economic, Social and Environmental Dimensions*, 289-312. https://doi.org/DOI: 10.1007/978-981-19-3540-4_11

Poullikkas, A. (2024). Strategic objectives for Cyprus' energy transition, In Business news, https://inbusinessnews.reporter.com.cy/article/2024/9/19/796476/strategikoi-stokhoi-gia-tenenergeiake-metabase-tes-kuprou/

Poullikkas, A. (2020). Long-term Sustainable Energy Strategy: Cyprus' Energy Transition to Hydrogen Economy. https://doi.org/DOI: 10.31219/osf.io/js5hu

Poullikkas, A. (2021). Perspectives for the development of energy strategies—Challenges towards a hydrogen economy in Cyprus. *Green Energy Sustain*, *1*(0004), 0004. Advance online publication. DOI: 10.47248/HKOD902101020004

Powys Whyte, K., McGregor, D., & Talaga, A. (2019). Indigenous lessons about sustainability are not just for "all humanity.". *Nature Communications*, *10*(1), 1–3. DOI: 10.1038/s41467-019-11971-9 PMID: 30602773

Prajapati, P., Guo, R., Cai, A., & Prasad, R. (2025). *Navigating the Energy Transition in India: Challenges and Opportunities Towards Sustainable Energy Goal.* Water-Energy Nexus.

Pratiwi, A. P., & Edeh, F. O. (2024). The Role of ESG Disclosure in Corporate Performance and Investment Decision-Making. *Sinergi International Journal of Accounting and Taxation*, *2*(1), 1–14. DOI: 10.61194/ijat.v2i1.476

Press Information Bureau. (2024). *PM-Surya Ghar: Muft Bijli Yojana launched.* Government of India.

Priano, F. H., Armas, R. L., & Guerra, C. F. (2016). A model for the smart development of island territories. In *Proceedings of the 17th international digital government research conference on digital government research* (pp. 465-474). DOI: 10.1145/2912160.2912187

Prodromou, A., Papademetriou, C., & Savvides, D. (2025). The Strategic Role of Green HRM in Hospitality Industry for Reaching Sustainable Development Goals. In K. Ragazou, A. Garefalakis, C. Papademetriou, & A. Samara (Eds.), *Sustainability Through Green HRM and Performance Integration* (pp. 1-38). IGI Global Scientific Publishing. https://doi.org/DOI: 10.4018/979-8-3693-5981-5.ch001

Psarros, G. N., Papakonstantinou, A. G., & Papathanassiou, S. A. (2020, November). Integration of storage into large island power systems: The case of Cyprus. In *The 12th Mediterranean Conference on Power Generation, Transmission, Distribution and Energy Conversion (MEDPOWER 2020)* (Vol. 2020, pp. 270-275). IET. https://doi.org/DOI: 10.1049/icp.2021.1219

Psarros, G. N., & Papathanassiou, S. A. (2022). Electricity storage requirements to support the transition towards high renewable penetration levels–Application to the Greek power system. *Journal of Energy Storage*, *55*, 105748. DOI: 10.1016/j.est.2022.105748

Psillaki, M., Apostolopoulos, N., Makris, I., Liargovas, P., Apostolopoulos, S., Dimitrakopoulos, P., & Sklias, G. (2023). Hospitals' energy efficiency in the perspective of saving resources and providing quality services through technological options: A systematic literature review. *Energies*, *16*(2), 755. DOI: 10.3390/en16020755

Psyllos, G., Hadjigeorgiou, I., Dimitrakopoulos, P. G., & Kizos, T. (2022). Grazing land productivity, floral diversity, and management in a semi-arid Mediterranean landscape. *Sustainability (Basel)*, *14*(8), 4623. DOI: 10.3390/su14084623

Purnomo, V. D., Hamid, A., & Tedjosaputro, L. (2024). The Influence of Economic Law on a Just Energy Transition. *International Journal of Business and Applied Economics*, *3*(6), 1077–1090. DOI: 10.55927/ijbae.v3i6.11950

Purwanti, I., Abadi, M. D., Suyanto, U. Y., & Ahmad, S. T. I. E. K. (2019). Green marketing: Strategy for gaining sustainable competitive advantage in industry 4.0. *Manajemen Bisnis*, *9*(2), 176–193. DOI: 10.22219/jmb.v9i2.10039

Qadir, S. A., Al-Motairi, H., Tahir, F., & Al-Fagih, L. (2021). Incentives and strategies for financing the renewable energy transition: A review. *Energy Reports*, *7*, 3590–3606. DOI: 10.1016/j.egyr.2021.06.041

Queku, Y. N., & Seidu, B. A. (2025). *Targeting in Africa: Readiness and Challenges*. In *Clean Energy and Development*. https://books.google.com/books?id=LSVEEQAAQBAJ

Quijano, A. (2000). Coloniality of power, Eurocentrism, and Latin America. *International Sociology*, *1*(3), 533–580. DOI: 10.1177/0268580900015002005

Qureshi, M., Arbab, M. A., & Rehman, S. U. (2024). Deep learning-based forecasting of electricity consumption. *Scientific Reports*, *14*(1), 6489. https://www.nature.com/articles/s41598-024-56602-4. DOI: 10.1038/s41598-024-56602-4 PMID: 38499617

Qu, S. Q., & Dumay, J. (2011). The qualitative research interview. *Qualitative Research in Accounting & Management*, *8*(3), 238–264. DOI: 10.1108/11766091111162070

Ragnitz, J., Markwardt, G., Schwartzkopff, J., Reitzenstein, A., Wehnert, T., Kurwan, J., & Beutel, J. (2022). Analysis of the historical structural change in the German lignite mining area of Lusatia (case study).

Raichand, M. (2024, February 23). Understanding green marketing: How to include sustainability in your marketing strategy. LinkedIn. https://www.linkedin.com/pulse/understanding-green-marketing-how-include-your-mehek-raichand-auhbf/

Raina, G., & Sinha, S. (2019). Outlook on the Indian scenario of solar energy strategies: Policies and challenges. *Energy Strategy Reviews*, *24*, 331–341.

Rajaeian, E., & Davarani, R. Z. (2023). Supplying the energy of islands with geopolitical positions using distributed generations to increase energy security. *International Journal of Hydrogen Energy*, *48*(42), 16057–16071. DOI: 10.1016/j.ijhydene.2023.01.114

Ram, M., Bogdanov, D., Aghahosseini, A., Gulagi, A., Oyewo, A. S., Mensah, T. N. O., & Breyer, C. (2022). Global energy transition to 100% renewables by 2050: Not fiction, but much needed impetus for developing economies to leapfrog into a sustainable future. *Energy*, *246*, 123419.

Ram, M., Osorio-Aravena, J. C., Aghahosseini, A., Bogdanov, D., & Breyer, C. (2022). Job creation during a climate compliant global energy transition across the power, heat, transport, and desalination sectors by 2050. *Energy, 238*, 121690. DOI: 10.1016/j.energy.2021.121690

Ramos-Real, F. J., Barrera-Santana, J., Ramírez-Díaz, A., & Perez, Y. (2018). Interconnecting isolated electrical systems. The case of Canary Islands. *Energy strategy reviews, 22*, 37-46.

Rane, N., Choudhary, S., & Rane, J. (2023). Metaverse marketing strategies: enhancing customer experience and analysing consumer behaviour through leading-edge Metaverse technologies, platforms, and models. *Platforms, and Models (November 3, 2023).*

Raworth, K. (2017). *Doughnut Economics: Seven Ways to Think Like a 21st-Century Economist*. Chelsea Green Publishing.

Rechsteiner, R. (2021). German energy transition (Energiewende) and what politicians can learn for environmental and climate policy. *Clean Technologies and Environmental Policy, 23*(2), 305–342. DOI: 10.1007/s10098-020-01939-3 PMID: 33041746

Reddy, R., Monisha, V., Ramya, P., Deeksha, H. C., & Lavanya, A. (2023, June). Earthquake magnitude prediction using deep learning techniques. In *2023 International Conference on Sustainable Computing and Smart Systems (ICSCSS)* (pp. 159-164). IEEE. https://ieeexplore.ieee.org/abstract/document/10169746/

Rediske, G., Burin, H. P., Rigo, P. D., Rosa, C. B., Michels, L., & Siluk, J. C. M. (2021). Wind power plant site selection: A systematic review. *Renewable & Sustainable Energy Reviews, 148*, 111293. DOI: 10.1016/j.rser.2021.111293

Rehman, I. H., Kar, A., Raven, R., Singh, D., Tiwari, J., Jha, R., & Mirza, A. (2010). Rural energy transitions in developing countries: A case of the Uttam Urja initiative in India. *Environmental Science & Policy, 13*(4), 303–311.

Renewable Energy Policy Network for the 21st Century. (2024). Renewables global status report. Paris: REN21.

Republic of Cyprus (2021). Territorial Just Transition Plan 2021-2027 (in Greek)

Republic of Cyprus Presidency. (2021). Joint Declaration of the 9th Greece-Cyprus-Egypt Trilateral Summit, in Athens, https://www.presidency.gov.cy/cypresidency/cypresidency.nsf/All/CB82614EC33A6A51C22587740031F9E0?OpenDocument

Republic of Cyprus, (2018). Plan for the production of electricity from renewable energy sources with the final outcome of the integration of the projects within the competitive market, (2018)

Republic of Cyprus. (2020). Cyprus' Integrated National Energy and Climate Plan, Nicosia, Cyprus

Republic of Cyprus. (2024). Final update of Cyprus' integrated energy and climate plan for 2021-2030, (in Greek), https://e-consultation.gov.cy/wp-content/uploads/2024/11/NECP_v.1.5.pdf

Riady, S. R., & Apriani, R. (2023, February). Multivariate time series with Prophet Facebook and LSTM algorithm to predict the energy consumption. In *2023 International Conference on Computer Science, Information Technology and Engineering (ICCoSITE)* (pp. 805-810). IEEE. https://ieeexplore.ieee.org/abstract/document/10127735

Rial, R. C. (2024). Biofuels versus climate change: Exploring potentials and challenges in the energy transition. *Renewable & Sustainable Energy Reviews, 196*, 114369. DOI: 10.1016/j.rser.2024.114369

Ridzwan, N. S. M., & Yusoff, S. H. M. (2023). Machine learning for earthquake prediction: A review (2017–2021). *Earth Science Informatics, 16*(2), 1133–1149. https://link.springer.com/article/10.1007/s12145-023-00991-z. DOI: 10.1007/s12145-023-00991-z

Ring, E., Wallgren, M., Marald, E., Westerfelt, P., Djupstrom, L., Davidsson, A., & Sonesson, J. (2024). Forest roads in Sweden-infrastructure with multiple uses and diverse impacts. *Silva Fennica, 58*(4). Advance online publication. DOI: 10.14214/sf.24021

Ringel, M., Schlomann, B., Krail, M., & Rohde, C. (2016). Towards a green economy in Germany? The role of energy efficiency policies. *Applied Energy, 179*, 1293–1303. DOI: 10.1016/j.apenergy.2016.03.063

Ripple, W. J., Wolf, C., Newsome, T. M., Barnard, P., & Moomaw, W. R. (2020). Corrigendum: World scientists' warning of a climate emergency. *Bioscience, 70*(1), 100–100. DOI: 10.1093/biosci/biz152

Risteska, L. (2023). Green marketing mix and promotion of responsible consumption. *Knowledge-International Journal, 57*(1), 121–126.

Ritchie, J., Ormston, R., McNaughton Nicholls, C., & Lewis, J. (2013). Qualitative research practice: A guide for social science students and researchers.

Rodríguez-Garavito, C. (2019). Ethnicity.gov: Global governance, indigenous peoples, and the right to prior consultation in social minefields. *Indiana Journal of Global Legal Studies, 18*(1), 263–305. DOI: 10.2979/indjglolegstu.18.1.263

Rotmans, J., Kemp, R., & Van Asselt, M. (2001). More evolution than revolution: transition management in public policy. *foresight, 3*(1), 15-31.

Roy, B., & Schaffartzik, A. (2021). Talk renewables, walk coal: The paradox of India's energy transition. *Ecological Economics, 180*, 106871.

Saeed, M. A., & Kersten, W. (2020). Sustainability performance assessment framework: A cross-industry multiple case study. *International Journal of Sustainable Development and World Ecology, 27*(6), 496–514. DOI: 10.1080/13504509.2020.1764407

Safarzadeh, S., Hafezalkotob, A., & Jafari, H. (2022). Energy supply chain empowerment through tradable green and white certificates: A pathway to sustainable energy generation. *Applied Energy, 323*, 119601. https://doi.org/https://doi.org/10.1016/j.apenergy.2022.119601. DOI: 10.1016/j.apenergy.2022.119601

Saha, S., Hiremath, R. B., Prasad, S., & Kumar, B. (2021)... *Journal of Infrastructure Development, 13*(2), 107–128.

Sahoo, S. K. (2016). Renewable and sustainable energy reviews solar photovoltaic energy progress in India: A review. *Renewable & Sustainable Energy Reviews, 59*, 927–939.

Sahu, B. K. (2016). Solar energy developments, policies and future prospectus in the state of Odisha, India. *Renewable & Sustainable Energy Reviews, 61*, 526–536.

Sakka, E. G., Bilionis, D. V., Vamvatsikos, D., & Gantes, C. J. (2020). Onshore wind farm siting prioritization based on investment profitability for Greece. *Renewable Energy, 146*, 2827–2839. DOI: 10.1016/j.renene.2019.08.020

Salami, I. (2021). Challenges and Approaches to Regulating Decentralized Finance. *AJIL Unbound, 115*, 425–429. DOI: 10.1017/aju.2021.66

Sánchez, A. S., Junior, E. P., Gontijo, B. M., de Jong, P., & dos Reis Nogueira, I. B. (2023). Replacing fossil fuels with renewable energy in islands of high ecological value: The cases of Galápagos, Fernando de Noronha, and Príncipe. *Renewable & Sustainable Energy Reviews, 183*, 113527. DOI: 10.1016/j.rser.2023.113527

Santopietro, L., & Scorza, F. (2021). The Italian experience of the covenant of mayors: A territorial evaluation. *Sustainability (Basel), 13*(3), 1289. DOI: 10.3390/su13031289

Sarabdeen, M., Elhaj, M., & Alofaysan, H. (2024). Exploring the influence of digital transformation on clean energy transition, climate change, and economic growth among selected oil-export countries through the panel ARDL approach. *Energies*, *17*(2), 298. DOI: 10.3390/en17020298

Sarangi, G. K. (2018). ADBI Working Paper No. 863.

Saratun, M. (2016). Performance management to enhance employee engagement for corporate sustainability. *Asia-Pacific Journal of Business Administration*, *8*(1), 84–102. DOI: 10.1108/APJBA-07-2015-0064

Sarbjit, K., & Sanjay, K. B. (2017). Opportunities and challenges of Green Marketing in India.

Sarkar, N., Gupta, R., Keserwani, P. K., & Govil, M. C. (2022). Air quality index prediction using an effective hybrid deep learning model. *Environmental Pollution*, *315*, 120404. https://www.sciencedirect.com/science/article/pii/S0269749122016189. DOI: 10.1016/j.envpol.2022.120404 PMID: 36240962

Sau, A., & Singh, A. (2024). Forecasting Energy Consumption Patterns: Weather-Driven Insights from Time Series Analysis. In *2024 IEEE Students Conference on Engineering and Systems (SCES)* (pp. 1-6). IEEE. https://ieeexplore.ieee.org/abstract/document/10652321

Sawant, A. V. (2020). challenge And Opportunities Before Green Marketing in India. Journal of marketing Economics, 7(4).

Schaltegger, S., & Wagner, M. (2006). Integrative management of sustainability performance, measurement and reporting. *International Journal of Accounting. Auditing and Performance Evaluation*, *3*(1), 1–19. DOI: 10.1504/IJAAPE.2006.010098

Schipper, L. (2004). International Comparisons of Energy End Use: Benefits and Risks. In Cleveland, C. J. (Ed.), *Encyclopedia of Energy* (pp. 529–555). Elsevier., https://doi.org/https://doi.org/10.1016/B0-12-176480-X/00169-8 DOI: 10.1016/B0-12-176480-X/00169-8

Schlosberg, D., & Collins, L. B. (2014). From environmental to climate justice: Climate change and the discourse of environmental justice. *Wiley Interdisciplinary Reviews: Climate Change*, *5*(3), 359–374. DOI: 10.1002/wcc.275

Schmidt, D., Reyment, J., Webster, E., Kirby, S., & Lyle, D. (2020). Workplace-based health research training: A qualitative study of perceived needs in a rural setting. *Health Research Policy and Systems*, *18*(1), 67. DOI: 10.1186/s12961-020-00580-2 PMID: 32539865

Scott, D. N. (2020). Fossil capital and the law: Toward a just transition. *Osgoode Hall Law Journal*, *57*(3), 845–878.

Scott, J. C. (2017). *Against the grain: A deep history of the earliest states*. Yale University Press. DOI: 10.2307/j.ctv1bvnfk9

Seidl, R., Moser, C., & Blumer, Y. (2017). Navigating behavioral energy sufficiency. Results from a survey in Swiss cities on potential behavior change. *PLoS One*, *12*(10), e0185963. DOI: 10.1371/journal.pone.0185963 PMID: 29016642

Sekar, A., Varghese, G. K., & Varma, R. (2022). Exposure to volatile organic compounds and associated health risk among workers in lignite mines. *International Journal of Environmental Science and Technology*, •••, 1–14. DOI: 10.1007/s13762-022-04056-4

Sen, T. (2024)The influence of green marketing mix on green purchase decision: the mediating role of green brand image.

Sendstad, L. H., Hagspiel, V., Mikkelsen, W. J., Ravndal, R., & Tveitstøl, M. (2022). The impact of subsidy retraction on European renewable energy investments. *Energy Policy*, *160*, 112675. DOI: 10.1016/j.enpol.2021.112675

Shaban, F., Siskos, P., & Tjortjis, C. (2023). Electromobility Prospects in Greece by 2030: A Regional Perspective on Strategic Policy Analysis. *Energies*, *16*(16), 6083. DOI: 10.3390/en16166083

Shafi, R. M., Shahid, W., Osman, I., & Arifin, N. A. M. (2025). Mapping the Landscape: A Bibliometric Analysis of Global Trends in Sustainable Performance. *International Review of Management and Marketing*, *15*(3), 228–245. DOI: 10.32479/irmm.18298

Sharma, Y. (2011). changing consumer behaviour with respect to green. international journal of multidisciplinary research, 152- 162.

Sharma, N. K., Tiwari, P. K., & Sood, Y. R. (2012). Solar energy in India: Strategies, policies, perspectives and future potential. *Renewable & Sustainable Energy Reviews*, *16*(1), 933–941.

Sheng, C. (2020). Not just the state: The role of entrepreneurs in China's energy transition. *Energy Research & Social Science*, *70*, 101814. DOI: 10.1016/j.erss.2020.101814

Shetty, N. (2019). Green Marketing Strategy: Opportunities And Challenges For Marketers. Advance and Innovative Research, 421.

Shrote, P., Dasarwar, P., Dongre, S., & Khobragade, P. (2024). Earthquake Prediction Through Machine Learning Approach. In *2024 4th International Conference on Technological Advancements in Computational Sciences (ICTACS)* (pp. 542-546). IEEE. https://ieeexplore.ieee.org/abstract/document/10840652

Sidiropoulou, A., Chouvardas, D., Mantzanas, K., Stefanidis, S., & Karatassiou, M. (2022). Impact of Transhumant Livestock Grazing Abandonment on Pseudo-Alpine Grasslands in Greece in the Context of Climatic Change. *Land (Basel), 11*(12), 2126. DOI: 10.3390/land11122126

Silaban, S. A., Inulingga, S., & Fadli, M. (2021). The effect of green marketing on purchase decisions and brand image as intervening variables: Case Study of Pt. Sari Coffee Indonesia- Starbucks Focal Point Medan. *International Journal of Research and Review, 8*(1), 159–169.

Simpson, L. B. (2017). *As we have always done: Indigenous freedom through radical resistance*. University of Minnesota Press. DOI: 10.5749/j.ctt1pwt77c

Singh, B. R., & Singh, O. (2016). Future scope of solar energy in India. *SAMRIDDHI: A Journal of Physical Sciences. Engineering and Technology, 8*(01), 20–25.

Singh, K., & Jana, S. K. (2024a). Determinants of Carbon Emission-A Study in South Asian Countries. *International Journal of Innovative Research in Engineering & Management, 11*(3), 71–76.

Singh, K., & Jana, S. K. (2024b). Trends of Energy Security Status in India-A Study for the Period 1970-2022. *International Journal of Innovative Research in Engineering and Management, 11*(2), 89–93.

Singh, S. (2012). Green marketing: Challenges and strategy in the changing scenario. *International Journal of Advanced Research in Management and Social Sciences, 1*(6), 164–172.

Sithole, K. D.Khumbulani Derrick Sithole. (2024). Employing Internet of Things (IoT) devices for Monitoring and Controlling Energy Management Systems-A Review. *Journal of Electrical Systems, 20*(11s), 753–760. DOI: 10.52783/jes.7273

Skjærseth, J. B. (2018). Implementing EU climate and energy policies in Poland: Policy feedback and reform. *Environmental Politics, 27*(3), 498–518. DOI: 10.1080/09644016.2018.1429046

Skjølsvold, T. M., Ryghaug, M., & Throndsen, W. (2020). European island imaginaries: Examining the actors, innovations, and renewable energy transitions of 8 islands. *Energy Research & Social Science, 65*, 101491. DOI: 10.1016/j.erss.2020.101491

Slattery, M. C., Lantz, E., & Johnson, B. L. (2011). State and local economic impacts from wind energy projects: Texas case study. *Energy Policy, 39*(12), 7930–7940. DOI: 10.1016/j.enpol.2011.09.047

Smith, J. A., Langenhove, L. V., & Harre, R. (1995). Rethinking methods in psychology. https://doi.org/DOI: 10.4135/9781446221792

Sohail, M. S. (2017). Green marketing strategies: How do they influence consumer-based brand equity? *Journal for Global Business Advancement, 10*(3), 229–243. DOI: 10.1504/JGBA.2017.084607

Solaiman, M., Osman, A., & Halim, M. S. B. A. (2015). Green marketing: A marketing mix point of view. *International Journal of Business and Technopreneurship, 5*(1), 87–98.

Soraya, B., Nurrochmah, A., & Hwihanus, H. (2024). Business Transformation Towards Sustainability: The Role of Green Accounting in Sustainability Management. *Journal of Environmental Economics and Sustainability, 1*(3), 1–8. DOI: 10.47134/jees.v1i3.342

Sotiriou, S. A. (2025). Connecting security with sustainable development in the Eastern Mediterranean and generating pay-offs for the European Union. *Mediterranean Politics, 30*(1), 208–217. DOI: 10.1080/13629395.2023.2279384

Soukissian, T. H., Karathanasi, F. E., & Zaragkas, D. K. (2021). Exploiting offshore wind and solar resources in the Mediterranean using ERA5 reanalysis data. *Energy Conversion and Management, 237*, 114092. DOI: 10.1016/j.enconman.2021.114092

Soukissian, T., Papadopoulos, A., Skrimizeas, P., Karathanasi, F., Axaopoulos, P., Avgoustoglou, E., Kyriakidou, H., Tsalis, C., Voudouri, A., Gofa, F., & Katsafados, P. (2017). Assessment of offshore wind power potential in the Aegean and Ionian Seas based on high-resolution hindcast model results. *AIMS Energy, 5*(2), 268–289. DOI: 10.3934/energy.2017.2.268

South Aegean Region, (2021). South Aegean Regional Programme 2021-2027 (in Greek)

South Aegean Region. (2021). *Strategic Operational Program for the South Aegean Region*. South Aegean Region. (in Greek)

Southeastern European Energy Institute. (2020). *Just Development Transition Plan: Current situation and perspectives for regions in energy transition in Greece*. Southeastern European Energy Institute. (in Greek)

Sovacool, B. K. (2021). Who are the victims of low-carbon transitions? Towards a political ecology of climate change mitigation. *Energy Research & Social Science*, *73*, 101916. DOI: 10.1016/j.erss.2021.101916

Sovacool, B. K., Burke, M., Baker, L., Kotikalapudi, C. K., & Wlokas, H. L. (2020). New frontiers and conceptual frameworks for energy justice. *Energy Policy*, *105*, 677–691. DOI: 10.1016/j.enpol.2017.03.005

Sovacool, B. K., Hook, A., Martiskainen, M., & Baker, L. (2020). The decarbonisation divide: Contextualizing landscapes of low-carbon exploitation and toxicity in Africa. *Global Environmental Change*, *60*, 102028. DOI: 10.1016/j.gloenvcha.2019.102028

Spanidis, P. M., Roumpos, C., & Pavloudakis, F. (2020). A multi-criteria approach for the evaluation of low risk restoration projects in continuous surface lignite mines. *Energies*, *13*(9), 2179. DOI: 10.3390/en13092179

Spilanis, I., Kondili, J., & Gryllaki, S. (2003). Measuring the attractiveness of small islands. A tool for sustainability. In *International conference on sustainability indicators. Malta.*–http://www. aegean. gr/lid/internet/elliniki_ekdosi/Dimosieuseis/ Malta_paper. pdf *[2015 02 21].*

Spilanis, I., Kizos, T., & Petsioti, P. (2012). Accessibility of Peripheral Regions: Evidence from Aegean Islands (Greece). *Island Studies Journal*, *7*(2), 199–214. DOI: 10.24043/isj.268

Spilanis, I., Kizos, T., Vaitis, M., & Koukourouvli, N. (2013). Measuring the economic, social and environmental performance of European island regions: Emerging issues for European and regional policy. *European Planning Studies*, *21*(12), 1998–2019. DOI: 10.1080/09654313.2012.722970

Stauropoulou, A., & Sardianou, E. (2019, November). Understanding and measuring sustainability performance in the banking sector. []. IOP Publishing.]. *IOP Conference Series. Earth and Environmental Science*, *362*(1), 012128. DOI: 10.1088/1755-1315/362/1/012128

Stergiopoulou, E. (2025). Presentation at the event organized by the Waste, Energy and Water Regulatory Authority in Thessaloniki in the framework of the conference 'Helexpo Dialogues', the timetable for the electrical interconnection of the islands, March 2025, (in Greek)

Stern, N. (2007). *The Economics of Climate Change: The Stern Review*. Cambridge University Press. DOI: 10.1017/CBO9780511817434

Strambo, C. (2020). *Just transition and the geopolitics of decarbonization in the EU*. Mistra Geopolitics.

Stratakis, A. M., & Pelagidis, T. (2020). *The Competitive Advantage of the Forthcoming Mediterranean Energy Hub and its Implementation Alternatives: the EastMed Pipeline or an LNG Terminals Network?.*

Streimikiene, D., Kyriakopoulos, G. L., Lekavicius, V., & Siksnelyte-Butkiene, I. (2021). Energy poverty and low carbon just energy transition: Comparative study in Lithuania and Greece. *Social Indicators Research, 158*(1), 319–371. DOI: 10.1007/s11205-021-02685-9 PMID: 33941997

Sullivan, J. R. (2012). Skype: An appropriate method of data collection for qualitative interviews? *The Hilltop Review, 6*(1), 10.

Sultan, M. F., Tunio, M. N., Shaukat, G., & Asim, M. (2023). Green Marketing Mix From the Perspective of Service Sector: Leveraging Marketing of Services With Green-SIVA Marketing Mix Elements. In Entrepreneurship and Green Finance Practices (pp. 267-283). Emerald Publishing Limited.

Suman, S. K., & Ahamad, J. (2018). Solar energy potential and future energy of India: An overview. *International Journal of Engineering Science, 8*(5), 17575–17579.

Sun, B., Zhang, Q., Mao, H., & Li, Z. (2024). Validation of a statistical-dynamic framework for predicting energy consumption: A study on vehicle energy conservation equation. *Energy Conversion and Management, 307*, 118330. https://www.sciencedirect.com/science/article/pii/S0196890424002711. DOI: 10.1016/j.enconman.2024.118330

SUNEER. S. (2024). *PREDICTIVE ANALYSIS OF BIRTH RATES IN INDIA* (Doctoral dissertation, St. Teresa's College (autonomous) Ernakulam). http://117.239.78.102:8080/jspui/bitstream/123456789/4484/1/SIFNA.pdf

Suthar, S., & Pindoriya, N. M. (2020). *Blockchain and Smart Contract based Decentralized Energy Trading Platform.* DOI: 10.1109/NPSC49263.2020.9331883

Taliotis, C., Fylaktos, N., Partasides, G., Gardumi, F., Sridharan, V., Karmellos, M., & Papanicolas, C. N. (2020a). The effect of electric vehicle deployment on renewable electricity generation in an isolated grid system: The case study of Cyprus. *Frontiers in Energy Research, 8*, 205. DOI: 10.3389/fenrg.2020.00205

Taliotis, C., Giannakis, E., Karmellos, M., Fylaktos, N., & Zachariadis, T. (2020b). Estimating the economy-wide impacts of energy policies in Cyprus. *Energy Strategy Reviews, 29*, 100495. DOI: 10.1016/j.esr.2020.100495

Taliotis, C., Rogner, H., Ressl, S., Howells, M., & Gardumi, F. (2017). Natural gas in Cyprus: The need for consolidated planning. *Energy Policy, 107*, 197–209. DOI: 10.1016/j.enpol.2017.04.047

Tandon, A. (2021) Transition finance: Investigating the state of play: A stocktake of emerging approaches and financial instruments. Report no. 179, OECD, 3 August. Paris: OECD Publishing

Telli, A., Erat, S., & Demir, B. (2021). Comparison of energy transition of Turkey and Germany: Energy policy, strengths/weaknesses and targets. *Clean Technologies and Environmental Policy*, *23*(2), 413–427. DOI: 10.1007/s10098-020-01950-8

Temple, B., & Young, A. (2004). Qualitative research and translation dilemmas. *Qualitative Research*, *4*(2), 161–178. DOI: 10.1177/1468794104044430

The Energy and Resources Institute. (2024). *About TERI*. TERI.

Theodoropoulou, S., Sabato, S., & Akgüç, M. (2025). National eco-social policies in the framework of EU just transition: The cases of Greece, Italy, Portugal and Spain. *Global Social Policy*, *25*(1), 86–111. DOI: 10.1177/14680181241291122

Therapontos, P., Tapakis, R., Nikolaidis, A., & Aristidou, P. (2022). Increasing RES penetration in the Cyprus power system: current and future challenges. In *13th Mediterranean Conference on Power Generation, Transmission, Distribution and Energy Conversion (MEDPOWER 2022)* (Vol. 2022, pp. 313-318). IET.

Thomas, D. R. (2006). A general inductive approach for analyzing qualitative evaluation data. *The American Journal of Evaluation*, *27*(2), 237–246. DOI: 10.1177/1098214005283748

Thomas, M., Maheswari, G. S., & Kavitha, M. (2025). PERFORMANCE MANAGEMENT AND SUSTAINABILITY GOALS. *Cuestiones de Fisioterapia*, *54*(3), 1138–1156. DOI: 10.48047/4jfe3865

Thompson, C. (1999). Qualitative research into nurse decision making: Factors for consideration in theoretical sampling. *Qualitative Health Research*, *9*(6), 815–828. DOI: 10.1177/104973239900900609 PMID: 10662261

Thompson, M. P., Gannon, B. M., & Caggiano, M. D. (2021). Forest roads and operational wildfire response planning. *Forests*, *12*(2), 110. DOI: 10.3390/f12020110

Tietenberg, T. (2010). *Emissions trading: principles and practice*. Routledge. DOI: 10.4324/9781936331284

Tilt, B. (2015). *Dams and development in China: The moral economy of water and power*. Columbia University Press.

Tkachuk, O. (2024). The European Union in its pursuit of being a global leader in climate neutrality. *Global Policy*, *15*(S8), 69–74. DOI: 10.1111/1758-5899.13480

Torres, J. F., Martínez-Álvarez, F., & Troncoso, A. (2022). A deep LSTM network for the Spanish electricity consumption forecasting. *Neural Computing & Applications, 34*(13), 10533–10545. https://link.springer.com/article/10.1007/s00521-021-06773-2. DOI: 10.1007/s00521-021-06773-2 PMID: 35153386

Tranberg, B., Corradi, O., Lajoie, B., Gibon, T., Staffell, I., & Andresen, G. B. (2019). Real-time carbon accounting method for the European electricity markets. *Energy Strategy Reviews, 26*, 100367. DOI: 10.1016/j.esr.2019.100367

Tranoulidis, A., Sotiropoulou, R. E. P., Bithas, K., & Tagaris, E. (2022). Decarbonization and transition to the post-lignite era: Analysis for a sustainable transition in the region of Western Macedonia. *Sustainability (Basel), 14*(16), 10173. DOI: 10.3390/su141610173

Tranoulidis, A., Sotiropoulou, R. E. P., Bithas, K., & Tagaris, E. (2024). Lessons from European Union just transition toolkits: A regional investment framework for Greece and Germany. *Energy Research & Social Science, 114*, 103603. DOI: 10.1016/j.erss.2024.103603

Trinh, H. H., Trinh, K., Trinh, N., & Haouas, I. (2023). Energy Technology RD&D Budgets, sustainable transition, carbon risk, and climate change: Fresh insights from IEA member countries under policy uncertainty. *Available at SSRN* 4475237. DOI: 10.2139/ssrn.4475237

Tripathy, S. N. (2025). *Energy transitions and sustainability in India: Decarbonization, nuclear futures, renewable growth and environmental justice*. Abhijeet Publications.

Trombulak, S. C., & Frissell, C. A. (2000). Review of ecological effects of roads on terrestrial and aquatic communities. *Conservation Biology, 14*(1), 18–30. DOI: 10.1046/j.1523-1739.2000.99084.x

Trotta, G. (2018). Factors affecting energy-saving behaviours and energy efficiency investments in British households. *Energy Policy, 114*, 529–539. https://doi.org/https://doi.org/10.1016/j.enpol.2017.12.042. DOI: 10.1016/j.enpol.2017.12.042

Trotta, G. (2020). Assessing drivers of energy consumption and progress toward energy targets in Italy. *Energy Sources. Part B, Economics, Planning, and Policy, 15*(3), 137–156. DOI: 10.1080/15567249.2020.1778817

Troulis, M. (2019). The East Med pipeline project: Europe's fourth energy artery. *Civitas Gentium, 7*(2), 153–168.

Trudeau, N., Tam, C., Graczyk, D., & Taylor, P. (2011). *Energy transition for industry: India and the global context.*

Tsagkari, M. (2020). Local energy projects on islands: Assessing the creation and upscaling of social niches. *Sustainability (Basel), 12*(24), 10431. DOI: 10.3390/su122410431

Tsagkari, M., & Roca Jusmet, J. (2020). Renewable energy projects on isolated islands in Europe: A policy review. *International Journal of Energy Economics and Policy, 10*(5), 21–30. DOI: 10.32479/ijeep.9683

Tsai, P. H., Lin, G. Y., Zheng, Y. L., Chen, Y. C., Chen, P. Z., & Su, Z. C. (2020). Exploring the effect of Starbucks' green marketing on consumers' purchase decisions from consumers' perspective. *Journal of Retailing and Consumer Services, 56*, 102162. DOI: 10.1016/j.jretconser.2020.102162

Tsakiris, T. (2017). Cyprus's natural gas strategy: Geopolitical and economic preconditions. *Mediterranean Quarterly, 28*(1), 29–57. DOI: 10.1215/10474552-3882786

Tsangas, M., Zorpas, A. A., Jeguirim, M., & Limousy, L. (2018). Cyprus energy resources and their potential to increase sustainability. In *2018 9th International Renewable Energy Congress (IREC)* (pp. 1-7). IEEE. https://doi.org/DOI: 10.1109/IREC.2018.8362480

Tsangas, M., & Zorpas, A. A. (2023). Cyprus: Energy Policy. In *Encyclopedia of Mineral and Energy Policy* (pp. 205–212). Springer Berlin Heidelberg., DOI: 10.1007/978-3-662-47493-8_293

Tserkezos, D., Apostolopoulos, S., & Prodromou, A. (2025). Sustainable Entrepreneurship in Under-Resourced Rural Areas in Relation to The EU Rural Vision: The Greek Case. In Entrepreneurial Opportunities in Disadvantaged Rural Communities (pp. 1-22). IGI Global Scientific Publishing.

Tserkezos, D., Apostolopoulos, S., & Prodromou, A. (2025). Sustainable Entrepreneurship in Under-Resourced Rural Areas in Relation to The EU Rural Vision: The Greek Case. In Apostolopoulos, S., Paunovic, I., Macaulay, B., & Kakouris, A. (Eds.), *Entrepreneurial Opportunities in Disadvantaged Rural Communities* (pp. 1–22). IGI Global Scientific Publishing., DOI: 10.4018/979-8-3373-0214-0.ch001

Tsoukala, O. (2022). Energy Diplomacy in the Eastern Mediterranean: The Cases of EuroAsia and EuroAfrica Interconnector Pipelines of Electricity Interconnection. *HAPSc Policy Briefs Series, 3*(1). Advance online publication. DOI: 10.12681/hapscpbs.31005

Tuli, F. (2010). The basis of distinction between qualitative and quantitative research in social science: Reflection on ontological, epistemological and methodological perspectives. *Ethiopian journal of education and sciences, 6*(1): 97-108 https://doi.org/DOI: 10.4314/ejesc.v6i1.65384

Türk, B., & Erciş, A. (2017). Elements Of The Sustainable Marketing Mix. *Studies on Sustainability Research*, 105-125.

Twinn, S. (1997). An exploratory study examining the influence of translation on the validity and reliability of qualitative data in nursing research. *Journal of Advanced Nursing*, *26*(2), 418–423. DOI: 10.1046/j.1365-2648.1997.1997026418.x PMID: 9292378

Tzanes, G., Zafirakis, D., Makropoulos, C., Kaldellis, J. K., & Stamou, A. I. (2023). Energy vulnerability and the exercise of a data-driven analysis protocol: A comparative assessment on power generation aspects for the non-interconnected islands of Greece. *Energy Policy*, *177*, 113515. DOI: 10.1016/j.enpol.2023.113515

Tzitiridou-Chatzopoulou, M., Zournatzidou, G., & Kourakos, M. (2024). Predicting future birth rates with the use of an adaptive machine learning algorithm: A forecasting experiment for scotland. *International Journal of Environmental Research and Public Health*, *21*(7), 841. https://www.mdpi.com/1660-4601/21/7/841. DOI: 10.3390/ijerph21070841 PMID: 39063418

Tzogopoulos, G. N. (2022). *EastMed Pipeline Could Be a Giant Step Towards Enhancing Regional Security*. Begin-Sadat Center for Strategic Studies.

Uche-Soria, M., & Rodríguez-Monroy, C. (2018). Special regulation of isolated power systems: The Canary Islands, Spain. *Sustainability (Basel)*, *10*(7), 2572. DOI: 10.3390/su10072572

Umar, A., Kumar, D., Ghose, T., Alghamdi, H., & Abdelaziz, A. Y. (2024). Decentralized Community Energy Management: Enhancing Demand Response through Smart Contracts in a Blockchain Network. *IEEE Access : Practical Innovations, Open Solutions*, *12*, 80781–80798. DOI: 10.1109/ACCESS.2024.3409706

UNEP. (2023). *Making Peace with Nature*. United Nations Environment Programme. https://www.unep.org/resources/making-peace-nature

United Nations Environment Programme. (2022). Emissions gap report 2022. https://www.unep.org/resources/emissions-gap-report-2022

United Nations Framework Convention on Climate Change (UNFCCC). (2015). *The Paris Agreement*. UNFCCC.

United Nations. (2007). *United Nations Declaration on the Rights of Indigenous Peoples*. United Nations General Assembly.

Vakulchuk, R., & Overland, I. (2024). The failure to decarbonize the global energy education system: Carbon lock-in and stranded skill sets. *Energy Research & Social Science*, *110*, 103446–103446. DOI: 10.1016/j.erss.2024.103446

Van Eck, N. J., & Waltman, L. (2013). VOSviewer manual. *Leiden. Univeristeit Leiden*, *1*(1), 1–53.

Van Nes, F., Abma, T., Jonsson, H., & Deeg, D. (2010). Language differences in qualitative research: Is meaning lost in translation? *European Journal of Ageing*, *7*(4), 313–316. DOI: 10.1007/s10433-010-0168-y PMID: 21212820

Vanhercke, B., & Verdun, A. (2021). The European semester as goldilocks: Macroeconomic policy coordination and the recovery and resilience facility. *Journal of Common Market Studies*, *60*(1), 204–223. DOI: 10.1111/jcms.13267

Varshney, N., Sreetharan, V., Vennila, C., Ahila, R., Mane, D. T., Kshirsagar, K., & Pragathi, B. (2024). Blockchain Technology in Energy Management Systems: Enhancing Security and Transparency. *E3S Web of Conferences*, *591*, 01007–01007. DOI: 10.1051/e3sconf/202459101007

Vasconcelos, P. N., Silva, G. S., De Souza, A. C. Z., & Fuly, B. I. (2022, December 1). *Decentralized Governance and Energy Carbon Footprint: Opportunity for Re-Empowering People?* IEEE Xplore. DOI: 10.1109/IHTC56573.2022.9998383

Vasev, N. (2017). Governing energy while neglecting health–The case of Poland. *Health Policy (Amsterdam)*, *121*(11), 1147–1153. DOI: 10.1016/j.healthpol.2017.09.008 PMID: 28958707

Veers, P., Dykes, K., Basu, S., Bianchini, A., Clifton, A., Green, P., Holttinen, H., Kitzing, L., Kosovic, B., Lundquist, J. K., Meyers, J., O'Malley, M., Shaw, W. J., & Straw, B. (2022). Grand Challenges: Wind energy research needs for a global energy transition. *Wind Energy Science Discussions*, *7*(6), 2491–2505. DOI: 10.5194/wes-7-2491-2022

Velasco-Fernández, R., Dunlop, T., & Giampietro, M. (2020). Fallacies of energy efficiency indicators: Recognizing the complexity of the metabolic pattern of the economy. *Energy Policy*, *137*, 111089. https://doi.org/https://doi.org/10.1016/j.enpol.2019.111089. DOI: 10.1016/j.enpol.2019.111089

Venizelou, V., & Poullikkas, A. (2025). Navigating the Evolution of Cyprus' Electricity Landscape: Drivers, Challenges and Future Prospects. *Energies (19961073)*, *18*(5). https://doi.org/DOI: 10.3390/en18051199

Vijai, C., & Anitha, P. (2020). The importance of green marketing. *International Journal of Future Generation Communication and Networking*, *13*(3), 4137–4142.

Vlachou, A., & Pantelias, G. (2022). Energy transitions: The case of Greece with a special focus on the role of the EU ETS. *Science and Society*, *86*(4), 516–545. DOI: 10.1521/siso.2022.86.4.516

Vlados, C., Chatzinikolaou, D., & Kapaltzoglou, F. F. (2021). Energy market liberalisation in Greece: Structures, policy and prospects. *International Journal of Energy Economics and Policy*, *11*(2), 115–126. DOI: 10.32479/ijeep.10804

Vlami, V., Danek, J., Zogaris, S., Gallou, E., Kokkoris, I. P., Kehayias, G., & Dimopoulos, P. (2020). Residents' views on landscape and ecosystem services during a wind farm proposal in an island protected area. *Sustainability (Basel)*, *12*(6), 2442. DOI: 10.3390/su12062442

Vo, D. H., & Ho, C. M. (2024). Understanding the characteristics of the household energy transition in a developing country. *Heliyon*, *10*(1).

von Hirschhausen, C. (2018a). German Energy and Climate Policies: A Historical Overview. In von Hirschhausen, C., Gerbaulet, C., Kemfert, C., Lorenz, C., & Oei, P.-Y. (Eds.), *Energiewende "Made in Germany": Low Carbon Electricity Sector Reform in the European Context* (pp. 17–44). Springer International Publishing., DOI: 10.1007/978-3-319-95126-3_2

von Hirschhausen, C. (2018b). Nuclear Power in the Twenty-First Century [Part I]. *Assessment*.

Vona, F. (2019). Job losses and political acceptability of climate policies: Why the 'job-killing' argument is so persistent and how to overturn it. *Climate Policy*, *19*(4), 524–532. DOI: 10.1080/14693062.2018.1532871

Vora, D., Zade, N., Patwa, P., Kushwaha, A., Agrawal, G., & Sharma, A. (2024). Temporal Energy Consumption Exploration and Time Series Forecasting of Electricity Dynamics. In *2024 IEEE Pune Section International Conference (PuneCon)* (pp. 1-6). IEEE. https://ieeexplore.ieee.org/abstract/document/10895847

Votsi, N., Mazaris, A., Kallimanis, A., Zomeni, M., Vogiatzakis, I., Sgardelis, S., & Pantis, J. (2012). Road effects on habitat richness of the Greek Natura 2000 network. *Nature Conservation*, *1*, 53–71. DOI: 10.3897/natureconservation.1.2086

Wallace, C., Farmer, J., & McCosker, A. (2019). Boundary spanning practices of community connectors for engaging 'hardly reached' people in health services. *Social Science & Medicine*, *232*, 366–373. DOI: 10.1016/j.socscimed.2019.05.034 PMID: 31132544

Wanday, J., & Ajour El Zein, S. (2022). Higher expected returns for investors in the energy sector in Europe using an ESG strategy. *Frontiers in Environmental Science*, *10*, 1031827. DOI: 10.3389/fenvs.2022.1031827

Wang, C., Li, C., Yong, S., Wang, X. A., & Yang, C. (2022). Time Series and Non-Time Series Models of Earthquake Prediction Based on AETA Data: 16-Week Real Case Study. *Applied Sciences (Basel, Switzerland)*, *12*(17), 8536. https://www.mdpi.com/2076-3417/12/17/8536. DOI: 10.3390/app12178536

Wang, H., & Wang, M. (2020). Effects of technological innovation on energy efficiency in China: Evidence from dynamic panel of 284 cities. *The Science of the Total Environment*, *709*, 136172. https://doi.org/https://doi.org/10.1016/j.scitotenv.2019.136172. DOI: 10.1016/j.scitotenv.2019.136172 PMID: 31905588

Wang, J., Li, X., Jin, L., Li, J., Sun, Q., & Wang, H. (2022). An air quality index prediction model based on CNN-ILSTM. *Scientific Reports*, *12*(1), 8373. https://www.nature.com/articles/s41598-022-12355-6. DOI: 10.1038/s41598-022-12355-6 PMID: 35589914

Wang, Q., Gontikaki, E., Stenzel, P., Louca, V., Küpper, F. C., & Spiller, M. (2024a). How to decarbonize Greece by comparing wind and PV energy: A land eligibility analysis. *Energies*, *17*(3), 567. DOI: 10.3390/en17030567

Wang, R., Mirza, N., Vasbieva, D. G., Abbas, Q., & Xiong, D. (2020). The nexus of carbon emissions, financial development, renewable energy consumption, and technological innovation: What should be the priorities in light of COP 21 Agreements? *Journal of Environmental Management*, *271*, 111027. https://doi.org/https://doi.org/10.1016/j.jenvman.2020.111027. DOI: 10.1016/j.jenvman.2020.111027 PMID: 32778307

Wang, X., Ali, D. A., & Zhao, W. (2024b). Urban energy transition and quality of life: A comprehensive statistical study on the nexus of wind clean energy adoption, job creation, and livability. *Sustainable Cities and Society*, *108*, 105436. DOI: 10.1016/j.scs.2024.105436

Wang, Y. D., Yin, F. K., Shen, L., & Wu, C. Z. (2023). A Review of Road Design for Wind Farms in China. *Applied Sciences (Basel, Switzerland)*, *13*(7), 4075. DOI: 10.3390/app13074075

Watters, J. K., & Biernacki, P. (1989). Targeted sampling: Options for the study of hidden populations. *Social Problems*, *36*(4), 416–430. DOI: 10.2307/800824

WEC. (2023). WEC Trilemma: Cyprus. World Energy Council, https://trilemma.worldenergy.org/#!/country-profile?country=Cyprus&year=2023

Weiss, A., Hernandez, D., & Gruenewald, D. (2024), "The electricity demand in Europe: growing or going? https://www.mckinsey.com/industries/electric-power-and-natural-gas/our-insights/electricity-demand-in-europe-growing-or-going, *accessed 9/7/2025*

Weisser, D. (2004). Power sector reform in small island developing states: What role for renewable energy technologies? *Renewable & Sustainable Energy Reviews*, *8*(2), 101–127. DOI: 10.1016/j.rser.2003.10.002

Welford, R. (2000). *Hijacking Environmentalism*. Earthscan.

Whyte, K. (2018). Indigenous science (fiction) for the Anthropocene: Ancestral dystopias and fantasies of climate change crises. *Environment and Planning. E, Nature and Space*, *1*(1–2), 224–242. DOI: 10.1177/2514848618777621

Whyte, K. P. (2017). Indigenous climate change studies: Indigenizing futures, decolonizing the Anthropocene. *English Language Notes*, *55*(1–2), 153–162. DOI: 10.1215/00138282-55.1-2.153

Wohlfarth, K., Worrell, E., & Eichhammer, W. (2020). Energy efficiency and demand response–two sides of the same coin? *Energy Policy*, *137*, 111070. DOI: 10.1016/j.enpol.2019.111070

Wolf, F., Surroop, D., Singh, A., & Leal, W. (2016). Energy access and security strategies in Small Island Developing States. *Energy Policy*, *98*, 663–673. DOI: 10.1016/j.enpol.2016.04.020

World Bank. (2022). *Climate-Resilient and Inclusive Cities*. https://documents.worldbank.org/en/publication/documents-reports/documentdetail/503201643479926123

World Bank. (2023). *World Development Indicators: Energy Use and Access*. World Bank.

Wu, C., Zhang, X. P., & Sterling, M. J. (2021). Economic analysis of power grid interconnections among Europe, North-East Asia, and North America with 100% renewable energy generation. *IEEE Open Access Journal of Power and Energy*, *8*, 268–280. DOI: 10.1109/OAJPE.2021.3085776

Xanthakis, M., Katsimanis, N., & Antonopoulos, N. (2022). Impact of a wind farm on the avifauna of a Mediterranean mountainous environment. *Environmental Sciences Proceedings*, *21*(1), 11. DOI: 10.3390/environsciproc2022021011

Xayasouk, T., Lee, H., & Lee, G. (2020). Air pollution prediction using long short-term memory (LSTM) and deep autoencoder (DAE) models. *Sustainability (Basel), 12*(6), 2570. https://www.mdpi.com/2071-1050/12/6/2570. DOI: 10.3390/su12062570

Xiang, L., Xie, T., & Xie, W. (2020). Prediction model of household appliance energy consumption based on machine learning. In *Journal of Physics: Conference Series* (Vol. 1453, No. 1, p. 012064). IOP Publishing. https://iopscience.iop.org/article/10.1088/1742-6596/1453/1/012064/meta

Xie, J. J., Martin, M., Rogelj, J., & Staffell, I. (2023). Decarbonizing the US power system presents diverse challenges and opportunities in the changing employment landscape. *Research Square (Research Square).* DOI: 10.21203/rs.3.rs-2402526/v1

Xu, X., Liu, Z., Liu, W., Pei, C., Wu, X., & Nie, Z. (2024). A sustainable development benchmarking framework for energy companies based on topic mining and knowledge graph: The case of oil and gas industry. *Renewable & Sustainable Energy Reviews, 196*, 114350. DOI: 10.1016/j.rser.2024.114350

Yang, G., Zhang, G., Cao, D., Zha, D., Gao, X., & Su, B. (2024). China's provincial-level sustainable energy transition requires accelerating renewable energy technological innovation. *Energy, 288*, 129672. DOI: 10.1016/j.energy.2023.129672

Yang, Y., Xia, S., Huang, P., & Qian, J. (2024). Energy transition: Connotations, mechanisms and effects. *Energy Strategy Reviews, 52*, 101320. DOI: 10.1016/j.esr.2024.101320

Yenneti, K. (2016). The grid-connected solar energy in India: Structures and challenges. *Energy Strategy Reviews, 11*, 41–51.

Yeşil, Y. P. K. İ. K., & Projeleri, L. (2019). Global Green Logistics Projects In Green Marketing Mix. *The Sciences, 5*(21), 1269–1288.

Yousefi, S., Emami, S. N., Nekoeimehr, M., Rahmati, O., Imaizumi, F., Gomez, C., & Valjarevic, A. (2024). A hot-spot analysis of forest roads based on soil erosion and sediment production. *Land (Basel), 13*(10), 1583. DOI: 10.3390/land13101583

Yuan, B. (2019). Effectiveness-based innovation or efficiency-based innovation? Trade-off and antecedents under the goal of ecological total-factor energy efficiency in China. *Environmental Science and Pollution Research International, 26*(17), 17333–17350. DOI: 10.1007/s11356-019-05082-5 PMID: 31012077

Yu, C., Wang, F., Wang, Y., Shao, Z., Sun, T., Yao, D., & Xu, Y. (2025). Mgsfformer: A multi-granularity spatiotemporal fusion transformer for air quality prediction. *Information Fusion, 113*, 102607. https://www.sciencedirect.com/science/article/pii/S1566253524003853. DOI: 10.1016/j.inffus.2024.102607

Zachmann, G., Fredriksson, G., & Claeys, G. (2018). The distributional effects of climate policies. *Bruegel. Blueprint Series, 28.*

Zafeiratou, E., & Spataru, C. (2017). Potential economic and environmental benefits from the interconnection of the Greek islands. *International Journal of Global Warming, 13*(3-4), 426–458. DOI: 10.1504/IJGW.2017.087211

Zervas, E., Vatikiotis, L., & Gareiou, Z. (2021a). Proposals for an environmental and social just transition for the post-lignite era in Western Macedonia, Greece. In *IOP Conference Series: Earth and Environmental Science* (Vol. 899, No. 1, p. 012049). IOP Publishing. https://doi.org/DOI: 10.1088/1755-1315/899/1/012049

Zervas, E., Vatikiotis, L., Gareiou, Z., Manika, S., & Herrero-Martin, R. (2021b). Assessment of the Greek national plan of energy and climate change—Critical remarks. *Sustainability (Basel), 13*(23), 13143. DOI: 10.3390/su132313143

Zhang, H., Sun, Z., Wang, Z., Zhang, Q., Pan, R., & Cao, S. (2022, August). Analysis of Influencing Factors of Birth Rate and Prediction of Policy Scenario. In *China National Conference on Big Data and Social Computing* (pp. 309-331). Singapore: Springer Nature Singapore. DOI: 10.1007/978-981-19-7532-5_20

Zhang, G., Zhang, J., & Xie, T. (2020). A solution to renewable hydrogen economy for fuel cell buses–A case study for Zhangjiakou in North China. *International Journal of Hydrogen Energy, 45*(29), 14603–14613. DOI: 10.1016/j.ijhydene.2020.03.206 PMID: 32322133

Zhang, L., Mu, R., Zhan, Y., Yu, J., Liu, L., Yu, Y., & Zhang, J. (2022). Digital economy, energy efficiency, and carbon emissions: Evidence from provincial panel data in China. *The Science of the Total Environment, 852*, 158403. https://doi.org/https://doi.org/10.1016/j.scitotenv.2022.158403. DOI: 10.1016/j.scitotenv.2022.158403 PMID: 36057314

Zhang, Q., Han, Y., Li, V. O., & Lam, J. C. (2022). Deep-AIR: A hybrid CNN-LSTM framework for fine-grained air pollution estimation and forecast in metropolitan cities. *IEEE Access : Practical Innovations, Open Solutions, 10*, 55818–55841. https://ieeexplore.ieee.org/abstract/document/9780279. DOI: 10.1109/ACCESS.2022.3174853

Zhang, X., Zhang, Z., Liu, Y., Xu, Z., & Qu, X. (2024). A review of machine learning approaches for electric vehicle energy consumption modelling in urban transportation. *Renewable Energy, 234*, 121243. https://www.sciencedirect.com/science/article/pii/S0960148124013119. DOI: 10.1016/j.renene.2024.121243

Zhang, Z., & Wang, Y. (2023). A spatiotemporal model for global earthquake prediction based on convolutional LSTM. *IEEE Transactions on Geoscience and Remote Sensing, 61*, 1–12. https://ieeexplore.ieee.org/abstract/document/10210326. DOI: 10.1109/TGRS.2023.3302316

Zhan, S. (2023). ESG and corporate performance: A review. In *SHS web of conferences* (Vol. 169, p. 01064). EDP Sciences.

Zhao, X., Li, F., Yuan, Y., Ari, G., Yan, Y., Zhang, Q., Olhnuud, A., & Liu, P. (2025). Wind farms reduce grassland plant community diversity and lead to plant community convergence. *BMC Ecology and Evolution, 25*(1), 10. DOI: 10.1186/s12862-025-02350-6 PMID: 39815201

Zharfpeykan, R., & Akroyd, C. (2022). Factors influencing the integration of sustainability indicators into a company's performance management system. *Journal of Cleaner Production, 331*, 129988. DOI: 10.1016/j.jclepro.2021.129988

Zhu, W., Zhang, Z., Li, X., Feng, W., & Li, J. (2019). Assessing the effects of technological progress on energy efficiency in the construction industry: A case of China. *Journal of Cleaner Production, 238*, 117908. https://doi.org/https://doi.org/10.1016/j.jclepro.2019.117908. DOI: 10.1016/j.jclepro.2019.117908

Zikos, A., & Constantinidis, T. (2022). Plant diversity of Mts. Oligirtos and Farmakas (NE Peloponnisos, Greece) with emphasis on their endemic flora. *Plants, 11*(19), 2649. DOI: 10.3390/plants11192649 PMID: 36235515

Ziouzios, D., Karlopoulos, E., Fragkos, P., & Vrontisi, Z. (2021). Challenges and Opportunities of Coal Phase-Out in Western Macedonia. *Climate (Basel), 9*(7), 115. DOI: 10.3390/cli9070115

Zisopoulos, F. K., Noll, D., Singh, S. J., Schraven, D., De Jong, M., Fath, B. D., Goerner, S., Webster, K., Fiscus, D., & Ulanowicz, R. E. (2023). Regenerative economics at the service of islands: Assessing the socio-economic metabolism of Samothraki in Greece. *Journal of Cleaner Production, 408*, 137136. DOI: 10.1016/j.jclepro.2023.137136

Zoll, M. (2022). Energy governance in the Republic of Poland. In *Handbook of energy governance in Europe* (pp. 923–958). Springer. DOI: 10.1007/978-3-030-43250-8_22

Zorpas, A. A., Tsangas, M., Jeguirim, M., Limousy, L., & Pedreno, J. N. (2017). Evaluation of renewable energy sources (solar, wind, and biogas) established in Cyprus in the framework of sustainable development.

Zotos, A., Kosma, C., Triantafyllidis, V., Kakabouki, I., Kehayias, G., Roussis, I., Mavroeides, A., Tataridas, A., & Bilalis, D. (2021). Plant species diversity of the wet meadows under natural and anthropogenic interventions: The case of the Lakes Amvrakia and Ozeros (W. Greece). *Notulae Botanicae Horti Agrobotanici Cluj-Napoca*, *49*(3), 12435–12435. DOI: 10.15835/nbha49312435

Zou, Y., & Wang, M. (2024). Does environmental regulation improve energy transition performance in China? *Environmental Impact Assessment Review*, *104*, 107335. DOI: 10.1016/j.eiar.2023.107335

Zülfikaroğlu, S. (2024). AI-Driven Strategic Management and Decision Making for Energy Sector. *Next Frontier for Life Sciences and AI*, *8*(1), 91. DOI: 10.62802/q7rkdb54

About the Contributors

Panos Dimitrakopoulos is a Lecturer in the Department of Accounting and Finance at the University of the Peloponnese (Greece). He is holds a PhD degree from school of Management at University of Peloponnese. He is PhD Thesis focused on The Impact of Major Energy Indexes on the Economic Growth of EU Countries, According to the Contribution of Industry to GDP. His academic background includes an MBA in Financial Planning and Public Sector Management, a Bachelor's degree in Accounting, and a postgraduate certificate in Pedagogical and Teaching Competence. His main research interests lie in the fields of energy transition, sustainable development, and green entrepreneurship, with a particular emphasis on rural economies. He has published in peer-reviewed journals indexed in Scopus and Clarivate and has contributed chapters to collective academic volumes by Springer and IGI Global. His work includes field-based qualitative research, systematic literature reviews, and the analysis of institutional frameworks affecting economic and energy performance across European regions. He has presented his work at international academic conferences in Greece, Serbia, the UK, and Italy, and has served as a reviewer for academic book submissions and journals such as Interdisciplinary Environmental Review. He has also led and participated in funded research projects related to local economic registries and the energy efficiency of institutions.

Dimitrios Salampasis is the FinTech Capability Lead and Associate Professor of Emerging Technologies and FinTech at the Swinburne University of Technology in Australia. He is also a Senior Executive Fellow at the Digital Economist in Washington DC. Dimitrios is an award-winning and globally-recognized thought leader, educator, researcher, and sought-after speaker, passionate about innovation and business strategy, global affairs and geopolitics, digital economy (FinTech), sustainability and emerging technologies nexus. Dimitrios holds several international visiting professorship positions and fellowships. He is the recipient of the 2022

Innovation Excellence Award (Hellenic-Australian Chamber of Commerce and Industry), the 2021 Blockchain Educator Award (Blockchain Australia) and the 2021 Swinburne University Vice Chancellor's Industry Engagement Award (Individual). Dimitrios publishes in international peer-reviewed academic journals and books and his work is regularly presented in major international conferences, leading industry forums, executive education masterclasses and invited keynotes around the world. His research has been published in leading journals incl. Sustainability Accounting, Management and Policy Journal, British Journal of Management and IEEE Transactions on Engineering Management. Dimitrios has edited three books published by Edward Elgar Publishing (forthcoming) and World Scientific Publishing. Dimitrios' research interests revolve around the organizational, human, geopolitical and ESG sides of innovation and open innovation in emerging technologies and FinTech. His areas of specialist expertise, research, teaching, industry engagement, policy and advisory work revolve around the emergence and development of FinTech-enabled business models (incl. open banking, open finance and ethics of programmable money), Blockchain-enabled human rights due diligence (modern slavery reporting) and corporate environmental stewardship (greenwashing), synthetic identity fraud and online scams, programmable democracy, quantum and neuromorphic computing in the financial and the public sectors, together with, their relevant global regulatory, societal, ethical and policy implications. Dimitrios is a trusted partner to C-Level Executives, CEOs and Board of Directors of leading corporations, start-ups, and government officials

Ilias Makris is a professor in Financial Analysis. He received his PhD on Business Administration at University of Patras. He currently serves as a Professor at the Department of Accounting and Finance, of the University of the Peloponnese, Teaching Economics and Finance, in undergraduate and postgraduate level. He has served as Scientific Director of Innovation and Entrepreneurship Unit of Technological Educational Institute of Peloponnese (2009-2015), Head of the Dept of Accounting and Finance and Director of postgraduate program in Finance, dept. of Accounting & Finance. He has authored many refereed publications in international journals, book chapters, collective volumes and international conferences. His scientific interests focus on sustainable economics, energy economics, innovation, entrepreneurship, and performance measures in Micro and macro level. His recent work has been published at J. Open Innov. Technol. Mark. Complex., Energy Sources, Part B: Economics, Planning, and Policy, Sustainability, and Energies.

Sotiris Apostolopoulos is a Lecturer in Entrepreneurship and Development at the University of Neapolis (Paphos, Cyprus). He also acts as academic Director of the Undergraduate Programs in Business Administration & Digital Business of the Department of Economics and Business for the same University. He holds a PhD degree from the School of Management of University of the Peloponnese and two Master's degrees (MSc) in Healthcare Administration and in Finance as well. He specialises in rural entrepreneurship, EU policies and the socio-economic development of the rural areas with emphasis on private and social health enterprises. His research interests focus on local development, European policies, private and social enterprises as well as on innovation. He has many years' experience working in the management of private companies. He served as Head of Communication and Public Relations of the Marketing Department of the Rehabilitation & Recovery Center "Diaplasis" in Greece and Chief Operating Officer (COO) for the largest regional private hospital "City Hospital" of the Peloponnese Region. His research articles have been published in major scientific journals by publishers such as Elsevier, Wiley, Emerald, Springer, MDPI, while he has participated in studies that were included and published as book chapters in edited volumes (IGI Global, Eburon, etc.). Specifically, towards IGI-GLOBAL publishing, he also served as co-editor in two books recently released namely "Real-World Tools and Scenarios for Entrepreneurship Exploration" (2024) & Entrepreneurial Opportunities in Disadvantaged Rural Communities (2025). Simultaneously, he has taken part through research work in international and European conferences on Economics, Entrepreneurship and Innovation held in Greece, Cyprus, UK, Switzerland, Italy, Spain, Portugal and Montenegro. Recently, the World Health Organization included in its official database two research papers of Sotiris (co-author), concerning the rural health enterprises during the Covid-19 pandemic.

***

Ayyoob A is a Senior Research Scholar at the School of Business Studies, University of Calicut, specializing in sustainability, corporate governance, and strategic corporate social responsibility (CSR). A seasoned academic, Ayyoob has served as a resource person for numerous workshops on research methodology and data analysis. His scholarly contributions include publishing book chapters and articles in reputed platforms and presenting research papers at prestigious national and international conferences, both in India and abroad, organized by renowned universities. Additionally, Ayyoob actively reviews manuscripts for esteemed journals, showcasing his expertise and commitment to advancing research in his fields of interest.

Eleftherios Aggelopoulos is an Associate Professor of Accounting in the Department of Business Administration, University of Patras, Greece. He has worked for many years in a systemic Greek bank in various managerial positions. He holds a master's degree from Cass Business School, City University London and a Ph.D. in Accounting from the Department of Business Administration, University of Patras. His research has been published in prestigious academic journals (such as European Journal of Operational Research, Accounting Forum and Journal of Accounting and Public Policy). His work has been presented at the Annual Conferences of the European Accounting Association (EAA). The courses he teaches (Management, Financial and Tax accounting) are accredited by the Association of Certified Chartered Accountants (ACCA).

Nikolaos Apostolopoulos is an Associated Professor in European Policies in Sustainability, Development and Entrepreneurship at the University of Peloponnese and Director of the Laboratory of European Sustainable Development Policies. He also acts as a Scientific Advisor at the Labour Institute of the General Confederation of Greek Employees (INE-GSEE) and he participates at the European Trade Unions Institute. He has participated in more than 10 funding initiatives of the EU Structural Funds. Among these initiatives, he was co-author of a report on the impact of the Europe 2020 strategy on the Greek economy funded by the EU Structural Funds. Moreover, he is a key researcher of a Jean Monnet Centre of Excellence on Governance at the University of Peloponnese. He has published numerous articles in leading academic journals. He is co-editor of 3 edited volumes and co-author of 4 books.

Christodoulos Daoutis is a graduate of the Department of Forestry and Environmental and Natural Resources Management of the School of Agriculture and Forestry of the Democritus University of Thrace. He holds three postgraduate degrees. He is a PhD candidate at the Department of Forestry and Environmental and Natural Resources Management of the Democritus University of Thrace since 2022. Since April 2022, he has been working at the Department of Forestry and Environmental and Natural Resources Management of the Democritus University of Thrace as a Laboratory Teaching Staff. He has published scientific papers in Greek and foreign scientific journals and at conferences.

Vasileios Giannopoulos serves as an Associate Professor in the Department of Accounting and Finance at the University of Peloponnese. He obtained a BSc from the University of Piraeus, Department of Banking and Finance, and a Master's Degree in Business Administration (New Principles of Business Administration - MBA) from the University of Patras. He also holds a PhD in Financial Management and Accounting from the University of Patras, Department of Business Administration. He worked for 14 years as an executive at a large Greek bank, focusing on financing evaluation and providing facilities to small businesses

(Small Business Banking). Since April 2019, he has been an Assistant Professor at the University of Peloponnese, teaching undergraduate and graduate courses such as Financial Accounting, Management Accounting, Greek Accounting Standards, Tax Accounting, Financial Statement Analysis, and Computerized Accounting. His research and teaching interests include Financial Accounting, Management Accounting, and Financial Management using machine learning models, Banking Management, Sustainable Accounting, Internal Audit, and Franchising. He has published articles in international journals and presented his research at numerous national and international conferences.

Gagan Gulati is serving as an assistant professor in the Department of Management (CDOE) at Manipal University Jaipur. His research interest area includes green marketing.

J. Manokaran received the bachelor's degree in electronics and communication engineering and the master's degree in applied electronics from Anna University, in 2009 and 2011, respectively. He is currently pursuing the Ph.D. degree with the SRM Institute of Science and Technology, Kattankulathur, Chennai, India. His list of academic achievements includes five publications, two journals, and three conference papers. His area of interests include the Internet of Things, computational intelligence, machine learning, deep learning, edge computing, cluster computing, and intrusion detection. During his post-graduation, he received the Best Technical Paper Award in the national level conference on ''Computer Intelligent and Application'' organized by the CARE College of Engineering, Tiruchirappalli.

Sebak Kumar Jana is currently Professor of Economics and former Head of the Department of Economics in Vidyasagar University, Midnapore, West Bengal, India. His major areas of research include resource and environmental economics, rural economics and economics of education. He has a good number of publications in national and international journals and books.

Mahesh Chandra Joshi is currently Professor of Marketing in the Department of Management at Manipal University Jaipur. His research interests include marketing and consumer behavior.

Maria Karamanoli is a PhD candidate in European Economic Governance at the University of Aegean (Mytilene, Lesvos). She is a Graduate in International and European Economic Studies of the University of Macedonia and Graduate in Sociology of the University of the Aegean. She possesses two Masters a) MBA in Accounting and Auditing and b) Master in European Societies and European Integration, both from the University of Aegean. She specializes in Fiscal Policy,

Tax Policy, Economic Crises and European policies. She has published research on fiscal policy, business management, tax fraud and human rights. She has many years' experience working for European Agencies (asylum) and Greek Public Authorities (taxes).

Iordanis Katemiadis is a Lecturer of Destination Marketing and Tourism of UCLan Cyprus University and Course Leader of the MA Hospitality, Tourism And Events Management. He holds a Phd in Business Administration in the area of Destination Marketing and Management performance from the University of the Aegean. He also holds a MSc in Hospitality Management from the University of Strathclyde and a BA in Tourism and Hospitality Management from the International University of Greece. His fields of expertise include Destination Marketing and Management, sustainable development, tourism and hospitality marketing and management and tourism policy. He has published scientific papers in International Journals and presented his work in various conferences. He has years of experience in luxury hotel properties and he managed his own business in the area of education.

Faisal Khan, Dr., is a faculty member in the Department of Commerce at Brainware University, Kolkata. His research interests include international trade, sustainable business practices, and the evolving dynamics of commerce in emerging economies. He has guided and contributed to publications that address the intersection of trade policy, institutional development, and sustainable growth.

Amit Kumar is currently serving as assistant professor (senior scale) in the department of business administration at Manipal University Jaipur. His research interest area includes consumer behavior and sustainability.

Ioannis Lampropoulos holds a Ph.D. in Managerial Accounting from the Department of Business Administration, University of Patras and a Master's Degree in Primary Health Care from the Department of Medicine of the University of Thessaly. He is a Postdoctoral Researcher of the University of Thessaly in the Department of Medicine. He also holds the Pedagogical Competence from the Hellenic Open University. He has been working since 2009 as a Retail Business Executive in various managerial positions. His research interests include, among others, Management & Financial Accounting, Performance Management, Health Economics and Big Data analysis. He has taught since 2008 in the Hellenic Open University the course Greek Accounting Standards in theory and practice, in the Department of Accounting of the TEI of Patras the courses Computerized Accounting I & Special Topics in Computerized Accounting, in the Department of Accounting of the TEI of Larissa the courses Computerized Accounting I, II & III, Accounting for Firms, Cost Accounting I & II, at the Department of Animal Production of the TEI of Larissa the course Livestock Accounting & Agricultural Accounting & Techno-economic Analysis, at the Department of Informatics Engineering TE Larissa of the TEI of Larissa the courses Operational Research, Business Economics, at the Department of Accounting of the University of Thessaly the courses Advanced Accounting, International Financial Reporting Standards, Business Responsibility, at the Department of Animal Production Sciences of the University of Thessaly the courses Educational Technology & Multimedia & Educational Psychology, at the Department of Agrotechnology of the University of Thessaly the course Agricultural Policy & International Trade.

Aimilia Lempesi received her Forester degree from the Department of Forestry & Environmental Management of Karpenisi in 2010 and her Master's degree in Ecology and Management of Forest Ecosystems from the Department of Forestry and Environmental Management and Natural Resources of the Democritus University of Thrace in 2013. In 2023, she was awarded a PhD from the Department of Forestry and Management of Environment and Natural Resources of the Faculty of Agriculture and Forestry of the Democritus University of Thrace, in the field of Range Science. Her research interests include the effect of grazing intensity on vegetation and the functional characteristics of understory plant species in grazed oak forests, the recording, analysis and evaluation of flora and vegetation.She has also published scientific papers in Greek and foreign journals and conferences. As part of the fellowship, in the year 2021-2022 she carried out auxiliary teaching work at the Democritus University of Thrace in the Cognitive Subject of Range Science. Since the spring semester of 2023 she has been working as a Part-Time Lecturer at the Democritus University of Thrace in the Cognitive Subject of Range Science

Andreas Masouras is the Head of the Department of Communication & Creativity, Member of the Senate, and Associate Professor of Communication and Marketing at Neapolis University Pafos. He also serves as MBA Coordinator, including the specialization in Tourism. As a full-time faculty member, he plays a key role in academic governance, program development, and the strategic academic direction of the university.

Aaliya Nashat is associated with the Department of Hospital Management at Brainware University, Kolkata. Her academic interests focus on healthcare administration, sustainable hospital practices, and the integration of management strategies to improve service delivery. She has contributed to research on the role of healthcare institutions in the context of energy transition and sustainability, and is actively involved in teaching and guiding students in hospital operations and policy studies.

Ananya Pandey is a Research Scholar in School of Law at Christ University, Bangalore. She has her research interests in the field of Constitutional Law, Environment Law, Human Rights, Food Security, Intellectual Property Law and Jurisprudence.

Georgios Papadopoulos is a Graduate in Economics at the University of the Peloponnese. He possesses Master's in Economic and Business Strategy and in Bioeconomy, Circular Economy and sustainable development from the University of the Piraeus. He also is PhD candidate at the Department of Management Science and Technology at the University of the Peloponnese with a scholarship from the Labour Institute of the General Confederation of Greek Employees (INE-GSEE)

Amvrosios Prodromou holds a DBA degree with specialisation in SMEs, MBA in Marketing and a Bachelors in Business Administration. He is a lecturer at Neapolis University Pafos (NUP). He is a member of the Cyprus Chamber Commerce and Industry (CCCI), and he runs his own business consulting firm. He has published several articles in academic journal and participated in a large number of European projects.

Aparna Sajeev is an Assistant Professor at the Department of Commerce and Management Studies, School of Business Studies, University of Calicut. She pursued her doctoral degree in macro-policy management at the Faculty of Management Studies, University of Delhi. Her key research areas include sustainable Development and Climate Change, Sustainable Finance, Psychometric Analysis, and Impact Assessment. She has three years of Education industry experience as a Senior Research Manager - Impact Assessment and continues her work in education sector engagements as a consultant. She collaborated with the Berkeley Evaluation and Assessment Research Center. With over a decade of research experience, she is presently a research supervisor at the Department of Commerce, University of Calicut. She received a 2 million Won - Selected Research Paper award from Kyunghee University, Seoul (2020).

Ankit Saxena is currently working as associate professor at Institute of Business Management, GLA University, Mathura. He is a Director Medalist from Dayalbagh Educational Institute, Agra. He qualified NET-JRF (UGC). He has excellent academic background and a rich experience of 17 years in research and academics. He has published number of research papers and contemporary business case studies. His area of interest comprises Strategic Financial Management and behavior of market participants.

Khushboo Singh is Ph. D. scholar in the Department of Economics, Vidyasagar University, Midnapore, India

Nicholas Theodorakopoulos is Professor of Entrepreneurship Development at the School of Economics, Administration & Computer Science of Neapolis University Pafos. His specialist areas of teaching, research, and industry engagement include strategic entrepreneurship, learning & development, change management & innovation, and diversity & inclusion. Nicholas Theodorakopoulos has received substantial financial support from various bodies, including the European Commission, Economic & Social Research Council, Department for Business, Innovation and Skills, Leicestershire Learning and Skills Council, Centre for Enterprise, and UK Trade and Investment, to examine entrepreneurship in different contexts and deliver impact.

Index

F

G

H

I

J

L

M

N

P

R

S

T

W

www.ingramcontent.com/pod-product-compliance
Lightning Source LLC
LaVergne TN
LVHW062335300926
849161LV00007B/1700